Nuclear Fission

Nuclear Fission

ROBERT VANDENBOSCH / JOHN R. HUIZENGA

Department of Chemistry
University of Washington
Seattle, Washington

Departments of Chemistry and Physics
University of Rochester
Rochester, New York

ACADEMIC PRESS New York and London 1973
A Subsidiary of Harcourt Brace Jovanovich, Publishers

ACADEMIC PRESS, INC.
111 Fifth Avenue, New York, New York 10003

United Kingdom Edition published by
ACADEMIC PRESS, INC. (LONDON) LTD.
24/28 Oval Road, London NW1

Library of Congress Cataloging in Publication Data

Vandenbosch, Robert, DATE
Nuclear fission.

Bibliography: p.
1. Nuclear fission. I. Huizenga, John Robert, DATE joint author.
QC790.V34 539.7′62 72-9340
ISBN 0-12-710850-5

PRINTED IN THE UNITED STATES OF AMERICA

Contents

Preface xi

I. Introduction

A. Discovery of Fission 1
B. Scope of the Book 3
References 13

II. The Fission Barrier

A. Liquid Drop Model 15

1. Introduction 15
2. Potential Energy Mapping and the Fission Barrier 17
   a. Symmetric Deformations 17
   b. Refinements of the Liquid Drop Model 20
   c. Stability of Equilibrium Shapes with Respect to Asymmetric Distortions 22
3. Barrier Curvature 22

B. Single Particle Model 24

1. Introduction 24
2. Single Particle Calculation of Deformation Energy 25

C. Strutinsky Hybrid Model 27

1. The Shell Correction Method 27
2. Predicted Barriers for Superheavy Elements 34
3. Extension of the Strutinsky Method to Reflection-Asymmetric Saddle-Point Shapes 36

D. Fission Barrier Height Systematics 37

E. Excitations at the Saddle Point 40

References 42

III. Spontaneous Fission

A. Introduction 45

B. Systematics of Ground State Spontaneous Fission Half Lives 46

C. Penetration of the Fission Barrier 51

D. Spontaneously Fissionable Isomers 59

References 77

IV. Fission Widths from Neutron Resonance Studies

A. Resonance Spin Determinations 81

1. Neutron Scattering 82
2. Interference between Levels 82
3. Nuclear Orientation 83
4. Transmission of Polarized Neutrons through Polarized Targets 85
5. Gamma-Ray Multiplicities 85
6. Relative Intensity of Specific Capture Gamma Rays 88

B. Average Fission Widths 88

C. Width Distributions 92

D. Fission Characteristics Which Apparently Depend on Resonance Spin 94

1. Mass Asymmetry 94
2. Kinetic Energy Release 95
3. Neutron Multiplicity 95
4. Ratio of Ternary to Binary Fission 95

E. Intermediate Levels in the Two Potential Wells of Heavy Nuclei 96

F. The n, γf Process 104

References 107

V. Properties of Low-Lying Levels of Transition Nuclei Determined from Reaction Studies

A. INTRODUCTION 109

B. PHOTOFISSION 112

1. Even Nucleus Targets 112
2. Odd Mass Targets 130
   a. Dipole Photofission of ^{239}Pu 130
   b. Dipole Photofission of Targets with Spin of $\frac{5}{2}$ 135
   c. Dipole Photofission of Targets with Spin of $\frac{7}{2}$ 139

C. NEUTRON-INDUCED FISSION 139

1. Odd-Mass Transition Nuclei (Formed by Neutron Fission of Even-Even Targets) 139
2. Odd-Mass Transition Nuclei (Two-Humped Barrier Penetration) 152
3. Even-Even Transition Nuclei (Formed by Neutron Fission of Odd-N Targets) 159

D. FISSION OF NUCLEI EXCITED IN DIRECT REACTIONS 163

1. Introduction 163
2. Fission of Even-Even Transition Nuclei 169
3. Fission of Odd-A Transition Nuclei 175

REFERENCES 176

VI. Fission-Fragment Angular Distributions at Moderate to High Excitation Energies

A. INTRODUCTION 179

B. THEORY OF ANGULAR DISTRIBUTIONS FOR TRANSITION STATES DESCRIBED BY A STATISTICAL MODEL 180

1. Exact Theoretical Expression 180
2. Approximate Theoretical Expressions 183
3. Comparisons of the Various Theoretical Expressions 185

C. EXPERIMENTAL FISSION-FRAGMENT ANGULAR DISTRIBUTIONS 189

D. MODIFICATION IN THEORY TO ACCOUNT FOR NUCLEAR PAIRING 197

E. SHAPE OF THE TRANSITION NUCLEUS 203

F. ANGULAR ANISOTROPY OF FISSION FRAGMENTS OF PARTICULAR MASS 209

G. ANGULAR CORRELATION BETWEEN FISSION FRAGMENTS 212

REFERENCES 215

VII. Competition between Fission and Neutron Emission

A. EXPERIMENTAL DETERMINATION OF Γ_n/Γ_f VALUES AT LOW AND MODERATE EXCITATION ENERGY 216

1. Fission Cross Sections 216
2. Photofission and Photoneutron Cross Sections 218
3. Mean Values of Γ_n/Γ_f Deduced from Higher Energy Neutron Fission Cross Sections 220
4. Mean Values of Γ_n/Γ_f Derived from Charged Particle-Induced Spallation Cross Sections 222
5. Summary of Experimental Values of Γ_n/Γ_f 225

B. THEORETICAL EXPECTATIONS FOR Γ_n/Γ_f 227

1. The Neutron and Fission Widths at Low Energies 227
2. Excitation Energy Dependence of Γ_n/Γ_f 232
3. Deformation Dependence of the Level Density Parameter a 240

C. EFFECT OF ANGULAR MOMENTUM ON Γ_n/Γ_f 244

1. Theoretical Considerations 244
2. Comparison with Experiment 250
3. Direct Fission 250

REFERENCES 252

VIII. Fission Barrier Heights

A. EXPERIMENTAL VALUES OF FISSION BARRIER HEIGHTS 254

B. METHODS OF DETERMINING FISSION BARRIER HEIGHTS FROM REACTIONS 254

1. Barrier Heights from Excitation Energy Dependence of Γ_n/Γ_f 254
2. Barrier Heights from (n, f), (t, pf), (t, df), and (d, pf) Excitation Functions 256
3. Barrier Heights from Photofission Excitation Functions 257
4. Barrier Heights for Nuclei Exhibiting a Double Barrier 257

REFERENCES 258

IX. Motion from Saddle to Scission: Theories of Mass and Energy Distributions

A. INTRODUCTION 259

B. ADIABATIC MODELS 263

1. The Liquid Drop Model 263
2. Single Particle Effects on the Potential Energy Surface and on the Dynamics of the Descent from Saddle to Scission 273
   - *a. Potential Energy Surface Mapping with Inclusion of Odd-Multipole Shape Distortions* 273
   - *b. Potential Energy Surface and Mass Asymmetry* 280
   - *c. Dynamics of Descent on a Particular Shell Structure-Corrected Potential Surface* 282

C. Nonadiabatic Models of Fission 283
1. Statistical Theory of Fission 283
2. A Model of Kinetic Dominance 286
References 287

X. Kinetic Energy Release in Fission

A. Dependence of the Total Kinetic Energy on the Charge and Mass of the Fissioning Nucleus 289
B. Dependence of the Total Kinetic Energy on the Mass Division 294
C. Dependence of the Total Kinetic Energy on Excitation Energy 300
D. Dependence of the Total Kinetic Energy on Angular Momentum 301
E. Dependence of the Total Kinetic Energy Release on Charge Division 303
References 303

XI. Distribution of Mass and Charge in Fission

A. Mass Distributions 304
1. Introduction 304
2. Spontaneous Fission 305
3. Particle-Induced Fission 307
4. Fine Structure in Primary Fission-Fragment and Fission Product Mass Distributions 317
5. Mass Distribution for Spontaneous Fission Isomers 320
B. Charge Distributions 322
1. Charge Distributions in Low Energy Fission 322
2. Dependence of Charge Division on Excitation Energy 331
References 333

XII. Prompt Neutrons from Fission

A. Introduction 335
B. Total Neutron Yield in Low Energy Fission 335
1. Time Scale of Neutron Emission 335
2. Average Neutron Yield $\bar{\nu}$ for Various Fissioning Species 337
3. Distribution of Neutron Emission Numbers 337
4. Dependence of Neutron Yield on Excitation Energy 340
5. The Neutron Energy Spectrum 344
C. Dependence of Neutron Yield on Fragment Mass 345
D. Correlation between Neutron Yields of Complementary Fragments 351
E. Dependence of Neutron Kinetic Energy on Fragment Mass 352
References 355

XIII. Gamma Rays from Primary Fission Products

A. Introduction 357

B. Gamma Rays from Fragments of All Masses 358

1. The Time of Emission 358
2. The Gamma-Ray Energy Spectrum 359
3. Angular Distribution of Fission Gamma Rays 361

C. Dependence of Gamma-Ray Yield and Energy on Fragment Mass 363

D. *K* x Rays and Conversion Electrons 367

E. Isomeric Yield Ratios and Rotational State Populations as a Measure of Fragment Angular Momentum 368

1. Isomeric Yield Ratios 368
2. Gound State Band Populations 369
3. Anisotropies of Specific Deexcitation Gamma Rays 370

F. Theoretical Estimates of Fission-Fragment Angular Momentum 371

References 372

XIV. Ternary Fission

A. Light-Particle-Accompanied Fission 374

1. Mechanism for Charged Particle Emission during Fission 375
2. Dependence of Yield on Nuclear Species and Excitation Energy 376
3. Relative Yields of the Different Light Particles 378
4. Angular and Energy Distributions of the Light Charged Particles 382
5. Dependence of the Probability of Alpha-Particle-Accompanied Fission on Mass Division 396
6. Energy Balance in Alpha-Accompanied Fission 399

B. Fission in Which Three Fragments of Comparable Mass Are Produced 400

References 405

Author Index 409

Subject Index 419

Preface

The intention governing the scope of this book was to provide a comprehensive account of our present understanding of nuclear fission. D. H. Wilkinson* remarked five years ago that

> Fission is a process of deadly fascination; had nature chosen her constants just a little differently, we should have been deprived of its potential for social good and spared its power for social evil. Despite the former and despite the undeniable fact that the latter is responsible for nuclear and particle physics being decades in advance of what would otherwise have been their time, I know what my own choice for the constants would have been. But we have always hoped that fission would teach us something about nuclear physics rather than just about itself—and ourselves. It has remained a phenomenon very much of its own, finding its explanation in the rest of nuclear physics but giving little in return. It now seems that an installment on the debt may be forthcoming.

Hence, a book on fission is timely in view of the recent intensification of interest in fission resulting from the discovery of the shape isomers, intermediate structure, and the double humped fission barrier. Advances in data collection and analysis techniques have enabled researchers to test

* From D. H. Wilkinson, *Comments on Nuclear and Particle Physics* **2**, 146 (1968).

detailed theoretical predictions in rapid sequence. This continual confrontation between experiment and theory over the last few years has led to our present enlightened era where fission is giving us new insights into nuclear physics. Theoretical methods developed recently to obtain potential energy surfaces as a function of deformation parameters relevant to fission are now being applied to the field of heavy ion reactions. One can expect further cross fertilization between fission and nuclear physics as the dynamics of both fission and reactions between very heavy ions continue to be explored.

This volume is written at a level to introduce students to the exciting field of the physics and chemistry of fission. At the same time, we have attempted to cover recent developments in nuclear fission at a sufficient depth to make the volume valuable also to research scientists. The theoretical framework for understanding fission is developed in a systematic way and discussed in terms of the most recent experimental observations.

We are much indebted to a large number of our colleagues for numerous helpful comments and discussions on fission over the years, and their enthusiastic willingness to allow us to make use of their data in the form of figures and tables. Our warm thanks go to S. Bjørnholm, I. Halpern, U. Mosel, J. R. Nix, J. Pedersen, W. N. Reisdorf, T. D. Thomas, and J. P. Unik for reading preliminary versions of portions of the manuscript. The typing of such material is a difficult task and we acknowledge J. Hutchison, L. Edwards, and B. Jaeger for their expert work.

CHAPTER

I

Introduction

A. Discovery of Fission

Following the discovery of the neutron, Fermi (1934) and co-workers initiated a series of experiments in which they irradiated natural uranium with neutrons. It was realized that such irradiations might lead to the production of element 93, and possibly even of elements with greater atomic numbers, by one or more successive β disintegrations following neutron capture. These early experiments catalyzed a considerable effort by a large number of investigators to further characterize the radioactivities which were produced on irradiating uranium and thorium with neutrons. The apparent discovery of transuranic elements by Fermi was, of course, of great interest to chemists. Different investigators associated the new radioactive species with various transuranic elements such as eka-rhenium, eka-osmium, and eka-iridium, whereas others associated them with radium and actinium. The fact that natural uranium bombarded with neutrons leads to the production of so many different radioactive species raised many questions and doubts about the interpretation of the results in terms of the discovery of transuranium elements. Confusion reigned in the field of transuranic elements from 1934 through most of 1938. An appreciation of this situation may be gained by reading the review article on transuranic chemistry written at that time by Quill (1938).

In 1938 Hahn and Strassman published a paper in which they confirmed earlier results showing that by using both Ba and La as carriers they could precipitate, with the Ba, radioactive species having half lives of 25 min, 110 min, and several days. From these grew daughter substances, precipitable with La, having half lives of 40 min, 4 hr, and 60 hr, respectively. The activities precipitating with Ba were attributed to isomeric states of ^{231}Ra and the activities precipitating with La were attributed to isomeric states of ^{231}Ac. The nuclear reaction producing these activities was assumed to be ^{238}U(n, 2α)^{231}Ra. The fact that the apparent (n, 2α) process of production of Ra from U was found to occur by thermal neutrons, as well as by fast neutrons, was cause for considerable concern that the reaction process had been erroneously assigned. Hence, Hahn and Strassman (1939) performed an elaborate series of experiments to prove rigorously the chemical identification of some of the so-called "radium" and "actinium" activities. The active "radium" activities in question were so similar chemically to barium that these two elements were the only ones to which the activities might be ascribed. In order to make sure beyond all doubt which of the two was correct, special experiments for distinguishing between them were performed. Much to the investigators' surprise, the activities were found to associate themselves with barium. This fact made it reasonable to infer that the daughter "actinium" activities were really isotopes of lanthanum, as was indeed found to be the case.

These unexpected and startling results indicated that uranium nuclei irradiated with neutrons could split into fragments of intermediate mass. The authors gave this explanation of their experimental findings with much reserve because such a conclusion seemed to be incompatible with already known properties of nuclei. However, other experiments produced additional evidence to support their interpretation of the experimental findings. In a second conclusive paper Hahn and Strassman (1939) showed beyond a doubt that the "radium" and "actinium" activities reported previously were isotopes of barium and lanthanum, respectively.

Meitner and Frisch (1939) recognized that if a nucleus divided into two fragments of comparable mass, the mutual Coulombic repulsion of the fragments would result in a total kinetic energy of about 200 MeV, an amount of energy available from the difference in the masses of the original nucleus and the two products. The predicted high kinetic energies of the fragments were very quickly confirmed experimentally by Frisch (1939), who observed the very large pulses produced by the fragments in an ionization chamber.

The discovery that the capture of a thermal or low energy neutron by a heavy nucleus resulted in the rupture of the nucleus into fragments of intermediate mass raised new theoretical problems. Meitner and Frisch (1939) were the first to suggest a theoretical explanation on the basis of a nuclear liquid drop model. They pointed out that just as a drop of liquid which is set into vibration may split into two drops, so might a nucleus divide into two smaller nuclei. These authors treated the stability of nuclei in terms of cohesive nuclear forces of short range, analogous to a surface tension, and an electrostatic energy of repulsion. They went on to estimate that nuclei with $Z \approx 100$ would immediately break apart. Since uranium had only a slightly smaller charge, they argued that it was plausible that this nucleus would divide into two nuclei upon receiving a moderate amount of excitation energy supplied by the neutron binding energy. To describe this exciting new process, Meitner and Frisch proposed the term nuclear "fission" in analogy to the process of division of biological cells. The stability of nuclei to fission was discussed by several authors in papers published in rapid succession, and in great detail in a comprehensive and classic paper on the theory of fission by Bohr and Wheeler (1939). An excellent review of the exciting events leading to the discovery of fission and the outburst of scientific activity immediately following the discovery has been published by Turner (1940). References to the early literature covering the period 1934–1939 are given at the end of this review.

B. Scope of the Book

Nuclear fission is an extremely complex reaction. In the fission process we are dealing with a cataclysmic rearrangement of a single nucleus into two nuclei. More than three decades have elapsed since the discovery of fission and an enormous number of articles have been published on the experimental and theoretical aspects of fission. In later chapters, we intend to cover most aspects of nuclear fission in a systematic and coherent way; however, our treatment will be in somewhat general terms, and we make no effort to give a comprehensive and exhaustive account of all fission work. Only those developments are treated in any detail which we feel give special insight into the understanding of fission.

Several compendiums on fission are available in the literature. Much of the earlier fission data is summarized by Hyde (1964). In addition, the proceedings of the International Atomic Energy Agency Symposiums on the Physics and Chemistry of Fission held in Salzburg (1965) and Vienna (1969) give detailed accounts of fission data and theory. A number of reviews of various aspects of fission have been published by Halpern (1959), Huizenga and Vandenbosch (1962), Wheeler (1963), Huizenga (1965), Wilets (1964), Fraser and Milton (1966), Gindler and Huizenga (1967), Fong (1969), and Brack *et al.* (1972).

Since fission is such a complex process, we attempt to describe it as a sequence of events starting with the formation of the excited compound nucleus (of course, spontaneous fission occurs from the ground state of the original nucleus) and ending with the decay of the radioactive fission products. Although a discussion of the fission process may be quite naturally divided into three major subjects concerned with events associated with the intermediate transition state nucleus, the scission configuration, and postscission phenomena, such a division must be considered quite arbitrary and of pedagogical value only.

Chapters II–VIII deal with the transition nucleus. In Chapter II, theoretical models of the fission barrier are discussed. Although investigators have used the charged liquid drop model of the nucleus with great success in describing the general features of the barrier in nuclear fission, experimental evidence shows that the barrier, at least for some nuclei, has structure. The experimental discoveries of fission isomers and subthreshold neutron fission resonances initiated a new wave of interest in nuclear physics and in fission in particular. Much of the pioneering work in the liquid drop model of fission and in the application of nuclear shell corrections to nuclidic masses was done by Swiatecki and collaborators (Cohen and Swiatecki, 1962, 1963; Myers and Swiatecki, 1966). The general dependence of the potential energy on the fission coordinate for a heavy nucleus like ^{240}Pu is shown in Fig. I-1. The expanded scale used in this figure shows the large decrease in energy of about 200 MeV as the fragments separate to infinity. It is known that ^{240}Pu is deformed in its ground state, which is represented by the lowest minimum of -1813 MeV

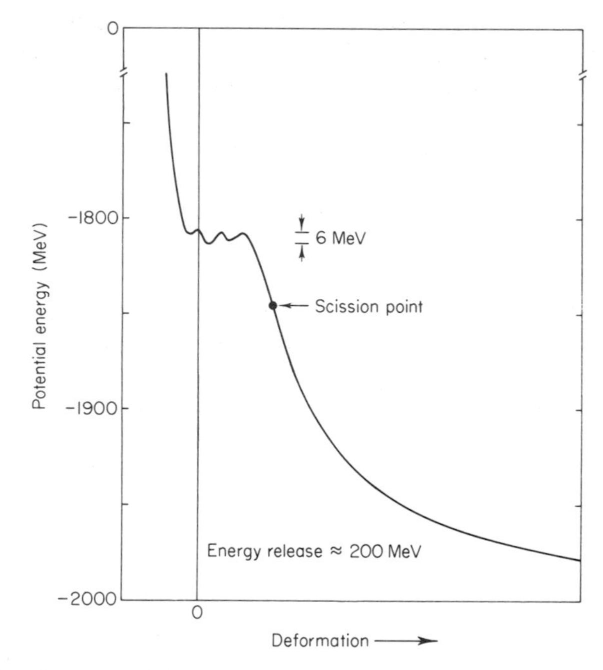

Fig. I-1. Plot of the potential energy in MeV as a function of deformation for the nucleus ^{240}Pu. The local fluctuations are obtained from a nonself-consistent microscopic model, whereas the smooth trends are obtained from a macroscopic (liquid drop) model. [After Bolsterli *et al.* (1972).]

near zero deformation. This energy represents the total nuclear binding energy where the zero of potential energy is the energy of the individual nucleons at a separation of infinity. The second minimum to the right of zero deformation illustrates the fission barrier structure introduced by shell corrections to the liquid drop mass. Although shell corrections introduce small wiggles in the potential energy surfaces as a function of deformation, it is important to remember that the gross features of potential energy surfaces are reproduced by the liquid drop model. This is illustrated in Fig. I-2, where the experimentally determined ground state masses of the β-stable nuclei are compared with liquid drop theory. The quantities actually plotted are the mass decrements from the whole number masses in units of energy (MeV). It can be seen that the liquid drop theory reproduces the overall trend of the masses throughout the periodic table to within a few MeV. There are, however, marked deviations from the predictions of the liquid drop model in the total nuclear binding energies in the vicinity of the well-known neutron and proton shells. The first method for quantitatively introducing a shell correction to the liquid drop potential energy as a function of deformation was introduced by Strutinsky (1967, 1968). This procedure is now widely used to calculate potential energy surfaces and has had considerable success in explaining phenomena associated with the two-humped fission barrier. Extension of this procedure to reflection-asymmetric shapes may finally be giving convincing insights into the origin of the long-standing puzzle of the strong preference of heavy elements to fission asymmetrically.

The calculation of the potential energy curve illustrated in Fig. I-1 may be summarized as follows. The smooth potential energy obtained from a macroscopic (liquid drop) model is added to a fluctuating potential energy, representing the shell and pairing energy corrections, derived from a non-self-consistent microscopic model. The calculation of these corrections requires several steps, namely (1) specification of the geometrical shape of the nucleus, (2) generation of a single particle potential related to this shape, (3) solution of the Schrödinger equation, and (4) calculation from these single particle energies of the shell and pairing energies.

The subject of barrier penetration and spontaneous fission is discussed in Chapter III. The lifetime for spontaneous fission from the ground state gives a measure of the width and height of the total barrier. The discovery of spontaneously fissionable isomers has added a new dimension to such studies in that these lifetimes give a measure of the properties of the outer barrier. Theoretical calculations for simple barrier shapes and assumptions about the inertial parameter are performed and compared with experimental results.

In Chapter IV we discuss the subject of fission widths. For fission of ^{239}Pu with resonance neutrons, it is well known that many combinations of fission fragments result. From these results we might conclude erroneously that fission

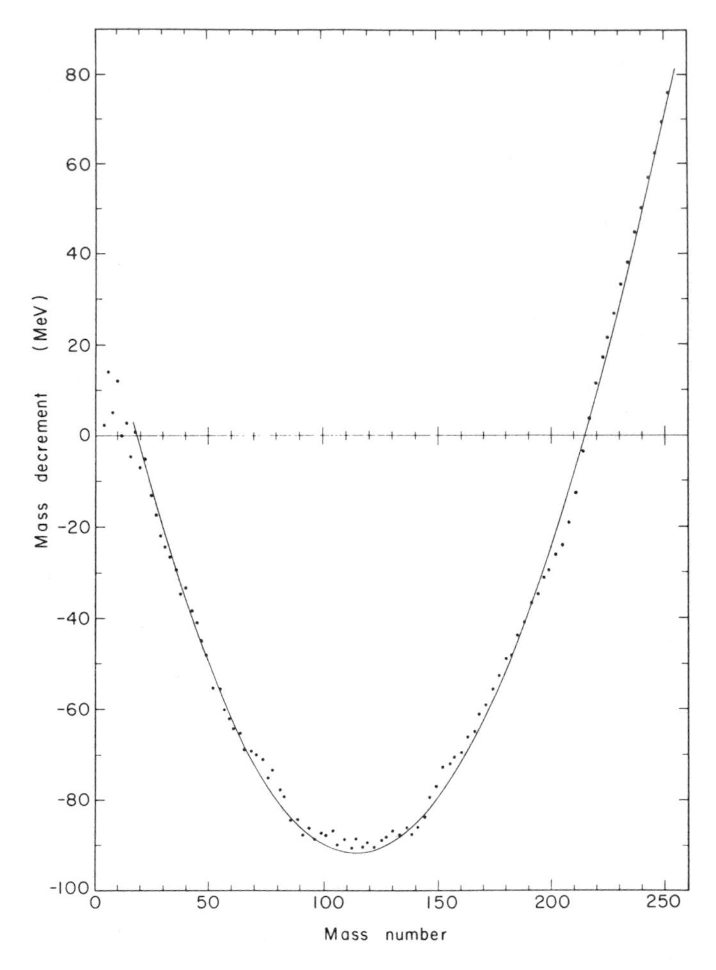

Fig. I-2. Comparison of the experimentally determined ground state masses of β-stable nuclei with masses from the liquid drop theory. The points represent the measured masses and the smooth curve the liquid drop part of the first mass formula of Myers and Swiatecki (1966). The quantities actually plotted are the mass decrements from the whole number masses in units of energy (MeV).

is a many channel process. It turns out, however, that an examination of the fission width distributions for both the $J = 0^+$ and $J = 1^+$ compound resonances in ^{240}Pu, for example, shows a chi-squared distribution with one or two open channels. A similar result is obtained if the number of open channels is calculated from the statistical formula $\sum T = 2\pi \langle \Gamma_f{}^J \rangle / \langle D_J \rangle$, where $\langle D_J \rangle$ is the average spacing between levels in the compound nucleus with spin J. Such results give supporting evidence to the concept of a transition or intermediate nucleus which has very low excitation energy and correspondingly few available levels. Hence, even though a large array of pairs of fission fragments eventually are formed, the fission process is a few channel process for low energy neutron fission.

Recent measurements of the subthreshold neutron fission cross sections of several nuclei have revealed groups of fissioning resonance states with wide energy intervals between each group where no fission occurs. Such a spectrum is illustrated in Chapter IV. In between the groups of fissioning resonance states are many other resonance states known from data on the total neutron cross section, but these have negligible fission cross sections. Such data can be explained in terms of a two-humped fission barrier. The mechanism for slow neutron subbarrier fission is postulated to proceed in the following way. The interaction of a slow neutron with the target nucleus produces a compound nucleus with a deformation comparable to that of the target and with a particular excitation energy E (see Fig. IV-3). This energy is the sum of the neutron binding energy and the neutron kinetic energy. By variation of the neutron kinetic energy, different compound resonance states are reached. Each of the compound resonance states decays by one or more of the available exit channels, such as emission of γ rays, emission of neutrons, and fission. However, if and only if the energy of a state at the initial deformation accidentally is almost equal to the energy of a level in the second potential well at another deformation will the nucleus undergo a transition from the compound nuclear level to the intermediate state, which, in turn, may fission. Such a mechanism gives rise to the experimentally observed groups of fissioning resonance states.

The concept of the transition nucleus was first applied by Bohr (1956) to explain fission-fragment angular distributions. The theoretical angular distribution of the fission fragments is based on two assumptions. First, the two fission fragments are assumed to separate along the direction of the nuclear symmetry axis, so that the angle θ (see Fig. V-1) represents the angle between the body-fixed axis and the space-fixed axis. Second, it is assumed that the transition from the saddle point (or transition nucleus) to scission is so fast that the Coriolis forces do not change the values of K established at the saddle point. The validity of these assumptions has been well established by experiments.

In Chapter V we discuss the low lying levels of transition nuclei as measured by photofission, neutron-induced fission, and direct reactions exemplified by the (d, pf) and $(\alpha, \alpha'f)$ reactions. It has been known for some time that the fission-fragment angular distributions may have structure at energies near the fission barrier. This structure changes for isotopes of some elements over energy intervals of the order of 100 keV and has usually been interpreted in terms of the Bohr model with different K values of specific fission channels contributing near the barrier. In the case of two barriers, the question arises which of the two barriers is responsible for the structure. If the second barrier is the higher, then it is postulated that the structure is associated with the second barrier. On the other hand, if the first barrier is the higher, some ambiguity exists as to which barrier controls the low energy angular distributions. If the second well is deep, it can be argued that the nucleus will pass through this region slowly enough so that the K value with which it passed over the first barrier is altered. The angular distribution may then reflect the K distribution at the second barrier. If this second barrier is sufficiently lower than the first barrier, many channels may be open, thus resulting in a statistical distribution of K. Although some experimental data for very heavy nuclei are consistent with this picture, such evidence for the double-humped barrier is rather indirect and sometimes unconvincing.

Experimental fission-fragment angular distributions show the presence of collective levels in even-even transition nuclei near the barrier. Photofission experiments have located, for example, the $J = 1^-$, $K = 0$ and $J = 2^+$, $K = 0$ levels in several even-even nuclei. Near-barrier-induced fission of even-even targets by direct reactions, such as the $(\alpha, \alpha'\mathrm{f})$ reaction, has clearly shown the presence of the $K = 0, J = 0^+, 2^+, 4^+, \ldots$ rotational band with excitation of high spin members in the transition nucleus. Excitation of Nilsson states in odd-A transition nuclei by neutron fission and by direct reaction followed by fission, such as the (d, pf) reaction, is well established also.

In Chapter VI we examine fission-fragment angular distributions at energies in excess of the barrier. At these energies we expect angular distributions given by statistical theory with a Gaussian K distribution. The standard deviation of the Gaussian K distribution is directly related to the deformation of the transition nucleus. Measurements of the angular distributions of fragments over a large range of nuclei have demonstrated that the shape of the saddle point changes in the direction predicted according to the liquid drop model of fission. As indicated previously, for low excitation energies the minimum number of levels in the transition nucleus corresponds to deformations associated with either the first or second barrier. For medium to high excitation energies, however, the minimum number of levels is thought to correspond to the liquid drop model barrier shape.

The competition between fission and neutron emission is treated in Chapter VII. In this section we assume the compound nuclear model with its basic assumption that the entrance and exit channels are independent. Hence, the reaction cross section of a particular process is separable into a product composed of two quantities, one quantity depending only on the entrance channels and the other quantity depending only on the exit channels. Except for conservation of the "constants of motion," including energy, angular momentum, parity, etc., the decay mode is completely independent of the formation mode.

The fission width for levels of all spins and both parities is calculated with the standard expression given previously, where the effective number of open channels is given by an integral over the level density in the transition nucleus. The total energy is partitioned between the fission and nonfission degrees of freedom and the energy in the nonfission degrees of freedom is the important quantity in the count of available channels at the saddle configuration.

The competition between fission and neutron emission depends on several factors, including the fission barrier, neutron binding energy, excitation energy, angular momentum, and the level density parameter of both the residual nucleus formed in neutron emission and the deformed intermediate transition nucleus. Since the deformation and the associated nuclear shell energies of these two nuclei can be very different, the effective values of the level density parameters may be very different. This in practice means that the measured fission barriers may in some mass regions deviate markedly from the liquid drop fission barriers.

Experimental information on fission barriers is discussed in Chapter VIII. Barrier heights are deduced from the excitation energy dependence of the neutron width to fission width ratios and from different types of excitation functions. Although experimental evidence confirms the double-humped fission barriers in the actinide region, certain details of the barrier structure are difficult to obtain from such measurements.

In the scission stage of fission the highly deformed nucleus, composed of a single nuclear potential, divides into at least two fragments, each with a separate nuclear potential. At the time of scission the nuclear deformation has increased beyond that of the transition state nucleus. As the mass of the fissioning nucleus decreases, the deformation of the transition nucleus approaches the scission deformation. At the instant of separation the primary fragments possess, in some cases, considerable deformation energy in addition to their other forms of energy.

Heavy nuclei near uranium in the periodic table fission asymmetrically to give light and heavy fragments. This early experimental result has been one of the more difficult puzzles for fission theorists to explain. Although various explanations based on the effects of nuclear shells either in the transition state nucleus or in the primary fragments have been proposed, no completely satisfactory theory has been developed to date. In Chapter IX we deal with theories of mass and energy distributions.

With the availability of large computers, extensive calculations based on the liquid drop model have been performed over the last few years in order to map out the potential energy surfaces of heavy nuclei as a function of their deformation. Cohen and Swiatecki (1962, 1963) and Strutinsky *et al.* (1963) have located the saddle points or transition state deformations for different nuclei. These early calculations gave symmetric shapes for the saddle points.

Recently, more realistic theoretical calculations have been performed. These calculations incorporate shell structure into the liquid drop potential energy surface and include odd-multipole shape distortions. With the inclusion of P_3 (pear-shaped deformation) and higher order asymmetric deformation (P_5), there is theoretical evidence that the potential surface is lower for asymmetric shapes at large deformation for some nuclei. The validity of these results rests on the degree of accuracy obtained in the potential energy surface when it is calculated as a function of deformation by simply summing the shape-dependent liquid drop and shell correction energies. The potential energy surface as a function of deformation between the saddle point and the scission shape is presently not well known but of extreme importance in fission theory. For heavy nuclei the scission deformation is much larger than the saddle deformation and it is necessary to consider the dynamics of the descent from saddle to scission.

One of the crucial questions in the passage from saddle to scission is the extent to which this process is adiabatic with respect to the particle degrees of freedom. As the nuclear shape changes through the value where a filled independent particle level is crossed by an unfilled level, we are interested in knowing the probability for the nucleons to remain in the lowest energy orbital. If the collective motion toward scission is very slow, then the single particle degrees of freedom continually readjust to each new deformation as the distortion proceeds. In this case, the adiabatic model is a good approximation. Motion along the lowest potential energy surface involving such a continual readjustment of nucleons, so that the lowest energy orbitals are always occupied, gives a substantial contribution to the collective inertia. In the adiabatic approximation the decrease in potential energy from saddle to scission appears in the collective degrees of freedom at scission, primarily as kinetic energy associated with the relative motion of the nascent fragments. On the other hand, if the collective motion between saddle point and scission is so rapid that equilibrium is not attained, then there will be transfer of collective energy into nucleonic excitation energy. Motion along a potential energy surface in which the nucleons do not readjust their orbitals gives a smaller inertial parameter. Such a nonadiabatic model in which collective energy is transferred to the single particle degrees of freedom during the descent from saddle to scission is often discussed in terms of the statistical theory of fission.

Present experimental evidence does not allow us to decide unequivocally in favor of the adiabatic or the nonadiabatic model. There is some experimental indication that the adiabatic inertial parameter is much larger than the irrotational value; in addition, the prescission kinetic energies of some heavy elements have been estimated to be sizeable. The latter piece of evidence indicates that the adiabatic approximation has some validity. However, as discussed in Chapter IX, the descent from saddle to scission is complicated. By this is meant that some degrees of freedom may be nearly adiabatic while others may be nearly nonadiabatic.

A number of light charged particles have been observed to occur in fission with low probability. These particles are believed to be emitted very near the time of scission; hence, they are of extreme interest. The evidence for these particles being emitted near the scission stage comes mainly from detailed studies of the α particles accompanying fission as discussed in Chapter XIV. Available evidence also indicates that neutrons are emitted at or near scission with considerable frequency. Experimental studies of the kinetic energy and angular distributions of the light charged particles have contributed to our understanding of the scission configuration. Additional information has come from measured correlations between the energy and emission angle of the light particle, between the fragment and particle energies, and between the angle of particle emission and the fragment mass ratio. Calculations aimed at reproducing these experimental observations support the evidence already mentioned that the fragments are moving apart at scission with appreciable kinetic energy.

In Chapter X we discuss the kinetic energy release in fission. A large fraction of the total energy release in fission goes into the kinetic energy of the fission fragments. The origin of the kinetic energies of the fragments is twofold. The major fraction of this energy is a result of the Coulomb repulsion between the nascent fragments which are created at the scission point. As discussed earlier, the remaining kinetic energy consists of the translational kinetic energy acquired prior to scission. The relative fraction of the total kinetic energy which is acquired before the scission point depends critically on the charge and mass of the fissioning system. Nuclei with highly deformed saddle shapes (i.e., shapes approaching the scission shape) are expected to have only a very small fraction of their kinetic energies acquired prior to scission, while nuclei with slightly distorted saddle shapes may have a significant fraction of their kinetic energies acquired prior to scission.

The overall average kinetic energy as a function of the charge and mass of a fissioning nucleus may be correlated as a function of the nuclear parameter $Z^2/A^{1/3}$. However, for a fixed fissioning nucleus, there are sizable variations in the kinetic energy release from event to event. In Chapter X we discuss these variations in terms of the dependence of the kinetic energy on mass division, excitation energy, angular momentum, and charge division.

The distribution of mass and charge in fission is discussed in Chapter XI. Heavy nuclei ($Z \geqslant 90$) fission asymmetrically to give light and heavy fragments. As the mass of the fissioning nucleus increases, the mass of the heavy fragment remains approximately the same, while the mass of the lighter fragment increases. Very recent measurements for the heaviest targets indicate that the two peaks are starting to merge, resulting in low energy symmetric fission. Although considerable experimental information about asymmetric fission has been known for a long time, this process has been a difficult puzzle for fission theorists to explain. As the excitation energy increases, the probability for division into two equal fragments increases. At high energies, the mass distribution is peaked about a symmetric mass division. The mass distributions for nuclei in the vicinity of radium are known to have three peaks. The question arises whether this effect is due to the potential energy surface of these nuclei or to multichance fission. Lighter nuclei with $Z \leqslant 84$ fission symmetrically at all excitation energies, although the width of the mass distribution increases with excitation energy.

The extent to which the potential energy surface at the saddle point (including the single particle energy corrections) can account for the trends in the mass distribution is a subject under active investigation. A number of authors have calculated that the second barrier for actinides has an asymmetric shape. Hence, at low energies an actinide nucleus starts its descent from the saddle point to scission in an asymmetric shape. It should be pointed out, however, that for actinide nuclei, it is a long way to scission. The degree to which the mass division is correlated with saddle shape is a subject requiring much more extensive investigation.

Unlike the mass division, the qualitative aspects of the charge division are explainable in terms of present fission theories. There is fairly clear evidence that a small redistribution of charge occurs prior to scission. The magnitude of the redistribution in charge is about that expected on the basis of statistical equilibrium at scission, and supports the conclusion of statistical equilibrium in the degree of freedom associated with nuclear polarizability.

Many experiments have been designed to investigate postscission phenomena. After the fragments are separated at scission, they are further accelerated as a result of the large Coulomb repulsion. The fragments reach at least 9/10 of their final kinetic energies in approximately 10^{-20} sec. The initially deformed primary fragments are thought to collapse to their equilibrium shapes in a time period shorter than this. The point is that the time for neutron evaporation from the fragments is long relative to the time for fragment acceleration; hence, the evaporated neutrons are emitted from fully accelerated fragments. These evaporated neutrons comprise most of the neutron yield in fission. (The scission neutrons are estimated to constitute as much as 30% of the neutron yield of some nuclei, whereas the delayed neutrons produced after β decay make up only 0.2–0.6% of the neutron yield. Isomeric fission offers a new source of delayed neutrons of very low intensity.) The fragments, after neutron emission, lose the remainder of their energy by γ radiation with a lifetime of about 10^{-11} sec. Emission of neutrons and γ rays from fission is discussed in Chapters XII and XIII, respectively.

The variation of the number of emitted neutrons with fragment mass has been studied rather extensively because of its relation to the fragment excitation energy. The resulting functions are "saw-toothed" and asymmetric about the symmetric fragment mass. Minimum neutron yields are observed for nuclei near closed shells because of the resistance of nuclei with closed shells to deformation. Maximum neutron yields occur for fragments that are "soft" toward nuclear deformation. Hence, at the scission configuration the fraction of the deformation energy stored in each fragment depends on the shell structure of the individual fragments. After scission, this deformation energy is converted to excitation energy; hence, the neutron yield is directly correlated with the fragment shell structure. This conclusion is further supported by the correlation between the neutron yield and the final kinetic energy. Closed shells result in a larger Coulomb energy at scission for fragments that have a smaller deformation energy and a smaller number of evaporated neutrons.

The relative yield of prompt γ rays as a function of fragment mass is very similar to the relative yield of neutrons. The high multiplicity of γ rays from one fragment is connected with a low multiplicity in the complimentary fragment. The average γ-ray energy for fission is considerably smaller than that for neutron capture. In addition to the information on γ rays, some information on the x rays and conversion electrons from fission is discussed also in Chapter XIII.

References

Bohr, A. (1956). *Proc. Int. Conf. Peaceful Uses At. Energy, Geneva*, **2**, p.151. United Nations, New York.

Bohr, N., and Wheeler, J. A. (1939). *Phys. Rev.* **56**, 426.

Brack, M., Damgaard, J., Jensen, A.S., Pauli, H.C., Strutinsky, V.M., and Wong, C.Y. (1972). *Rev. Mod. Phys.* **44**, 320.

Bolsterli, M., Fiset, E. O., Nix, J. R., and Norton, J. L. (1972). *Phys. Rev. C.* **5**, 1050.

Cohen, S., and Swiatecki, W. J. (1962). *Ann. Phys. (New York)* **19**, 67.

Cohen, S., and Swiatecki, W. J. (1963). *Ann. Phys. (New York)* **22**, 406.

Fermi, E. (1934). *Nature (London)* **133**, 898.

Fong, P. (1969), "Statistical Theory of Nuclear Fission." Gordon & Breach, New York.

Fraser, J. S., and Milton, J. C. D. (1966). *Ann. Rev. Nuc. Sci.* **16**, 379.

Frisch, O. R. (1939). *Nature (London)* **143**, 276.

Gindler, J. E., and Huizenga, J. R. (1967). *In* "Nuclear Chemistry" (L. Yaffe, ed.), Vol. II, Chapter 7. Academic Press, New York.

Hahn, O., and Strassmann, F. (1938). *Naturwissenschaften* **26**, 755.

Hahn, O., and Strassmann, F. (1939). *Naturwissenschaften* **27**, 11.

Hahn, O., and Strassmann, F. (1939). *Naturwissenschaften* **27**, 89.

Halpern, I. (1959). *Ann. Rev. Nuc. Sci.* **9**, 245.

Huizenga, J. R., and Vandenbosch, R. (1962). *In* "Nuclear Reactions" (P. M. Endt and P. B. Smith, eds.), Chapter II. North-Holland Publ., Amsterdam.

Huizenga, J. R. (1965). *In* "Nuclear Structure and Electromagnetic Interactions" (N. MacDonald, ed.). Plenum, New York.

Hyde, E. K. (1964). "Nuclear Properties of the Heavy Elements," Vol. III, Fission Phenomena. Prentice-Hall, Englewood Cliffs, New Jersey.
Meitner, L., and Frisch, O. R. (1939). *Nature* (*London*) **143**, 239.
Myers, W. D., and Swiatecki, W. J. (1966). *Nucl. Phys.* **81**, 1.
Quill, L. L. (1938). *Chem. Rev.* **23**, 87.
Salzburg (1965). *Proc. IAEA Symp. Phys. Chem. of Fission.* IAEA. Vienna.
Strutinsky, V. M., Lyashchenko, N. Ya, and Popov, N. A. (1963). *Nucl. Phys.* **46**, 639.
Strutinsky, V. M. (1967). *Nucl. Phys. A* **95**, 420.
Strutinsky, V. M. (1968). *Nucl. Phys. A* **122**, 1.
Turner, L. A. (1940). *Rev. Mod. Phys.* **12**, 1.
Vienna (1969). *Proc. IAEA Symp. Phys. Chem. of Fission 2nd.* IAEA, Vienna.
Wheeler, J. A. (1963). *In* "Fast Neutron Physics" (J. B. Marion and J. L. Fowler, eds.). Wiley (Interscience), New York.
Wilets, L. (1964). "Theories of Nuclear Fission." Oxford Univ. Press (Clarendon), London and New York.

CHAPTER

The Fission Barrier

A. Liquid Drop Model

1. Introduction

The uniformly charged liquid drop model is of interest in providing a first orientation of the energetics associated with the nuclear distortion encountered in nuclear fission. The rationale for this model is similar to that for the Weizsacker semiempirical mass equation. The first and dominant term in the mass equation is proportional to the nuclear volume and expresses the fact that the nuclear binding energy is proportional to the number of nucleons. In discussing the distortion of nuclei, volume conservation is assumed, reflecting the saturation property of the nuclear density. Thus, this term is independent of distortion and does not appear in the liquid drop model of fission. The second term (of opposite sign) in the mass equation is proportional to the surface area and takes into account the reduction in binding associated with nucleons on the nuclear surface. This term is smallest for a sphere, and any distortion away from a sphere is associated with larger potential energy from this effect. The third term is the Coulomb term describing the repulsive force between the protons. For prolate distortion the Coulomb energy is decreased due to the larger average separation between the charge elements. A fourth composition-dependent term, proportional to $(N-Z)^2/A$, is designated the symmetry energy term. As expressed here, it depends only on the nuclear volume, hence does not depend on distortion. (The surface energy term is also made composition dependent in some mass equations.) Thus, the liquid drop model in its simplest form describes the potential energy changes associated with shape distortions in terms of the interplay between surface and Coulomb effects. In this idealization of the nucleus no account is taken of shell effects or residual interactions arising from independent particle motion in the nucleus. As will be shown later, the changes in Coulomb and surface energy with distortion are of comparable magnitude but of opposite sign: the resulting near-cancellation of these two effects can result in a significant modification of the potential energy surface by higher order contributions, such as single particle effects which are small compared to the Coulomb or surface terms individually.

Let us start our discussion of the potential energy changes associated with nuclear distortions by considering small distortions of a sphere. For small, axially symmetric distortions the radius can be written as

$$R(\theta) = R_0[1+\alpha_2 P_2(\cos\theta)] \qquad \text{(II-1)}$$

where θ is the angle of the radius vector, the coefficient α_2 is a parameter describing the amount of quadrupole distortion, and R_0 is the radius of the undistorted sphere. The distortions described by this expression are not exactly spheroidal, as for spheroidal distortions additional terms starting with $\alpha_4 P_4$ are required. For small spheroidal distortions α_4 is considerably less than α_2. The surface and Coulomb energies for small distortions are given by (Bohr and Wheeler, 1939)

$$E_s = E_s^{\,0}(1+\tfrac{2}{5}\alpha_2^{\,2}), \qquad E_c = E_c^{\,0}(1-\tfrac{1}{5}\alpha_2^{\,2}) \qquad \text{(II-2)}$$

where $E_s^{\,0}$ and $E_c^{\,0}$ are the surface and Coulomb energies of undistorted spheres. In order for the charged liquid drop to be stable against small distortions, the decrease in Coulomb energy $\Delta E_c = -\tfrac{1}{5}\alpha_2^{\,2} E_c^{\,0}$ must be smaller than the increase in surface energy $\Delta E_s = \tfrac{2}{5}\alpha_2^{\,2} E_s^{\,0}$. The drop will become unstable when $|\Delta E_c|/\Delta E_s = 1$, or when $E_c^{\,0}/2E_s^{\,0} = 1$. Following Bohr and Wheeler (1939), we will define the fissility parameter x to be equal to this ratio or

$$x = E_c^{\,0}/2E_s^{\,0} \qquad \text{(II-3)}$$

For x less than unity the drop will be stable with respect to small distortions. For x greater than unity there will be no potential energy barrier to inhibit spontaneous division of the drop.

Let us now evaluate the surface and Coulomb energies for the idealized spherical nucleus. From electrostatics

$$E_c^{\,0} = \tfrac{3}{5}(Ze)^2/R_0$$

where Z is the number of protons and e the protonic charge. The surface energy is the product of the spherical area and the surface tension Ω, giving

$$E_s^{\,0} = 4\pi R_0^{\,2}\,\Omega$$

These equations may be expressed in terms of the mass number A and the nuclear charge Z by fitting experimental nuclear masses with the semiempirical mass equation. Such an analysis has been performed by Green (1954), whose results give in units of MeV

$$E_c^{\,0} = 0.7103Z^2/A^{1/3} \qquad \text{and} \qquad E_s^{\,0} = 17.80A^{2/3}$$

(More recent analyses are also available, but usually include higher order terms or have had constants adjusted to fit fission barriers. For the moment, we prefer to consider the simpler form and see what predictions can be made.) Substituting these values into (II-3), we have $x = Z^2/50.13A$. Some typical values of Z^2/A and x are 35.56 and 0.71 for ^{238}U and 38.11 and 0.76 for ^{252}Cf. For Z larger than about 125, the simple liquid drop model would predict no barrier; hence, these nuclei would be expected to fission in a time comparable to a nuclear vibrational period.

2. Potential Energy Mapping and the Fission Barrier

a. Symmetric Deformations

In order to accurately describe distortions as large as are encountered at the top of the fission barrier, or the saddle point, we must include higher order polynomials than are given in Eq. (II-1). It is convenient to describe the drop shape in terms of an expansion in Legendre polynomials

$$R(\theta) = (R_0/\lambda)\left[1 + \sum_{n=1} \alpha_n P_n(\cos\theta)\right] \qquad \text{(II-4)}$$

The parameter λ is a scale factor required to ensure that the volume remains constant at the value for a sphere of radius R_0. An illustration of the shapes associated with different combinations of the leading coefficients α_2 and α_4 is given in Fig. II-1.

For a nucleus like ^{252}Cf, with a value of the fissility parameter x of about 0.76 (assuming the parameters of Green), the saddle occurs for $\alpha_2 = 0.636$ and $\alpha_4 = 0.158$ (the α_n's for higher n are almost an order of magnitude smaller than the leading terms). A cut through the potential energy surface corresponding to $dE/d\alpha_4 = 0$ is given in Fig. II-2. This cut follows the bottom of the valley of the two-dimensional potential energy surface described by the coefficients α_2 and α_4. The surface and Coulomb energies are also illustrated separately to show the large cancellation between the two contributions to the distortion energy. At the saddle, ΔE_s, the increase in surface energy relative to the sphere, is 84.5 MeV, while ΔE_c, the decrease in Coulomb energy, is 77.0 MeV. The net fission barrier is only 7.5 MeV. This illustration shows how carefully the surface and Coulomb energies must be calculated in order to

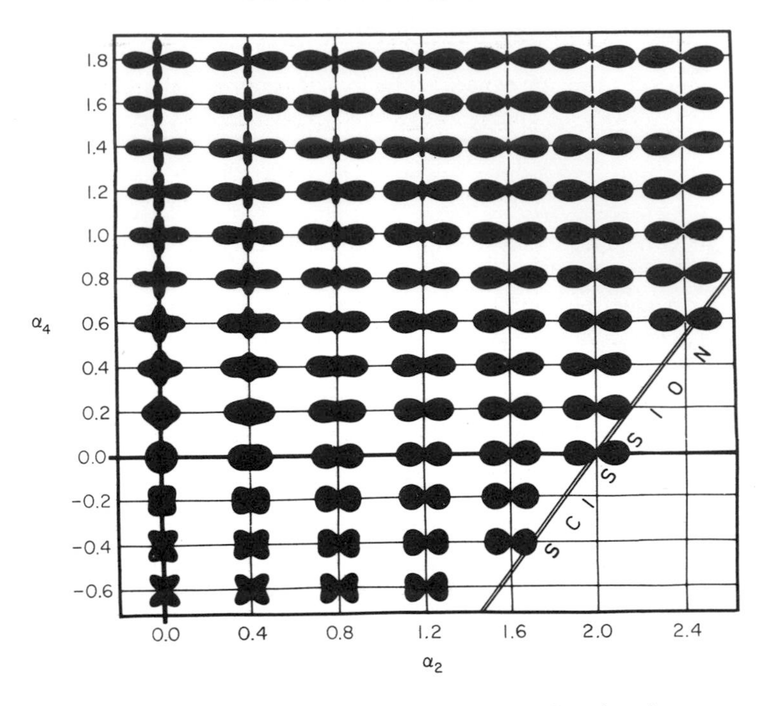

Fig. II-1. Shapes of nuclear surfaces corresponding to various locations on an α_2–α_4 map. The figures possess rotational symmetry about the horizontal axis. [From Cohen and Swiatecki (1962).]

obtain reasonably accurate distortion energies from the liquid drop model; it also alerts us to the likelihood that single particle and other effects not included in the liquid drop model may be important.

Rather extensive investigations of the equilibrium configuration have been performed by Cohen and Swiatecki (1963) and by Strutinsky *et al.* (1963). The mathematical methods used by the two groups are somewhat different, but in general the results are in good agreement. These saddle point shapes and energies can be fairly well reproduced for $x \geqslant 0.7$ by a simpler two-parameter representation given by Lawrence (1965). The saddle point shapes for various values of x are illustrated in Fig. II-3. As x decreases from unity down to 0.7, the saddle configuration becomes stretched to a cylindrical shape. As x decreases below 0.7, the saddle configuration develops a well-defined neck, and for x less than 0.6 the maximum elongation actually decreases. The transition from stretching to necking-in with decreasing x is rather sharp, occurring at $x \sim 0.67$. For x values slightly larger than this value, the total potential energy varies rather slowly with deformation near the equilibrium configuration, and the barrier is rather flat. This is illustrated in Fig. II-4.

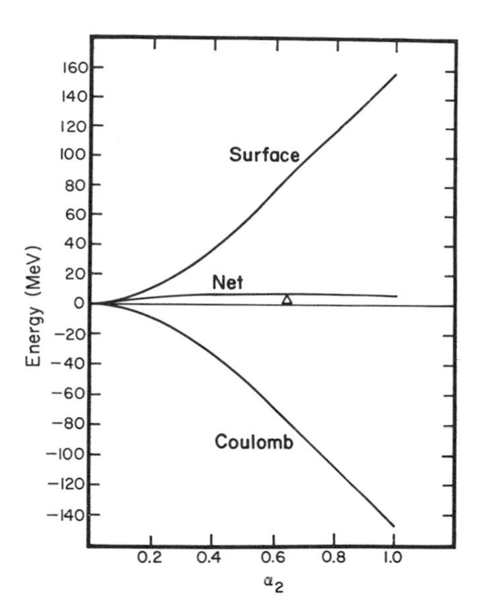

Fig. II-2. Surface, Coulomb, and net deformation energies (in MeV) shown as a function of α_2 for a cut through the potential energy surface corresponding to $dE/d\alpha_4 = 0$. All the coefficients higher than α_4 are zero. The triangle indicates the position of the saddle point. The fissility parameter x is 0.76 in this example, corresponding approximately to ^{252}Cf. [Constructed from a table of Cohen and Swiatecki (1962) based on a figure given by Frankel and Metropolis (1947).]

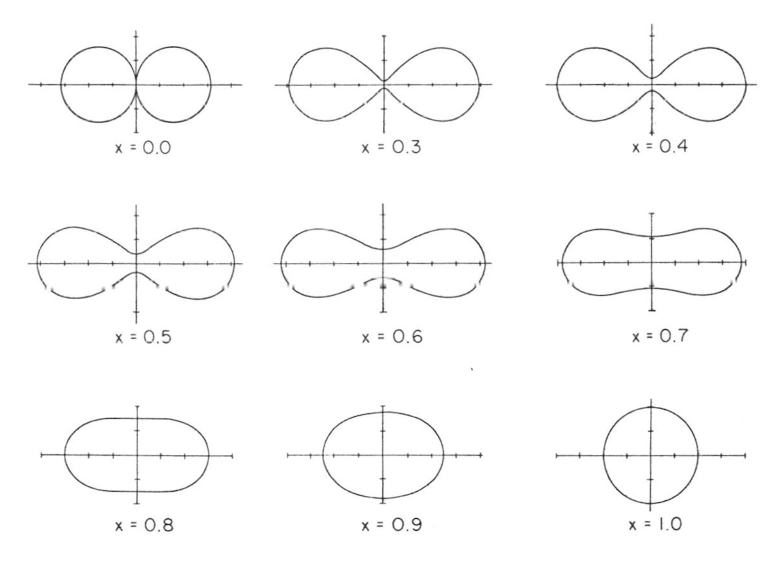

Fig. II-3. Saddle point shapes for various values of x. [From Cohen and Swiatecki (1963).]

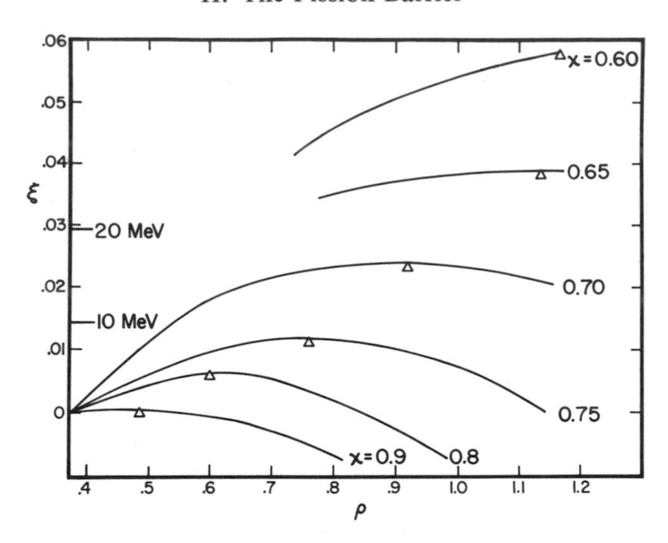

Fig. II-4. Potential energy of a uniformly charged liquid drop for selected values of the fissility parameter x. The left-hand scale on the ordinate gives the energy in units of the surface energy of a sphere, while the right-hand scale is the approximate energy in MeV for a nucleus like ^{238}U. The abscissa represents the distortion as characterized by a distortion parameter ρ, defined as the distance of the mass center of one of the halves of the nucleus from the center of the nucleus. ρ is given in units of the radius of a sphere of equal volume. The triangles are located at the saddle point for each x value. [Adapted from Strutinsky *et al.* (1963).]

The dependence of the liquid drop saddle point energies of Cohen and Swiatecki (1963) on fissility x is illustrated in Fig. II-5. The fission barrier heights observed in the heavy element region show a weaker dependence on x than predicted. Since the experimental barrier heights are the differences between the ground state and saddle masses, the deviations may reflect failure of the liquid drop model to describe the ground state masses as well as the saddle masses.

A simple formula (Cohen and Swiatecki, 1963) that approximately reproduces the calculated saddle point energies is

$$E = 0.38(0.75 - x) \qquad \text{for} \quad \tfrac{1}{3} < x < \tfrac{2}{3}$$

$$E = 0.83(1 - x)^3 \qquad \text{for} \quad \tfrac{2}{3} < x < 1$$

The energies in this formula are given in units of the surface energy of a sphere.

b. Refinements of the Liquid Drop Model

Apart from the obvious need to include some type of single particle effects in the description of nuclei, there are other possible refinements of the liquid drop model which might be included, at least in principle. These refinements include modifications for a diffuse surface, nonuniform charge distribution, compressibility, and a curvature dependent surface tension. A composition dependence of the surface as well as the volume term has also been used. Of particular relevance to fission are those corrections which are dependent on distortion. Strutinsky (1964a) has investigated some of the effects of a curvature-dependent surface tension which takes into account the deformation of the diffuse layer of the nucleus. The surface energy was assumed to have the form

$$E_s = [4\pi(1-\Gamma)]^{-1} \int dS(1-\Gamma H)$$

where H is the mean curvature ($H = 1$ for a sphere) and Γ is a constant expected to be negative and of the order of magnitude of the ratio of the thickness of the

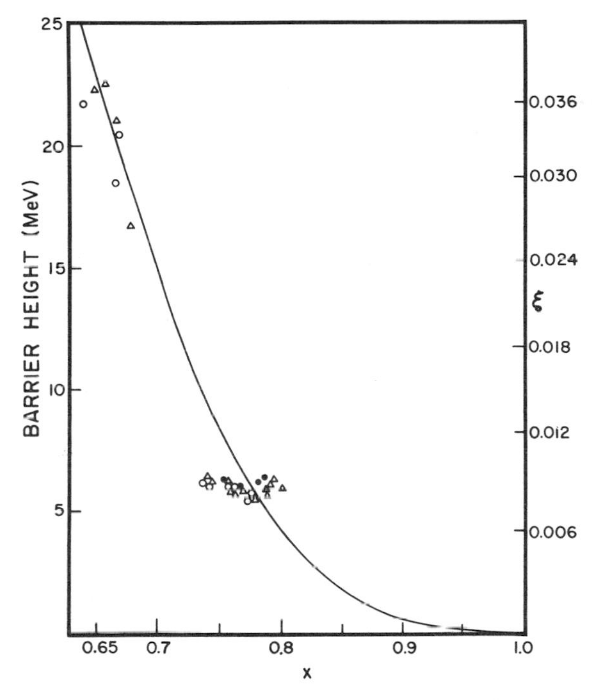

Fig. II-5. The simple liquid-drop model prediction of the fission barrier height as a function of the fissility parameter x (Cohen and Swiatecki, 1963). In order to present the results in MeV, the ordinate corresponds at each x value to typical values of Z and A for known nuclei, using the surface energy constant of Green (1954). The experimental barrier heights were plotted at x values assuming $(Z^2/A)_{crit}$ is equal to 50.13. The ratio of the barrier height to the surface energy of a sphere is indicated by the slightly nonlinear scale at the right. ○ Even-even, △ odd-A, ● odd-odd.

diffuse layer to the radius of the nucleus, which is approximately equal to $A^{-1/3}$. Inclusion of this correction increases the saddle point energy, but this variation in the energy is smaller than the absolute value of the correction to the surface energy of the sphere. The latter effect decreases $(Z^2/A)_{\mathrm{crit}}$ hence, it increases the x value for a particular nucleus. Thus the change in saddle point energy with Γ is less than the change in saddle point energy for the two different x values of a given nucleus corresponding to the different values of $(Z^2/A)_{\mathrm{crit}}$ before and after the curvature correction. The overall result is that the predicted barriers are lower, hence in better agreement with experiment. No quantitative statement can be made, however, unless a redetermination of all the constants of the semiempirical mass formula with the inclusion of the curvature correction is performed to obtain a consistent value of $(Z^2/A)_{\mathrm{crit}}$. This has not yet been done.

c. Stability of Equilibrium Shapes with Respect to Asymmetric Distortions

Historically one of the incentives for investigation of the liquid drop model was the hope that the tendency toward asymmetric mass division might reflect some simple instability of a liquid drop. Unfortunately, the liquid drop model has not shed a great deal of light on this problem.

No asymmetric *equilibrium* configurations with lower energy than the symmetric equilibrium configurations have been found. It is well established that for $x \leqslant 0.39$ the saddle point configuration becomes unstable with respect to $\lambda = 3$ deformation as well as $\lambda = 2$ deformation. This is not particularly relevant to the problem at hand, not only because nuclear fission has only been studied for much larger x values, but also because this instability apparently corresponds to the sucking up of one fragment by the other. Cohen and Swiatecki have investigated the stability of symmetric equilibrium shapes for $x > 0.39$ and found them to be stable against asymmetric distortions. Strutinsky (1964b) has reported that for x values between approximately 0.75 and 0.85 the saddle configuration is unstable with respect to asymmetric deformation. The investigations of Cohen and Swiatecki (1963) and Nix (1967) do not substantiate this claim. The difficulty may be that Strutinsky's measure of the asymmetry depends on the curvature of the neck. At $x \sim 0.8$, the curvature goes to zero (cylinderlike neck shapes) and his asymmetric displacement becomes a shift of the drop's center of mass which has a zero restoring force.

3. Barrier Curvature

Besides the barrier height, other properties of the barrier of interest in fission are the curvature of the barrier at the saddle point and the total width of the barrier. The curvature is of importance in determining quantum mechanical penetration and reflection properties which determine the rate at which the barrier can be overcome as a function of energy. The total width of the barrier, or more specifically its full shape, is of interest in determining the barrier penetration rates which determine spontaneous fission lifetimes. Unfortunately the liquid drop model has been of limited usefulness for determining spontaneous fission lifetimes. There are two reasons for this. In the first place, the potential energy surface beyond the saddle point has not been adequately explored, partly because of difficulties in the expansions employed to describe the shape. In the second place, the liquid drop model predicts spherical ground states, whereas most fissionable nuclei are known to have appreciable deformations in their ground states. Thus, the liquid drop model barriers are probably too broad at their base.

The liquid drop model might be expected to have more success in predicting the barrier curvature at the top of the barrier. In order to obtain a quantity which can be compared with experiment, it is necessary to calculate both the potential and kinetic energies in the neighborhood of the saddle point. From the potential energy E the elements of the stiffness (or elastic) matrix K can be calculated. An investigation of the kinetic energy requires an assumption about the nature of the nucleonic matter flow, and it is usually assumed that the flow is perfectly irrotational. The calculation of the inertial matrix M is rather complicated and will not be discussed further (see Nix, 1967). Once the stiffness and inertial matrices have been obtained, we transform from the original α_n coordinates to the normal coordinates.

The frequency of the normal mode oscillation is given by $\omega_n = (\mathsf{K}_n/\mathsf{M}_n)^{1/2}$, and for the fission mode $n = 2$ is purely imaginary. It is customary to multiply the absolute value of ω_2 by $\hbar$ to obtain a quantity $\hbar\omega$ in energy units. We shall designate this quantity the barrier curvature energy; it is shown as a function of x in Fig. II-6. The magnitude of $\hbar\omega_2$ determines the ease with which a nucleus with energy less than the barrier will penetrate the barrier and one with energy more than the barrier will be reflected by the barrier. Small values of $\hbar\omega_2$ correspond to relatively flat barriers with small penetrabilities, whereas large values correspond to relatively thin barriers with large penetrabilities. For small values of $\hbar\omega_2$ the observed fission threshold will be very sharp, whereas for large values of $\hbar\omega_2$ the fission cross section will vary rather slowly with excitation energy in the neighborhood of the fission barrier. Experimental excitation functions and angular distributions can be analyzed to obtain values of $\hbar\omega$. Unfortunately, most of the available experimental information on the barrier curvature energy is for elements with $Z \sim 92$, where single particle effects have qualitatively changed the character of the barrier (see Section C). For such nuclei a second minimum in the potential energy surface is present, resulting in a double barrier. The energy dependence of the penetration of such a barrier is rather complicated and depends on the

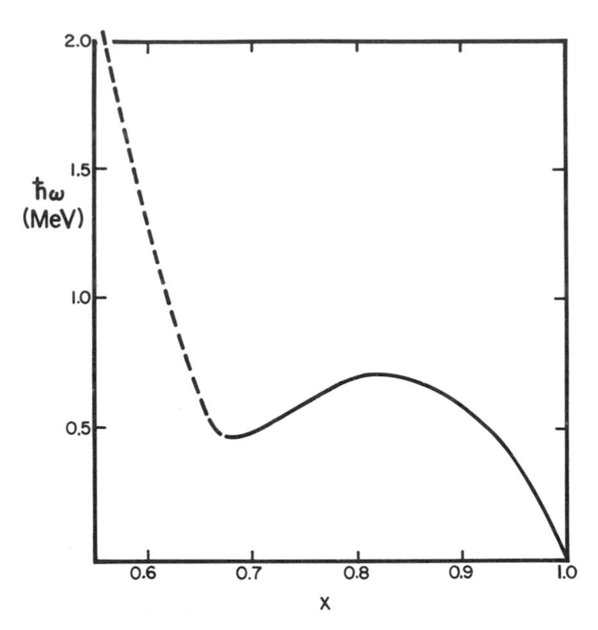

Fig. II-6. Barrier curvature energy as a function of x. The full curve is the liquid drop model prediction of the barrier curvature energy as calculated by Nix (1967). The dashed portion is of uncertain accuracy.

curvature of both barriers as well as on their separation and relative heights (see Section V,C,2).

B. Single Particle Model

1. Introduction

It is well known that the liquid drop model is an inadequate model for predicting many properties of nuclear structure, particularly effects associated with the shell structure of nuclei. The single particle or shell model has had considerable success in explaining the shell gaps and in describing the spin and parity of nuclei with an odd number of neutrons or protons. The single particle model assumes that each nucleon moves in a potential well that is an approximate representation of the interaction of that nucleon with all the other nucleons. If we calculate the eigenstates of a spherical potential whose dimensions are consistent with known properties of nuclei, we find that the correct ordering of levels is not obtained unless we also introduce a spin-orbit force.

With inclusion of the spin-orbit force, the spherical single particle model has been very successful in accounting for the N and Z numbers associated with regions of special stability corresponding to the filling of shells and subshells. It has also been successful in accounting for the spin and parity of many nuclei in the region close to the shell closures. However, it is known that nuclei far away from closed shells exhibit rotational spectra and are permanently deformed. If we extend the single particle model by allowing the potential to be nonspherical, it is possible to account for the spins and parities of the deformed nuclei as well. It has even been possible to use the single particle model to calculate the equilibrium deformations of deformed nuclei by calculating the total energy as a function of deformation and finding the deformation at which the energy is a minimum. In the simplest of such calculations the total energy is obtained by summing the lowest N neutron and the lowest Z proton level energies for each deformation (Mottelson and Nilsson, 1959) and adding a contribution due to the change in Coulomb energy (Bes and Szymanski, 1961) associated with that deformation. Such calculations have been fairly successful in accounting for the experimental deformations of heavy rare earth and transuranic nuclei. The parameters characterizing the deformed potential were adjusted, however, so as to reproduce the observed level structure of these nuclei. In principle, it is possible to extend these calculations to larger deformations and, therefore, obtain fission barriers.

2. Single Particle Calculation of Deformation Energy

The first attempt in this direction was made by Primack (1966). He simply summed approximate single particle energies as a function of deformation for a spheroidal potential, and added a correction for the Coulomb energy. The single particle energies were obtained in the "asymptotic approximation" of Nilsson (1955). The resulting potential energy curve gave a ground state equilibrium deformation which was too small and a saddle deformation which was much smaller than that indicated by experiment. The barrier was also too narrow. The failure of this calculation is not surprising in view of the limitation to spheroidal deformations and the use of quite approximate energy levels. It should be recognized that this type of calculation suffers from a problem very similar to that of the liquid drop model, where there is a large cancellation between the surface and Coulomb energies. In the single particle model calculation the sum of the single particle energies, which includes the surface energy, is a very large number, and we are looking for small variations in this sum with deformation.

A more elaborate calculation has been performed by Nilsson and collaborators (1969). The potential used is a deformed harmonic oscillator with both P_2 and P_4 coefficients to account for spheroidal and neck-constriction degrees of freedom in describing the shape. The potential also includes terms proportional to the spin-orbit coupling $\mathbf{l}\cdot\mathbf{s}$ and to l^2. The latter term is included to correct for the well-known tendency of the harmonic oscillator potential to underestimate the binding of particles of large l due to the unrealistic shallowness of the harmonic oscillator well at large distances. The coefficients of the $\mathbf{l}\cdot\mathbf{s}$ and l^2 terms are chosen to reproduce the experimental level ordering for nuclei in their ground states (both spherical and deformed). An attempt to impose a constraint of constant density (volume) as a function of deformation has been formulated by requiring that the volume enclosed by any equipotential surface be conserved. The Coulomb energy has also been included, assuming a charge distribution for a spheroid of the same shape as the potential deformation.

The energy can now be mapped as a function of the distortion parameters ε and ε_4 employed by Nilsson. (To lowest order, $\alpha_2 = \frac{2}{3}\varepsilon$ and $\alpha_4 = -\varepsilon_4$.) We can then find the energy minimum from a variation of ε_4 for each ε value. The energy for this path as a function of ε is illustrated in Fig. II-7. Although a strong suggestion of a saddle appears at $\varepsilon = 0.5$, at larger ε the energy starts to rise again quite rapidly. This is believed to arise primarily from the failure of the constraint of constant volume within any equipotential surface to conserve constant density. That is, the matter distribution need not follow the

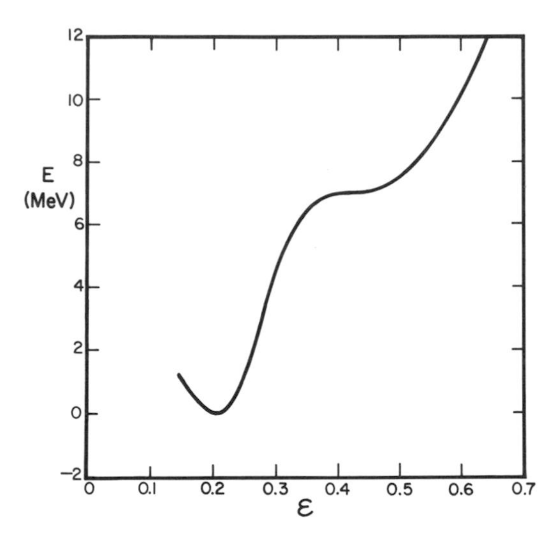

Fig. II-7. Single particle model calculation of energy as a function of the deformation parameter ε after minimization of the energy with respect to variation of ε_4. The rise in energy at large deformations illustrates the failure of the single particle model when the surface to volume ratio has increased greatly. [After Nilsson (1969).]

potential distribution. This can be expected in general for a deformed potential and is further complicated by the inclusion of the l^2 correction term.

C. Strutinsky Hybrid Model

1. The Shell Correction Method

In view of the failure of single particle models to predict reasonable deformation energies at large deformations, it is desirable to find some way of incorporating nucleon shell effects without requiring the single particle model to also properly predict the surface energy. Strutinsky (1967) has proposed a method whereby shell effects are considered as a small deviation from a uniform single particle energy level distribution. This deviation is then treated as a correction to the liquid drop model energy which contains the dominant surface and Coulomb effects. The dependence of the pairing strength on deformation can also be treated as a correction in a similar manner.

The total energy is then written as a sum of the liquid drop model (LDM) energy E_{LDM}, and the shell and pairing corrections $\delta U + \delta P$

$$E = E_{LDM} + \sum_{p,n} (\delta U + \delta P) \tag{II-5}$$

where the corrections for neutrons and protons are treated separately. All of the quantities are functions of the deformation. Let us consider the shell correction δU first. It is the *difference* between the sum of single particle energies for two different single particle models: a realistic shell model exhibiting nonuniform energy spacings and level degeneracies; and a "uniform" distribution. The shell correction is then given by

$$\delta U = U - \tilde{U} \tag{II-6}$$

The sum U for the shell model is given by

$$U = \sum_{\nu} 2\varepsilon_\nu n_\nu \tag{II-7}$$

where ε_ν are single particle energies of nucleon levels in a realistic shell model potential and the n_ν are the occupation numbers of these levels. For the "uniform" distribution

$$\tilde{U} = 2 \int_{-\infty}^{\lambda} \varepsilon \tilde{g}(\varepsilon)\, d\varepsilon \tag{II-8}$$

where $\tilde{g}(\varepsilon)$ is a "uniform" distribution of nucleon states, λ the corresponding chemical potential defined by $N = 2\int_{-\infty}^{\lambda} \tilde{g}(\varepsilon)\, d\varepsilon$, and N the total number of particles. The principle of this correction method is that any systematic errors arising from the general problem of calculating the total energy from a single particle model will cancel, and only effects associated with special degeneracies and splitting of the levels in the particular shell model potential will remain as a shell correction. It is therefore important that the "uniform" distribution be as consistent as possible with the overall shell distribution.

This is accomplished by taking the shell distribution and averaging it over a sufficiently large energy interval to wash out the shell effects. This averaging is performed by using a weighting function yielding

$$\tilde{g}(\varepsilon) = (\pi\gamma)^{-1/2} \sum_{\nu} \exp[\gamma^{-2}(\varepsilon-\varepsilon_{\nu})^{2}] \tag{II-9}$$

The sum in Eq. (II-9) is the number of levels in the energy interval $(\pi\gamma)^{1/2}$, which is centered at the energy ε. A correction factor, usually in the form of a polynomial multiplying the Gaussian, is introduced to avoid distortions in $\tilde{g}(\varepsilon)$ due to averaging over a finite energy interval. It can be shown that if γ is taken to be of the order of the energy difference between major shells, $\hbar\omega_0 \approx 7$ MeV, and if the above-mentioned correction factor is included, $\tilde{g}$ is not sensitive to the exact value of γ. Although this averaging procedure seems to work satisfactorily for single particle levels from a harmonic oscillator potential, difficulties are encountered if a finite potential of Woods–Saxon form is used (Lin, 1970). Somewhat arbitrary truncations have to be employed to avoid problems with the continuum states.

Some examples of the small correction δU as a function of deformation are shown in Fig. II-8a. The single particle energies used are from the Nilsson

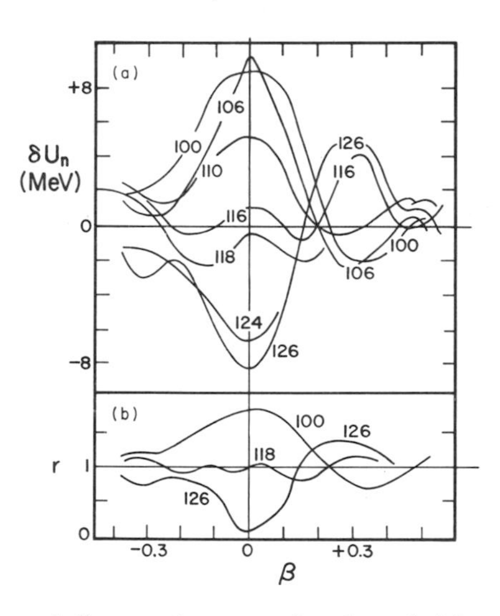

Fig. II-8. (a) Neutron shell corrections as a function of deformation using Nilsson single particle energies. Curves are shown for selected neutron number at and below the $N = 126$ shell. The energy scale is only approximate. (b) The ratio of the shell single particle level density to the uniform one at the Fermi energy. [From Strutinsky (1967).]

model. For spherical nuclei the correction is negative (corresponding to a lower total energy or stronger binding) for nuclei at or near closed shells, while for mid-shell nuclei the correction is positive. At some finite deformation β of the order of 0.3 the situation is typically reversed, with a positive correction for nucleon numbers close to the shell closure and a negative one for mid-shell nuclei. This behavior accounts for the observed equilibrium deformations of nuclei in their ground states. The shell correction which favors nonspherical shapes for mid-shell nuclei is large enough to override the favoring of the spherical shape from the LDM part of the total energy.

It is of interest to try to understand the oscillations exhibited in Fig. II-8a in somewhat more general terms. The shell correction δU is very sensitive to the density of single particle levels near the Fermi energy. These density variations reflect the shell-type compression and thinning out of levels as a function of deformation and energy or nucleon number. A "shell" level density $g_{\text{shell}}(\varepsilon)$ can be defined by using the weighting function of Eq. (II-9), but with a smaller averaging interval of the order of the distance between subshells, i.e., $\hbar\omega_0 A^{-1/3} \approx 1$ MeV. In principle, the sum $\sum_\nu 2\varepsilon_\nu n_\nu$ of Eq. (II-7) could be replaced by

$$\int_{-\infty}^{\lambda} 2\varepsilon g_{\text{shell}}(\varepsilon)\, d\varepsilon$$

although in practice, expression (II-7) is used when computing the total energy as a function of deformation. The "shell" level density is of interest in showing the direct connection between the shell correction δU and the fluctuation in the "shell" level density about the uniform level density. Strutinsky (1967) has found a strong correlation between the shell correction δU and the ratio of the "shell" density to the uniform density, as can be seen by comparing Figs. II-8a and II-8b. Qualitatively the correlation is due to the fact that since the Fermi energy is essentially independent of deformation, a decrease of level density at the Fermi surface is associated with the unfilled levels being pushed up (relative to the uniform density distribution) and the filled levels being pushed down. The sum of the single particle energies for the filled levels is therefore decreased relative to that for the uniform distribution. Similarly, compression of the levels at the Fermi surface leads to the opposite effect.

Myers and Swiatecki (1967) and Strutinsky (1968) have shown how the reversals of the shell correction as a function of deformation can be understood qualitatively by looking at the way bands of bunched levels become debunched with deformation. This is shown schematically in Fig. II-9. At deformations where levels from two shell bunches start to cross, regions of relatively higher level density appear at the same energy at which there were shell gaps for the spherical configuration. Similarly, for this same deformation at other energies there are regions of lower density where for the corresponding spherical case

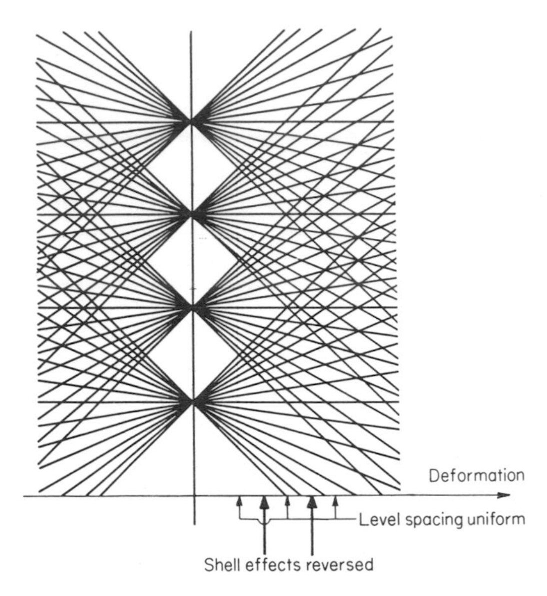

Fig. II-9. A highly schematic Nilsson level diagram with fans of levels radiating from equally spaced, completely degenerate bunches. The overlap of fans leads to regions of increased densities of levels. Every so often the level spacing becomes uniform, followed by a reversal of bunched and rarefied regions. [From Myers and Swiatecki (1967).]

the density is highest. Strutinsky has shown that this same qualitative behavior can be shown from an idealized schematic diagram which includes subshells. Both diagrams indicate that the level density exhibits further oscillations at larger deformations, although in the Myers and Swiatecki diagram the oscillations are quite rapidly damped. That these regularities are not washed out in a realistic energy level diagram can be seen by examining the Nilsson diagram for protons in Fig. II-10.

A highly schematic example of how crossings in the Nilsson diagram can be related to irregularities in the deformation energy is given in Fig. II-11. This example also illustrates how a deformation energy surface calculated from a single particle model (results of such a calculation are presented in Fig. II-7) is pulled down so as to make the average trend agree with the liquid drop model.

It should be mentioned that although the Strutinsky prescription (1967, 1968) is intuitively attractive and appears quite successful, there exists no complete theory to justify it. The problem of obtaining the potential energy as a function of deformation has been discussed by Bassichis and Wilets (1969). The validity of the procedure has been reinforced by recent comparisons (Bassichis *et al.*, 1972) of the deformation energy surface from constrained Hartree–Fock calculations with calculations based on the Strutinsky method

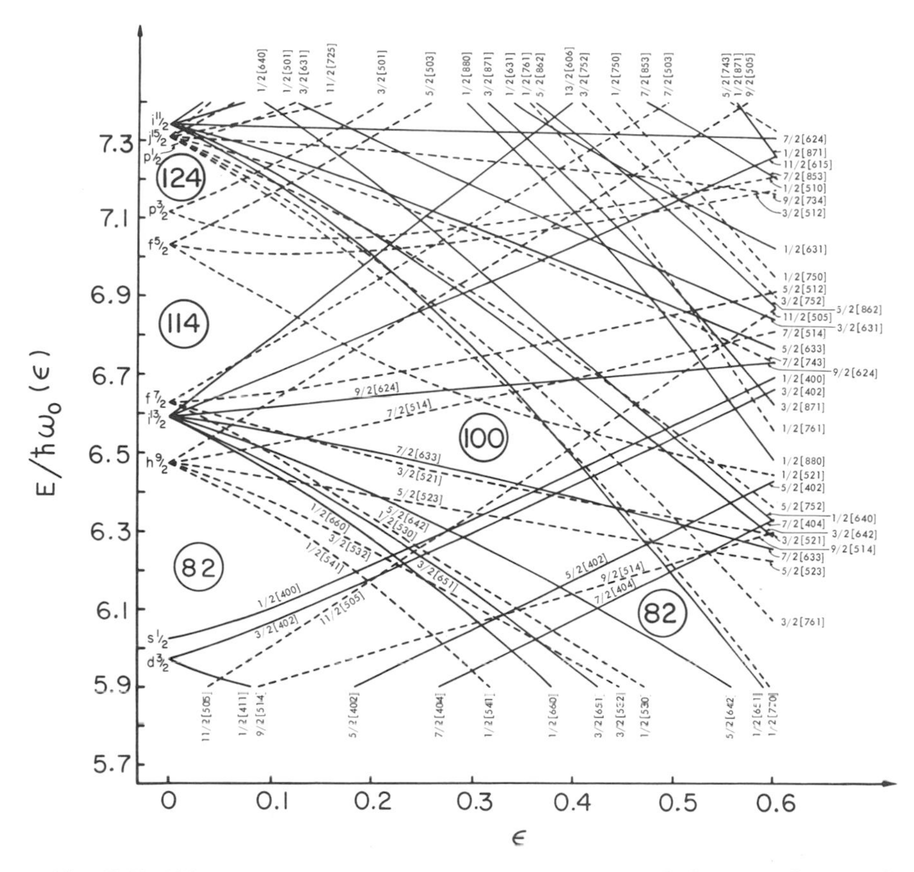

Fig. II-10. Nilsson diagram for odd protons, $82 \leqslant Z \leqslant 126$. [After Gustafson *et al.* (1967).]

using the Hartree–Fock eigenenergies. The test was performed for the heaviest nucleus, $A = 108$, for which Hartree–Fock calculations are presently practical. Both the deformation at which minima occurred and the approximate depth of the minima in the Hartree–Fock calculations were reproduced in the calculations based on the Strutinsky method.

Returning to our discussion of the calculation of the total energy with Eq. (II-5), we note that the pairing energy correction δP can be calculated in the usual approximation used in the application of the Bardeen, Cooper, and Schrieffer (BCS) theory to nuclei. The pairing energy is found to vary by several MeV with deformation, exhibiting modulated oscillations which are out of phase with the shell correction δU. The reason for this is that the pairing energy is also determined by the density of single particle states close to the

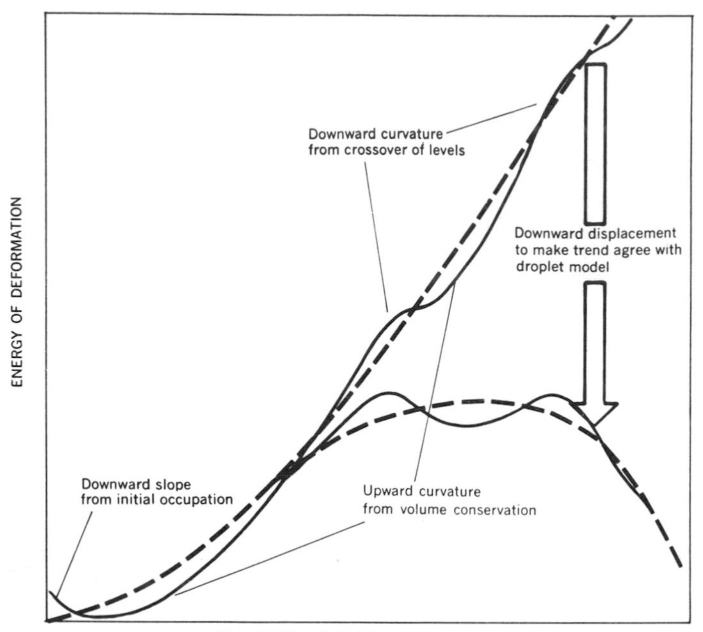
ENERGY OF DEFORMATION
Downward curvature
from crossover of levels
Downward displacement
to make trend agree with
droplet model
Downward slope
from initial occupation
Upward curvature
from volume conservation
ELONGATION OF DEFORMED NUCLEUS, δ

(a)

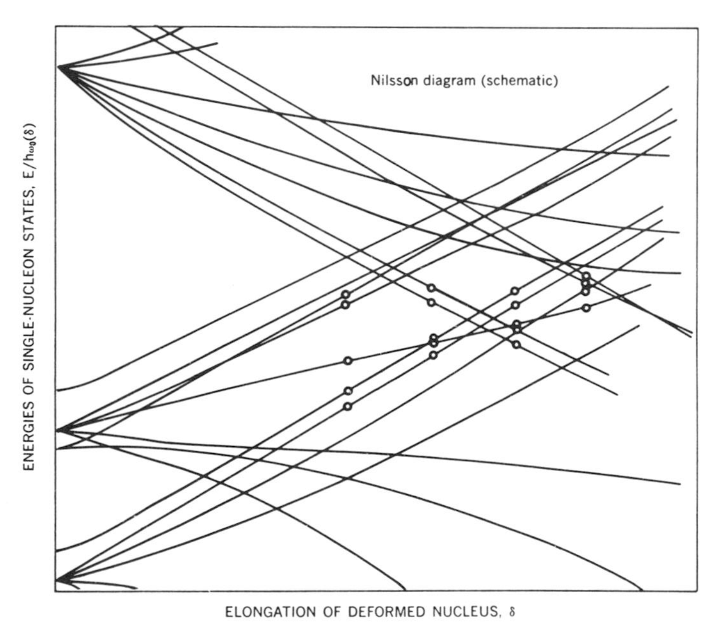
Nilsson diagram (schematic)
ENERGIES OF SINGLE-NUCLEON STATES, E/ħω₀(δ)
ELONGATION OF DEFORMED NUCLEUS, δ

(b)

Fig. II-11. Schematic extension of the Nilsson diagram shows (b) features that help to explain the energy curves in (a). The upper heavy curve in (a) is based on volume conservation within equipotential surfaces of the potential from which the single particle energies are calculated. The lower heavy curve is that obtained by requiring the general trend to be consistent with the liquid drop model as suggested by Strutinsky. Small open circles indicate the topmost occupied levels; they may refer to either neutrons or protons. Initial downward curvature of energy in (a) corresponds to the first hump of the fission barrier; thereafter, the curve wavers upward and downward about an average curve shown in dashed form. [From Inglis (1969).]

Fermi surface. In the case of the pairing correction, however, the total energy is lowered when there is an unusually high level density, since the higher the level density, the larger the number of levels into which pairs can be scattered. The pairing energy correction and the shell energy correction therefore reach their maximal and minimal values simultaneously, so that there is some

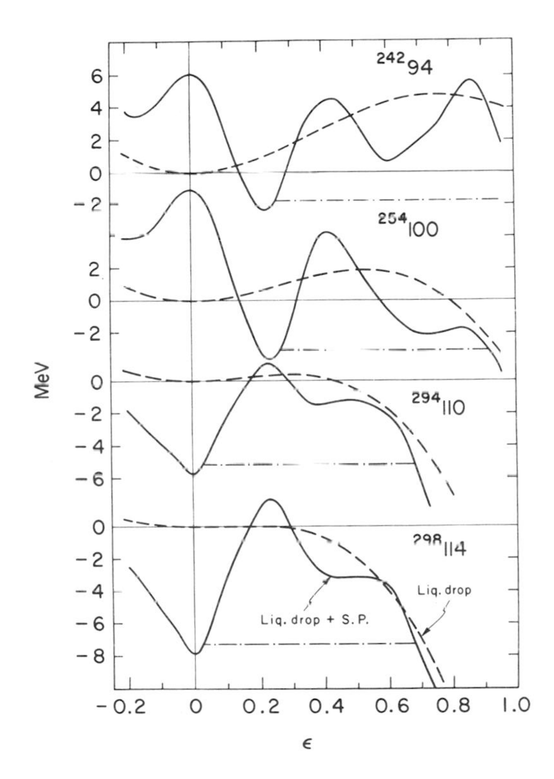

Fig. II-12. Potential energy minimized with respect to ε_4 as a function of ε for various nuclei illustrating the effect of shell structure on a liquid drop background. --- Liquid drop fission barriers; —— barriers after inclusion of shell and pairing effects. [From Tsang and Nilsson (1970).]

compensation from the pairing energy correction to the shell correction. As a result, the total deformation energy is somewhat less modulated than would be given by the shell correction alone.

Examples of the total nuclear deformation energy for heavy elements obtained from Eq. (II-5) are shown in Fig. II-12. The first minimum of the deformation energy occurs at approximately the known deformation of the nuclear ground state. Of special interest is the second minimum in the deformation energy in the vicinity of the saddle point. This minimum is caused by a second reversal in the oscillatory shell correction at this deformation. The shell correction is particularly large for neutrons when $N \sim 150$. There is evidence for the presence of such a second minimum in several nuclei from the known existence of spontaneously fissioning isomers (see Section III, D) and intermediate structure observed in resonance neutron fission studies (see Section IV, E). The second minimum begins to disappear for Cf and heavier nuclei, partly because the liquid drop energy falls off very steeply at a smaller deformation as illustrated in Fig. II-12.

2. Predicted Barriers for Superheavy Elements

An interesting application of this model is the prediction of fission barriers of superheavy nuclei as yet undiscovered. This problem has been investigated by a number of groups, using the Strutinsky renormalization method. The first step is to find potential parameters which reproduce the single particle energies for both deformed and spherical nuclei in order to be able to extrapolate to higher nucleon numbers with greater confidence. The single proton and single neutron level diagrams obtained by Nilsson *et al.* (1969) are illustrated in Fig. II-13. In this representation a larger shell gap occurs at $Z = 114$ than at $Z = 126$, a result indicated previously in calculations with a Woods–Saxon potential. The neutron shell does appear to have a moderately large gap at the expected nucleon number $N = 184$. The nuclide with $Z = 114$ and $N = 184$ is predicted to have a fission barrier of about 9 MeV. Deformation energies for several nuclei with $Z = 114$ are illustrated in Fig. II-14. Results from similar calculations of Muzychka *et al.* (1969) are illustrated in Fig. II-15. As the neutron number decreases, the barrier decreases. Unfortunately, the more stable isotopes of $Z = 114$ with $N \sim 184$ may be difficult to synthesize artificially because of their large neutron excess. However, some of the lighter $Z = 114$ isotopes may have spontaneous fission half lives long enough for observation. It is also interesting to note that the more neutron-rich species with large barriers are predicted to have half lives longer than the age of the universe. Thus, such nuclides might exist in nature providing they were produced in nucleosynthesis and that they exhibit sufficient stability with respect to α and β decay.

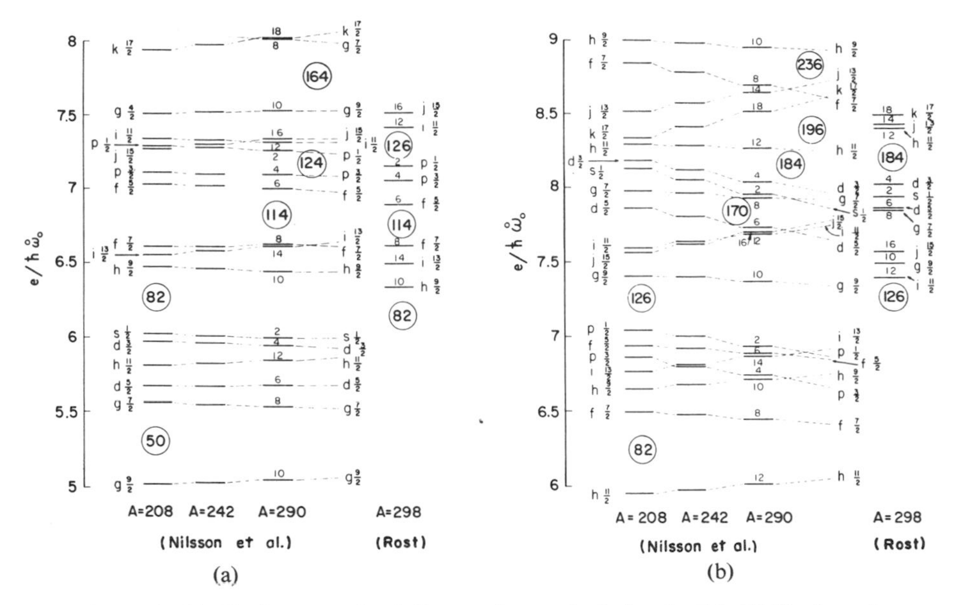

Fig. II-13. (a) Single proton level diagram for a spherical potential. (b) Single neutron level diagram for a spherical potential. [From Nilsson *et al.* (1969).]

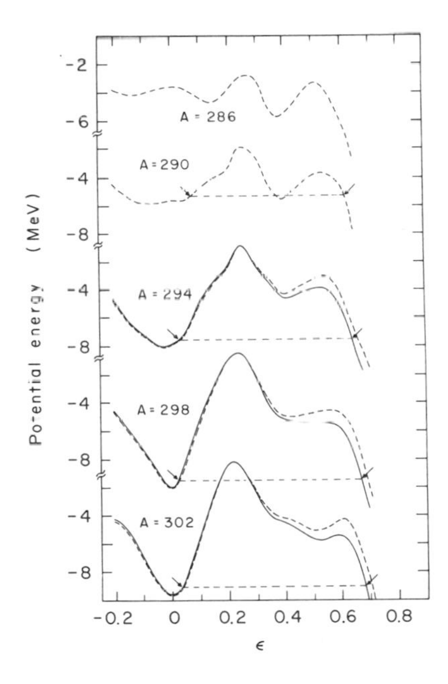

Fig. II-14. Minimum deformation energy projection along the ε axis for $Z = 114$, $A = 286$–302. Each point along the curve corresponds to an energy minimum with respect to ε_4 [From Nilsson *et al.* (1969).]

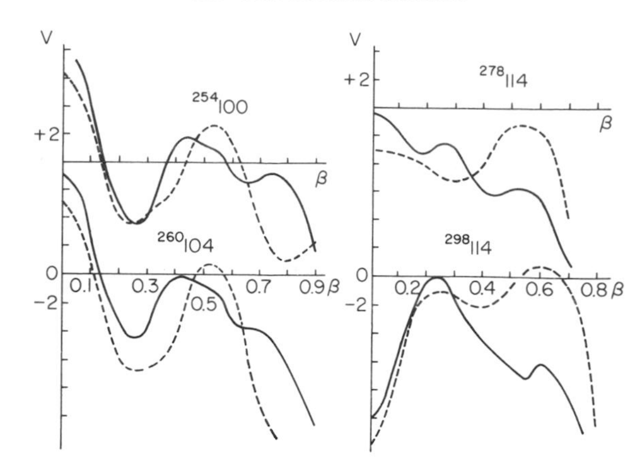

Fig. II-15. Deformation energies (in MeV) for different nuclei calculated with the Nilsson level scheme (——) and with the approximate level scheme for the deformed Woods–Saxon potential (– – –). [From Muzychka *et al.* (1969).]

3. Extension of the Strutinsky Method to Reflection-Asymmetric Saddle-Point Shapes

The shell corrected deformation energy surface calculations discussed thus far have been for reflection-symmetric shapes with only the $n = 2$ and $n = 4$ terms in Eq. (II-4). Although the liquid drop energy surface is stable with respect to asymmetric distortions, for large P_2 distortions the liquid drop energy surface becomes quite soft with respect to distortions that involve a certain combination of P_3 and P_5 deformations (Cohen and Swiatecki, 1963; Nix, 1967). It is thus possible that at certain deformations shell effects might overcome the liquid drop stability with respect to asymmetric distortions. Möller *et al.* (1969) have shown that inclusion of shell effects for the P_3 term alone does not destroy the stability of the liquid drop against asymmetric distortions. Inclusion of the P_5 term, as well, in such a way as to exploit the softness of the liquid drop energy surface, does result in the loss of stability with respect to asymmetric distortions for shapes with symmetric distortions corresponding to the top of the second barrier. The result is to lower the height of the second barrier by approximately 2 MeV for ^{236}U and ^{240}Pu (Möller and Nilsson, 1970; Pauli *et al.*, 1971; Bolsterli *et al.*, 1972). As a consequence, the theoretical barrier heights are in better agreement with experimental barrier heights (see Chapter VIII). The specific nucleonic orbitals primarily responsible for the instability have been discussed by Gustafson *et al.* (1971). The first barrier and second minimum are still stable with respect to asymmetric distortions. The decrease in the height of the second barrier due to the asymmetric degrees of freedom is less for the heavier actinides, amounting to only about

0.5 MeV for Cf and Fm. For the lighter elements the situation is unclear, with Möller and Nilsson (1970) reporting ^{210}Po ($Z=84$) stable with respect to asymmetric distortions, while Bolsterli *et al.* (1972) find it unstable. Pashkevich (1971) finds the nearby nucleus ^{208}Pb ($Z=92$) unstable with respect to asymmetric distortions. The dependence of the potential energy surface on asymmetric distortions is discussed in greater detail in Chapter IX in connection with mass asymmetry. Contour maps of the potential energy surface for such distortions are illustrated in Figs. IX-17 and IX-18.

D. Fission Barrier Height Systematics

A comparison of the experimental barrier heights with the simplest a priori expectations of the liquid drop model has already been given in Fig. II-5. A much more successful fit has been obtained by introducing semiempirical shell corrections to the ground state energy together with adjustment of the mass-equation coefficients to fit one of the barriers (Myers and Swiatecki, 1967) We concentrate our attention here on the barriers for the heavy elements, particularly their dependence on Z and A. We also examine any odd-even dependences of the barrier heights.

The fission barrier heights for the even-even and even-odd nuclides with $Z \geqslant 90$ are plotted as a function of the fissility parameter in the lower part of Fig. II-16. Two qualitative features of the data are apparent in this figure. The barrier heights depend rather weakly on the fissility parameter x, with curium isotopes having even higher barriers than the plutonium isotopes. The other feature is the near equality of the barrier heights for even-even and even-odd nuclei.

The weak dependence of the barrier heights on fissility and the larger-than-expected barriers for the curium isotopes presumably are a manifestation of the shell effects discussed in previous sections. It is possible to largely isolate the shell effects on the ground state energy from which the barriers are measured by correcting the observed barrier heights by an amount equal to the energy difference between the experimental ground state mass, which includes shell effects, and the mass predicted by the liquid drop part of a semiempirical mass equation. Swiatecki (1955) has found a similar correction to be very successful in accounting for the deviations in spontaneous fission lifetimes from a smooth dependence on fissility x (see Section III, B). Fission barrier heights corrected in this way are shown in the upper part of Fig. II-16. Also shown is the relative dependence of the barrier height on fissility implied by the simple liquid drop model as given by $E_f \propto (1-x)^3$. Although a stronger dependence on fissility is now observed, there is still a significant difference between the expected and observed dependence of the barrier heights on fissility. These differences are

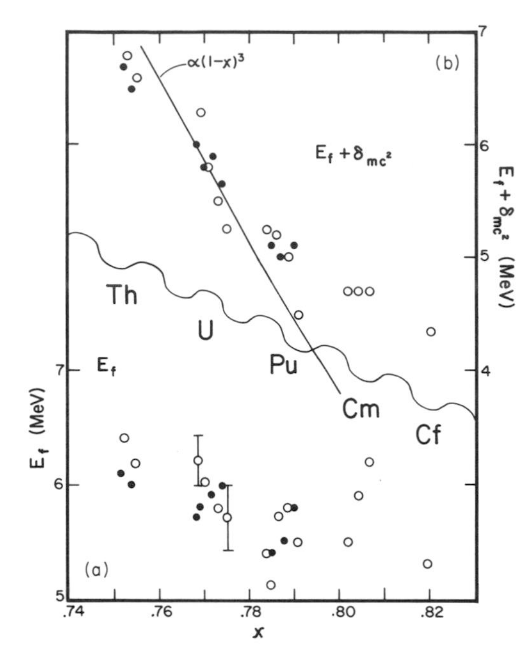

Fig. II-16. (a) Experimental fission barrier heights for the even-even and even-odd isotopes of heavy elements plotted as a function of the Myers–Swiatecki (1966) fissility parameter. Typical uncertainties in the barrier heights are indicated. (b) Plot of the sum of the experimental barrier heights and the energy difference between the experimental ground state mass and the liquid drop model ground state mass (Myers and Swiatecki, 1967). ● Even-even, ○ even-odd.

presumably due to shell effects at the deformation corresponding to the top of the barrier. Comparison with Fig. III-12 suggests that this is due to a large shell effect for the higher Z nuclei at the deformation corresponding to the first barrier. Comparison with the dependence of ground state spontaneous fission lifetimes on fissility, Fig. III-2, where the curium isotopes follow the trend of the lighter elements, supports this suggestion. If the barriers for the curium isotopes were more or less uniformly larger at all deformations, the penetrability implied by the spontaneous fission lifetimes would also be anomalously large.

We have already remarked that the fission barriers for even-odd nuclides do not seem to be much larger than for the neighboring even-even nuclei. The average differences in barrier height between different nuclear types is given in Table II-1. These small differences are of some interest in connection with suggestions that the fission barriers of odd-A and odd-odd nuclei might be appreciably larger than for even-even nuclei because of an increase of the pairing energy gap with deformation (Griffin, 1963). A theoretical indication

TABLE II-1

Average Differences between Heavy Element ($Z \geqslant 90$) Fission Barrier Heights of Even-Z, Odd-N (eo), Odd-Z, Even-N (oe), and Odd-Odd (oo) Nuclei as Compared to Even-Even (ee) Nuclei (first row)[a]

	$E_f(\text{eo}) - E_f(\text{ee})$	$E_f(\text{oe}) - E_f(\text{ee})$	$E_f(\text{oo}) - E_f(\text{ee})$
Experiment ($Z \geqslant 90$)	0.25 ± 0.10	0.2 ± 0.10	0.25 ± 0.10
Level density effect on pairing at first barrier	0.25	0.25	0.5
Level density effect on pairing at second barrier	0.1	0.1	0.2
Surface area effect on pairing at first barrier	0.1	0.1	0.2
Surface area effect on pairing at second barrier	0.45	0.45	0.9

[a] The theoretical estimates in the second through fifth rows are based on the calculations of Nilsson *et al.* (1969) for the nucleus ^{254}Fm with $A = 265$ parameters. They appear, however, from other results presented in that paper to be typical of the values expected for heavy elements.

of a dependence of the pairing strength on the surface-to-volume ratio has been obtained by Kennedy *et al.* (1964). Indeed, at one time there was also thought to be experimental evidence for such an effect from the dependence of the fission cross section and fragment angular anisotropy on excitation energy (Britt *et al.*, 1965). This experiment has been reinterpreted and the bump in the cross section at low energy originally identified as the threshold for the 0+ transition state has now been identified as a vibrational resonance due to the double barrier (Back *et al.*, 1969). If there were to be an effect due to pairing, it would be expected to be twice as large for odd-odd nuclei as for odd-even or even-odd nuclei. It would also lead to a systematically greater retardation of odd-odd spontaneous fission lifetimes than for odd-A spontaneous fission lifetimes (see Section III, D). Neither of these effects is indicated by the data (Ignatyuk and Smirenkin, 1969). The hindrance factor for spontaneous fission of odd-odd ^{242}Am is less than the hindrance factor for many odd-A nuclei. In addition, the fission barrier height differences for odd-odd nuclei are not larger than for odd-A, although there are considerable uncertainties in the estimates in Table VIII-1.

How large are the effects of deformation on the pairing energy expected to be? There are actually two possible contributions to the increase in the pairing energy gap with deformation. The first of these is due to the increase in the gap parameter with single particle level density. The strong correlation of the shell correction with single particle level density implies that there will be an unusually low level density at the equilibrium deformation and an unusually high level density at barrier peaks arising from the shell corrections. Estimates of the magnitude of this effect for the first (inner) and second (outer) barrier as compared to the equilibrium ground state deformation are given in the second and third rows of Table II-1. This effect is weaker at the second barrier than at the first due to the damping of the shell fluctuations with deformation. The second possible contribution to the dependence of the pairing gap on deformation is that due to the varying surface-to-volume ratio. It has been suggested (Kennedy *et al.*, 1964; Nilsson *et al.*, 1969) that pairing is predominantly a surface effect. If the pairing strength is assumed to be proportional to the surface area S, the gap parameter Δ is approximately proportional to S^3. The slab model of Kennedy *et al.* (1964) gives a weaker dependence, $\Delta \propto S^{3/2}$. If the pairing strength is assumed to be proportional to the surface area (Stepien and Szymanski, 1968; Nilsson *et al.*, 1969), the contributions indicated in the fourth and fifth rows of the table are obtained. Since the experimental barrier differences are determined by the higher barrier, the assumption of a linear dependence of the pairing strength on surface area is inconsistent with experiment unless the second barrier is always lower in energy than the first barrier. In fact, the small increases in the experimental barrier heights for odd-A and odd-odd nuclei compared to even-even nuclei may simply be a residual specialization energy effect; the spin state with the lowest barrier may not have been significantly populated in the reaction from which the threshold was deduced. There is also evidence from the lighter elements, where the surface area at the saddle is quite large, that the assumption of pairing strengths proportional to the surface area predicts fission barriers which are considerably lower (Mosel, 1972) than indicated by experiment. (See also discussion of deformation dependence of pairing energy in Section III, B in connection with spontaneous fission lifetimes.)

E. Excitations at the Saddle Point

We have concentrated our attention thus far on the lowest energy state of the nucleus at the fission barrier. In order to discuss fission at higher energy, we must consider excited states of the saddle point configuration. Our concern will be primarily with even-even nuclei. We can perhaps best get oriented by considering the excitation spectrum of an even-even permanently deformed nucleus in its equilibrium ground state configuration. The equilibrium ground state, being nonspherical, exhibits a rotational band with energy levels given by $E_J = (\hbar^2/2\mathscr{I})[J(J+1)]$. For heavy elements $\hbar^2/2\mathscr{I}$ is typically 7 keV. In addition to the rotational degree of freedom, there are a number of vibrational modes which can be excited. The simplest of these is a breathing mode described by the distortion parameter α_2 or β. This type of excitation is called a β vibration. The lowest asymmetric mode corresponds to inversions of a pear-shaped configuration, or a "sloshing" vibration. The projection of the total angular momentum on the nuclear symmetry axis K is zero for the rotations and vibrations considered thus far. There are also vibrational modes corresponding to oscillations which destroy axial symmetry. These are known as γ vibrations, the lowest γ-vibrational mode having $K = 2$. All three of these types of vibrational states have been observed in the excitation spectrum of even-even nuclei, and typically have energies of 1 MeV or slightly less.

Finally, we expect particle excitations, corresponding to the breaking of a nucleon pair, at about $2\Delta = 1.2$–1.4 MeV.

Let us now consider how these excitations depend on deformation, and what might be expected at the saddle deformation. The β-vibrational mode corresponds to motion in the fission direction and has a purely imaginary frequency at the saddle, and does not contribute to the excitation spectrum of the saddle point. The $K = 2$ γ-vibration is expected on the basis of liquid drop-model considerations to have a higher energy at the saddle than at the equilibrium deformation. It is not clear what is to be expected for the mass asymmetry or sloshing vibration, as the frequency for this mode of excitation at the ground state deformation varies quite drastically over a narrow range in mass number. Liquid drop model calculations indicate that the energy of the mass asymmetry vibration occurs at about 1.5 MeV (Nix, 1967) to 2.5 MeV (Griffin, 1965). However, the energy of this vibrational mode is expected to be appreciably less in real nuclei, as the liquid drop model usually overestimates energies of collective motions. Shell effects can drastically alter these expectations based on the liquid drop model. It has been shown (Pashkevich, 1969; Larsson *et al.*, 1972; Götz *et al.*, 1972) that the inner barrier of nuclei with $Z \gtrsim 94$ becomes unstable with respect to the degree of freedom associated with the $K = 2$ γ-vibration. We have also remarked that the outer barrier can be unstable with respect to the mass asymmetry degree of freedom.

There is another type of collective state which has not been identified in the excitation spectrum of nuclei in their ground state, but would be expected to be lower in energy at the saddle point. This is the bending mode, with $K = 1$ and negative parity. The liquid drop model prediction of the energy for this mode is quite large, about 3 MeV, but again this estimate may be too high for real nuclei.

For each of these vibrational excitations, it is possible to have an associated sequence of rotational excitations, with a rotational energy constant somewhat smaller than for the ground state because of the larger moment of inertia of the saddle configuration. It is also possible to have combination excitations where more than one vibrational mode is excited simultaneously. The combination of

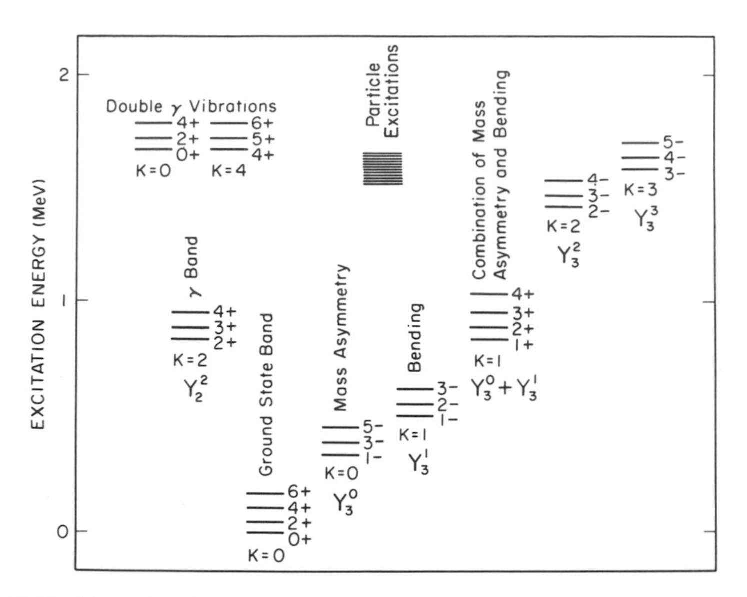

Fig. II-17. Schematic diagram of some possible collective band structures for an even-even transition nucleus with a stable quadrupole deformation. The β-vibration band (Y_2) is missing because it is a vibration in the fission direction. The excitation energies of the various collective bands are *not* to be interpreted literally. The energy scale is meant in a schematic sense only. [After Huizenga (1967).]

the bending and sloshing vibrations is of special interest as it is probably least costly energetically for obtaining a 1^+ collective state.

We also expect to find particle excitations at about $2\Delta_f$, where Δ_f, the gap parameter at the saddle deformation, may be larger than Δ, the gap parameter at the equilibrium ground state deformation, if there is a large dependence of the pairing energy on surface area. The evidence discussed in the previous section suggests that $2\Delta_f$ is not much larger than 2Δ.

A schematic diagram of a possible collective band spectrum is shown in Fig. II-17.

References

Back, B. B., Bondorf, J. P., Otroshenko, G. A., Pedersen, J. and Rasmussen, B. (1969). *Proc IAEA Symp. Phys. Chem. of Fission, 2nd*, p. 351. IAEA, Vienna.

Bassichis, W. A., and Wilets, L. (1969). *Phys. Rev. Lett.* **22**, 799.

Bassichis, W. A., Kerman, A. K., Tsang, C. F., Tuerpe, D. R., and Wilets, L. (1972). *In* "Magic without Magic: John Archibald Wheeler." Freeman, San Francisco.

Bes, D., and Szymanski, Z. (1961). *Nucl. Phys.* **28**, 42.

Bohr, N., and Wheeler, J. A. (1939) *Phys. Rev.* **56**, 426.

Bolsterli, M., Fiset, E. O., Nix, J. R., and Norton, J. L. (1972) *Phys. Rev. C* **5**, 1050.
Britt, H. C., Gibbs, W. R., Griffin, J. J., and Stokes, R. H. (1965). *Phys. Rev.* **139**, *B* 354.
Cohen, S., and Swiatecki, W. J. (1962). *Ann. Phys.* (*N.Y.*) **19**, 67.
Cohen, S., and Swiatecki, W. J. (1963). *Ann. Phys.* (*N.Y.*) **22**, 406.
Frankel, S., and Metropolis, N. (1947). *Phys. Rev.* **72**, 914.
Götz, U., Pauli, H. C., and Junken, K. (1972). *Phys. Lett.* **39B**, 436.
Green, A. E. S. (1954). *Phys. Rev.* **95**, 1006.
Griffin, J. J. (1963). *Phys. Rev.* **132**, 2204.
Griffin, J. J. (1965). "Physics and Chemistry of Fission," Vol. I, p. 23. Salzburg. IAEA, Vienna.
Gustafson, C., Lamm, I. L., Nilsson, B., and Nilsson, S. G. (1967). *Proc. Int. Symp.* "*Why and How should We Investigate Nuclides Far off the Stability Line*," *Lysekil, Sweden, August 21–27, 1966*, p. 613. Almquist and Wiksell, Stockholm; *Ark. Fys.* **36**, 613.
Gustafson, C., Möller, P., and Nilsson, S. G. (1971). *Phys. Letts. B* **34**, 349.
Huizenga, J. R. (1967). *Int. Nucl. Phys. Conf.* (R. L. Becker, C. D. Goodman, P. H. Stelson, and A. Zucker, eds.), p. 721. Academic Press, New York.
Hyde, E. K. (1964). "Nuclear Properties of the Heavy Elements," Vol. III, "Fission Phenomena." Prentice-Hall, Englewood Cliffs, New Jersey.
Ignatyuk, A. V., and Smirenkin, G. N. (1969). *Phys. Lett. B* **29**, 159.
Inglis, D. R. (1969). *Phys. Today* **22**, 29.
Johansson, S. A. E. (1961). *Nucl. Phys.* **22**, 529.
Kennedy, R. C., Wilets, L., and Henley, E. M. (1964). *Phys. Rev. Lett.* **12**, 36.
Larsson, S. E., Ragnarsson, I., and Nilsson, S. G. (1972). *Phys. Lett.* **38B**, 269.
Lawrence, J. N. P. (1965). *Phys. Rev. B* **139**, 1227.
Lin, W. (1970). *Phys. Rev. C* **2**, 871.
Möller, P., and Nilsson, S. G. (1970). *Phys. Lett. B* **31**, 283.
Möller, P., Nilsson, S. G., Sobiczewski, A., Szymanski, Z., and Wycech, S. (1969). *Phys. Lett. B* **30**, 223.
Mosel, U. (1972). *Phys. Rev. C* **6**, 971.
Mottelson, B. R., and Nilsson, S. G. (1959). *Kgl. Dan. Vidensk. Selsk. Mat. Fys. Skr.* **1**, No. 8.
Muzychka, Yu. A., Pashkevich, V. V., and Strutinsky, V. M. (1969). *Sov. J. Nucl. Phys.* **8**, 417.
Myers, W. D., and Swiatecki, W. J. (1966). *Nucl. Phys.* **81**, 1.
Myers, W. D., and Swiatecki, W. J. (1967). *Proc. Int. Symp.* "*Why and How Should We Investigate Nuclides Far off the Stability Line*," *Lysekil, Sweden, August 21–27, 1966*, p. 343. Almqvist and Wiksell, Stockholm; *Ark. Fys.* **36**, 343.
Nilsson, S. G. (1955). *Kgl. Dan. Vidensk. Selsk. Mat. Fys. Medd.* **29**, No. 16.
Nilsson, S. G. (1969). *Proc. Int. School Phys.* "*Enrico Fermi*" **40**. Academic Press, New York.
Nilsson, S. G., Nix, J. R., Sobiczewski, A., Szymanski, Z., Wyceck, S., Gustafson, G., and Möller, P. (1968). *Nucl. Phys. A* **115**, 545.
Nilsson, S. G., Tsang, C. F., Sobiczewski, A., Szymanski, Z., Wycech, S., Gustafson, C., Lamm, I. L., Möller, P., and Nilsson, B. (1969). *Nucl. Phys. A* **131**, 1.
Nix, J. R. (1967). *Ann. Phys.* (*N.Y.*) **41**, 52.
Pashkevich, V. V. (1969). *Nucl. Phys.* **A 133**, 400.
Pashkevich, V. V. (1971). *Nucl. Phys. A* **169**, 275.
Pauli, H. C., Ledergerber, T., and Brack, M. (1971). *Phys. Lett. B* **34**, 264.
Primack, J. R. (1966). *Phys. Rev. Lett.* **17**, 539.
Stepien, W., and Szymanski, Z. (1968). *Phys. Lett. B* **26**, 181
Strutinsky, V. M. (1964a). *Sov. Phys. JETP* **18**, 1298.

Strutinsky, V. M. (1964b). *Sov. Phys. JETP* **18**, 1305.
Strutinsky, V. M. (1967). *Nucl. Phys. A* **95**, 420.
Strutinsky, V. M. (1968). *Nucl. Phys. A* **122**, 1.
Strutinsky, V. M., Lyashchenko, N. Ya., and Popov, N. A. (1963). *Nucl. Phys.* **46**, 639.
Swiatecki, W. J. (1955). *Phys. Rev.* **100**, 937.
Tsang, C. F., and Nilsson, S. G. (1970). *Nucl. Phys. A* **140**, 275.

CHAPTER

Spontaneous Fission

A. Introduction

It is possible for a nucleus in its ground state to quantum mechanically tunnel through the fission barrier discussed in the previous chapter. As will be shown later, the probability for tunneling through a barrier with a given width is expected to depend exponentially on the square root of the barrier height. Indeed, for elements lighter than thorium the barrier heights are so high and the penetrability so low that spontaneous fission has not been observed. For the heaviest elements known, spontaneous fission becomes a very prominent mode of decay, and the limitation on the existence of elements heavier than those presently known arises primarily from instability with respect to spontaneous fission.

For nuclides with relatively long half lives spontaneous fission lifetimes are usually determined by measuring the fission disintegration rate of a sample of known weight. Gas ionization counters, solid state detectors, or dielectric track detectors are used to identify the fission fragments. Because of the large energy of the fission fragments, counter backgrounds are very low and rather long-lived species can be investigated. Since, however, spontaneous fission rates vary by many orders of magnitude, it is necessary to be careful about the chemical and isotopic purity of the samples. For shorter-lived nuclides the spontaneous fission half life is determined by measuring the total half life and the branching ratio for spontaneous fission. This is most difficult for those nuclides which decay by electron capture, since the determination of absolute decay rates for this process is more difficult than for α, β, and spontaneous fission decay.

B. Systematics of Ground State Spontaneous Fission Half Lives

A rather complete tabulation of spontaneous fission half lives is given by Hyde (1964). A summary including more recent data is given in Table III-1. The logarithms of the spontaneous fission half lives for even-even and even-odd

TABLE III-1

Half Lives for Spontaneous Fission[a]

Nuclide	Half life	Reference
^{230}Th	$\geqslant 1.5\times10^{17}$ yr	Segrè (1952)
^{232}Th	$>10^{20}$ yr	Podgurskaya *et al.* (1955)
	$>10^{21}$ yr	Flerov *et al.* (1958)
^{232}U	$(8\pm5.5)\times10^{13}$ yr	Jaffey and Hirsch (1951)
^{233}U	$(1.2\pm0.3)\times10^{17}$ yr	Aleksandrov *et al.* (1966a, b)
^{234}U	1.6×10^{16} yr	Ghiorso *et al.* (1952)
^{235}U	1.8×10^{17} yr	Segrè (1952)
	$(3.5\pm0.9)\times10^{17}$ yr	Aleksandrov *et al.* (1966a, b)
^{236}U	2×10^{16} yr	Jaffey and Hirsch (1949)
^{238}U	$(5.9\pm0.14)\times10^{15}$ yr	Kuroda and Edwards (1957)
	$(5.8\pm0.5)\times10^{15}$ yr	Gerling *et al.* (1959)
	$(1.3\pm0.2)\times10^{16}$ yr	Perfilov (1947)
	$(6.5\pm0.3)\times10^{15}$ yr	Kyz'minov *et al.* (1960)
	$(1.01\pm0.03)\times10^{16}$ yr	Fleischer and Price (1964)
^{237}Np	$>10^{18}$ yr	Druin *et al.* (1961a)
^{236}Pu	3.5×10^{9} yr	Ghiorso *et al.* (1952)
^{238}Pu	$(5\pm0.6)\times10^{10}$ yr	Druin *et al.* (1961a)
^{239}Pu	5.5×10^{15} yr	Segrè (1952)
^{240}Pu	1.2×10^{11} yr	Chamberlain *et al.* (1954)
	1.22×10^{11} yr	Barclay *et al.* (1954)
	$(1.32\pm0.03)\times10^{11}$ yr	Kinderman (1953)
	$(1.34\pm0.02)\times10^{11}$ yr	Watt *et al.* (1962)
	1.20×10^{11} yr	Mikheev *et al.* (1960)
	$(1.45\pm0.02)\times10^{11}$ yr	Malkin *et al.* (1963)
^{242}Pu	$7.06\pm0.19\times10^{10}$ yr	Mech *et al.* (1956), Jones *et al.* (1955)
	$(6.64\pm0.10)\times10^{10}$ yr	Butler *et al.* (1956a)
	$(6.5\pm0.7)\times10^{10}$ yr	Druin *et al.* (1961a)
	$(7.45\pm0.17)\times10^{10}$ yr	Malkin *et al.* (1963)
^{244}Pu	$(2.5\pm0.8)\times10^{10}$ yr	Fields *et al.* (1955)

TABLE III-1 (continued)

Nuclide	Half life	Reference
^{241}Am	$(2.3 \pm 0.8) \times 10^{14}$ yr	Druin *et al.* (1961b)
^{242}Am	$(9.5 \pm 3.5) \times 10^{11}$ yr	Caldwell *et al.* (1967)
^{243}Am	$(3.3 \pm 0.3) \times 10^{13}$ yr	Aleksandrov *et al.* (1966a, b)
^{240}Cm	1.9×10^{6} yr	Ghiorso *et al.* (1952)
^{242}Cm	7.2×10^{6} yr	Ghiorso and Robinson (1947), Hanna *et al.* (1951)
^{244}Cm	1.4×10^{7} yr	Ghiorso *et al.* (1952)
	$(1.346 \pm 0.006) \times 10^{7}$ yr	Metta *et al.* (1965)
	$(1.25 \pm 0.01) \times 10^{7}$ yr	Barton and Koontz (1970)
^{246}Cm	$(2 \pm 0.8) \times 10^{7}$ yr	Fried (1956)
	$(1.66 \pm 0.10) \times 10^{7}$ yr	Metta *et al.* (1965)
	$(1.80 \pm 0.01) \times 10^{7}$ yr	Metta *et al.* (1969)
^{248}Cm	$(4.6 \pm 0.5) \times 10^{6}$ yr	Butler *et al.* (1956b)
	$(4.22 \pm 0.12) \times 10^{6}$ yr	Metta *et al.* (1969)
^{250}Cm	2×10^{4} yr	Huizenga and Diamond (1957)
^{249}Bk	$(1.65 \pm 0.17) \times 10^{9}$ yr	Vorotnikov *et al.* (1970)
^{246}Cf	$(2.1 \pm 0.3) \times 10^{3}$ yr	Hulet *et al.* (1953)
^{248}Cf	7×10^{3} yr	Hulet (1953)
	4.1×10^{4} yr	Skobelev *et al.* (1968)
^{249}Cf	1.5×10^{9} yr	Ghiorso *et al.* (1954a)
	$\geqslant 2 \times 10^{9}$ yr	Vorotnikov *et al.* (1970)
^{250}Cf	$(1.5 \pm 0.5) \times 10^{4}$ yr	Ghiorso *et al.* (1954b), Fields *et al.* (1954a), Magnusson *et al.* (1954)
	$(1.73 \pm 0.06) \times 10^{4}$ yr	Phillips *et al.* (1962)
^{252}Cf	66 ± 10 yr	Magnusson *et al.* (1954), Ghiorso *et al.* (1954b)
	82 ± 6 yr	Eastwood *et al.* (1957)
	85.5 ± 0.5 yr	Metta *et al.* (1965)
^{254}Cf	56.2 ± 0.7 day	Huizenga and Diamond (1957)
	61.9 ± 1.1 day	Metta *et al.* (1965)
	55 day	Fields *et al.* (1956)
	60.5 ± 0.2 day	Phillips *et al.* (1963)
^{253}Es	3×10^{5} yr	Fields *et al.* (1954b), Ghiorso *et al.* (1954a)
	$(7 \pm 3) \times 10^{5}$ yr	Jones *et al.* (1956)
	6.3×10^{5} yr	Metta *et al.* (1965)
^{254}Es	$> 2.5 \times 10^{7}$ yr	Fields *et al.* (1967)
^{255}Es	2440 ± 140 yr	Fields *et al.* (1967)
^{244}Fm	$\geqslant 3.3 \pm 0.5$ msec	Nurmia *et al.* (1967)
^{246}Fm	15 ± 5 sec	Nurmia *et al.* (1967)
	~ 20	Druin *et al.* (1971)
^{248}Fm	10 ± 5 hr	Nurmia *et al.* (1967)
	~ 60 hr	Druin *et al.* (1971)

TABLE III-1 (continued)

Nuclide	Half life	Reference
^{250}Fm	~10 yr	Druin *et al.* (1971)
^{252}Fm	115 yr	Nucl. Data Sheets (1969)
^{254}Fm	200 day	Choppin *et al.* (1954)
	220 ± 40 day	Fields *et al.* (1954b)
	246 day	Jones *et al.* (1956)
^{255}Fm	$(1.2 \pm 0.6) \times 10^4$ yr	Brandt *et al.* (1963)
^{256}Fm	3 hr	Choppin *et al.* (1955)
	170 min	Hoff *et al.* (1968)
^{257}Fm	120 yr	Nucl. Data Sheets (1969)
^{258}Fm	380 ± 60 μsec	Hulet *et al.* (1971)
^{257}Md	$\geqslant 30$ hr	Sikkeland *et al.* (1965)
^{252}No	~7.5 sec	Sikkeland *et al.* (1968), Oganesian *et al.* (1970)
^{254}No	$\geqslant 9 \times 10^4$ sec	Dubna work quoted in Ghiorso and Sikkeland (1967)
^{256}No	~1500 sec	Ghiorso and Sikkeland (1967), Kuznetsov *et al.* (1967)
^{258}No	1.2×10^{-3} sec	Nurmia *et al.* (1969); Nurmia (1970)
^{256}Lw	$> 10^5$ sec	Flerov *et al.* (1971)
^{257}Lw	$> 10^5$ sec	Flerov *et al.* (1971)
^{258}Lw	$\geqslant 20$ sec	Flerov *et al.* (1971)
^{261}Rf	$\geqslant 650$ sec	Ghiorso *et al.* (1970)
^{261}105	8 sec	Flerov *et al.* (1971)

[a] This table is an amended and updated version of a table given by Hyde (1964).

species have been plotted in Fig. III-1 versus the fissility parameter x defined in the previous chapter. We expect at least a qualitative correlation with this parameter in view of the smooth dependence of the barrier heights with x predicted by the liquid drop model. It can be seen that there are two types of systematic deviations from this expectation. In the first place, the half lives for even-even isotopes of a given element *decrease* with x for the heavier isotopes of each element. This effect becomes much more dramatic for nuclides such as ^{250}Cm, ^{252}Cf, ^{254}Cf, ^{254}Fm, ^{256}Fm, and ^{258}Fm with $N > 152$. This is presumably related to the apparent subshell closure at $N = 152$. In the second place, even-odd nuclides appear to have abnormally long half lives, as do the few odd-even and odd-odd nuclides for which spontaneous fission half lives are known (but not plotted in Fig. III-1).

Swiatecki (1955) has shown that there is a close correlation between the deviations in spontaneous fission half lives from a smooth trend with Z^2/A

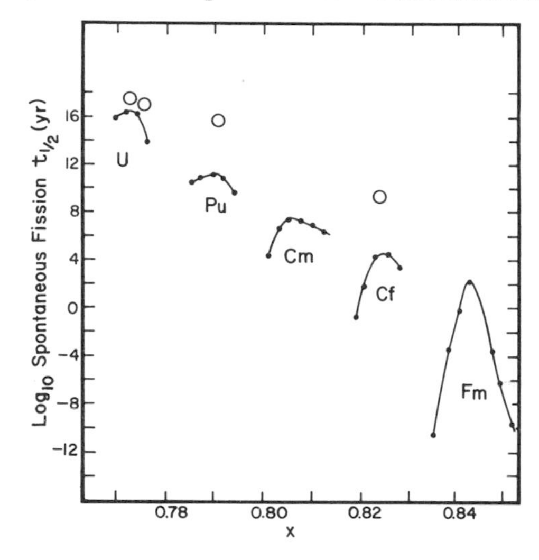

Fig. III-1. Spontaneous fission half lives of even-even (●) and even-odd (○) nuclides as a function of the fissility parameter x appropriate to the Myers–Swiatecki (1967) mass formula. (In this formula x is not simply proportional to Z^2/A, since there is a composition-dependent correction term proportional to $[(N-Z)/A]^2$ in the surface energy expression.)

and the deviations of the ground state masses δm from a smooth liquid drop model reference surface. A nucleus with less than average stability in its ground state is found to have a shorter half life than that given by the overall trend with Z^2/A. Each MeV of extra ground state instability was found to correspond to approximately 10^5 times shorter lifetime. If we then plot $\log t_{1/2}(\text{exp}) + 5\,\delta m$ versus the fissility parameter x, a dramatic tightening of the correlation is found to occur (see Fig. III-2). Thus both the systematic deviation in half lives of the heavier isotopes for a given element and the anomalously short half lives for nuclides with $N > 152$ are attributed to instabilities in the ground state masses associated with their shell structure.

Half lives for odd-even, even-odd, and odd-odd nuclides, corrected in the same manner, are also shown in Fig. III-2. It is seen that these nuclides still exhibit "corrected" half lives that are longer than those of even-even nuclides. It is possible to obtain a measure of this retardation of spontaneous fission rates for odd-A and odd-odd nuclei by defining a hindrance factor (HF) analogous to that employed in α-decay systematics. The log HF is then given by the ratio of the corrected logarithm of the observed half life to that expected for an even-even nucleus of the same fissility x:

The average log hindrance factor for eight odd-A nuclides is 4.2, while the odd-odd nuclide ^{242}Am has a log hindrance factor of 1.8, and ^{254}Es (for which

$$\log \text{HF} = \log[t_{1/2}(\text{obs})/t_{1/2}(\text{e–e})] + k[\delta m(\text{obs}) - \delta m(\text{e–e})] \quad \text{(III-1)}$$

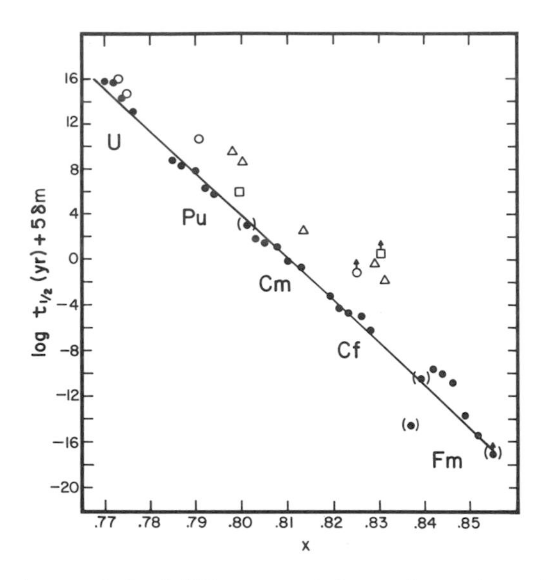

Fig. III-2. Spontaneous fission half lives corrected according to the method of Swiatecki. The ordinate is $\log t_{1/2}\,(\mathrm{yr}) + k\,\delta m$, where $k = 5$ and δm is in MeV. The abscissa is the fissility parameter appropriate to the Myers–Swiatecki (1967) mass formula (see caption to Fig. III-1). Points enclosed in parentheses have corrections based on masses estimated from mass systematics rather than experimental masses. ● Even-even, ○ even-odd, △ odd-even, □ odd-odd.

only a lower limit on the half life is known) has a log hindrance factor greater than 9.2.

There have been several suggestions as to the origin of the retardation in spontaneous fission rates associated with the presence of an odd number of neutrons or protons. Newton (1955) and Wheeler (1955) have observed that conservation of the spin and parity of the odd nucleon with deformation may lead to an increased barrier for this type of nuclide. This idea will be pursued later. Griffin (1963) suggests that a more important factor may be a possible increase in the pairing energy at the saddle point compared to the ground state configuration. There are theoretical indications that the pairing interaction is predominately a surface effect (Kennedy *et al.*, 1964) and would be expected to be proportional in some way to the surface area. It does not appear however that this can be a large effect, as it would be expected to be a systematic effect for all nuclei, whereas for several nuclei the HF observed is rather small. This model would also predict a systematic increase in the hindrance factor with decreasing x due to the increasing difference between the surface area of the saddle configuration relative to the equilibrium configuration with decreasing x. No evidence for such a trend has been observed. (See further discussion of the dependence of the pairing energy on deformation in Section II, D.) A third factor influencing the penetrability for odd-A nuclei is related to the mass or inertial parameter in the penetration integral discussed below. The larger the effective mass, the longer the half life. It has been shown that pairing reduces the mass parameter given by an independent particle model by several orders of magnitude for even-even nuclei. The presence of an odd particle blocks one level from contributing to the pairing correlations, resulting in an increase in the effective mass parameter. Estimates of an increase of 20 to 30% in the mass parameter have been given (Urin and Zaretsky, 1966; Sobiczewski, 1970).

C. Penetration of the Fission Barrier

We expect the penetration of the fission barrier to be the dominant factor in determining the probability for spontaneous fission. The penetrability p is given in the WKB approximation by

$$p = \exp\left\{-(2/\hbar)\int_{r_1}^{r_2}\left[2(V(r)-T)\sum_i m_i(dx_i/dr)^2\right]^{1/2} dr\right\} \qquad \text{(III-2)}$$

where r is the distance between the centers of gravity of the nascent fragments, $(V(r)-T)$ is the "negative kinetic energy" at the separation r, $V(r)$ is the potential energy of deformation, T is any kinetic energy in the fission degree of freedom, and x_i is the coordinate of the ith nucleon of mass m_i. The integral extends from the point r_1 of stable equilibrium over the fission saddle and down the other side of the barrier to the point r_2 where $(V(r)-T)$ is again zero.

If we ignore the changes which take place in the future fragments during the barrier penetration and hence assume each nucleon moves the same distance as the fragment center of mass, then $\sum_i m_i(dx_i/dr)^2 = M$. In principle we should also take into account the zero-point energies of the ground state configuration, but these will be neglected now. For nuclei in their ground state, $T=0$, and we have the penetrability

$$p = \exp\left\{-(2/\hbar)\int_{r_1}^{r_2}[2V(r)M]^{1/2}\,dr\right\} \qquad \text{(III-3)}$$

With this expression we can now understand more clearly the origin and nature of the Swiatecki correction for variations in ground state masses. If a nucleus has an unusually low ground state mass (see dashed line in Fig. III-3), the potential energy as measured from the stable ground state will be larger at every deformation. It can also be seen that a shell effect in the vicinity of the top of the barrier (as illustrated by dotted line in Fig. III-3) will have a smaller effect on the integral in the exponential and hence on the penetrability. It is

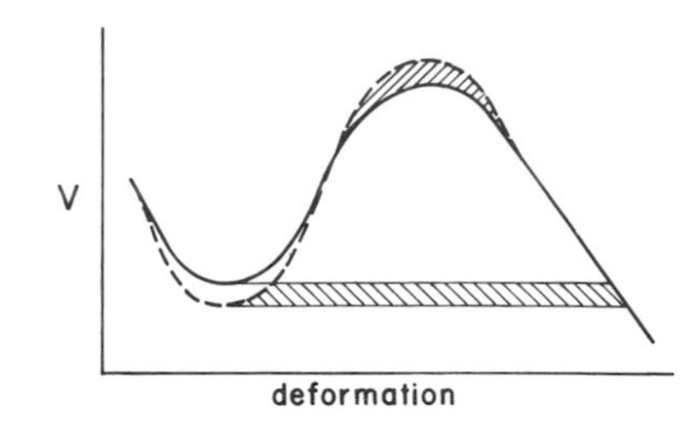

Fig. III-3. Origin of the greater effect on the barrier penetrability for a shell deviation in the ground state energy compared to a comparable deviation in the saddle point energy. The zero-point energy has been neglected.

therefore not possible to conclude from the success of the Swiatecki correlation that shell effects are not present in the potential energy surface in the neighborhood of the saddle point.

If we recognize the nucleonic structure of the nuclei, then the total energy as a function of deformation is not expected to be smooth, but will instead exhibit discontinuities when level crossings occur at the Fermi surface as the deformation is changed. Thus on a Nilsson diagram the lowest unfilled energy level for a particular nucleon will change with deformation as level crossings occur. This effect is shown very schematically in Fig. III-4.

We must of course ask what happens as the nucleus deforms and level crossings occur. If the nucleus is even-even and there is a pair of particles in a given level, as the deformation changes and a level crossing is encountered, it might be possible either to continue on the same level or to transfer the two nucleons to the crossing level so as to lower the total energy. The latter transfer will conserve spin and parity of the system since the nucleons will still be paired. It is generally thought that it is possible for the nucleus to change its nucleonic configuration during the change in deformation from stable equilibrium to the saddle point. In the case of spontaneous fission of even-even nuclei it is possible that this adiabatic approximation may be valid and that nucleonic configuration changes will occur with deformation so that the total energy is minimized. It is important to recognize that the consequences will not be greatly changed if the probability for exploiting a level crossing is not 100%. If the transfer does not occur, the barrier will be increased and fission is much less likely to occur from this configuration. Thus unless the transfer probability at a crossing is very small, most fission will proceed via the path of minimum energy.

Even if the path of minimum energy is followed, this path will depend on the number and positions of the level crossings in the Nilsson diagram. Johansson (1959) has attempted to improve the systematics of spontaneous fission half lives by taking into account single particle effects as a correction to the liquid drop model (LDM) barrier. In practice he takes a barrier of the general form

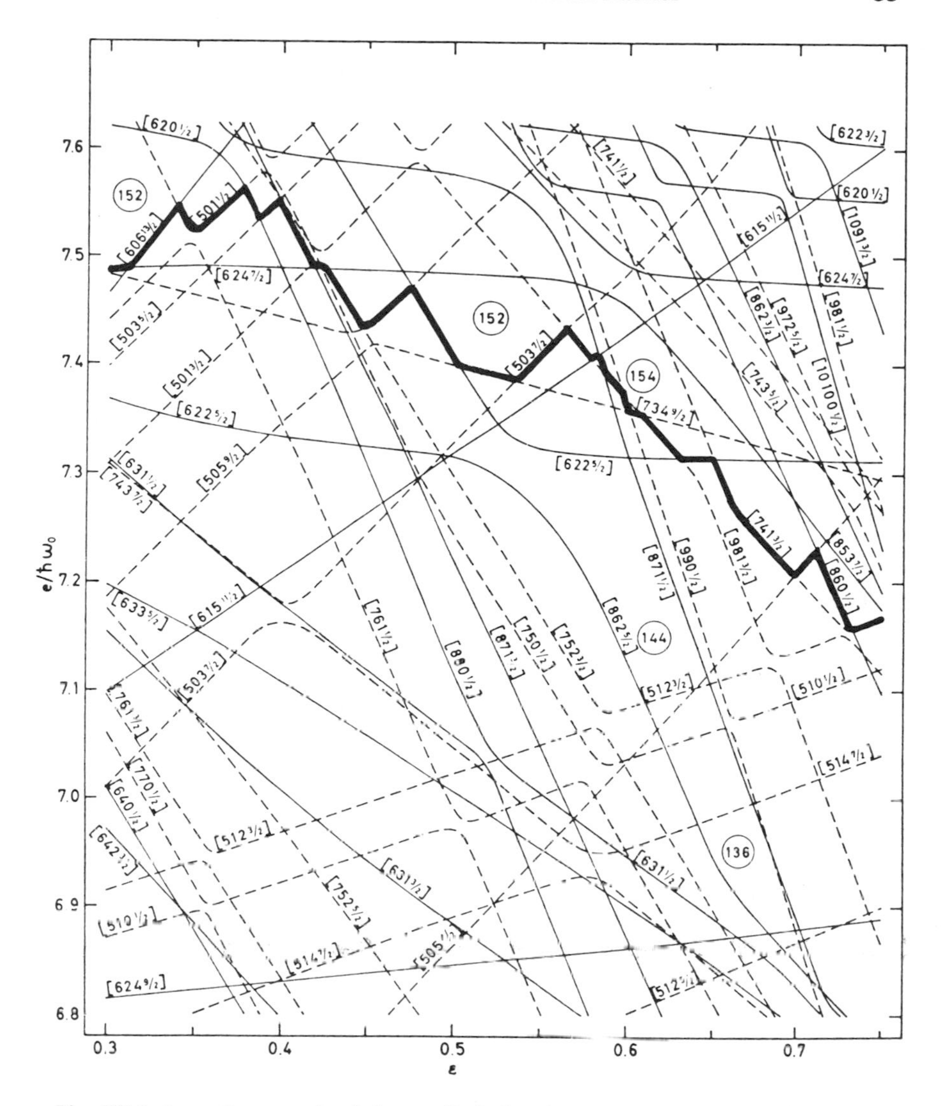

Fig. III-4. Lowest energy level (heavy line) for the 152nd neutron as a function of deformation. The parameters used are those for $A=242$ with $\varepsilon_4=0.04$. [Nilsson diagram from Nilsson *et al.* (1969).]

predicted by the LDM and finds an empirical parameterization to account for observed deformations, fission thresholds, and the general trend of spontaneous fission half lives. He then corrects the penetrability by a factor

$$\exp\left\{-(k/2)\int_{\varepsilon_1}^{\varepsilon_2}[\Delta V(\varepsilon)/(V(\varepsilon))^{1/2}]\,d\varepsilon\right\} \qquad \text{(III-4)}$$

where ΔV is the difference between the single particle curve and the drop model curve. He has therefore corrected the observed half lives for the variations from nucleus to nucleus in the difference between the smooth liquid drop barrier and a single particle model barrier derived from the envelope of the highest occupied level at each deformation. For reasons that are not clear, the upper limit in the integral has been taken as the top of the barrier rather than as the point where one emerges from the other side of the barrier. The Nilsson levels used at that time were based on extrapolations from smaller deformations guided by the asymptotic behavior. The corrected half lives for even-even nuclei when plotted against Z^2/A no longer show the systematic deviations exhibited in Fig. III-1, although the scatter of points is considerably larger than in the Swiatecki correlation.

We might wonder why two correlation schemes, one based only on ground state mass deviations and the other based on single particle effects at all deformations up to the saddle point, are both fairly successful. The two findings can be at least partially reconciled by recognizing (as was pointed out earlier) that deviations at the ground state equilibrium point have a larger effect on the penetrability than do deviations at any other point on the potential energy curve. This is reflected by the factor $(V(\varepsilon))^{1/2}$ in the denominator of the expression in the exponential of Eq. (III-4).

In the case of a nucleus with an odd neutron or proton, the situation is usually quite different. It is not always possible for the odd nucleon to transfer to another level at a crossing and still conserve the total spin and parity. When such transfers to the lowest levels cannot take place, the fission barrier for a nucleus with an unpaired nucleon will be higher and wider than for an even-even nucleus. This "specialization energy" associated with the increase in the barrier has been suggested as the origin (Newton, 1955; Wheeler, 1955) of the fact that nuclei with an odd number of neutrons or protons have considerably longer spontaneous fission half lives than do neighboring even-even nuclei. The detailed path of the odd nucleon in the level diagram depends on whether K, the projection of the nuclear spin on the nuclear symmetry axis, is a good quantum number. A nucleus with an odd nucleon will usually have in its lowest energy state the spin and parity of the odd nucleon. In addition, deformed odd-A nuclei will have a rotational band with spin values I, $I+1$, $I+2$, etc. The projection of the total nuclear spin on the nuclear symmetry axis is $K = I$ and the members of a particular rotational band all have the K value of the single particle state on which the band is based. The extent to which the nucleus can follow the lowest energy state at level crossings depends on whether K remains a good quantum number as the deformation proceeds. If K can freely change at each crossing, then a nucleon can make the transition to the new level if both levels have the same parity and if the K value of the new level is not greater than the total angular momentum of the odd nucleon. If the

K value of the new level is lower than the total angular momentum of the odd particle, then there will be a low lying rotational level with the same angular momentum as the initial particle, and transfer can occur with conservation of angular momentum. The number of crossings where transfer can occur may be reduced if K is a good quantum number (i.e., must be conserved at crossings). There is considerable evidence that K is conserved to a high degree in decay processes for low lying states in heavy deformed nuclei. K is not a constant of the motion, however, as Coriolis forces can couple states of different K. Wolf (1969) has concluded that the available data is consistent with the assumption that only $\Delta K = 0$ and 1 transitions can occur at level crossings for nuclei with an odd nucleon. The problem of the retardation of spontaneous fission in nuclei with an odd number of neutrons and/or protons should be reexamined with the improved level schemes now available. The consequences of reflection-asymmetric shapes at the larger deformations must be considered. Odd-even effects on the effective inertial mass, which also affect the penetrability, should also be explored.

Thus far we have only used the penetration formula to obtain *relative* corrections to decay rates. To obtain an absolute penetrability we must specify the shape of the barrier, and to obtain an absolute decay rate we must specify the number of barrier assaults. Let us for the moment choose a particularly simple shape for the barrier, namely, an inverted parabola of base width Δr and height E_f, as sketched in Fig. III-5. For this barrier the exponent is simply

$$-(\pi \Delta r/2\hbar)(2ME_f)^{1/2} \tag{III-5}$$

The number of assaults on the barrier can be estimated from vibrational frequencies for the equilibrium configuration. For a vibrational frequency corresponding to $\hbar\omega = 1$ MeV, we obtain a number of barrier assaults per

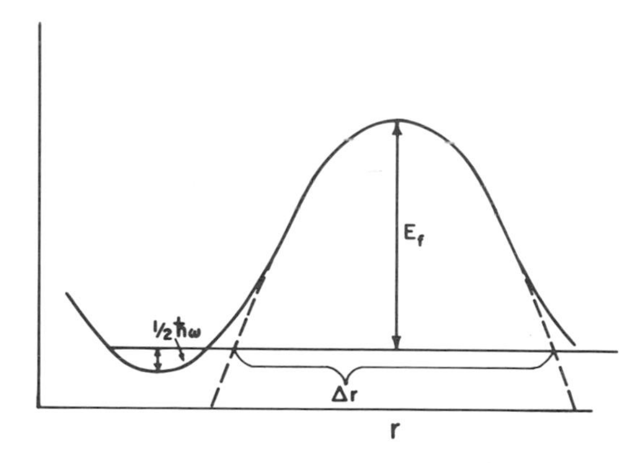

Fig. III-5. Approximation of fission barrier by inverted parabola (---) of width Δr and height E_f, measured from the ground state with zero-point energy $\frac{1}{2}\hbar\omega$.

second, $n = \omega/(2\pi)$, of 2.5×10^{20}. The spontaneous fission rate $\Gamma/\hbar$ is then equal to np, and the mean life is given by

$$\tau = \hbar/\Gamma = 1/np \qquad \text{(III-6)}$$

and the half life by

$$t_{1/2} = \ln(2)/np \qquad \text{(III-7)}$$

We may now ask whether the observed spontaneous fission rates give a reasonable value of the width of the barrier if the barrier is assumed to be parabolic. For ^{238}U, whose spontaneous fission half life is approximately 2×10^{23} sec and fission barrier height is approximately 6 MeV, we obtain a barrier width of approximately 15 F if the mass parameter is taken as the reduced mass of the separating fragments (Halpern, 1959). This is a fairly reasonable value, considering the approximations made in describing the mass motion and the shape of the barrier.

It is also possible to use this expression to determine the approximate sensitivity of the half life to a given change in barrier height, assuming a constant barrier width at the base. For the foregoing width of 15 F, we find that a change in barrier height of 1 MeV is associated with a change in half life of a factor of 10^4. This corresponds to a k value of 4 for the coefficient of the mass deviation term in the Swiatecki correlation, which is fairly close to the empirically determined coefficient of 5.

This factor of 10^4 change in half life for a change in barrier height of 1 MeV for a fixed-width barrier must not be confused with the change in half life expected for a change in barrier height of 1 MeV for a barrier of fixed curvature. The latter quantity is of interest in discussing the change in half life expected in a given nucleus with a change in excitation energy, and is approximately 10^8 per MeV. This distinction is illustrated in Fig. III-6.

The exponential dependence of the spontaneous fission half life on the square root of the fission barrier height is not unique to a parabolic barrier. If we transform Eq. (III-3) to a coordinate system characterized by the distortion

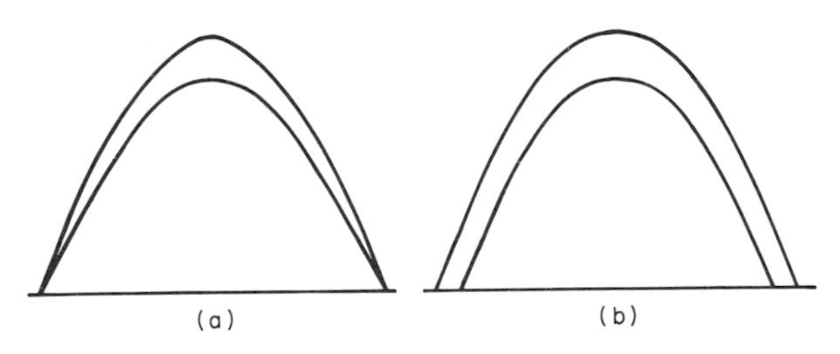

Fig. III-6. (a) Two parabolic barriers of constant width but differing in height by 1 MeV, for which the penetration differs by a factor of 10^4. (b) Two barriers of constant width (curvature) again differing by 1 MeV in height. For these barriers the penetration differs by a factor of approximately 10^8.

parameter ε, we obtain

$$p = \exp\left\{-(2/\hbar)\int_{\varepsilon_1}^{\varepsilon_2}[2B(V(\varepsilon)-T)]^{1/2}\,d\varepsilon\right\} = \exp(-K) \qquad \text{(III-8)}$$

where B is the inertial mass associated with the displacement along the distortion coordinate ε. Since ε is dimensionless, B will have dimensions of the moment of inertia. Assuming B to be a constant with deformation, we can evaluate the integral for several simple limiting barrier shapes:

(a) a parabola centered around the saddle point $\varepsilon_s = (\varepsilon_2+\varepsilon_1)/2$,
(b) a rectangular barrier of width $(\varepsilon_2-\varepsilon_1)$,
(c) a triangular barrier with its peak at the saddle point.

The exponent for these three examples is (Nilsson, 1969)

$$\text{(a)} \quad K = \pi(\varepsilon_2-\varepsilon_1)(BE_f/2\hbar^2)^{1/2} \qquad \text{(III-9)}$$

$$\text{(b)} \quad K = 4(\varepsilon_2-\varepsilon_1)(BE_f/2\hbar^2)^{1/2} \qquad \text{(III-10)}$$

$$\text{(c)} \quad K = (8/3)(\varepsilon_2-\varepsilon_1)(BE_f/2\hbar^2)^{1/2} \qquad \text{(III-11)}$$

demonstrating the common functional dependence on the barrier width and the square root of the barrier height.

For small barrier heights the WKB approximation breaks down, giving $p = 1$ for $E_f = 0$. Fortunately there is an exact expression for a parabolic barrier due to Hill and Wheeler (1953),

$$p = [1+\exp(2\pi E_f/\hbar\omega_f)]^{-1} \qquad \text{(III-12)}$$

where $\hbar\omega_f = (|C|/B)^{1/2}$, which for a parabolic barrier is

$$\hbar\omega_f = \{8E_f\hbar^2/[B(\varepsilon_1-\varepsilon_2)^2]\}^{1/2}$$

We have designated $\hbar\omega_f$ the barrier curvature energy as discussed in Section II, A. This expression can also be used to calculate the change in penetrability of a given barrier with energy E

$$p = \{1+\exp[2\pi(E_f-E)/\hbar\omega_f]\}^{-1} \qquad \text{(III-13)}$$

and gives the expected quantum mechanical reflection for energies above the barrier as well as penetration for energies below the barrier. The penetrability is equal to 0.5 at $E = E_f$.

A more realistic shape for the fission barrier can be obtained with a cubic expression for the dependence of the potential energy on deformation. Nix (1967) has suggested a way to parameterize this barrier so that the coefficients of the quadratic and cubic terms do not require a new parameter. The penetrability for this barrier is given to a high degree of accuracy by

$$p = \{1+\exp(36/5\hbar\omega_f)[E_f-E-(1-[10\pi/36]E\ln(E_f/E))]\}^{-1} \qquad \text{(III-14)}$$

where $\hbar\omega_f$ is the curvature energy appropriate to the top of the barrier.

For nuclides which exhibit a double-peaked barrier due to shell effects, a more complicated representation of the barrier is required. One such representation for which an exact expression for the penetrability can be obtained is that of two parabolic peaks connected with a third inverted parabola forming the intermediate well (Wong and Bang, 1969; Cramer and Nix, 1970). (The method of Cramer and Nix ensures that the parabolas are joined smoothly.) The penetrability is obtained in terms of the standard parabolic cylinder functions and their derivatives. For energies of interest in spontaneous fission the WKB approximation is adequate. The penetrability as a function of energy, which exhibits resonance behavior, is discussed in Chapter V.

Thus far our discussion of barrier penetration has emphasized the height and shape of the barrier. A calculation of the absolute penetrability requires a knowledge of the mass parameter M in Eq. (III-3), or equivalently, the inertial parameter B in Eq. (III-8) or the barrier curvature energy $\hbar\omega_f$ in Eq. (III-12). It will be most convenient for the present purposes to work in deformation space, where the inertial parameter B is appropriate. If we consider an independent particle model without residual interactions, deformation changes usually result in crossings of the single particle levels. If the nuclear configuration follows the deformation by retaining the occupation of the original single particle levels, the inertial parameter is expected to be given approximately by that for irrotational flow. For a spherical liquid drop the irrotational inertial parameter is given by $B_{irr}(0) = (2/15)\,MAR_0^2$, which, with $R_0 = 1.2A^{1/3}$ F, can be expressed as

$$B_{irr}(0) = 0.004632A^{5/3}\hbar^2 \quad \text{MeV}^{-1}$$

The irrotational inertial parameter depends somewhat on distortion, and for quadrupole distortions is given by (Sobiczewski *et al.*, 1969)

$$B_{irr}(\varepsilon) = B_{irr}(0)\{1-\tfrac{1}{3}\varepsilon^2-\tfrac{2}{27}\varepsilon^3\}^{-4/3}[1+\tfrac{2}{9}\varepsilon^2]/[1-\tfrac{2}{3}\varepsilon]^2$$

Griffin (1969, 1971) has pointed out, however, that if rearrangement occurs at level crossings, so that the lowest energy orbitals are occupied at every deformation (as might be expected in spontaneous fission of even-even nuclei), the inertial parameter can become several orders of magnitude larger than the inertia given by irrotational motion. If we take into account pairing effects, the inertial parameter is reduced from that for an independent particle model (including the contribution of crossings), resulting in a final inertial parameter which is approximately an order of magnitude larger than the irrotational value (Wilets, 1964; Urin and Zaretsky, 1966; Griffin, 1971). Detailed numerical calculations have been performed for heavy nuclei using single particle states of harmonic oscillator potentials (Sobiczewski *et al.*, 1969; Damgaard *et al.*, 1969). The values of the inertial mass parameter obtained depend on the deformation and are approximately five to ten times that for irrotational flow.

These calculations based on one-center oscillator potentials become unrealistic for large deformations as the neck develops and the scission shape is approached. The inertial mass is expected with increasing distortion to approach asymptotically the value appropriate to that of separated fragments.

D. Spontaneously Fissionable Isomers

In 1962 investigators in the Soviet Union reported on the discovery of a nuclide (later identified as ^{242m}Am) exhibiting an unusually short spontaneous fission half life ($t_{1/2} \sim 0.02$ sec) (Polikanov *et al.*, 1962). This half life was much shorter than predicted by systematics of known spontaneous fission half lives. To account for the short half life they suggested that the observed species might be an isomeric state and hence not have to penetrate the whole fission barrier. Since this report, a number of such species have been observed. For a considerable period of time all of the isomers observed were odd-odd, but more recently isomers of all nuclear types have been observed.

The experimental methods used to investigate these short-lived isomeric states are of three types, depending primarily on the lifetimes one wishes to investigate. The first method makes use of mechanical devices, such as rotating wheels or moving belts, to transport the nuclear reaction products, which recoil out of a thin target to a detector (Polikanov *et al.*, 1962). The nuclear reactions are induced by energetic particles from an accelerator, and most of the momentum associated with the projectile ends up in the heavy recoil products. The detector may be a photographic film, a mica or polycarbonate resin sheet, or a gas or solid state ionization counter, and it records fission fragments associated with delayed fission of the recoil products. The half life can be determined by observing the decrease in the number of events as a function of the distance of the detector along the wheel or belt, or by varying the speed of the wheel or belt.

A second method utilizes the principle of decay in flight (Gangrsky *et al.*, 1968; Lark *et al.*, 1969). The reaction products are allowed to recoil from the target. Detectors, shielded from the target itself, are placed along the path of the reaction products. The detectors are often polycarbonate plastic sheets which are chemically etched and then sparked to reveal the tracks of individual fragments. The lifetimes accessible by this method depend on the momentum of the beam projectiles, but typical recoil energies are 0.06 MeV and typical velocities are 2×10^8 cm/sec. Thus species with half lives in the 1- to 1000-nsec range can be investigated with this method.

The third method is based on an electronic measurement of the time difference between the induction of the nuclear reaction and the spontaneous fission decay of the reaction product (Vandenbosch and Wolf, 1969). The beam projectiles must be bunched in time, and a time signal related to each beam burst must be available. A semiconductor detector is used to produce a signal upon detection of a fission fragment. Pulsed beams may be obtained either by pulsing an external beam from an accelerator or by using the inherent pulsed structure associated with the internal beams of some accelerators. This method can be used over a wide lifetime range with appropriate pulse width and spacing. In some experiments the recoil and electronic methods have been combined, with solid state detectors oriented to detect fission fragments from decay of recoils stopped on a recoil catcher placed downstream from the target (Polikanov and Sletten, 1970).

A summary of the available data is given in Table III-2. The half lives for

TABLE III-2

Half Lives, Excitation Energies, and Isomer Ratios (at Peak of Isomer Excitation Function) for Spontaneous Fission Isomers

	$t_{1/2}$(sec)	Ref.	Excitation energy	Ref.	(Isomer/ground state) ratio × 10^4	Ref.
^{234}U	†					
^{235}U	†					
^{236}U	$(105 \pm 20) \times 10^{-9}$	*r, i*				
^{238}U	$(195 \pm 30) \times 10^{-9}$	*r, l*			1.8	*t*
^{235}Pu	$(30 \pm 5) \times 10^{-9}$	*j, q*	2.4	*j*		
^{236}Pu	$(34 \pm 8) \times 10^{-9}$	*i*	4.0	*j,i*	0.1	*j*
^{237}Pu	$(82 \pm 8) \times 10^{-9}$	*m, i, h*			1.6	*m*
	$(1120 \pm 80) \times 10^{-9}$	*m, l, h*			1.4	*m*
^{238}Pu	$(6.5 \pm 1) \times 10^{-9}$	*j*	3.7	*j*	0.4	*j*
	0.5×10^{-9}	*k*	2.4	*k*		
^{239}Pu	$(8 \pm 1) \times 10^{-6}$	*l, n*	2.8	*j*	5 ± 1.5	*b*
^{240}Pu	$(3.8 \pm 0.3) \times 10^{-9}$ ‡	*j, b, i, q*	2.6	*j*	8.6 ± 2.4	*b*
^{241}Pu	$(23 \pm 1) \times 10^{-6}$ §	*p, l*	2.6	*j, p*		
^{242}Pu	28×10^{-9}	*i*			0.95	*t*
^{243}Pu	33×10^{-9}	*i, l*				
^{237}Am	5×10^{-9}	*l*	2.4	*j, l*		
^{238}Am	35×10^{-6}	*v*	2.7	*j, k*	5	*a*
^{239}Am	$(160 \pm 40) \times 10^{-9}$	*i, q*	2.5	*j, i*	1.5	
^{240}Am	$(0.91 \pm 0.07) \times 10^{-3}$	*e, n*	3.0	*j, e, o*	5	*a*
^{241}Am	$(1.5 \pm 0.6) \times 10^{-6}$	*i*	2.2	*j, i*	3	
^{242}Am	$(14.0 \pm 0.4) \times 10^{-3}$	*g, f*	2.9	*j, d, o*	4	*c*
					1.6	*t*
^{243}Am	$(6 \pm 1) \times 10^{-6}$	*l, n*			2.7	*t, n*
^{244}Am	$(1.1 \pm 0.2) \times 10^{-3}$	*f, n*				
^{245}Am	$(640 \pm 60) \times 10^{-9}$	*n*				
^{246}Am	$(73 \pm 10) \times 10^{-6}$	*n*				

TABLE III-2 (continued)

	$t_{1/2}$(sec)	Ref.	Excitation energy	Ref.	(Isomer/ ground state) ratio × 10^4	Ref.
^{241}Cm	$(15 \pm 1) \times 10^{-9}$	*j, l, q, s*	2.3	*j*		
^{242}Cm	180×10^{-9}	*s, j*	3.2	*j*	0.1	*j*
^{243}Cm	$(42 \pm 5) \times 10^{-9}$	*n, l, s*	2.3	*j*	0.65	*n*
^{244}Cm	$\geqslant 50 \times 10^{-9}$	*n, q*	3.7	*j*	0.1	*j*
^{245}Cm	$(13 \pm 2) \times 10^{-9}$	*j, n, q*	2.7	*j*	0.5	*n*
^{242}Bk	$(9.5 \pm 2.0) \times 10^{-9}$	*n*			0.4	*n*
	$(600 \pm 100) \times 10^{-9}$	*n*			0.48	*n*
^{244}Bk	$(820 \pm 60) \times 10^{-9}$	*n*			0.37	*n*
^{245}Bk	2×10^{-9}	*s*				

† Several isomers have been reported in a study of neutron-induced reactions using a pulsed neutron source (Elwyn and Ferguson, 1970). Among them is a 35-nsec activity attributed to ^{234}U. This isomer has been unsuccessfully searched for in other reactions where it would be expected to be produced in good yield (see Refs. *l, j, r*). In view of this result and the surprisingly large yields, these neutron studies are believed to be in error due to experimental difficulties. The similarities in the decay curves obtained from ^{233}U and ^{239}Pu targets, which have very similar fast neutron excitation functions, suggest complications due to scattered (time-delayed) neutrons.

‡ In the work referred to in footnote † a second isomer in ^{240}Pu was reported. No evidence has been found for this isomer in the work described in Ref. *j*.

§ A 35-nsec isomer in ^{241}Pu has been reported in Ref. *i*. Later work (Ref. *l*) indicates the half life is shorter (7–12 nsec) and that the isomer may be in either ^{241}Pu or ^{242}Pu.

[a] Borggreen *et al.* (1967)
[b] Vandenbosch and Wolf (1969)
[c] Flerov *et al.* (1968)
[d] Flerov *et al.* (1967)
[e] Bjørnholm *et al.* (1967)
[f] Brenner *et al.* (1966)
[g] Flerov *et al.* (1965)
[h] Temperley *et al.* (1971)
[i] Lark *et al.* (1969)
[j] Britt *et al.* (1971)
[k] Sletten and Limkilde (1973)
[l] Polikanov and Sletten (1970)
[m] Russo *et al.* (1971)
[n] Wolf and Unik (1972)
[o] Gangrsky *et al.* (1970a)
[p] Gangrsky *et al.* (1970b)
[q] Metag *et al.* (1969a, b)
[r] Wolf *et al.* (1970)
[s] Repnow *et al.* (1971)
[t] Gangrsky *et al.* (1970c)
[u] Back *et al.* (1971)
[v] Jorgenson *et al.* (1972)

some Pu isomers are compared with the spontaneous fission half lives of some Pu isotopes in their ground states in Fig. III-7. It can be seen that the isomeric half lives are some 10^{26} times shorter than for the same nuclide in its ground state. Only the fission decay mode has been observed for these isomers. A few searches for α- and γ-ray decay have been made, but due to the higher background encountered when looking for such decay modes, sizeable γ or α branching cannot be excluded.

Recently, confirmation of the view that the isomer has a much larger deformation than the ground state has been obtained. Specht *et al.* (1972) have identified the conversion lines of the rotational band built on the shape isomeric

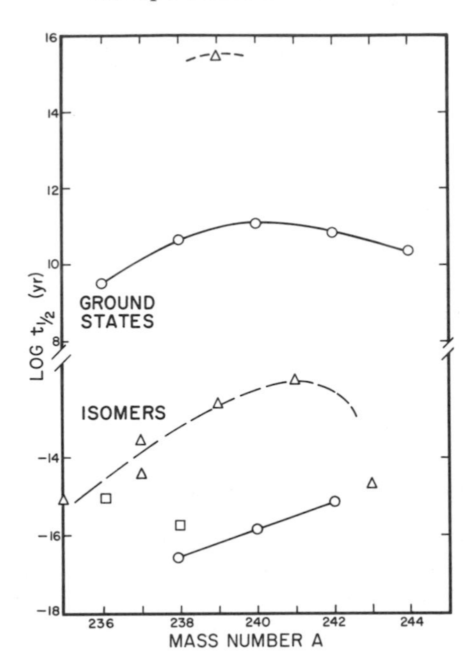

Fig. III-7. Spontaneous fission half lives for both ground and isomeric states of various plutonium isotopes. Even-even isomers with anomalously large excitation energies are designated as two quasi-particle states. ○ Even-even; △ even-odd; □ even-even, two quasi particles.

state of ^{240}Pu. Their results are illustrated in Fig. III-8. The moment of inertia of the isomeric state band is more than twice as large as that of the ground state band. A similar result has been obtained by Heffner *et al.* (1972) for ^{236}U. These results for the moment of inertia of the isomeric state are in good agreement with the theoretical calculations of Damgaard *et al.* (1969). It can also be seen from Fig. III-8 that the transition energies of the shape isomer follow the $J(J+1)$ rotational energy dependence much more closely than do those of the ground state band. This is consistent with expectations for deviations due to centrifugal stretching.

The origin of the greatly enhanced spontaneous fission rates and the implicit γ-ray retardation is of considerable interest. One possibility that had to be considered was that these isomers might simply be high spin metastable states whose decay with respect to γ emission was retarded due to spin forbiddenness. However, threshold measurements showed (Bjørnholm *et al.*, 1967) that these isomeric states had excitation energies of approximately 3 MeV, much too large to sustain hindrance with respect to γ-ray decay unless the spins of these states were very large. Measurements of the isomer ratio in reactions where different amounts of angular momentum were brought in by the

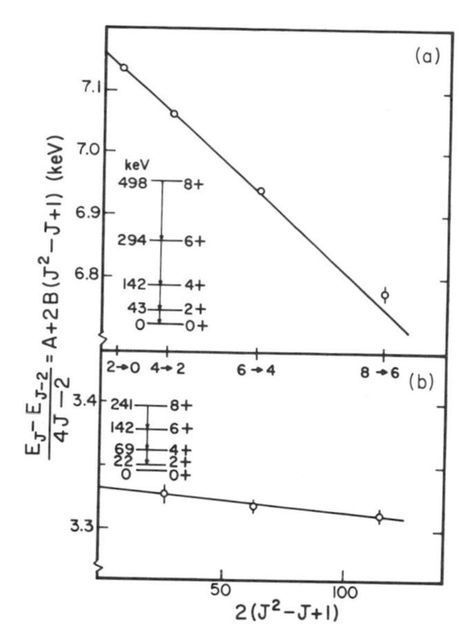

Fig. III-8. Fit of the transition energies to the energy expression $E = AJ(J+1) + BJ^2(J+1)^2$ (a) for the ground state band of the first well ($A = 7.156$ keV, $B = -3.55$ eV) and (b) for the isomeric state band of the second well ($A = 3.331 \pm 0.008$ keV, $B = 0.17 \pm 0.10$ eV). The intercept gives the rotational constant A and the slope the non-adiabaticity parameter B. [After Specht *et al.* (1972).]

inducing particle showed that the angular momentum could not be very large (Flerov *et al.*, 1968). The picture that then emerged is that these isomers are associated with the second minimum in the potential energy surface predicted by Strutinsky and described previously, and are hence to be considered shape isomers. The short spontaneous fission lifetime is attributed to the fact that the barrier is narrower and only approximately one half as high for the isomer as for the ground state.

Besides the half life, another property of isomers which can be deduced in a fairly straightforward manner is the isomer excitation energy E_i. Ideally, the isomer formation excitation function is measured to an energy as close to the threshold as experimentally feasible, and then a theoretically guided extrapolation is performed to deduce the actual threshold. A suitable yet simple way to perform the extrapolation is to use a modification (Jägare, 1970; Vandenbosch, 1972) of the Jackson (1956) evaporation model. (The application of this model to the extraction of Γ_n/Γ_f branching ratios from excitation functions of normal ground states is discussed in Chapter VII.) The model must be modified to take into account that for the last neutron emitted there is a competition between populating states in the first and in the second well that is dependent on the excitation energy of the intermediate nucleus.

For a reaction such as a (p, 2n) reaction the probability of emitting two neutrons leading to a residual excitation energy sufficient to form the isomer is proportional to

$$\int_0^{E-B_1-B_2-E_i} \int_0^{E-B_1-B_2-E_i-\varepsilon_i} (\varepsilon_1/T^2)\exp(-\varepsilon_1/T)$$

$$\times (\varepsilon_2/T^2)\exp(-\varepsilon_2/T)\, d\varepsilon_1\, d\varepsilon_2$$

$$= I(\Delta_2{}^i, 3) = 1 - \exp(-\Delta_2{}^i) \sum_{n=0}^{3} (\Delta_2{}^i)^n/n!$$

for an initial excitation energy E below the (p, 3n) or (p, 2nf) threshold. ε_1, ε_2 and B_1, B_2 are, respectively, the kinetic energies and binding energies of the first and second neutrons, T is the nuclear temperature, and $\Delta_2{}^i = (E-B_1-B_2-E_i)/T$. It is assumed that the competition between neutron emission and prompt fission is independent of excitation energy, and that γ emission does not compete if neutron emission or prompt fission is energetically possible. The expression above differs from the usual expression for a (p, 2n) reaction below the (p, 3n) threshold, for which only a single integral is necessary

$$\int_0^{E-B_1-B_2} (\varepsilon_1/T^2)\exp(-\varepsilon_1/T)\, d\varepsilon_1 = I(\Delta_2, 1) = 1 - \exp(-\Delta_2) \sum_{n=0}^{2} (\Delta_2)^n/n!$$

where $I(\Delta_2, 1)$ is Pearson's incomplete gamma function and

$$\Delta_2 = (E-B_1-B_2)/T$$

In this case it is only necessary to know the fraction of the first neutrons emitted leaving the intermediate nucleus with sufficient energy to emit a second neutron, whereas in the case of isomer formation we must take into account those nuclei in which a second neutron is emitted but for which the residual excitation energy is too low for isomer formation.

It is assumed that the relative probability of emitting neutrons to states in the second well, leading to formation of the isomer, as compared to the probability of emitting neutrons to states in the first well, leading to formation of the ground state, is simply proportional to the relative density of states in the two wells for a given neutron energy. For a constant temperature level density $\rho(E) \propto \exp(E/T)$, the ratio of states in the second well compared to those in the first well is a constant independent of the neutron emission energy for all energies above the isomer threshold.

A fit to the ^{240}Pu(p, 2n)^{240m}Am reaction is shown in Fig. III-9. It is important to use a realistic nuclear temperature in the fit to the data, or erroneously low isomer excitation energies will be obtained. The nuclear temperature appropriate to the modified Jackson model as applied to production of an

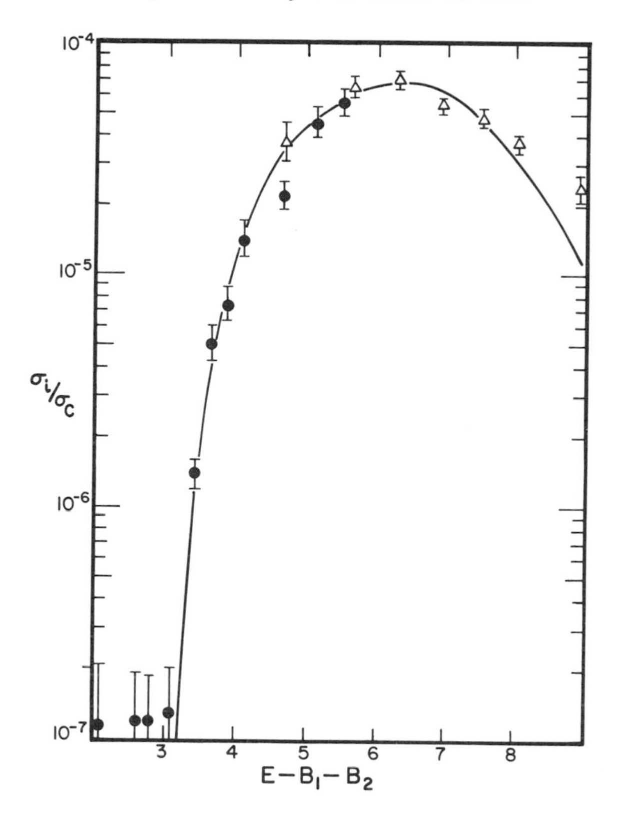

Fig. III-9. Evaporation model fit to the excitation function for production of the ^{240m}Am fission isomer. The (p, 2n) data (●) is that of Bjørnholm *et al.* (1967) and the (d, 2n) data (△) is that of Britt *et al.* (1971). The quantity σ_i/σ_c represents the ratio of the cross section for the production of the isomer by the (p, 2n) [or (d, 2n)] reaction to the corresponding total compound nucleus cross section as a function of projectile bombarding energy.

isomer is expected to be lower than that appropriate to a ground state (p, 2n) reaction, as the former involves an integration over the neutron emission spectrum of both the first and second neutrons, whereas the latter only involves the first neutron emission spectrum. The average excitation energy is higher for the first emission than for the second emission. Britt *et al.* (1971) have reanalyzed a considerable body of excitation function data to extract isomer excitation energies. Although the model employed was considerably more complicated than the simple modified evaporation model just presented, the isomer excitation energies obtained by the two approaches are in good agreement if an appropriate temperature is used in the evaporation model analysis. These values have been included in Table III-2.

Before we attempt to develop any systematics for the spontaneous fission half lives given in Table III-2, a number of points should be made. In the first place, we must be aware of experimental limitations which prevent observation of isomers with half lives much shorter than 10^{-9} sec. Since a large fraction of the observed isomers have half lives between 10^{-7} and 10^{-9} sec, it is quite likely that some isomers have not been observed simply because of their short half lives. Second, the half lives listed are the total half lives, so that the partial half lives for spontaneous fission may be longer if appreciable γ branching is present. A third complication is that occasionally there is more than one isomer observed for a given species. This is illustrated in Fig. III-7, where several of the plutonium isotopes are seen to exhibit multiple isomerism.

Let us consider the even-even Pu isomers first. The ^{236}Pu isomer and the longer-lived ^{238}Pu isomers appear to have longer half lives than would be expected on the basis of the short-lived ^{238}Pu isomer and the ^{240}Pu isomer. The excitation energies of ^{236}Pu and long-lived ^{238}Pu are also anomalous, as is illustrated in Fig. III-10. The excitation energies of these isomers is approximately 1 MeV larger than for the majority of the isomers. It has been suggested that these isomers correspond to two-quasi-particle excitations of the normal shape isomer, which would be expected to occur at an energy corresponding to the pairing gap 2Δ, which is approximately 1 MeV for these nuclei. (Two-quasi particle isomeric states with excitation energies of $\sim 2\Delta$ have been observed in a number of even-even hafnium nuclei and in ^{244}Cm.) The yields of these anomalous isomers are also an order of magnitude less than for the isomers with more typical excitation energies. That they have a longer half life than the lowest state in the second minimum is attributed to a specialization energy similar to that operative in determining the spontaneous fission lifetimes of odd-A and odd-odd nuclei (see Section C). It is not clear whether the

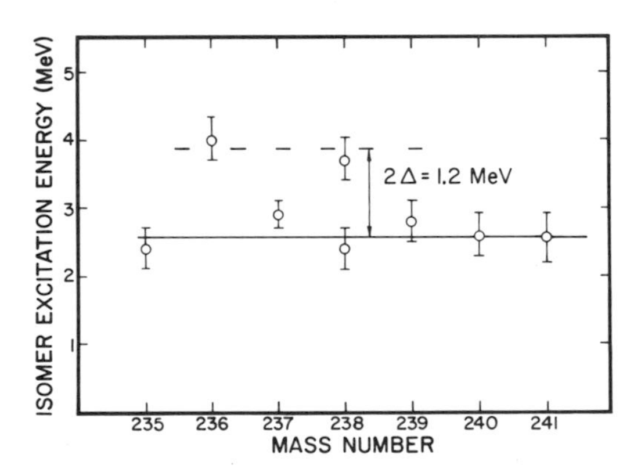

Fig. III-10. Isomer excitation energies for plutonium spontaneous fission isomers. Values are from Table III-2.

observed lifetime is due to a direct fission branch or γ decay to the lowest energy level of the second minimum followed by spontaneous fission.

The occurrence of two isomers in the same odd-A nucleus is easily accounted for by the possibility of having two close lying single particle levels of sufficiently different spin to make the γ decay from the higher to the lower energy state an isomeric transition. The spontaneous fission lifetimes may be rather different, even for two single particle states fairly close in energy, due to the possibility of very different specialization energies for the two states. Such a situation occurs for ^{237}Pu, where isomers of 80 and 1100 nsec have been observed. A study of the dependence of the relative yields on angular momentum deposition in different reactions has shown that the two isomers differ in spin, with the short-lived isomer having the lower spin (Russo *et al.*, 1971). On the basis of a comparison of the absolute yields with statistical model calculations, the high spin isomer is believed to have $I = \frac{11}{2}$. A $\frac{11}{2} - [505]$ single particle state and a $\frac{11}{2} + [615]$ single particle state are located close to the Fermi level in the Nilsson diagram at the deformation expected for the second minimum. The presence of these high spin single particle states close to the Fermi level makes them likely candidates for one of the quasi particles in the neighboring ^{236}Pu and ^{238}Pu two-quasi-particle isomeric states suggested above.

A plot of the spontaneous fission half lives versus neutron number is given in Fig. III-11. The lines for even-even nuclei have been drawn ignoring the lifetimes for the "anomalous" isomers. The values for the odd-A and odd-odd nuclides vary more smoothly with neutron number than perhaps we would have reason to expect in view of the different specialization energies possible. Such a plot also emphasizes how all the known isomers have neutron numbers between 141 and 149. This is in good agreement with the calculations of Strutinsky (1968), who finds a minima in the energy due to shell effects at $N = 146$ or 148. The N and Z boundaries are also in qualitative agreement with the calculations of Nilsson *et al.* (1969) illustrated in Fig. III-12. For nuclides with $Z = 98$ or greater the outer barrier is too small to support isomers of measurable half life. For $Z \sim 92$ the inner barriers become appreciably lower than the outer barriers, so that the lifetime for γ decay to the ground state may be very short.

When considering a possible γ-decay branch back through the inner barrier it must be recognized that the ratio of fission decay to γ decay for the isomer is not simply the ratio of the penetrability of the outer barrier $\bar{p}_B$ compared to the penetrability of the inner barrier $\bar{p}_A$ (Nix and Walker, 1969). For purposes of this and later discussions it will be helpful to designate the barriers and minima in the potential energy surface as indicated in Fig. III-13. As indicated in Section C, the spontaneous fission lifetime for the ground state is simply the product of the number of barrier assaults times the penetrability for the

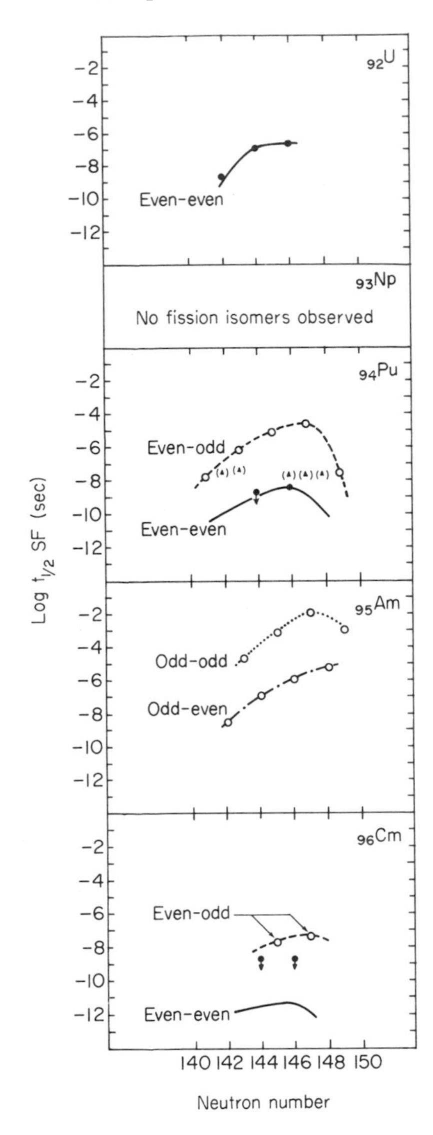

Fig. III-11. Spontaneous fission half lives for isomers as a function of neutron number. o odd systems, ● even-even systems. Points with an arrow represent upper limits. The triangles plotted off the curves for Pu represent excited states in the second potential well. [From Polikanov and Sletten (1970).]

barrier. The same expectation holds for isomeric fission decay, if we consider the penetration through the outer barrier $\bar{p}_B$. Implicit in this relation is the reasonable assumption that once the barrier is penetrated, fission is certain.

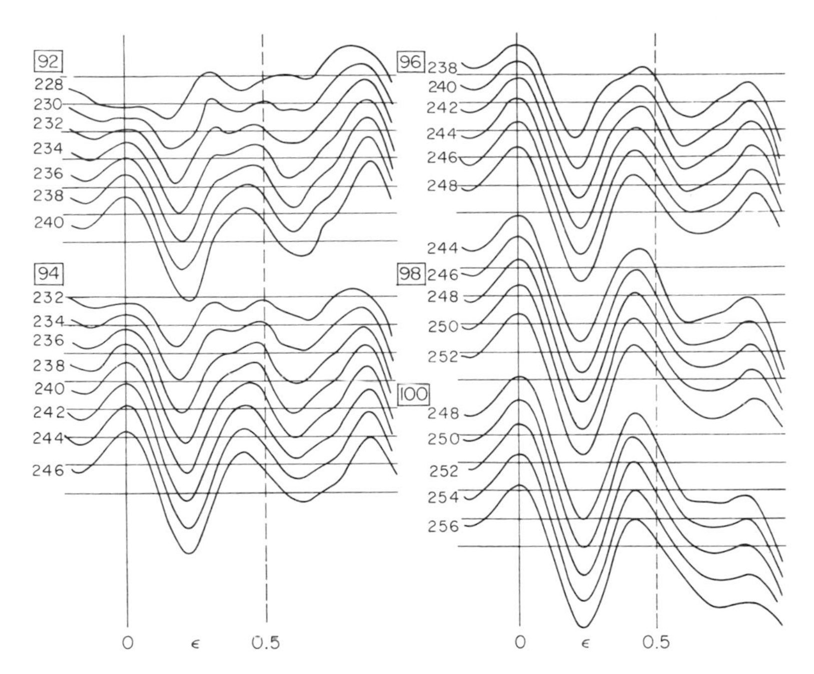

Fig. III-12. Calculated fission barriers for $Z = 92$–100. Only reflection-symmetric shapes were considered. The calculated height of the second barrier is expected to be decreased by several MeV for $Z \sim 94$ if reflection-asymmetric shapes are allowed. Separation between successive horizontal lines is 2 MeV. [From Nilsson *et al.* (1969).]

The situation is somewhat different for penetration through the inner barrier, as the decay takes place through the much weaker (and slower) electromagnetic radiation process. Loosely speaking, the nucleus after reaching the first minimum may return to the second minimum before it has had time for radiative decay. Since the energy of the nucleus exactly matches an energy level of the second minimum (the isomeric level), the penetration back is resonant and

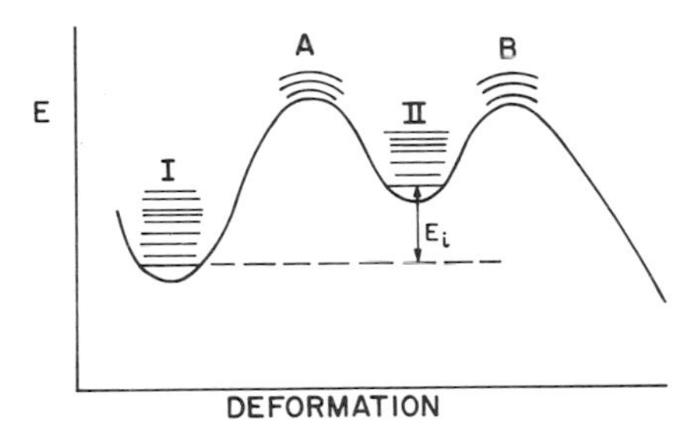

Fig. III-13. Designation of the barriers and potential minima for a double-peaked barrier.

close to unity. The probability for γ decay is therefore reduced by the ratio of the lifetime for a normal γ decay τ_0^{γ}, estimated to be of the order of 10^{-14} sec, to the mean time between barrier assaults $1/n$, approximately 10^{-20} sec. This problem can be looked at more rigorously in terms of an expansion of the wave function for the isomeric state in terms of the component situated in the region of the first (ground state) minimum and the component situated in the region of the secondary (isomer) minimum. We can designate the amplitudes of these components by a and b, respectively. If the wave function for the isomeric state is concentrated in the second well, as will be the case unless there is an energy level in the first well nearly coincident with the energy in the second well, the ratio $|a|^2/|b|^2$ will be simply the penetrability in going from the region of the second minimum back to the region of the first minimum. The decay constant for γ decay of the isomer λ_2^{γ} can therefore be written as the product of the penetrability $\bar{p}_A$ times the frequency of barrier assaults n multiplied by the retardation factor $(1/n\tau_0^{\gamma})$, giving

$$\lambda_2^{\gamma} = \bar{p}_A n(1/n\tau_0^{\gamma}) = \bar{p}_A/\tau_0^{\gamma}$$

The lifetime for γ decay is then $\tau_2^{\gamma} = \tau_0^{\gamma}/\bar{p}_A$. The ratio of the lifetime for γ decay of the isomer to the lifetime for spontaneous fission of the isomer is therefore

$$\tau_2^{\gamma}/\tau_2^{\mathrm{f}} = (\tau_0^{\gamma}/\bar{p}_A)/(1/\bar{p}_B n)$$

or, according to our previous estimates for τ_{γ}^{0} and n,

$$\tau_2^{\gamma}/\tau_2^{\mathrm{f}} = 10^6(\bar{p}_B/\bar{p}_A)$$

The implication of this result is that the penetrability of the inner barrier must be approximately 10^6 larger than for the outer barrier in order to have γ decay compete with fission decay. This means that even if the inner barrier is lower in energy than the outer barrier, the predominate mode of decay may be spontaneous fission.

What can be said about the relative heights of the two barriers? Unfortunately, there is little direct experimental evidence on this point. There is evidence from the ^{241}Am(n, γ) reaction that in the case of ^{242}Am the first barrier is higher than the second barrier (Gangrsky *et al.*, 1970d). This is obtained from a comparison of the excitation functions for the ratio of the isomer cross section to the ground state cross section σ_i/σ_g and the ratio of the isomer cross section to the prompt fission cross section σ_i/σ_f. The higher of the two barriers is larger than the neutron binding energy, so that the prompt fission excitation function exhibits a threshold behavior at about 0.5 MeV. As one goes from energies below the threshold to energies well above the threshold, the ratio σ_i/σ_f does not change much, as is illustrated in Fig. III-14. On the other hand, the ratio σ_i/σ_g has increased several orders of magnitude. These

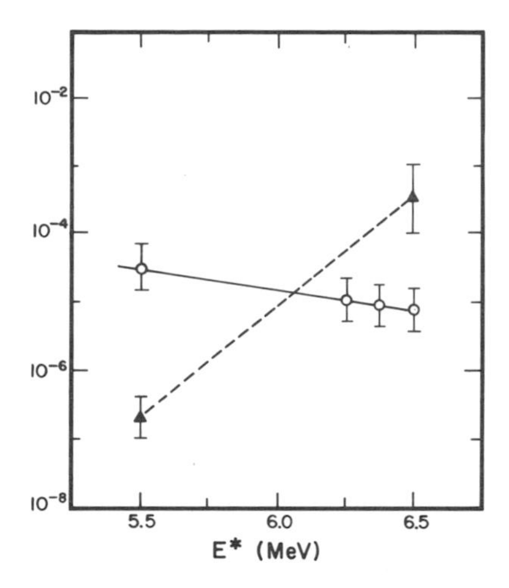

Fig. III-14. Isomeric to prompt fission cross section ratio σ_i/σ_f (○) and isomeric to ground state cross section ratio σ_i/σ_g (▲) as a function of excitation energy for the ^{241}Am(n, γ) reaction. [From Gangrsky *et al.* (1970d).]

results can be understood by assuming that the first barrier is higher in energy than the second barrier, so that the yields for both the isomer and prompt fission are determined by the penetration of this barrier, and hence behave similarly. Similar results have been obtained for the ^{243}Am(n, γ)^{244}Am reaction (Dalhsuren *et al.*, 1970). There is evidence from the angular distribution of fission fragments from neutron-induced fission that as one goes from the lighter ($A \lesssim 236$) to heavier ($A \gtrsim 240$) nuclei, the structure in the anisotropy as a function of neutron energy disappears (see Fig. V-24). This is consistent with a higher second barrier for the lighter elements and a lower second barrier for the heavier elements, as is discussed in further detail in Chapter V. This dependence of the relative heights of the two barriers on mass number is consistent with the general trend of the theoretical calculations displayed in Fig. III-12.

We have seen that the general trend of the spontaneous fission isomer half lives with N and Z illustrated in Fig. III-11 is qualitatively consistent with the trends in the theoretical double barrier heights. We will now attempt a somewhat more quantitative comparison. Pauli and Ledergerber (1971) have calculated the fission barrier in a shape representation that includes a reflection-asymmetric degree of freedom. They have tabulated the energies of the ground and isomeric state minima and of the first and second barriers for even-even nuclei from $Z = 90$ to 106. From these values we can determine the height of the barrier the isomer must penetrate to decay by spontaneous fission. To calculate the penetration of the barrier we must also know the shape of the barrier and the effective mass. In the absence of such information we can attempt a simplified description of the barrier in terms of an inverted parabola. The penetration of this barrier is given in Eq. (III-10). The dependence of the penetrability on the barrier shape and the effective mass is contained in the barrier curvature energy $\hbar\omega$. We have assumed this quantity to be the same for all nuclei, although not necessarily the same for the first and second barrier. It can be evaluated for the second barrier by fitting the known half life of ^{240m}Pu assuming the theoretical barrier height $(E_b - E_i)$. The half life is related to the penetrability by Eq. (III-7). Using our previous estimate that the number of barrier assaults n is 2.5×10^{20} per second, a barrier curvature energy $\hbar\omega_B = 0.72$ MeV is obtained. We can then calculate the expected spontaneous fission half lives for other even-even nuclei from their theoretical outer barrier heights. These half lives are shown by the solid lines in Fig. III-15. For thorium and uranium isotopes the predicted height of the first barrier is

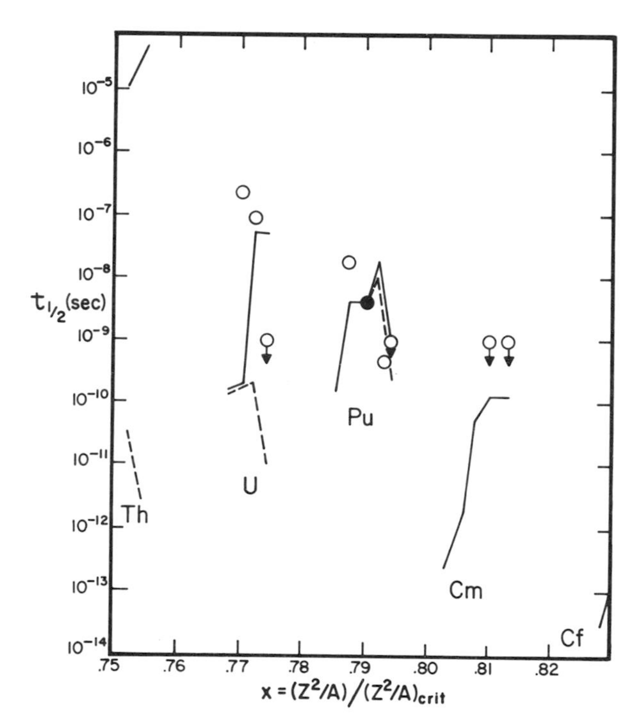

Fig. III-15. (—) Spontaneous fission isomer half lives for different even-even nuclei calculated from the theoretical outer barrier heights of Pauli and Ledergerber (1971) assuming a constant barrier curvature energy for all nuclei. The curvature energy $\hbar\omega_B = 0.72$ was determined from a fit to the ^{240}Pu isomer half life. (---) Expected contribution from γ decay, with the inner barrier curvature energy, $\hbar\omega_B = 1.0$, chosen so as to ensure that ^{240}Pu decays primarily by spontaneous fission. ○ Experimental half lives or upper limits on half lives for even-even nuclei. ● Normalization point.

sufficiently lower than that of the second barrier that γ decay of the isomer may become the predominant mode of decay. The predicted half life for γ decay can be obtained from our previous estimate that for a given penetrability, γ decay will be inhibited by a factor of 10^6 compared to fission decay. There remains the specification of the barrier curvature energy for the inner barrier. A larger curvature energy for the first barrier than for the second barrier has been obtained in previous empirical analyses (Bjørnholm, 1970; Vandenbosch and Wolf, 1969) and in the example in Fig. III-17. A value of $\hbar\omega_A = 1.0$ MeV was chosen for the present analysis. If a value appreciably larger than this is used, fission would no longer be the predominant mode of decay for the ^{240m}Pu isomer. The predicted total half lives, based on a constant curvature energy of 1.0 MeV for all nuclei and the theoretical first barrier heights of Pauli and Ledergerber, are indicated by the dashed lines in Fig. III-15. The experimental half lives or upper limits to half lives are also indicated. It can be seen that the predicted dependence of the half lives, on the basis of the theoretical barrier heights, reproduces the observations quite well. Only the uranium and plutonium even-even isomers are expected to have sufficiently long half lives for observation by present techniques. In fact, if it were not for hindrances associated with odd nucleons, the number of observable isomers might be much smaller.

The isomer excitation energies for even-Z nuclides are compared with the theoretical values of Pauli and Ledergerber in Fig. III-16. Experimental isomer excitation energies of isomers attributed to two-quasi-particle excited states have been omitted. The experimental and theoretical values are in quite good agreement.

We might wonder if it is possible to construct a one-dimensional barrier which is consistent with both the ground state and isomer state spontaneous

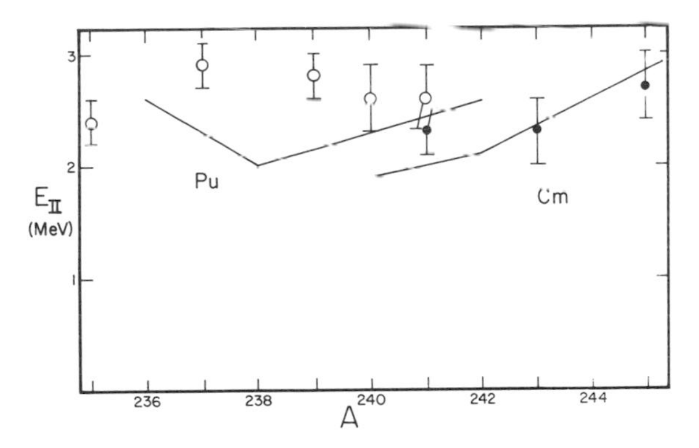

Fig. III-16. Comparison of the experimental isomer excitation energies of Britt *et al.* (1971) (○, ●) with the theoretical calculations of Pauli and Ledergerber (1971) (——).

fission half lives and with the isomer excitation energy. Such a comparison will have a well-defined meaning only if the isomer and ground state have the same spin and parity, which means in practice that we must restrict ourselves to even-even nuclei. Unfortunately there is only one even-even nucleus for which the foregoing experimental data is available, i.e., ^{240}Pu. It has been shown that it is indeed possible to construct such a barrier (Vandenbosch and Wolf, 1969) for ^{240}Pu. In fact, it is not possible to determine uniquely the barrier parameters from half life and isomer excitation energy data alone, as is indicated in Fig. III-17. All of the barriers shown are consistent with both the ground and isomer spontaneous fission half lives, the isomer excitation energy, and the location of the intermediate structure resonance (see Section V, D, 2) at 4.95 MeV. It can be seen that there are insufficient restraints to determine the barrier uniquely. We can discriminate somewhat among possible barriers by appeal to (d, pf) cross section data, which is quite sensitive to the height of the highest of the barriers. This analysis, however, does not allow us to determine which of the two barriers is higher in energy.

The success in accounting for the various data for ^{240}Pu in terms of the properties of a one-dimensional barrier cannot be entirely attributed to the number of barrier parameters which can be varied. A case in point is ^{242}Am.

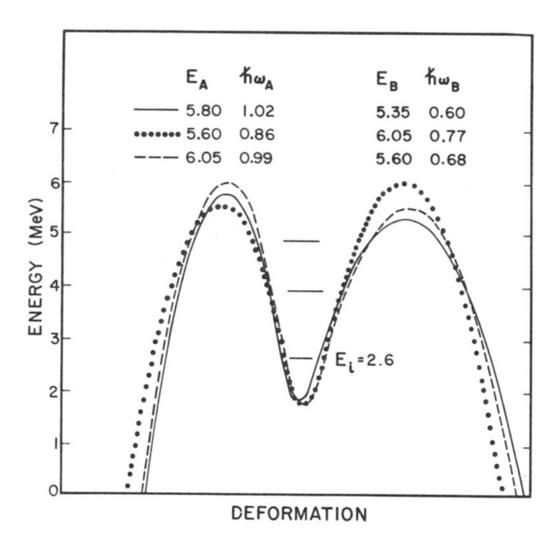

Fig. III-17. Illustration of various one-dimensional barriers which all reproduce the experimentally known ground and isomeric state half lives, isomer excitation energy, and intermediate structure resonance energy for ^{240}Pu. The broken line and dotted curves show two barriers with the same value of the height of the highest barrier, but with the inner barrier higher in one case and the outer barrier higher in the other case. The location of the second and third vibrational states, corresponding to intermediate structure resonances, are shown for the potential illustrated by the solid curve.

Nix and Walker (1969) have shown that it is not possible to account for the available data in this case. The difficulty arises in accounting for the relative shortness of the ground state spontaneous fission half life. If there is a sufficiently large barrier to the right of the second minimum to account for the spontaneous fission half life of the isomeric state, and a nearly comparable barrier on the left so that γ-ray decay of the isomer does not predominate, the total barrier for the ground state is too large to permit the observed spontaneous fission rate from the ground state. The most likely explanation for this apparent impasse is to recognize that the ground and isomeric states of this odd-odd nucleus likely have different spins; thus, the effective barriers seen by the two states are not expected to be the same because the specialization energies are likely to be different.

Another factor in the discrepancy could be a failure of the one-dimensional assumption about the fission barrier. It could well be that the position of the isomeric state on a two-dimensional potential energy surface is not located along the path of maximum penetration from the ground state. This is illustrated schematically in Fig. III-18. Even though the barrier may be a bit higher along the dashed path, if the path length is appreciably shorter than that along the dotted path (via the minimum responsible for the isomeric state), the penetrability will be greater for the dashed path. This is a consequence of the greater sensitivity of the penetrability integral to the width than to the height of the barrier. Recent calculations of the potential energy surface with inclusion of reflection-asymmetric degrees of freedom indicate that such an effect may arise. The ground state, first barrier, and second minimum are associated with reflection-symmetric shapes. The second barrier, however, corresponds to a shape with appreciable reflection asymmetry. It is difficult to quantitatively estimate the magnitude of such an effect, especially when we recognize that the effective mass probably depends on the distortion coordinates.

The yields of the fission isomers are usually very small, as can be seen from the entries in the last column of Table III-2. We can make a very qualitative estimate of the isomer ratio by assuming that the isomer ratio will be proportional to the ratio of the number of levels in each of the two wells. At the peak of the excitation function, i.e., at the highest energies where the nucleus will get "trapped" in one minimum or the other, this will be determined primarily by the ratio of the level densities at the top of the respective wells. Because of the lower level density at the top of the shallower well compared to the ground state minimum, the yield of the isomer will be much smaller than for the ground state. If we assume a level density expression

$$\rho(E^*) \propto (1/E^*) \exp[2(aE^*)^{1/2}]$$

and equal barrier heights $E_B = E_A$, the level density ratio is given by

$$\rho_{II}/\rho_I = [E_A/(E_A - E_i)] \exp\{2a^{1/2}[(E_A - E_i)^{1/2} - E_A^{1/2}]\}$$

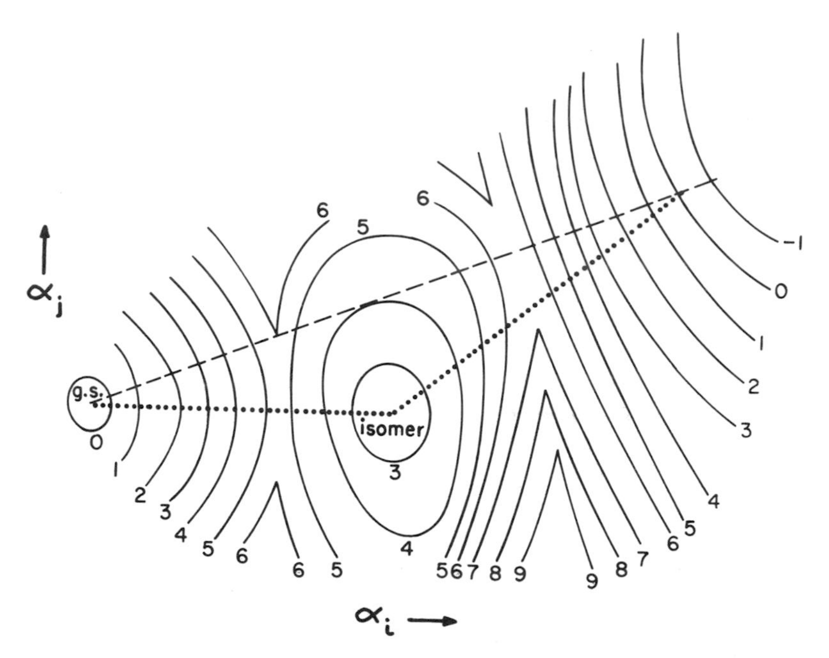

Fig. III-18. Hypothetical potential energy surface for two arbitrary distortion parameters α_i and α_j. Contour lines are at intervals of 1 MeV. The dotted path corresponds to the minimum potential energy as a function of motion in the α_i direction. The broken line path might, however, give an increased penetrability because the increase in potential height could be compensated for by the decreased width of the barrier, since the penetration depends more sensitively on the width than the height.

where E_i is the isomer excitation energy. For $E_A = E_B = 6$ MeV, $E_i = 3$ MeV, and $a = (A/10) = 24$, we obtain a ratio of level densities at the top of the barrier of 2×10^{-3}. If the barrier heights are unequal, an isomer ratio less than the foregoing estimate is expected. The observed isomer ratios are qualitatively consistent with these considerations. A much more detailed model for calculating isomer ratios has been discussed by Jägare (1969, 1970) and extended by Britt *et al.* (1971). Many of the parameters required in a more detailed calculation are usually not known, however. In principle, some of these parameters can be obtained from a fit to the data.

We might wonder whether the properties of the final fragments for fission originating from the second well exhibit any differences from "over the barrier" fission. The available evidence is that the mass and kinetic energy distributions are very similar for the two cases (Erkkila and Leachman, 1968; Ferguson *et al.*, 1970).

References

Aleksandrov, B. M. Krivokhatskii, L. S., Malkin, L. Z., and Petrzhak, K. A. (1966a). *Sov. At. Energy* **20**, 352.
Aleksandrov, B. M., Krivokhatskii, L. S., Malkin, L. Z., and Petrzhak, K. A. (1966b). *At. Energ.* **20**, 315.
Back, B. B., Bondorf, J. P., Otroshenko, G. A., Pedersen, J., and Rasmussen, B. (1969). *Proc. IAEA Symp. Phys. Chem. of Fission, 2nd, Vienna, 1969*, p. 351. IAEA, Vienna.
Back, B. B., Britt, H. C., and Erkkila, B. H. (1971). *Bull. Amer. Phys. Soc.* [II], **16**, 1149.
Barclay, F. R., Galbraith, W., Glover, K. M., Hall, G. R., and Whitehouse, W. J. (1954). *Proc. Phys. Soc. London Sect. A* **67**, 646.
Barton, D. M., and Koontz, P. G. (1970). *J. Inorg. Nucl. Chem.* **32**, 731.
Bjørnholm, S. (1970). *Proc. Robert A. Welch Found. Chem Res.* **13**, 447.
Bjørnholm, S., Borggreen, J., Westgaard, L., and Karnaukhov, V. A. (1967). *Nucl. Phys. A* **95**, 513.
Bohr, N., and Wheeler, J. A. (1939). *Phys. Rev.* **56**, 426.
Borggreen, J., Gangrsky, Yu. P., Sletten, G., and Bjørnholm, S. (1967). *Phys. Lett. B* **25**, 402.
Brandt, R., Gatti, R., Phillips, L., and Thompson, S. G. (1963). *J. Inorg. Nucl. Chem.* **25**, 1085.
Brenner, D. S., Westgaard, L., and Bjørnholm, S. (1966). *Nucl. Phys.* **89**, 267.
Britt, H. C., Burnett, S. C., Erkkila, B. H., Lynn, J. E., and Stein, W. E. (1971). *Phys. Rev. C* **4**, 1444.
Butler, J. P., Lounsbury, M., and Merritt, J. S. (1956a). *Can. J. Chem.* **34**, 253.
Butler, J. P., Eastwood, T. A., Jackson, H. G., and Schuman, R. P. (1956b). *Phys. Rev.* **103**, 965.
Caldwell, J. T., Fultz, S. C., Bowman, C. D., and Hoff, R. W. (1967). *Phys. Rev.* **155**, 1309.
Chamberlain, O., Farwell, G. W., and Segrè, E. (1954). *Phys. Rev.* **94**, 156.
Choppin, G. R., Thompson, S. G., Ghiorso, A., and Harvey B. G., (1954). *Phys. Rev.* **94**, 1080.
Choppin, G. R., Harvey, B. G., Thompson, S. G., and Ghiorso, A., (1955). *Phys. Rev.* **98**, 1519.
Cramer, J. D., and Nix, J. R. (1970). *Phys. Rev. C* **2**, 1048.
Dalhsuren, B., Flerov, G. N., Gangrsky, Yu. P., Lasarev, Yu. A., Markov, B. N., and Khanh, N. C. (1970). *Nucl. Phys. A* **148**, 492.
Damgaard, J., Pauli, H. C., Strutinsky, V. M., Wong, C. Y., Brack, B., and Stenholm Jensen, A. (1969). *Proc. IAEA Symp. Phys. Chem. of Fission, 2nd, Vienna, 1969*, p. 213. IAEA, Vienna.
Druin, V. A., Perelygin, V. P., and Khlebnikov, G. I. (1961a). *Sov. Phys. JETP* **13**, 913.
Druin, V. A., *et al.* (1961b). *Sov. Phys. JETP* **13**, 889.
Druin, V. A., Skobelev, N. K., and Rud, V. I. (1971). *Sov. J. Nucl. Phys.* **12**, 24.
Eastwood, T. A., *et al.* (1957). *Phys. Rev.* **107**, 1635.
Elwyn, A. J., and Ferguson, A. T. G. (1970). *Nucl. Phys. A* **148**, 337.
Erkkila, B. H., and Leachman, R. B. (1968). *Nucl. Phys. A* **108**, 689.
Ferguson, R. L., Plasil, F., Alam, G. D., and Schmitt, H. W. (1970). *Phys. Lett. B* **31**, 526.
Fields, P. R., *et al.* (1954a). *Nature (London)* **174**, 265.
Fields, P. R., *et al.* (1954b). *Phys. Rev.* **94**, 207.
Fields, P. R., Gindler, J. E., Harkness, A. L., Studier, M. H., Huizenga, J. R., and Friedman, A. M. (1955). *Phys. Rev.* **100**, 172.
Fields, P. R., *et al.* (1956). *Phys. Rev.* **102**, 180.

Fields, P. R., Diamond, H., Friedman, A. M. Milsted, J., Lerner, J. L., Barnes, R. F., Sjoblom, R. K., Metta, D. N., and Horwitz, E. P. (1967). *Nucl. Phys. A* **96**, 440.
Fleischer, R. L., and Price, P. B. (1964). *Phys. Rev. B* **133**, 63.
Flerov, G. N., *et al.* (1958). *Sov. Phys. Dokl.* **3**, 79.
Flerov, G. N., *et al.* (1965). *Proc. IAEA Symp. Phys. Chem. of Fission, Salzburg, 1965*, Vol. I, p. 307. IAEA, Vienna.
Flerov, G. N., Pleve, A. A., Polikanov, S. M., Tretyakova, S. P., Martologu, N., Poenaru, D., Sezon, M., Vilcov, I., and Vilcov, N. (1967). *Nucl. Phys. A* **97**, 444.
Flerov, G. N., Gangrsky, Yu. P., Markov, B. N., Pleve, A. A., Polikanov, S. M., and Jungclaussen, H. (1968). *Sov. J. Nucl. Phys.* **6**, 12.
Flerov, G. N., Oganesyan, Yu. Ts., Lobanov, Yu. V., Lasarev, Yu. A., and Tretiakova, S. P. (1971). *Nucl. Phys. A* **160**, 181.
Fried, S. (1956). *J. Inorg. Nucl. Chem.* **2**, 415.
Gangrsky, Yu. P., Markov, B. N., Polikanov, S. M., Kharisov, I. F., and Jungclaussen, H. (1968). *Izv. Akad. Nauk SSSR Ser. Fiz.* **32**, 1644.
Gangrsky, Yu. P., Markov, B. N., and Tsipenyuk, Yu. M. (1970a). *Sov. J. Nucl. Phys.* **11**, 30.
Gangrsky, Yu. P., Markov, B. N., and Tsipenyuk, Yu. M. (1970b). *Phys. Lett. B* **32**, 182.
Gangrsky, Yu. P., Nagy, T., Vinnay, I., and Kovacs, I. (1970c). Dubna Preprint P. 3–5528.
Gangrsky, Yu. P., Gavrilov, K. A., Markov, B. N., Khan, N. K., and Polikanov, S. M. (1970d). *Sov. J. Nucl. Phys.* **10**, 38.
Gerling, E. K., *et al.* (1959). *Radiokhimiya* **1**, 223.
Ghiorso, A., and Robinson, H. P. (1947). Unpublished results.
Ghiorso, A., and Sikkeland, T. (1967). *Phys. Today* **20**, 25.
Ghiorso, A., Higgens, G. H., Larsh, A. E., Seaborg, G. T. and Thompson, S. G. (1952). *Phys. Rev.* **87**, 163.
Ghiorso, A., *et al.* (1954a). Unpublished results.
Ghiorso, A., Thompson, S. G., Choppin, G. R., and Harvey, B. G., (1954b). *Phys. Rev.* **94**, 1081.
Ghiorso, A., Nurmia, M., Eskola, K., and Eskola, P. (1970). *Phys. Lett. B* **32**, 95.
Griffin, J. J. (1963). *Phys. Rev.* **132**, 2204.
Griffin, J. J. (1969). *Proc. IAEA Symp. Phys. Chem. of Fission, 2nd, Vienna, 1969*, p. 3. IAEA, Vienna.
Griffin, J. J. (1971). *Nucl. Phys. A* **170**, 401.
Halpern, I. (1959). *Annu. Rev. Nucl. Sci.* **9**, 245.
Hanna, G. C., *et al.* (1951). *Phys. Rev.* **81**, 466.
Heffner, R. H., Pedersen, J., and Swanson, H. H. (1972). Private communication.
Hill, D. L., and Wheeler, J. A. (1953). *Phys. Rev.* **89**, 1102.
Hoff, R. W., Evans, J. E., Hulet, E. K., Dupzyk, R. J., and Quaheim, B. J. (1968). *Nucl. Phys. A* **115**, 225.
Huizenga, J. R., and Diamond, H. (1957). *Phys. Rev.* **107**, 1087.
Hulet, E. K. (1953). Ph.D. Thesis, Univ. of California, Berkeley (Unclassified Rep. UCRL–2283).
Hulet, E. K., Thompson, S. G., and Ghiorso, A. (1953). *Phys. Rev.* **89**, 878.
Hulet, E. K., *et al.* (1971). *Phys. Rev. Lett.* **26**, 523.
Hyde, E. K. (1964). "Nuclear Properties of the Heavy Elements," Vol. III, "Fission Phenomena," pp. 75–77. Prentice-Hall, Englewood Cliffs, New Jersey.
Jackson, J. D. (1956). *Can. J. Phys.* **34**, 767.
Jaffey, A. H., and Hirsch, A. (1949). Unpublished results.
Jaffey, A. H., and Hirsch, A. (1951). Unpublished results.

Jägare, S. (1969). *Nucl. Phys. A* **137**, 241.
Jägare, S. (1970). *Phys. Lett. B* **32**, 571.
Johansson, S. A. E. (1959). *Nucl. Phys.* **12**, 449.
Jones, M., *et al.* (1955). Lab. Rep. KAPL–1378. Knolls At. Power Lab., Schenectady, New York.
Jones, M., Schuman, R. P., Butler, J. P., Cowper, G., Eastwood, T. A., and Jackson, H. G. (1956). *Phys. Rev.* **102**, 203.
Jorgenson, A. B., Polikanov, S. M., and Sletten, G. (1972). Unpublished results.
Kennedy, R. C., Wilets, L., and Henley, E. M. (1964). *Phys. Rev. Lett.* **12**, 36.
Kinderman, E. M. (1953). Declassified Rep. HW–27660. At. Energy Comm., Hanford.
Kuroda, P. K., and Edwards, R. R. (1957). *J. Inorg. Nucl. Chem.* **3**, 345.
Kuznetsov, V. I., Lobanov, Yu. V., and Perelygin, V. P. (1967). *Sov. J. Nucl. Phys.* **4**, 332.
Kyz'minov, B. D., Kutsaeva, L. S., Nesterov, V. G., Prokhorova, L. I., and Smirenken, G. P. (1960). *Sov. Phys. JETP* **10**, 290.
Lark, N. L., Sletten, G., Pedersen, J., and Bjørnholm, S. (1969). *Nucl. Phys. A* **139**, 481.
Magnusson, L. B. *et al.* (1954). *Phys. Rev.* **96**, 1576.
Malkin, L. Z., *et al.* (1963). *At. Energy.* **15**, 158.
Mech, J., *et al.* (1956). *Phys. Rev.* **103**, 340.
Metag, V., Repnow, R., von Brentano, P., and Fox, J. P. (1969a). *Z. Phys.* **226**, 1.
Metag, V., Repnow, R., von Brentano, P., and Fox, J. P. (1969b). *Proc. IAEA Symp. Phys. Chem. of Fission, 2nd, Vienna, 1969*, p. 449. IAEA, Vienna.
Metta, D., Diamond, H., Barnes, R. F., Milsted, J., Gray, Jr., J., Henderson, D. J., and Stevens, C. M. (1965). *J. Inorg. Nucl. Chem.* **27**, 33.
Metta, D. N., Diamond, H., and Kelly, F. R. (1969). *J. Inorg. Nucl. Chem.* **31**, 1245.
Mikheev, V. L., Skobelev, N. K., Druin, V. A. and Flerov, G. N. (1960). *Sov. Phys. JETP* **10**, 612.
Myers, W. D., and Swiatecki, W. J. (1966). *Nucl. Phys.* **81**, 1.
Myers, W. D., and Swiatecki, W. J. (1967). *Proc. Int. Symp. "Why and How Should We Investigate Nuclides Far off the Stability Line", Lysekil, Sweden, Aug. 21–27, 1966*, p. 343. Almqvist and Wiksell, Stockholm; *Ark. Fys.* **36**, 343.
Newton, J. O. (1955). *Progr. Nucl. Phys.* **4**, 234.
Nilsson, S. G. (1969). *Proc. Int. School Phys. "Enrico Fermi"* **40**.
Nilsson, S. G., *et al.* (1969). *Nucl. Phys. A* **131**, 1.
Nix, J. R. (1967). *Ann. Phys.* **41**, 52.
Nix, J. R., and Walker, G. E. (1969). *Nucl. Phys. A* **132**, 60.
Nucl. Data Sheets B (1969). **3**, No. 2.
Nurmia, M. (1970). *Proc. Int. Conf. Nucl. Reactions Induced by Heavy Ions, Heidelberg, July 1969* (R. Bock and W. K. Hering, eds.), p. 666. North-Holland Publ., Amsterdam.
Nurmia, M., Sikkeland, T., Silva, R., and Ghiorso, A. (1967). *Phys. Lett. B* **26**, 78.
Nurmia, M., Eskola, E., Eskola, P., and Ghiorso, A. (1969). Rep. No. UCRL–18667, 63. Univ. of California Radiation Lab., Berkeley, unpublished.
Oganesian, Yu. Ts., Lobanov, Yu. V., Tretiakova, S. P., Lasarev, Yu. A., Kolesov, I. V., Gavrilov, K. A., Plotko, V. M., and Poluboyarinov, Yu. F. (1970). *At. Energ.* **28**, 393.
Pauli, H. C., and Ledergerber, T. (1971). *Nucl. Phys. A* **175**, 545.
Perfilov, N. A. (1947). *Zh. Eksp. Teor. Fiz.* **17**, 476.
Phillips, L., Gatti, R., Brandt, R., and Thompson, S. G. (1962). Unpublished results.
Phillips, L., Gatti, R. C., Brandt, R., and Thompson, S. G. (1963). *J. Inorg. Nucl. Chem.* **25**, 1085.
Podgurskaya, A. V., *et al.* (1955). *Zh. Eksp. Teor. Fiz.* **28**, 503.
Polikanov, S. M., and Sletten, G. (1970). *Nucl. Phys. A* **151**, 656.

Polikanov, S. M., Druin, V. A., Karnaukhov, V. A., Mikheev, V. L., Pleve, A. A., Skobelev, N. K., Subbotin, V. G., Ter-Akopyan, G. M., and Fomichev, V. A. (1962). *Sov. Phys. JETP* **15**, 1016.

Repnow, R., Metag, V., and von Brentano, P. (1971). *Z. Phys.* **243**, 418.

Russo, P. A., Vandenbosch, R., Mehta, M., Tesmer, J. R., and Wolf, K. L. (1971). *Phys. Rev. C* **3**, 1595.

Segrè, E. (1952). *Phys. Rev.* **86**, 21.

Sikkeland, T., Ghiorso, A., Latimer, R., and Larsh, A. E. (1965). *Phys. Rev. B* **140**, 277.

Sikkeland, T., Ghiorso, A., and Nurmia, M. (1968). *Phys. Rev.* **172**, 1232.

Skobelev, N. K., Grogdev, B. A., and Druin, V. A. (1968). *Sov. J. At. Energy* **24**, 69.

Sletten, G., and Limkilde, P. (1973). *Nucl. Phys. A* **199**, 504.

Sobiczewski, A. (1970). *Proc. Robert A. Welch Found. Conf. Chem. Res.* **13**, 472.

Sobiczewski, A., Szymanski, Z., Wycech, S., Nilsson, S. G., Nix, J. R., Tsang, C. F., Gustafson, C., Möller, P., and Nilsson B. (1969). *Nucl. Phys. A* **131**, 67.

Specht, H. J., Weber, J., Konecny, E., and Heunemann, D. (1972). *Phys. Lett. B* **41**, 43.

Strutinsky, V. (1968). *Nucl. Phys. A* **112**, 1.

Swiatecki, W. J. (1955). *Phys. Rev.* **100**, 937.

Temperley, J. K., Morrissey, J. A., and Bacharach, S. L. (1971). *Nucl. Phys. A* **175**, 433.

Urin, M. G., and Zaretsky, D. F. (1966). *Nucl. Phys.* **75**, 101.

Vandenbosch, R. (1972). *Phys. Rev. C* **5**, 1428.

Vandenbosch, R., and Wolf, K. L. (1969). *Proc. IAEA Symp. Phys. Chem. of Fission, 2nd, Vienna, 1969*, p. 439. IAEA, Vienna.

Viola, V. E., Jr., and Wilkins, B. D. (1966). *Nucl. Phys.* **82**, 65.

Vorotnikov, P. E., Dubrovina, S. M. Otroschenko, G. A., Chistyakov, L. V., Shigin, V. A. and Shubka, V. M. (1970). *Sov. J. Nucl. Phys.* **10**, 419.

Watt, D. E., Bannister, F. J., Laidler, J. B. and Brown, F. (1962). *Phys. Rev.* **126**, 264.

Wheeler, J. A. (1955). *In* "Niels Bohr and the Development of Physics" (W. Pauli, ed.), p. 163. Pergamon, Oxford.

Wilets, L. (1964). "Theories of Nuclear Fission." Oxford Univ. Press (Clarendon), London and New York.

Wolf, K. L., (1969). Ph. D. Thesis, Univ. of Washington, Seattle.

Wolf, K. L., and Unik, J. P. (1972). *Phys. Lett. B* **38**, 405.

Wolf, K. L., Vandenbosch, R., Russo, P. A., Mehta, M. K., and Rudy, C. R. (1970). *Phys. Rev. C* **1**, 2096.

Wong, C. Y., and Bang, J. (1969). *Phys. Lett. B* **29**, 143.

CHAPTER

Fission Widths from Neutron Resonance Studies

At excitation energies corresponding to the binding energy of a neutron to a heavy element nucleus the average level spacing of the compound nuclear levels is of the order of or larger than the average level width. Typical level widths are 1 eV or less. It is therefore possible to observe resonance structure in excitation functions for partial and total cross sections for neutrons in the electron volt energy range.

Since the earliest days of the investigation of fission, information obtained from thermal and resonance energy neutron induced fission has played an important part in our understanding of fission. Indeed it was largely such information that led Bohr and Wheeler to conclude that ^{235}U and not ^{238}U was responsible for the fission observed in thermal neutron fission. Currently there is renewed interest in resonance fission as more detailed information becomes attainable. In particular, the dependence of various fission characteristics on the spin of the fissioning state has received considerable attention. We therefore will first consider experimental methods for determination of resonance spins.

A. Resonance Spin Determinations

We will consider here primarily only those methods which have been applied with some success to fissionable nuclides. Methods based on somewhat circular arguments involving fission theory or fission models will not be considered here. Correlations between certain fission characteristics and resonance spin will be considered in a later section.

1. Neutron Scattering

If the average resonance spacing is large compared to the average resonance width, the resonances are usually fairly well isolated and the analytic form of a resonance is given by the dispersion formula, also known as the Breit–Wigner single level formula. This formula for an isolated resonance can be written as

$$\sigma_i = \pi \hbar^2 g \frac{\Gamma_n \Gamma_i}{(\varepsilon - \varepsilon_0)^2 + (\Gamma/2)^2} \tag{IV-1}$$

where σ_i is the cross section for exit channel i, Γ_i is the partial width for the channel, Γ is the total width of the level, ε_0 is the neutron energy at the peak of the resonance, ε is the neutron energy, and

$$g = \frac{g_c}{g_n g_A} = \frac{(2J+1)}{(2s+1)(2I+1)} = \frac{2J+1}{2(2I+1)}$$

where J is the spin of the level, I the spin of the target, and s the spin of the neutron. Transmission measurements of the absolute total cross section ($\Gamma_i = \Gamma$) and the resonance width enable a determination of Γ and $g\Gamma_n$. Γ is determined from the shape of the excitation function, and this value of Γ together with the absolute cross section at the peak of the resonance, or the area of the resonance cross section, gives a value for $g\Gamma_n$. If the neutron scattering cross section ($\Gamma_i = \Gamma_n$) is measured, $g\Gamma_n^2/\Gamma$ can be obtained from Eq. (IV-1). We can now determine g from the previously determined $g\Gamma_n$ and Γ values from the relation

$$g = \frac{(g\Gamma_n)^2}{\Gamma(g\Gamma_n^2 \Gamma)}$$

This method has been applied primarily to ^{239}Pu ($I = \frac{1}{2}$), where the g values for the possible level spins of the compound nucleus for s-wave neutrons are rather different ($g = \frac{1}{4}$ for $J = 0$ and $g = \frac{3}{4}$ for $J = 1$). Resonance spins determined by this method and by the methods discussed in the following sections are summarized in Table IV-2 for ^{235}U and Table IV-3 for ^{239}Pu.

2. Interference between Levels

This method is based on the fact that a resonance of one spin can interfere with other resonances of the same spin but not with resonances of a different spin. In the case of ^{233}U and ^{235}U, the level spacings are small enough so that there is a great deal of interference between resonances. In the case of ^{239}Pu the level spacing is larger and interference effects are less dominant but still sufficient to be observed. This method requires that at least one and preferably several resonance spins be known from independent evidence, as the method gives only relative information as to whether two neighboring levels have the same or different spin. Multilevel analyses have been performed on cross section data for several nuclides, and in the case of ^{239}Pu a large number of spin assignments have been made by this technique.

3. Nuclear Orientation

It is possible to measure the angular distribution of fission fragments from uranium nuclei aligned by cooling single crystals of $UO_2Rb(NO_3)_2$. The nucleus aligns with its major axis in the plane perpendicular to the crystalline c axis (symmetry axis of the pseudo-hexagonal $UO_2Rb(NO_3)_2$ crystal). Results for thermal neutron fission were obtained some time ago, and are now being extended to the resonance region. Unfortunately the angular distribution is dependent not only on the spin of the resonance, but also on the K value of the saddle point states through which fission occurs. The angular distribution can be written as

$$\omega(\theta) = 1 + (A/T)\, P_2(\cos\theta) \qquad \text{(IV-2)}$$

where θ is the angle with respect to the c axis of the crystal. Table IV-1 gives theoretical values of A for different J and K combinations. Figure IV-1 shows anisotropy measurements in the neighborhood of the 0.28-eV resonance in ^{235}U. There is a strong correspondence of the anisotropy with the fraction of the fission cross section associated with one of the two spin states. If we take the calculated fraction of the cross section attributed to each spin state as a function of energy from the multilevel fit and deduce the values of A_2 for each spin state in the absence of the other, the spin state associated with the 0.28-eV

TABLE IV-1

Calculated A Values in Eq. (IV-2) (in °K) Corresponding to Various J, K Values[a]

K	$J = 3-$	$J = 4-$
0	+0.077	Parity forbidden
±1	+0.058	+0.065
±2	0	+0.031
±3	−0.096	−0.027
±4	Spin forbidden	−0.108

[a] The quadrupole coupling constant Phc/k was assumed to be 0.0154°K for ^{235}U. [From Dabbs *et al.* (1965).]

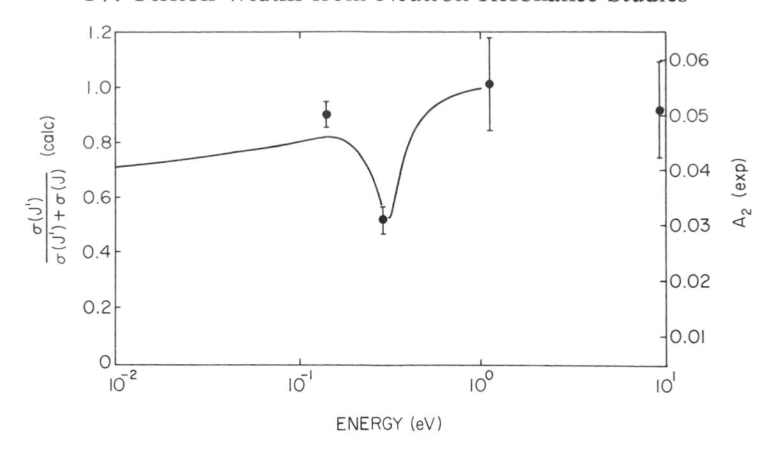

Fig. IV-1. (●) A values of Eq. (IV-2) as indicated by the scale at the right of the figure. (——) Curve derived from a multilevel fit and giving the fraction of the cross section due to resonances with spin J' different from that of the 0.28 resonance with spin J (scale at left of the figure). [After Dabbs *et al.* (1965).]

resonance would have $A \approx 0$, and the other spin state would have $A \approx 0.055$. It has been assumed in the multilevel fit that only one channel per spin value is operative, which may well not be the case. We now wish to deduce whether the 0.28-eV resonance has $J = 3$ or $J = 4$ from the anisotropy data. Let us first consider this assuming a Gaussian distribution of K values, $F(K) \propto \exp(-K^2/2K_0{}^2)$, although at this low excitation energy such a statistical assumption is not likely to be valid. It turns out that the predicted anisotropy varies from 0.065 for $K_0{}^2 = \frac{1}{2}$ to 0.021 for $K_0{}^2 = 5$, and differs by less than 10% for the two possible J values. It is therefore impossible to account for the observations with a statistical assumption about the distribution of K states. In view of the low excitation energy above the fission barrier, only a few K states are expected to be accessible. According to the discussion in Chapter II, the lowest collective state with $J\pi = 3-$ probably has $K = 0$, with perhaps a collective state with $K = 1$ somewhat higher in energy. The lowest collective state with $J\pi = 4-$ is likely to have $K = 1$, with a $K = 2$ state not too much higher in energy. Reference to Table IV-1 shows that neither spin state would be expected to give an A value as low as 0 for the K values just mentioned. Either higher K states are accessible, or the multilevel fit overestimates the contribution of other resonances at the peak of the 0.28-eV resonance. About the most that can be said is that the 0.28-eV resonance is more likely to be $4-$ than $3-$ on the expectation that the $K = 0$ state will be the principal contributor for $J\pi = 3-$.

Measurements with aligned nuclei have been extended to higher energy (Dabbs *et al.*, 1969: Pattenden and Postma, 1971) and to other targets (Kuiken *et al.*, 1972a, b). In general spin assignments have not been possible, although in the case of ^{237}Np it has been suggested that a group of resonances near 40 eV have $J = 2$. This group of resonances has been interpreted in terms of intermediate structure as discussed in Section IV, E.

4. Transmission of Polarized Neutrons through Polarized Targets

The total cross section for the interaction of polarized neutrons with a polarized nuclear target is dependent on the spin of the compound nuclear states involved. This spin dependence can be observed as a transmission effect. The experiment is rather difficult and as yet has only been applied to the nucleus ^{235}U over a limited energy region covering the first few low energy resonances (Schermer *et al.*, 1968). The results show that the 0.28-eV resonance is in the opposite spin state to the 1.14-eV resonance and to the major part of the thermal cross section. The 2.08-eV resonance is probably in the same spin state as the 0.28-eV resonance. There is also evidence that the thermal cross section is a mixture of spin states. In order to make absolute spin assignments for the ^{235}U resonances it is necessary to know the sign of the hyperfine interaction. There are no direct measurements of this quantity, but if we make certain assumptions about the chemical environment and assume that ^{235}U has a negative nuclear magnetic moment, then the resonance at 0.28 eV has $J = 3$ and the resonance at 1.14 eV has $J = 4$ (Schermer *et al.*, 1968). Unfortunately the spin assignment for the 0.28-eV resonance disagrees with that inferred from the nuclear orientation experiment (Dabbs *et al.*, 1965), and the spin assignment for the 1.14-eV resonance disagrees with that determined by the γ-ray multiplicity method to be discussed in the next section. It may be that the sign of the polarization has been incorrectly assumed.

5. Gamma-Ray Multiplicities

This method is based on the expectation that the number of γ rays in the cascade deexciting the compound nucleus to the ground state of an even-even spin zero nucleus should be larger for the higher spin state of the compound nucleus. It has been quite successful in determining spins for resonances in nonfissile nuclei. The principal difficulty in applying this method to fissile nuclei is that of distinguishing fission γ rays from capture γ rays. In principle this could be done by requiring an anticoincidence with a 4π fission counter so as to investigate only capture gammas, but this has not yet been done. Asghar *et al.* (1968) have reported on an experiment with ^{235}U where the ratio R of the coincidence rate for two γ detectors to the singles rate in one of the counters has been taken as a measure of the multiplicity. The R values for 24 resonances fall into two groups, one with $0.32 \leqslant R \leqslant 0.365$, and the other with $0.266 \leqslant R \leqslant 0.287$. The spin assignments are summarized in Table IV-2, where results from other methods are also listed.

TABLE IV-2

Spin Assignments for Neutron Resonances in $^{235}U + n$[a]

Resonance energy (eV)	Resonance fission width (eV)	Method for Spin Assignment: Neutron scattering							
		b	*c*	*d*	Polarized neutron[e]	γ-Ray multiplicity[f]	Multilevel analysis[g]	Discrete γ intensity[h]	Neutron multiplicity[i]
0.28	0.10				3				
1.135	0.115				4	3			3
2.04	0.010					4			4
2.84	0.16								
3.15	0.090					3			3
3.61	0.045					3			4
4.845	0.004					(4)			4
5.46	0.023								
5.84									
6.17	0.04								
6.39	0.009					4		4	4
7.08	0.027					3		3	3
8.79	0.08	3	3	3		3			3
9.29	0.115					3		3	4
10.20	0.046					3			3
11.65	0.004		4	4		4			4
12.39	0.028	4	3	(4)		4		4	4
15.40	0.053					3		3	3

TABLE IV-2 (continued)

Resonance energy (eV)	Resonance fission width (eV)	Method for Spin Assignment							
		Neutron scattering			Polarized neutron[e]	γ-Ray multiplicity[f]	Multilevel analysis[g]	Discrete γ intensity[h]	Neutron multiplicity[i]
		b	*c*	*d*					
16.08	0.018					(4)		3	3
16.69	0.097					3	3	3	4
18.07	0.116					3	3		4
18.97	0.065						4		
19.31	0.060		4	4		4	3	4	3
21.08								3	
22.94	0.052					3	3	3	3
23.42								4	
27.82	0.068					3	3	4	4
32.07	0.053					(4)	3	4	
33.53	0.023					4	3	4	4
34.39	0.038					4	4	3	
35.20	0.126					4	4	4	3
39.41	0.055					4	4	4	4

[a] Resonance energies and widths have been taken from Cao *et al.* (1968) or from BNL 325.
[b] Poortmans *et al.* (1967).
[c] Poortmans *et al.* (1970).
[d] Simpson *et al.* (1971).
[e] Schermer *et al.* (1968).
[f] Asghar *et al.* (1968).
[g] Cramer (1969).
[h] Weigmann *et al.* (1969).
[i] Ryabov *et al.* (1971).

6. Relative Intensity of Specific Capture Gamma Rays

The difference in initial spin for the two possible spin states is expected to manifest itself throughout the deexcitation cascade. It is somewhat hazardous to employ the intensity of γ rays originating from the initial compound nuclear state and decaying to a specific low lying state, as these are expected to fluctuate according to the Porter–Thomas distribution for 1 degree of freedom. If, however, we look at the relative intensity of a low lying transition whose parent state can be fed by many paths in the preceding cascade, the transition intensity may depend more on the spin of the capturing state than on any more specific properties of the state. An example is the 642-keV transition from the 687-keV level to the 45-keV 2+ level in ^{236}U. The spin of the 687-keV level is either 1− or 2−. In either case this level would be expected to be more strongly fed by 3− than by 4− resonances. The relative intensity of this γ ray has been measured for approximately 20 resonances by Weigmann *et al.* (1969) and a grouping of intensities has been observed. The spin assignments in most cases are consistent with those made by Asghar *et al.* (1968) on the basis of γ-ray multiplicity.

Another variant of this technique would be to measure the relative intensities of the deexcitation γ rays in the rotational cascade. It has been shown (Huizenga and Vandenbosch, 1960) that factors of 2 or more in relative intensity are expected for the two spin states from targets with spins of $\frac{5}{2}$ or $\frac{7}{2}$. This method has been successfully used in the rare earth region but has apparently not been used for fissile targets.

B. Average Fission Widths

If the fission cross section as well as the total cross section is measured as a function of neutron energy, and if the resonances are well separated, it is possible to determine the fission width for each resonance. These are found to fluctuate considerably from resonance to resonance. The distribution of widths will be considered later, but in this section we will be interested in the average width. Using the relation $2\pi\langle\Gamma_f\rangle/\langle D_J\rangle = \sum T$, where the right-hand term gives the sum of the transmission coefficients for all of the contributing channels, it is possible to obtain information about the number and "openness" of contributing channels from the average fission width and average spacing per spin state. We will designate the quantity $\overline{N} = 2\pi\langle\Gamma_f\rangle/\langle D_J\rangle$ as the effective number of fission channels, bearing in mind that for a given value it is not possible to know whether it corresponds to a larger number of slightly open channels or a smaller number of more open channels.

There are sufficient data for several nuclides for us to obtain average fission

TABLE IV-3

Spin Assignments for Neutron Resonances in ($^{239}Pu+n$)[a, b]

Energy (eV)	Direct determinations of spin *J*					Γ_f (meV)	Indirect determinations		
	c	*d*	*e*	*f*	*g*	*h*	*i*	*j*	*k*
7.82	1	1		1		47		1	1
10.93	1	1		1		143		1	1
11.89	1	1		1		24		1	1
14.31			(1)	(1)		67		1	1
14.69	0	1	1	1		30		1	1
15.42		(0)				650	(1)	1	0
17.66	1	1	1	1		34	1	1	1
22.28	0	1	1	1		62	1	1	1
26.29			1	1		55	1	(1)	0
32.38		(0)		(0)		110	(0)	0	0
35.43		(1)				5			
41.42						3			
41.66	1	1	1	1	1	54	1	(0)	
44.48	0	1	1	1	1	5	1	(1)	1
47.60		0	0	(1)		240	1	1	
49.71						690	0	(0)	
50.08			1	1	1	12	1		1
52.60		1	1	1	1	9	1	1	1
57.44				0		~500	0	1	1
58.84						~1100	0	1	0
59.22			0	1		133	1		
60.94						~6000	0		
65.71			1	1	1	74	1	0	
66.57						~140	0		1
74.05			1			32			
74.95		1	1	1	1	84	1	1	1
81.76						~2000	0		

[a] From Moore (1970), supplemented by more recent data.

[b] Those assignments shown in parentheses are less certain. The values in columns 2–6 represent direct determinations of the spins by resonance neutron scattering. Shown for comparison are values of the fission width as determined by Derrien *et al.* (Ref. *h*). At the right are assignments of resonance spins based on the assumptions that (i) the mass distribution is spin dependent (Ref. *i*), (ii) the fragment kinetic energy distribution is spin dependent (Ref. *j*), and (iii) the neutron multiplicity is spin dependent (Ref. *k*).

[c] Fraser and Schwartz (1962).

[d] Sauter and Bowman (1965).

[e] Asghar (1967).

[f] Simpson *et al.* (1971).

[g] King and Block (1969).

[h] Derrien *et al.* (1967).

[i] Cowan *et al.* (1966).

[j] Melkonian and Mehta (1965).

[k] Ryabov *et al.* (1972).

widths for each spin state. A summary of the available data is given in Table IV-4.

Farrell (1968) has performed a multilevel analysis of the ^{239}Pu fission cross section and reported that 15 levels with $J\pi = 0+$ have an average fission width of (1.0 ± 0.4) eV, and that 32 levels with $J\pi = 1+$ have an average fission width of (0.066 ± 0.016) eV. The latter number is probably not representative of the 1+ levels, as apparently a number of resonances with small fission widths were not detected. For the $J\pi = 0+$ levels $2\pi\langle\Gamma_f\rangle/\langle D_J\rangle = 1.3 \pm 0.5$, possibly indicating more than one contributing channel. This is consistent with the nature of the interference required in the multilevel analysis, which indicates that two or more channels are contributing for this spin state. At this energy, which is approximately 0.7 MeV above the barrier, the lowest transition state is expected to be fully open, and another 0+ state from a double vibrational excitation might be partly open. Ryabov *et al.* (1971) have reported a much lower average fission width for the 0+ levels, 0.29 eV, and a correspondingly low value of $2\pi\langle\Gamma_f\rangle/\langle D_J\rangle$. This apparently is because a multilevel analysis was not performed and the broad "background" resonances were not identified. For the 1+ levels $2\pi\langle\Gamma_f\rangle/\langle D_J\rangle = 0.24$. This number is somewhat uncertain and probably too large as a consequence of missed resonances with small widths. The average spacing has, however, been corrected for the missed levels, assuming there should be three times as many $J = 1$ as $J = 0$ resonances. Since the transmission coefficient is less than 0.5, the barrier for the lowest $J\pi = 1+$ transition state is higher in energy than the neutron binding energy. This would not be surprising, as the lowest collective state with $J\pi = 1+$ has been hypothesized as a combination of bending and mass asymmetry modes, and is therefore not expected at low excitation energies of the transition state nucleus.

In the case of ^{235}U, we again find some discrepancies between the results of different analyses. The average fission widths from the multilevel analysis are considerably larger than from the more conventional analysis. The effective number of channels is also larger for the 4− than for the 3− spin states in the multilevel analysis, contrary to the results of other analyses. There is evidence, from the fact that $\langle D_{J=3}\rangle/\langle D_{J=4}\rangle = 1.6$ rather than $\leqslant 1.3$ from the expected dependence of the spacing on J, that a number of 3− levels were missed in the analysis. The excitation energy for low energy fission of ^{235}U is only 0.4 MeV above the lowest 0+ transition state in the ^{236}U fissioning nucleus. The $\bar{N}$ values for the 3− state are consistent with the $K = 0$, $J = 3-$ reflection-asymmetric state being half open, corresponding to the penetrability of a state having an energy of approximately 0.5 MeV above the lowest $K = 0$, $J = 0+$ barrier. This is in agreement with an analysis of Britt *et al.* (1968) of (d, pf) and (t, pf) fission-fragment angular correlations. The large discrepancy between the $\bar{N}$ values for the $J = 4-$ levels makes the situation for this spin

TABLE IV-4

Summary of $2\pi \langle \Gamma_f \rangle / \langle D_J \rangle$ Where $\langle D_J \rangle$ Is the Average Spacing per Spin State

Target nucleus	$\langle \Gamma_f \rangle$ (eV)	$\langle D_J \rangle$ (eV)	$\bar{N} = 2\pi \langle \Gamma_f \rangle / \langle D_J \rangle$	Ref.
^{232}U	0.56	5.7	0.6	*b*
^{233}U	0.379	1.58	1.5[a]	*c*
^{235}U				
$J = 3-$	0.067	1.26	0.33	*d*
	0.114	1.75	0.4	*j*
	0.063			*k*
	0.087	1.26	0.42	*l*
$J = 4-$	0.027	1.26	0.13	*d*
	0.142	1.10	0.8	*j*
	0.038			*k*
	0.026	1.26	0.13	*l*
^{238}Pu			0.002	*e*
^{239}Pu				
$J = 0+$	1.0	5.0	1.3	*f*
	0.29	8.3	0.19	*l*
$J = 1+$	0.066	1.7	0.24	*f*
	0.045	3.8	0.088	*l*
^{242}Pu	0.00009	29	0.00002	*n*
^{241}Pu				
$J = 2+$	0.51	4.1	0.77	*g*
$J = 3+$	0.19	2.2	0.55	*g*
^{241}Am	0.00015	1.0	0.001	*h*
^{242m}Am			2.5[a]	*i*
^{244}Cm	0.00135	13	0.00062	*m*
^{245}Cm	0.6	9.2	0.41[a]	*m*
^{246}Cm	0.00048	38	0.00008	*m*
^{247}Cm	0.14	8.2	0.10[a]	*m*
^{248}Cm	0.0013	35	0.00023	*m*

[a] Two spin states are possible; the value of N is the average value per spin state.
[b] Auchampaugh *et al.* (1968).
[c] Bergen and Silbert (1968).
[d] The average fission widths for the two J states are based on the spin assignments of Asghar *et al.* (1968), the fission widths are from BNL 325, and the $\langle D_J \rangle$ are from Ref. *l*.
[e] Bowman *et al.* (1967).
[f] Farrell (1968).
[g] Sauter and Bowman (1968).
[h] BNL 325.
[i] Bowman *et al.* (1968).
[j] Cramer (1969).
[k] Weigmann *et al.* (1969).
[l] Ryabov *et al.* (1971).
[m] Moore and Keyworth (1971).
[n] Bergen and Fullwood (1971).

state less clear, although the expectations are that it should lie higher in energy and have a lower $\bar{N}$ value.

For neutron fission of ^{241}Pu the $\bar{N}$ values are 0.77 and 0.55 for the $J = 2+$ and $J = 3+$ states, respectively. Again the excitation energy is approximately $\frac{1}{2}$ MeV above the lowest barrier state, and the $2+$ state would be expected to be nearly fully open. The $3+$ state is expected to lie considerably higher in energy and to be less open.

Less extensive information is available for other nuclides, partly due to lack of spin assignments for enough resonances to determine the average widths for each spin state. In the case of ^{233}U a multilevel analysis yields $\langle D \rangle =$ 0.79 eV (or 1.58 eV per spin state) and $\langle \Gamma_f \rangle = 0.379$ eV (Bergen and Silbert, 1968). From these values we obtain $\bar{N} = 1.5$ per spin state. This is consistent with at least one fully open channel, probably the $2+$ rotational state of the ground state band.

A summary of $\langle \Gamma_f \rangle$, $\langle D_J \rangle$, and $\bar{N}$ values for these and other nuclides is given in Table IV-4. Many of these values are subject to large uncertainties associated with a failure to use multilevel fits (particularly crucial for ^{233}U and ^{235}U), an insufficient number of resonances studied, and inability to observe very narrow or very broad resonances. Lynn (1964) has emphasized the difficulties resulting from level-level interference effects. These effects can greatly perturb the apparent number and widths of resonances in a single level analysis, and even make it questionable whether a unique multilevel fit can be obtained. Historically the $2\pi\langle \Gamma_f \rangle / \langle D_J \rangle$ values have tended to increase as better data became available.

C. Width Distributions

One of the earliest indications that the probability for fission is determined by a relatively small number of saddle point states, rather than the very large number of possible final states associated with various mass, charge, and excitation energy divisions, was the observation that fission widths fluctuate considerably from resonance to resonance. In this respect the fission widths resembled the neutron widths, which also fluctuate a great deal, rather than the radiation widths, which vary little from resonance to resonance. There is a single exit channel for neutron emission, but many final states which can be reached by emission of a γ ray. The width distribution for N channels is given by the chi-squared distribution

$$P_N(x)\,dx = [\Gamma(N/2)]^{-1}(N/2\bar{x})^{N/2}\,x^{(N-2)/2}\exp(-Nx/2\bar{x})\,dx \qquad \text{(IV-3)}$$

where x is the fission width and $\Gamma(p)$ is the gamma function. The number of participating channels may be computed from

$$N = 2\langle\Gamma\rangle^2/(\langle\Gamma^2\rangle - \langle\Gamma\rangle^2) \tag{IV-4}$$

The combined error due to a constant uncertainty in the measurements, $\delta\Gamma/\Gamma$, and the finite sample size associated with n observed resonances has been given by Wilets (1964) as

$$\left(\frac{\delta N}{N}\right)^2 = \frac{1}{n(\langle\Gamma^2\rangle - \langle\Gamma\rangle^2)^2}\Bigg[\left(\frac{\delta\Gamma}{\Gamma}\right)^2(\langle\Gamma^4\rangle - 2\langle\Gamma^2\rangle\langle\Gamma^3\rangle/\langle\Gamma\rangle + \langle\Gamma^2\rangle^3/\langle\Gamma\rangle^2) + \tfrac{1}{4}(\langle\Gamma^4\rangle - \langle\Gamma^2\rangle^2) - \langle\Gamma^2\rangle\langle\Gamma^3\rangle/\langle\Gamma\rangle + \langle\Gamma^2\rangle^3/\langle\Gamma\rangle^2\Bigg] \tag{IV-5}$$

It is important to note that Eq. (IV–4) is obtained whether all participating channels are completely open or whether all of the channels are only partially (but equally) open. In the more general case we interpret N from Eq. (IV-4) as the effective number of channels. This estimate will always be equal to or greater than that obtained from $2\pi\Gamma/D$.

Another method of estimating the number of participating channels is by graphical comparison with the cumulative distributions derived from Eq. (IV-3). Such a comparison for ^{239}Pu fission widths is illustrated in Fig. IV-2.

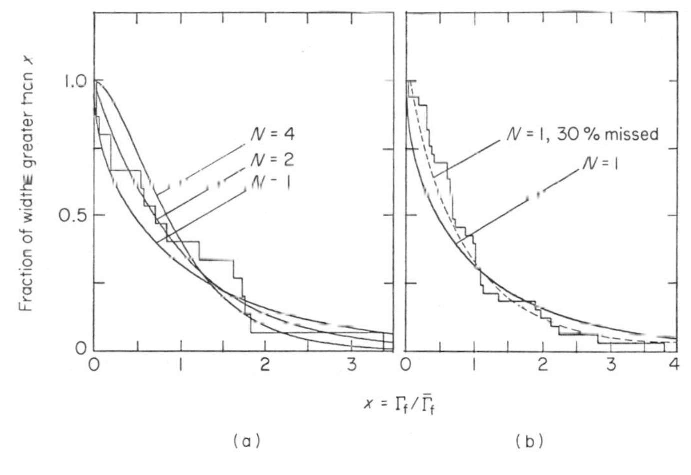

Fig. IV-2. The integral distributions of fission widths for the two spin states. (a) Porter–Thomas distributions for 1, 2, and 4 open channels are plotted for the 0^+ state ($\langle\Gamma_f\rangle = 1033$ meV). (b) The solid curve for the 1^+ state is the Porter–Thomas distribution for one channel ($\langle\Gamma_f\rangle = 66$ meV). The dashed curve is the truncated Porter–Thomas distribution with the 30% smallest widths left out. [From Farrell (1968).]

It is seen that for the $J = 0$ resonances $N = 2$ gives a better fit than $N = 1$ or 4. The situation for the $J = 1$ resonances is less clear, since many resonances with small widths may have been missed.

The width distributions for ^{233}U, ^{235}U, and ^{241}Pu seem to indicate several open channels. For example, $N = 3.6$ for ^{233}U, but this may be a result of the superposition of two $N = 1$ distributions with different average Γ. Unfortunately at this time there is insufficient information for us to determine the number of open channels for each spin state separately.

D. Fission Characteristics Which Apparently Depend on Resonance Spin

1. Mass Asymmetry

A number of experiments on the variation of mass asymmetry from resonance to resonance have been performed since the suggestion of Bohr (1956) and Wheeler (1963) that the symmetry of fission might depend on the spin and parity of the resonance. Such a dependence might arise if in some way the eventual mass division were influenced by the symmetry properties of the wave function describing the saddle point channel. Thus the lowest saddle point channel with $J = 0+$ has a symmetric wave function and might be expected to exhibit an increased probability for symmetric fission, while the lowest collective $3-$ state has a reflection-asymmetric wave function and might be expected to exhibit increased asymmetric fission. It is more difficult to predict the symmetry properties of the wave function for $4-$ and $1+$ states (the other spin states expected in neutron fission of ^{235}U and ^{239}Pu, respectively), as intrinsic excitations may lie as low in energy as collective excitations.

Variations in the ratio of symmetric to asymmetric fission from resonance to resonance have been observed, and sufficient data exist for both ^{235}U and ^{239}Pu to indicate a bimodal distribution for this ratio. For resonance fission of ^{239}Pu, the $0+$ resonances have both a higher probability for symmetric fission and larger widths (Cowan *et al.*, 1966). This observation is consistent with the foregoing proposal. It has previously been seen that the average fission width for $J = 0+$ resonances is appreciably larger than for $J = 1+$, due to the lower barrier for this spin state. In the case of ^{235}U, the group of resonances with increased fission symmetry also has a larger average fission width than the group with smaller ratios of symmetric to asymmetric fission (Cowan *et al.*, 1970). This is contrary to the above-mentioned expectations if the group with the larger average fission width corresponds to $J = 3-$, as would be expected on the basis of the relative barrier heights of $3-$ and $4-$ transition states. Although the grouping of the resonances according to mass ratio reveals a different average width for each group, the correlation of mass asymmetry with resonance spin is not at all clear. There is little correlation between the spin values listed in Table IV-2 and the mass asymmetry. If, however, we take the larger average fission width as evidence of a 3− assignment to the resonances in this group, we are left with a contradiction with the expectations based on the symmetry properties of the saddle point transition state. It has been noted by Cowan *et al.* (1966) that the correlation of greater mass symmetry with larger widths may indicate a stronger dependence on excess energy at the fission barrier than on the spin and parity of the channel. Another possibility is that the (n, γf) process may be playing a role, as discussed in Section IV, F.

2. Kinetic Energy Release

Melkonian and Mehta (1965) have searched for a correlation between kinetic energy release and resonance spin. Although slight variations from resonance to resonance do occur, it is not clear that any grouping occurs. For those resonances in ^{235}U for which the direct measurements of the spin do not disagree with one another, one resonance with lower than average kinetic energy has $J = 3$ and two have $J = 4$. There are two $J = 3$ and two $J = 4$ resonances with higher than average kinetic energy. The situation is a little better for ^{239}Pu, where most of the resonances with higher kinetic energy have $J = 1$. However, of the few resonances thought to have lower than average kinetic energy, half had $J = 0$ and half had $J = 1$. Further work is required to unambiguously establish the correlation of kinetic energy with spin.

3. Neutron Multiplicity

If the total energy release in fission is independent of resonance spin, any dependence of the kinetic energy release on spin would be expected to be compensated for by a change in the number of neutrons (or possibly γ rays) emitted in fission. Variations in neutron multiplicity from resonance to resonance have been observed, with some indication that the values fall into two groups. The results obtained in different laboratories are unfortunately contradictory, with Ryabov *et al.* (1972) concluding that for ^{239}Pu the $J = 0$ resonances have lower than average neutron multiplicities, whereas Weinstein *et al.* (1969) find the $J = 0$ resonances have the higher multiplicity. Similar discrepancies exist for ^{235}U.

4. Ratio of Ternary to Binary Fission

There is also some indication that the ratio of ternary to binary fission varies from resonance to resonance. Unfortunately there is very poor agreement between the results of Michaudon *et al.* (1965) and Melkonian and Mehta (1965). The variations from resonance to resonance are not large (Kvitek *et al.*, 1966).

E. Intermediate Levels in the Two Potential Wells of Heavy Nuclei†

Recently a rather surprising and dramatic effect has been observed in the excitation function for neutron-induced fission of ^{240}Pu. It is found that in the energy range between 0.5 and 3 keV most of the compound nuclear resonances have very small fission widths (usually too small to measure), whereas at periodic spacings of about 0.6 keV there are groups of resonances with greatly enhanced widths. This region of enhancement typically embraces four or five resonances within an energy interval of about 0.1 keV. Similar structure has been observed in neutron-induced fission of ^{237}Np as well as in other nuclides. This structure is related to a second minimum in the potential energy surface as a function of deformation. The implications of this second minimum will be discussed in some detail in this section.

The levels in the first potential well (equilibrium deformation β_1) will be designated class I levels and the levels in the second potential well (saddle point deformation β_2) will be designated class II levels. This classification of levels is illustrated in Fig. IV-3. If the excitation energy exceeds the barrier energy between the two potential wells (barrier A), the two groups of levels, class I and class II, will be quite thoroughly mixed. On the other hand, if the excitation energy is less than barrier A, the two groups of levels will be well separated and exhibit little, if any, mixing. In the intermediate case, there will be weak coupling through the barrier and class I levels with energies near the more widely spaced class II levels will mix to form the actual compound levels.

For fission induced with monoenergetic neutrons of subbarrier energies, we observe structure in the fission cross section as illustrated in Fig. IV-4 for the $^{237}Np+n$ reaction. At low energy resolution the fission excitation function exhibits structure with an average separation of about 50 eV between peaks. This spacing is too large to be due to the spacing between compound levels (class I) in the first well and is identified with the spacing of the levels in the second well (class II). These levels in the second well can be thought of as acting as intermediate states for the initial compound nucleus on the path of increasing deformation toward scission. This structure is therefore designated intermediate structure. At higher energy resolution the intermediate structure resonances can be resolved into more closely spaced resonances

† We follow closely the notation developed by Bjørnholm and Strutinsky (1969).

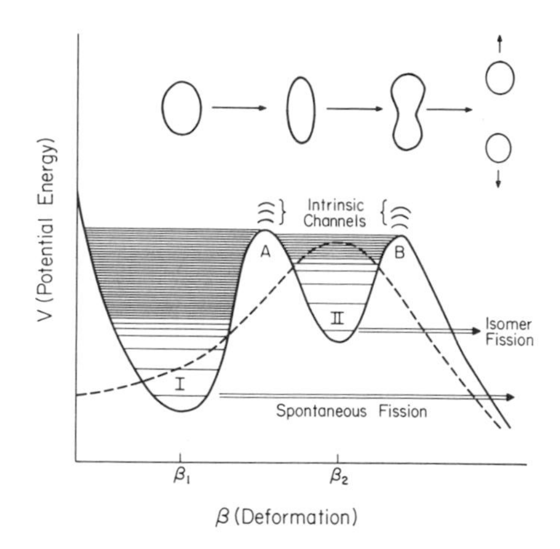

Fig. IV-3. Schematic illustrations of single-humped (---) and double-humped (—) fission barriers. Intrinsic excitations in the first and second wells at deformations β_1 and β_2 are designated class I and class II states, respectively. Intrinsic channels at the two barriers are also illustrated. The transition in the shape of the nucleus as a function of deformation is schematically represented in the upper part of the figure. Spontaneous fission of the ground state and isomeric state occurs from the lowest energy class I and class II states, respectively.

whose spacing is characteristic of the compound levels of the first well. If the width of the states in the second minimum is small compared to their spacing, we expect an enhancement of the fission widths of the compound nuclear levels if and only if the energy of the compound nuclear resonance by accident is almost equal to the energy of one of the intermediate levels. The number of compound levels whose fission widths will be enhanced depends on the width of the intermediate states.

A schematic representation of the observed fission cross section for fission induced with neutrons of subbarrier energies is shown in Fig. IV-5. The quantity $\gamma_2{}^{\mathrm{t}}$ represents the total width of levels in the second potential well, and is given by

$$\gamma_2{}^{\mathrm{t}} = \overleftarrow{\gamma}_2 + \gamma_2{}^{\mathrm{d}} \qquad \text{(IV-6)}$$

where $\overleftarrow{\gamma}_2$ represents the width of internal nonradiative transitions associated with a shape change from the second potential well to the first potential well and $\gamma_2{}^{\mathrm{d}}$ is the sum of decay widths for all partial decay widths which are energetically possible, $\gamma_2{}^{\mathrm{d}} = \gamma_2{}^{\mathrm{n}} + \gamma_2{}^{\gamma} + \gamma_2{}^{\mathrm{f}}$. The total width of the compound nucleus formed in neutron capture is given by Γ_{t}. The fission width Γ_{f} is given by the area of each resonance in Fig. IV-5. The quantities D_{I} and D_{II} are the average spacings of the levels in potential wells I and II, respectively. As can be seen in Fig. IV-5, the schematic illustration corresponds to a case where $\gamma_2{}^{\mathrm{t}}$

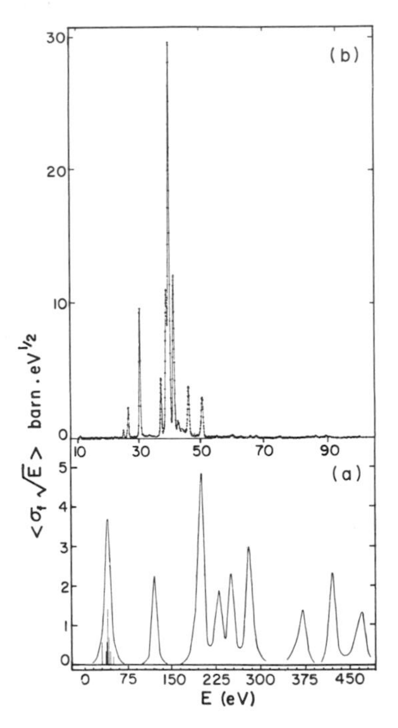

Fig. IV-4. (a) Fission cross section multiplied by $(E_n)^{1/2}$ averaged over 10 eV for ^{237}Np. (b) A smaller energy region at high resolution. [After Paya *et al.* (1968a).]

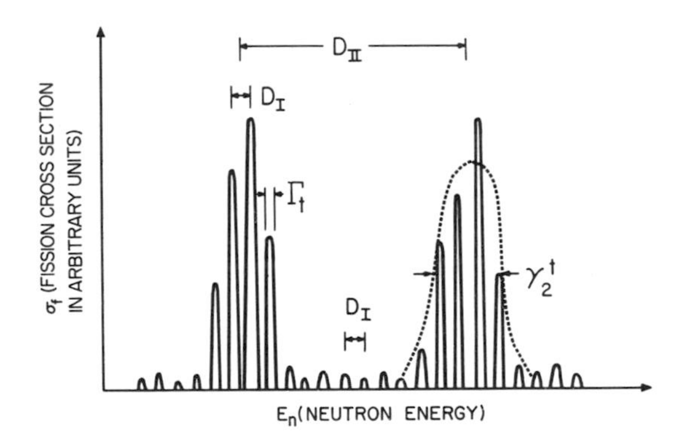

Fig. IV-5. Schematic diagram of the quantities that are important in the deformation of the resonance cross section for the fission of ^{240}Pu with subbarrier neutrons. The quantities D_I and D_{II} are the spacings between class I and class II states, respectively, Γ_t is the total width of the compound nucleus formed in neutron capture, and $\gamma_2{}^t$ is the total width of the class II state.

is considerable larger than Γ_t but smaller than the average spacing D_{II} between levels in the second well.

In the following derivations, we will assume that the excitation energy is sufficiently high so that the nucleus is thermodynamically hot at both deformations, ensuring the applicability of a statistical model. Indices 1 and 2 correspond to quantities in the first and second potential wells, respectively. Quantities denoted by Γ and γ refer to the compound nucleus as a whole and one of the potential wells, respectively. If the energy spread of the beam is large compared to the level spacings in both potential wells, the following decay processes

$$\xleftarrow{\gamma_1{}^d} n_1 \underset{\overleftarrow{\gamma}_2}{\overset{\vec{\gamma}_1}{\rightleftarrows}} n_2 \xrightarrow{\gamma_2{}^d}$$

give rise to the following differential equations

$$-\hbar\frac{dn_1}{d\tau} = (\gamma_1{}^d + \vec{\gamma}_1)n_1 - \overleftarrow{\gamma}_2 n_2 = \gamma_1{}^t n_1 - \overleftarrow{\gamma}_2 n_2 \tag{IV-7}$$

$$-\hbar\frac{dn_2}{d\tau} = -\vec{\gamma}_1 n_1 + (\gamma_2{}^d + \overleftarrow{\gamma}_2)n_2 = -\vec{\gamma}_1 n_1 + \gamma_2{}^t n_2 \tag{IV-8}$$

where n_1 and n_2 are the respective populations in the first and second potential wells at time τ. The total width of levels in the first potential well is given by

$$\gamma_1{}^t = \vec{\gamma}_1 + \gamma_1{}^d$$

where $\vec{\gamma}_1$ is the width of internal nonradiative transitions from the first to the second potential well and $\gamma_1{}^d = \gamma_1{}^n + \gamma_1{}^\gamma$, the sum of decay widths for all channels which are energetically open. Direct radiative transitions and neutron emission between the first and second wells are excluded. We assume that the first well is deeper than the second well.

The foregoing differential equations may be solved as follows. First, solve Eq. (IV-7) for n_2 (in the remaining discussion $\hbar$ is set equal to 1)

$$n_2 = \frac{1}{\overleftarrow{\gamma}_2}\left(\gamma_1{}^t n_1 + \frac{dn_1}{d\tau}\right) \tag{IV-9}$$

Differentiate Eq. (IV-9)

$$\frac{dn_2}{d\tau} = \frac{\gamma_1{}^t}{\overleftarrow{\gamma}_2}\frac{dn_1}{d\tau} + \frac{1}{\overleftarrow{\gamma}_2}\frac{d^2 n_1}{d\tau^2} \tag{IV-10}$$

Substitute (IV-9) and (IV-10) into (IV-8) and rearrange terms

$$\frac{d^2 n_1}{d\tau^2} + (\gamma_1{}^t + \gamma_2{}^t)\frac{dn_1}{d\tau} + (\gamma_1{}^t\gamma_2{}^t - \vec{\gamma}_1\overleftarrow{\gamma}_2)n_1 = 0 \tag{IV-11}$$

The two solutions of the second order differential equation are linearly independent and the general solution is composed of a linear combination of these two solutions with two constants, namely,

$$n_1 = c_1 \exp\{\tau[-\gamma_1{}^{\mathrm{t}} - \gamma_2{}^{\mathrm{t}} - ((\gamma_1{}^{\mathrm{t}} - \gamma_2{}^{\mathrm{t}})^2 + 4\vec{\gamma}_1\overleftarrow{\gamma}_2)^{1/2}]/2\}$$
$$+ c_2 \exp\{\tau[-\gamma_1{}^{\mathrm{t}} - \gamma_2{}^{\mathrm{t}} + ((\gamma_1{}^{\mathrm{t}} - \gamma_2{}^{\mathrm{t}})^2 + 4\vec{\gamma}_1\overleftarrow{\gamma}_2)^{1/2}]/2\} \qquad \text{(IV-12)}$$

For simplicity, let

$$\Delta = [(\gamma_1{}^{\mathrm{t}} - \gamma_2{}^{\mathrm{t}})^2 + 4\vec{\gamma}_1\overleftarrow{\gamma}_2]^{1/2} = \lambda_> - \lambda_<$$
$$= [R^2 + 4\vec{\gamma}_1(\gamma_1{}^{\mathrm{d}} - \gamma_2{}^{\mathrm{d}})]^{1/2}$$

where $R = \gamma_1{}^{\mathrm{t}} - \gamma_2{}^{\mathrm{t}} - 2\vec{\gamma}_1 = (\gamma_1{}^{\mathrm{d}} - \gamma_2{}^{\mathrm{d}}) - (\vec{\gamma}_1 + \overleftarrow{\gamma}_2)$ and

$$\lambda_> = (\gamma_1{}^{\mathrm{t}} + \gamma_2{}^{\mathrm{t}} + \Delta)/2, \qquad \lambda_< = (\gamma_1{}^{\mathrm{t}} + \gamma_2{}^{\mathrm{t}} - \Delta)/2$$

Equation (IV-12) may then be written

$$n_1 = c_1 \exp(-\lambda_> \tau) + c_2 \exp(-\lambda_< \tau) \qquad \text{(IV-13)}$$

The quantity n_2 is obtained by solving the first order differential equation (IV-9). The solution is

$$n_2 = c_1\left(\frac{\gamma_1{}^{\mathrm{t}}}{\overleftarrow{\gamma}_2} - \frac{\lambda_>}{\overleftarrow{\gamma}_2}\right)\exp(-\lambda_> \tau) + c_2\left(\frac{\gamma_1{}^{\mathrm{t}}}{\overleftarrow{\gamma}_2} - \frac{\lambda_<}{\overleftarrow{\gamma}_2}\right)\exp(-\lambda_< \tau) \qquad \text{(IV-14)}$$

The constants c_1 and c_2 are evaluated from the boundary conditions, namely, that at $\tau = 0$, $n_1 = 1$ and $n_2 = 0$. The results are

$$c_1 = (\gamma_1{}^{\mathrm{t}} - \lambda_>)/(\lambda_> - \lambda_<) \qquad \text{and} \qquad c_2 = -(\gamma_1{}^{\mathrm{t}} - \lambda_>)/(\lambda_> - \lambda_<)$$

If these values of c_1 and c_2 are substituted into Eqs. (IV-13) and (IV-14), the resulting equations for n_1 and n_2 are

$$n_1 = [(\gamma_1{}^{\mathrm{t}} - \gamma_2{}^{\mathrm{t}} + \Delta)/2\Delta]\exp(-\lambda_> \tau) + [(\gamma_2{}^{\mathrm{t}} - \gamma_1{}^{\mathrm{t}} + \Delta)/2\Delta]\exp(-\lambda_< \tau) \qquad \text{(IV-15)}$$

$$n_2 = (\vec{\gamma}_1/\Delta)[\exp(-\lambda_< \tau) - \exp(-\lambda_> \tau)] \qquad \text{(IV-16)}$$

The rate of any partial decay of the compound nucleus, as well as the probability for a specific excitation in the first or second potential well, may be calculated from the equation above. The fission rate is given by

$$\frac{dn_{\mathrm{f}}}{d\tau} = \vec{\gamma}_2 n_2 \qquad \text{(IV-17)}$$

where $\vec{\gamma}_2$ is equal to $\gamma_2{}^{\mathrm{f}}$. Substitution of n_2 from Eq. (IV-16) and integrating Eq. (IV-17) from $\tau = 0$ to ∞ gives

$$n_{\mathrm{f}}(\tau = \infty) = \vec{\gamma}_2\vec{\gamma}_1/(\gamma_1{}^{\mathrm{t}}\gamma_2{}^{\mathrm{t}} - \vec{\gamma}_1\overleftarrow{\gamma}_2) = \langle\Gamma_{\mathrm{f}}\rangle/\langle\Gamma_{\mathrm{t}}\rangle \qquad \text{(IV-18)}$$

where $n_{\mathrm{f}}(\tau = \infty)$ is the ratio of the average fission width to the average total width of the compound nucleus. The average width $\langle \Gamma_t \rangle$ is defined in terms of the remaining nuclei, $n_1(\tau)+n_2(\tau)$, at time τ (where $\tau = 1/\langle \Gamma_t \rangle$) being e times smaller than at $\tau = 1$, where

$$\begin{aligned} n_1(\tau) + n_2(\tau) &= [(\gamma_1{}^{t} - \gamma_2{}^{t} + \Delta - 2\vec{\gamma}_1)/2\Delta] \exp(-\lambda_> \tau) \\ &\quad + [(\gamma_2{}^{t} - \gamma_1{}^{t} + \Delta + 2\vec{\gamma}_1)/2\Delta] \exp(-\lambda_< \tau) \\ n_1(\tau) + n_2(\tau) &= (1/2\Delta)[(R+\Delta)\exp(-\lambda_> \tau) + (\Delta - R)\exp(-\lambda_< \tau)] \end{aligned} \tag{IV-19}$$

Weak Coupling to Second Well. If the first barrier is larger than both the second barrier and the neutron binding energy, $\vec{\gamma}_1 \ll \gamma_1{}^{d}$ and $\vec{\gamma}_1 \ll \gamma_2{}^{t}$, and there is weak coupling between the levels in potential wells I and II. Then

$$R \approx \gamma_1{}^{d} - \gamma_2{}^{t} \tag{IV-20}$$

The quantity Δ can be written

$$\Delta = \{[(\gamma_1{}^{d} - \gamma_2{}^{d} - \vec{\gamma}_2) + \vec{\gamma}_1]^2 + 4\vec{\gamma}_1\vec{\gamma}_2\}^{1/2}$$

Combining the terms, we obtain

$$\Delta = [(\gamma_1{}^{d} - \gamma_2{}^{d} - \vec{\gamma}_2)^2 + \vec{\gamma}_1{}^2 + 2\vec{\gamma}_1(\gamma_1{}^{d} - \gamma_2{}^{d} + \vec{\gamma}_2)]^{1/2}$$

Neglecting $\vec{\gamma}_1{}^2$ and using the binomial expansion, an approximate relation is derived, namely

$$\begin{aligned} \Delta &\approx (\gamma_1{}^{d} - \gamma_2{}^{d} - \vec{\gamma}_2) + \frac{1}{2}\frac{2\vec{\gamma}_1(\gamma_1{}^{d} - \gamma_2{}^{d} + \vec{\gamma}_2)}{|\gamma_1{}^{d} - \gamma_2{}^{d} - \vec{\gamma}_2|} + \cdots \\ &\approx |R| + [\vec{\gamma}_1(\gamma_1{}^{d} - \gamma_2{}^{d} + \vec{\gamma}_2)/|R|] \end{aligned} \tag{IV-21}$$

The total width is given by

$$\langle \Gamma_t \rangle \approx \gamma_1{}^{d} + [\vec{\gamma}_1(\gamma_1{}^{d} - \gamma_2{}^{d} + \vec{\gamma}_2)/(\gamma_1{}^{d} - \gamma_2{}^{t})] \tag{IV-22}$$

and making use of Eq. (IV-18), we find the fission width $\langle \Gamma_f \rangle$ to be

$$\langle \Gamma_f \rangle \approx \vec{\gamma}_1\vec{\gamma}_2\langle \Gamma_t \rangle/(\gamma_1{}^{t}\gamma_2{}^{t} - \vec{\gamma}_1\vec{\gamma}_2) \tag{IV-23}$$

Recalling that for weak coupling, $\gamma_1{}^{d} \gg \vec{\gamma}_1$, we can neglect the second term of Eq. (IV-22) and if, in addition, we reduce the denominator of Eq. (IV-23) to $\gamma_1{}^{d}\gamma_2{}^{t}$, then

$$\langle \Gamma_f \rangle \approx \vec{\gamma}_1\vec{\gamma}_2/\gamma_2{}^{t} \tag{IV-24}$$

Strong Coupling between First and Second Wells. If the first barrier is smaller than both the second barrier and the neutron binding energy, the nonradiative width $\vec{\gamma}_2$ is larger than both of the decay widths $\gamma_1{}^{d}$ and $\gamma_2{}^{d}$ and strong coupling exists between the two potential wells. If we expand the expression for Δ given after Eq. (IV-12), we obtain

$$\Delta \approx \pm R\{1 + [2\tilde{\gamma}_1(\gamma_1{}^{\mathrm{d}} - \gamma_2{}^{\mathrm{d}})/R^2] + \cdots\} \tag{IV-25}$$

where $2\tilde{\gamma}_1 = (\gamma_1{}^{\mathrm{d}} - \gamma_2{}^{\mathrm{d}}) + (\tilde{\gamma}_1 - \tilde{\gamma}_2) - R$ and further simplification gives

$$\Delta \approx -R - \{(\gamma_1{}^{\mathrm{d}} - \gamma_2{}^{\mathrm{d}})[(\tilde{\gamma}_1 - \tilde{\gamma}_2) + (\gamma_1{}^{\mathrm{d}} - \gamma_2{}^{\mathrm{d}})]/R\} + (\gamma_1{}^{\mathrm{d}} - \gamma_2{}^{\mathrm{d}})$$

Recalling that $R = (\gamma_1{}^{\mathrm{d}} - \gamma_2{}^{\mathrm{d}}) - (\tilde{\gamma}_1 + \tilde{\gamma}_2)$ and the earlier approximation about the nonradiative widths and the decay widths, we obtain

$$\Delta \approx (\tilde{\gamma}_1 + \tilde{\gamma}_2) + [(\tilde{\gamma}_1 - \tilde{\gamma}_2)(\gamma_1{}^{\mathrm{d}} - \gamma_2{}^{\mathrm{d}})/(\tilde{\gamma}_1 + \tilde{\gamma}_2)] \tag{IV-26}$$

The values of $\lambda_>$ and $\lambda_<$ now become

$$\lambda_> \approx \tilde{\gamma}_1 + \tilde{\gamma}_2, \qquad \lambda_< \approx (\tilde{\gamma}_1\gamma_2{}^{\mathrm{d}} + \tilde{\gamma}_2\gamma_1{}^{\mathrm{d}})/(\tilde{\gamma}_1 + \tilde{\gamma}_2)$$

and the total width $\langle\Gamma_{\mathrm{t}}\rangle$ reduces to

$$\langle\Gamma_{\mathrm{t}}\rangle \approx \lambda_< \tag{IV-27}$$

By substitution of $\lambda_<$ into Eq. (IV-18) for $\langle\Gamma_{\mathrm{t}}\rangle$, we derive

$$\langle\Gamma_{\mathrm{f}}\rangle \approx \lambda_<\tilde{\gamma}_1\tilde{\gamma}_2/(\gamma_1{}^{\mathrm{t}}\gamma_2{}^{\mathrm{t}} - \tilde{\gamma}_1\tilde{\gamma}_2) \tag{IV-28}$$

Hence

$$\langle\Gamma_{\mathrm{f}}\rangle \approx \tilde{\gamma}_1\tilde{\gamma}_2/(\tilde{\gamma}_1 + \tilde{\gamma}_2) \tag{IV-29}$$

Estimate of Fission Width and Number of Open Channels at Barriers A and B. If the spreading width $\gamma_2{}^{\mathrm{t}}$ of each state in the second well is large compared to the average distance D_{I} between states in the first well, then each of the class II states in the second well overlaps with many class I states in the first well, as illustrated in Fig. IV-5. In this case it is possible to estimate average nonradiative widths from the Bohr–Wheeler statistical model, namely

$$\langle\tilde{\gamma}_1\rangle \approx \frac{D_{\mathrm{I}}}{2\pi}N_A \tag{IV-30}$$

$$\langle\tilde{\gamma}_2\rangle \approx \frac{D_{\mathrm{II}}}{2\pi}N_A \tag{IV-31}$$

$$\langle\tilde{\gamma}_2\rangle \approx \frac{D_{\mathrm{II}}}{2\pi}N_B \tag{IV-32}$$

where N_A and N_B are the effective number of open channels over barriers A and B, respectively (see Fig. IV-3).

Structure has been observed in the fission of several even-mass target nuclei when bombarded with subbarrier monoenergetic neutrons. In such cases the fission probability is low and the total width Γ_{t} is determined almost entirely by the radiation width Γ_γ and the neutron width Γ_n. For example, in the case of resonance neutron fission of ^{240}Pu, illustrated in Fig. IV-6, a comparison of the fission resonance groups with the total number of radiation resonance groups leads us to conclude that the second well is some 1.9 MeV above the first well (see Fig. IV-3). This result is obtained with the assumption that the level density has the same energy dependence at the two deformations. The level spacing of the compound nucleus is determined by the total excitation energy in potential well I, $U_I^* = U - \Delta_I$, whereas the level spacing in the second minimum is determined by the available excitation energy in potential well II, $U_{II}^* = U - E_{\text{isomer}} - \Delta_{II}$. The level spacing ratios are given by

$$\frac{\langle D_I \rangle}{\langle D_{II} \rangle} = \frac{\rho_{II}}{\rho_I} = \left(\frac{U_I^*}{U_{II}^*} \right)^{3/2} \exp\{2a^{1/2}[(U_{II}^*)^{1/2} - (U_I^*)^{1/2}]\} \qquad \text{(IV-33)}$$

Inserting the values of 15 eV, 650 eV, 5.41 MeV, and 25 MeV^{-1} for $\langle D_I \rangle$, $\langle D_{II} \rangle$, U_I^*, and a gives a value of $E_{\text{isomer}} = 1.9$ MeV ($\Delta_I = \Delta_{II} = 0$ is assumed).

For the case of resonance neutron fission of ^{240}Pu the spacing D_{II} is much larger than D_I, and therefore

$$\langle \vec{\gamma} \rangle \ll \langle \overleftarrow{\gamma}_2 \rangle \qquad \text{(IV-34)}$$

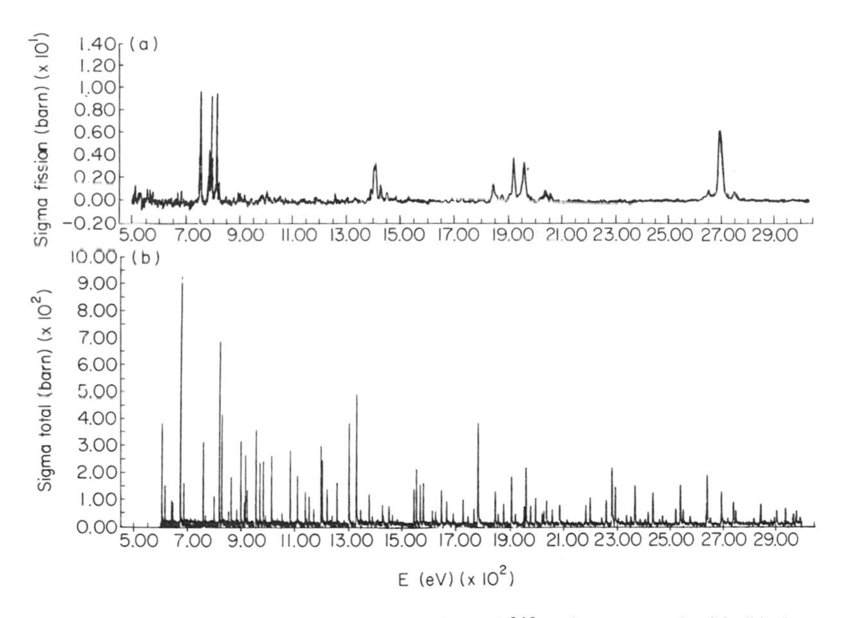

Fig. IV-6. (a) The neutron fission cross section of ^{240}Pu is compared with (b) the total neutron cross section to demonstrate the grouping of fission resonances. [After Strutinsky and Pauli (1969).]

The spreading width $\gamma_2{}^{\mathrm{t}}$ is given by the sum of $\overset{\uparrow}{\gamma}_2$ and $\vec{\gamma}_2$.

$$\langle \gamma_2{}^{\mathrm{t}} \rangle \approx \langle \vec{\gamma}_2 \rangle + \langle \overset{\uparrow}{\gamma}_2 \rangle \approx \frac{D_{\mathrm{II}}}{2\pi}(N_A + N_B) \tag{IV-35}$$

For the case of weak coupling the average fission width is given by

$$\langle \Gamma_{\mathrm{f}} \rangle \approx \frac{\langle \vec{\gamma} \rangle_1 \langle \vec{\gamma}_2 \rangle}{\langle \gamma_2{}^{\mathrm{t}} \rangle} \approx \frac{D_{\mathrm{I}} N_A N_B}{2\pi(N_A + N_B)} \tag{IV-36}$$

where for this case $N_B > N_A$ and Eq. (IV-36) reduces to

$$\langle \Gamma_{\mathrm{f}} \rangle \approx \frac{D_{\mathrm{I}} N_A}{2\pi} \tag{IV-37}$$

For the case of strong coupling the average fission width is given by Eq. (IV-29). For the case where the first well is considerably deeper than the second well, we see from Eq. (IV-34) that $\langle \overset{\uparrow}{\gamma}_2 \rangle \gg \langle \vec{\gamma}_1 \rangle$. The denominator of Eq. (IV-29) reduces then to $\overset{\uparrow}{\gamma}_2$ and the average fission width is given by

$$\langle \Gamma_{\mathrm{f}} \rangle \approx \frac{D_{\mathrm{I}} N_B}{2\pi} \tag{IV-38}$$

The experimental values of the quantities D_{I}, D_{II}, $\langle \gamma_2{}^{\mathrm{t}} \rangle$, and $\langle \Gamma_{\mathrm{f}} \rangle$ for resonance neutron fission of ^{240}Pu are 15, 650, 50, and 0.002 eV, respectively. If weak coupling is assumed, insertion of the foregoing values into Eqs. (IV-35) and (IV-37) gives values of $N_A = 0.001$ and $N_B = 0.48$. On the other hand, if strong coupling is assumed, then insertion of those values into Eqs. (IV-35) and (IV-38) gives values of $N_B = 0.001$ and $N_A = 0.48$. From the experimental width and spacing data it is not possible to decide which of the two foregoing numbers is related to barrier A and which to barrier B. If barrier A is the higher barrier ($N_A = 0.001$ and $N_B = 0.48$), then there is only weak coupling between the compound and intermediate states. In this case all the observed resonance states in the fission cross section are almost pure compound states and the intermediate state does not show up in the other reaction channels because of its large fission width. On the other hand, if barrier B is higher ($N_A = 0.48$ and $N_B = 0.001$), there is relatively strong mixing of each intermediate state with the compound nuclear states and, as a result, each intermediate state is distributed among the observed resonance states in each group.

F. The n, γf Process

When a compound nucleus is formed by capture of a low energy neutron, the compound nucleus can subsequently decay by either reemission of the neutron (elastic scattering), γ-ray emission, or if energetically possible, fission. It has been pointed out by Stavinsky and Shaker (1965) and by Lynn (1965) that it is possible for fission to occur subsequent to the emission of a low energy γ ray. If the residual nucleus following the emission of the γ ray has sufficient excitation energy relative to the appropriate fission barrier, it is possible that fission may successfully compete with secondary γ-ray emission. In general we would expect the contribution of this type of fission to be small. However, the fact that the intermediate compound nucleus following the emission of the primary γ ray may have a spin and parity different from that of the initial compound nucleus may allow fission following γ-ray emission through channels not open to the initial compound nucleus. A schematic diagram illustrating the relevant barriers and possible spin changes is given in Fig. IV-7.

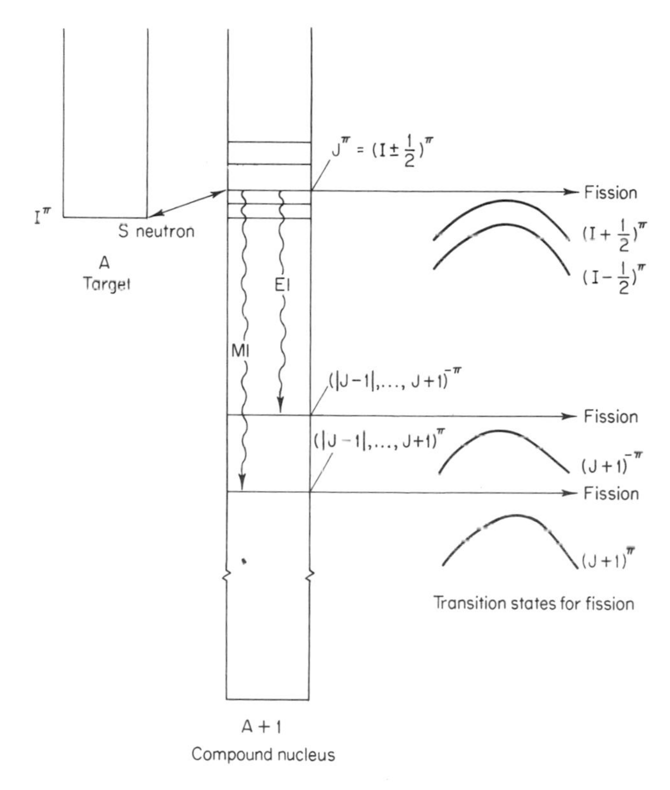

Fig. IV-7. Schematic diagram of the (n, γf) reaction. The inverted parabolic curves on the right indicate the transition states at the saddle point in the deformation energy through which the compound nucleus must pass in the process of fission. Only the lowest transition states for different spin and parity are schematically indicated. In a typical case the ground state of the compound nucleus will lie about 6 MeV below the neutron separation energy, while the very lowest transition state for fission will be up to 1.5 MeV below the neutron separation energy. [After Lynn (1965).]

Theoretical estimates for the magnitude of the contribution to the fission widths from the (n, γf) process have been made for several thermal neutron fissionable nuclides. It is dependent on the energy distribution of the primary γ rays, the relative amounts of electric and magnetic radiation, the relative amounts of dipole and quadrupole radiation, and the barrier heights for states of spin and parity accessible from the initial state. Since the most probable γ-ray energy is usually larger than the difference between the initial excitation energy and the transition state energies, we expect part of the contribution to be due to subbarrier penetration. Stavinsky and Shaker (1965) estimate a contribution of approximately half the radiation width, while in a more careful analysis Lynn (1965) concludes that the contribution will be appreciably smaller. Experimentally it is somewhat difficult to determine the contribution of this process to the total fission width. This process should manifest itself in the distribution of fission widths with respect to both a minimum value of the fission width to be expected and the fluctuations in the fission widths. For the (n, f) reaction only a few exit channels are generally open and a wide fluctuation in fission widths is expected. If fission is occurring subsequent to γ emission, the width fluctuations are greatly damped due to the large number of secondary compound nucleus states reached directly by γ emission from the initial state. Evidence for the (n, γf) reaction in slow neutron capture by ^{238}Pu has been reported. The fission widths for 10 resonances have been measured and have been found to fluctuate less strongly than would be expected for the (n, f) reaction. The ^{239}Pu compound nucleus is a good candidate for observation of the (n, γf) reaction since sizeable photofission cross sections have been observed below the neutron binding energy. Furthermore, for the initial compound nucleus to fission it must use the $K = \frac{1}{2}+$ fission saddle state, which is largely closed at this energy. If we assume that the smallest fission width observed corresponds to the average value of the nonfluctuating (n, γf) process, then the average width for the (n, f) process is estimated to be $0.00475 - 0.0012 = 0.00355$ eV. Thus, approximately 25% of the total fission widths may be attributed to the (n, γf) process in this case.

It has been suggested that the width distribution for neutron-induced fission of ^{235}U can be attributed to a superposition of the distribution for a few channels comprising most of the fission cross section, corresponding to (n, f) fission, and the distribution for a larger number ($\sim$12) of channels associated with the (n, γf) process and comprising a smaller fraction of the total fission probability (Lukyanov and Shaker, 1970).

In the case of odd mass number target nuclei the relative probability for compound nucleus states of the two possible spin values to contribute to the (n, γf) process will in general be different. This may have a number of consequences. A feature of the (n, γf) process is that there is a contribution to fission from nuclei with lower excitation energies than fission associated with the (n, f) reaction. This can give rise to a number of interesting effects with respect to the characteristics of fission observed as one goes from resonance to resonance. These effects include variation in mass asymmetry, kinetic energy release, and the probability for emission of ternary α particles. For example, if capturing states of one spin have a larger probability for fissioning prior to emitting a γ ray, fission associated with these states will be more symmetric than fission associated with states of the other spin because of the increased probability of symmetric fission at higher excitation energy. There is evidence that for both ^{239}Pu and ^{235}U the tendency for increased symmetry is associated with the spin state expected to have the smallest relative contribution from the (n, γf) reaction. However, it is not clear that the magnitude of the variations in mass asymmetry can be accounted for quantitatively on the basis of this effect.

References

Asghar, M. (1967). *Nucl. Phys. A* **98**, 33

Asghar, M., Michaudon, A., and Paya, D. (1968). *Phys. Lett. B* **26**, 664.

Auchampaugh, G. F., Bowman, C. D., and Evans, J. E. (1968). *Nucl. Phys. A* **112**, 329.

Bergen, D. W., and Fullwood, R. R. (1971). *Nucl. Phys. A* **163**, 577.

Bergen, D. W., and Silbert, M. G. (1968). *Phys. Rev.* **166**, 1178.

Bjørnholm, S., and Strutinsky, V. M. (1969). *Nucl. Phys. A* **136**, 1.

Bohr, A. (1956). *Proc. Int. Conf. Peaceful Uses At. Energy, Geneva 1955*, **2**, p. 151. United Nations, New York.

Bowman, C. D., Auchampaugh, G. F., Stubbins, W. F., Young, T. E., Simpson, F. B., and Moore, M. S. (1967). *Phys. Rev. Lett.* **18**, 15.

Bowman, C. D., Auchampaugh, G. F., Fultz, S. C., and Hoff, R. W. (1968). *Phys. Rev.* **166**, 1219.

BNL 325, 2nd ed., Supplement No. 2 (1965) Vol. III.

Britt, H. C., Rickey, F. A., Jr., and Hall, W. S. (1968). *Phys. Rev.* **175**, 1525.

Cao, M. G., Migneco, E., Theobald, J. P., Wartena, J. A. and Winter, J. (1968). *J. Nucl. Energy* **22**, 211.

Cao, M. G., Migneco, E., and Theobold, J. P. (1968). *Phys. Lett. B* **27**, 409.

Cowan, G. A., Bayhurst, B. P., Prestwood, R. J., Gilmore, J. S., and Knobeloch, G. W. (1966). *Phys. Rev.* **144**, 197.

Cowan, G. A., Bayhurst, B. P., Prestwood, R. J., Gilmore, J. S., and Knobeloch, G. W. (1970). *Phys. Rev. C* **2**, 615.

Cramer, J. D. (1969). *Nucl. Phys. A* **126**, 471.

Dabbs, J. W. T., Walter, F. J., and Parker, G. W. (1965). *Proc. IAEA Symp. Phys. Chem. of Fission, Salzburg, 1965*, **1**, p. 39. IAEA, Vienna.

Dabbs, J. W. T., Eggerman, C., Cauvin, B., Michaudon, A. and Sanche, M. (1969). *Proc. Symp. Phys. Chem. of Fission, 2nd, Vienna, 1969*, p. 321. IAEA, Vienna.

Derrien, H., Blons, J., Eggerman, C., Michaudon, A., Paya, D. and Ripon, P. (1967). *In* "Nuclear Data for Reactors," Vol. II, p. 195. IAEA, Vienna.

Farrell, J. A. (1968). *Phys. Rev.* **165**, 1371.

Fraser, J. S., and Schwartz, R. B. (1962). *Nucl. Phys.* **30**, 269.

Huizenga, J. R., and Vandenbosch, R. (1960). *Phys. Rev.* **120**, 1305.
James, G. D., and Rae, E. R. (1968). *Nucl. Phys. A* **118**, 313.
King, T. J., and Block, R. C. (1969). *Nucl. Phys. A* **138**, 556.
Kuiken, R., Pattenden, N. J., and Postma, H. (1972a) *Nucl. Phys. A* **190**, 401.
Kuiken, R., Pattenden, N. J., and Postma, H. (1972b) *Nucl. Phys. A* **196**, 389.
Kvitek, J., Popov, Yu. P., and Ryabov, Y. V. (1966). *Sov. J. Nucl. Phys.* **2**, 484.
Lukyanov, A. A., and Shaker, M. O. (1970). *Sov. J. Nucl. Phys.* **10**, 456.
Lynn, J. E. (1964). *Phys. Rev. Lett.* **13**, 412.
Lynn, J. E. (1965). *Phys. Lett.* **18**, 31.
Melkonian, E., and Mehta, G. K. (1965). *Proc. IAEA Symp. Phys. Chem. of Fission, Salzburg, 1965*, **2**, p. 355. IAEA, Vienna.
Michaudon, A. (1967). Determination of the spins of resonances of fissile nuclei. *In* "Nuclear Data for Reactors," Vol. II, p. 161. IAEA, Vienna.
Michaudon, A., Lottin, A., Paya, D., and Trochon, E. J. (1965). *Nucl. Phys.* **69**, 573.
Migneco, E., and Theobald, J. P. (1968). *Nucl. Phys. A* **112**, 603.
Moore, M. S. (1970). *In* "Neutron Resonance Spectroscopy" (J. A. Harvey, ed.), p. 348. Academic Press, New York.
Moore, M. S., and Keyworth, G. A. (1971). *Phys. Rev. C* **3**, 1656.
Pattenden, N. J., and Postma, H. (1971). *Nucl. Phys. A* **167**, 225.
Paya, D., *et al.* (1968a). *Proc. Colloq. Medium Mass and Heavy Nuclei, Bordeaux, 1967. J. Phys. (Paris)* **29**, Suppl. No. 1, p. 159.
Paya, D., *et al.* (1968b). *Proc. Int. Conf. Nucl. Struct., Dubna, 1968*, p. 185. IAEA, Vienna.
Poortmans, F., *et al.* (1967). *Proc. Int. Conf. Nucl. Data for React. Paris, 1967*, Vol. II, p. 211. IAEA, Vienna.
Poortmans, F., Ceulemans, H., Migneco, E., and Theobald, J. (1970). *Proc. Int. Conf. Nucl. Data for Reactions, 2nd, Helsinki, 1970.* To be published.
Ryabov, Y. V., So. Don Sik, Chikov, N., and Janeva, N. (1971). *Sov. J. Nucl. Phys.* **13**, 255.
Ryabov, Y. V., So. Don Sik, Chikov, N., and Janeva, N. (1972). *Sov. J. Nucl. Phys.* **14**, 519.
Sauter, G. D., and Bowman, C. D. (1965). *Phys. Rev. Lett.* **15**, 761.
Sauter, G. D., and Bowman, C. D. (1968). *Phys. Rev.* **174**, 1413.
Schermer, R. I., Passell, L., Brunhart, G., Reynolds, C. A., Sailor, L. V., and Shore, F. J. (1968). *Phys. Rev.* **167**, 1121.
Shapiro, F. L. (1968). *Proc. Int. Conf. Nucl. Struct. Dubna, 1968*, p. 283. IAEA, Vienna.
Simpson, F. B., Miller, L. G., Moore, M. S., Hockenbury, R. W., and King, T. J. (1971). *Nucl. Phys. A* **164**, 34.
Stavinsky, V., and Shaker, M. O. (1965). *Nucl. Phys.* **62**, 667.
Strutinsky, V. M., and Pauli, H. C. (1969). *Proc. IAEA Symp. Phys. Chem. of Fission, 2nd, Vienna, 1969*, p. 155. IAEA, Vienna.
Weigmann, H. (1968). *Z. Phys.* **214**, 7.
Weigmann, H., Winter, J., and Heske, M. (1969). *Nucl. Phys. A* **134**, 535.
Weinstein, S., Reed, R., and Block, R. C. (1969). *Proc. Symp. Phys. Chem. of Fission, 2nd, Vienna, 1969*, p. 477. IAEA, Vienna.
Wheeler, J. A. (1963). *In* "Fast Neutron Physics" (J. B. Marion and J. L. Fowler, eds.), pp. 2051–2184. Wiley (Interscience), New York.
Wilets, L. (1964). "Theories of Nuclear Fission." Oxford Univ. Press (Clarendon), London and New York.

CHAPTER

Properties of Low-Lying Levels of Transition Nuclei Determined from Reaction Studies

A. Introduction

In 1956, A. Bohr suggested that low energy fission may be understood in terms of a very few levels in the transition nucleus. Although the level spacing in the compound nucleus at an excitation energy of about 6 MeV is of the order of 1 eV or less, most of the excitation energy goes into deformation energy during the passage from the initially excited nucleus to the highly deformed transition state nucleus (or saddle point). Hence, the transition state nucleus is thermodynamically "cold" and is expected to have a spectrum of excited states analogous to those of a normal nucleus near its ground state.

When the excitation energy of the compound nucleus is approximately equal to the fission barrier, present evidence strongly supports the concept of Bohr (1956) that fission occurs through one or only a few channels. Information on the properties of these transition levels is obtained from a study of fission-fragment angular distributions. Extensive discussion of fission-fragment angular distributions for several different reactions at excitation energies near the fission barrier will be presented in this chapter. Although the fission barrier is double humped (see Chapter II), we assume in some parts of this chapter, for the sake of simplicity, that the second barrier has a higher potential energy than the first barrier. In particular, we make the assumption that the class I compound states are thoroughly admixed into the class II states and that the fission width is determined by the degree of penetration through the second barrier.

If we assume that the fission fragments separate along the nuclear symmetry axis and that K (the projection of J on the nuclear symmetry axis) is a good quantum number in the passage of a nucleus from its transition state to the configuration of separated fragments, the directional dependence of fission fragments resulting from a transition state with quantum numbers J, K, and M is uniquely determined. The quantum numbers J and M (projection of J on a space-fixed axis which is usually taken as the beam direction) are conserved in the entire fission process. Whereas J and M are fixed throughout the various extended shapes on the path to fission, no such restriction holds for the parameter K. In going from the original compound nucleus (with a shape which is the same as that of the ground state) to the saddle point or transition state, it is reasonable to suppose that the nucleus suffers many vibrations and changes in shape, and redistributes its energy and angular momentum in many ways. The K value (or values) of the transition nucleus are, therefore, unrelated to the initial K values of the compound nucleus. The assumption above postulates, however, that once the nucleus reaches the transition state deformation, K is a good quantum number beyond this point of the fission process. The relationship between J, M, and K is schematically illustrated in Fig. V-1. For the actimide nuclei where the first or inner barrier is the higher energy barrier, we assume that K mixing occurs prior to reaching the second or outer barrier. At the deformation of the inner barrier, heavy nuclei are very soft toward axial asymmetric deformations which mix K. Hence, on the basis of theoretical expectations and limited experimental evidence, we assume that the K distribution is frozen in at the second barrier for low-energy fission. On the other hand, at high excitation energies the K distribution of the liquid drop saddle is expected to be applicable. The probability of emitting fission fragments from a transition state with quantum numbers J, M, and K at an angle θ is given by (Wheeler, 1963)

$$P^J_{M,K}(\theta) = [(2J+1)/4\pi R^2]\,|d^J_{M,K}(\theta)|^2\, 2\pi R^2 \sin\theta\, d\theta \tag{V-1}$$

where $P^J_{M,K}(\theta)$ represents the probability of emitting fission fragments at angle θ into the conical volume defined by the angular increment $d\theta$. The normalization is such that the probability integrates to unity for limits 0 and π. The area of the annular ring on a sphere of radius R through which the fission fragments are passing is given by the width of the strip $R\, d\theta$ times the circumference of the ring $2\pi R \sin\theta$, which is $2\pi R^2 \sin\theta\, d\theta$. This annular ring area must be divided by the total area of the sphere $4\pi R^2$ in order to give the probability as defined by Eq. (V-1).

The foregoing probability distribution depends on the $d^J_{M,K}(\theta)$ function and is universal in the sense that it is independent of the polar angle, the angle of rotation about the symmetry axis, and the moments of inertia. Hence the probability distribution depends only on the angle θ between the space-fixed

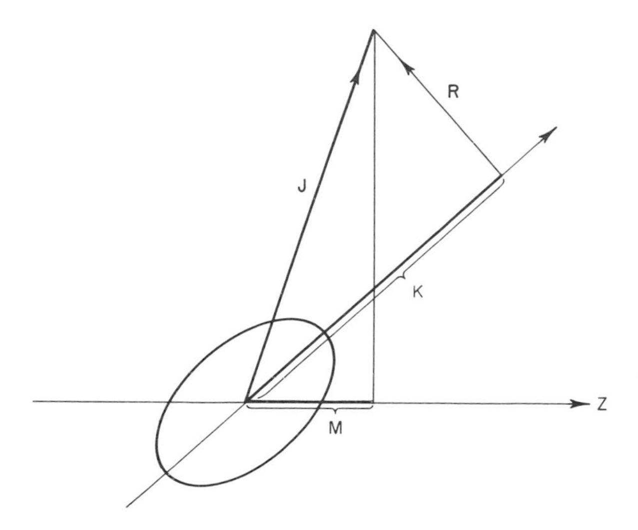

Fig. V-1. Angular momentum coupling scheme for a deformed nucleus. The vector J defines the total angular momentum. The quantity M is the component of the total angular momentum on the space-fixed Z axis. We define this direction as the beam direction. The quantity K is the component of the total angular momentum along the nuclear symmetry axis. The collective rotational angular momentum R is perpendicular to the nuclear symmetry axis; thus, K is entirely a property of the intrinsic motion. The angle θ in this chapter represents the angle between the nuclear symmetry axis and the space-fixed Z spin.

and body-fixed axes. The $d^J_{M,K}(\theta)$ functions† are defined by the following relation (Lamphere, 1962)

$$d^J_{M,K}(\theta) = \{(J+M)!(J-M)!(J+K)!(J-K)!\}^{1/2} \times \sum_X \frac{(-1)^X [\sin(\theta/2)]^{K-M+2X} [\cos(\theta/2)]^{2J-K+M-2X}}{(J-K-X)!(J+M-X)!(X+K-M)!X!} \quad \text{(V-2)}$$

where the sum is over $X = 0, 1, 2, 3, \ldots$ and contains all terms in which no negative value appears in the denominator of the sum for any one of the quantities in parentheses. There are a number of symmetry relationships among the d's, for example, $d^J_{M,K}(\theta) = (-)^{M-K} d^J_{K,M}(\theta)$. Tables of the rotational wave functions $d^J_{M,K}(\theta)$ are published for a range of J, M, and K values (Behkami, 1971).

The angular distribution $W^J_{M,K}(\theta)$ is obtained by dividing the probability for emitting fission fragments at angle (θ) as defined in Eq. (V-1) by $\sin\theta$,

$$W^J_{M,K}(\theta) = [(2J+1)/2]\, |d^J_{M,K}(\theta)|^2 \quad \text{(V-3)}$$

† A discussion of the connections among the various conventions is given by Preston (1962 Appendix A). We are using the convention of Rose (1957).

In analogy to Eq. (V-1), the normalization of Eq. (V-3) is such that $\int_0^\pi W^J_{M,K}(\theta)\sin\theta\, d\theta = 1$. The angular distribution function $W^J_{M,K}(\theta)$ is directly related to the differential fragment cross section for a particular channel (J, π, K, M) at angle θ by the relation

$$\frac{d\sigma_f}{d\Omega}(J, \pi, K, M, \theta) = \frac{W^J_{M,K}}{2\pi}(\theta)\,\sigma_f(J, \pi, K, M) \qquad \text{(V-4)}$$

where the differential fragment cross section is to be expressed in the same units per steradian as the total fission cross section. The factor 2π appears instead of the usual 4π due to the fact that two fragments separated by 180° arise from each fission event.

B. Photofission

1. Even Nucleus Targets

Bohr (1956) pointed out in his original paper on fission anisotropy that the case of photofission of even-even targets has some especially simple features. The spin and parity of an even-even target nucleus is $0+$, and the electric dipole absorption of a photon gives a compound state with spin and parity $1-$ and $M = \pm 1$ (space-fixed axis along the photon beam direction). At excitation energies only slightly in excess of the lowest fission barrier B_f, the available energy above this lowest barrier $(E - B_f)$ is insufficient for nucleon pairs to be broken, and the energy levels of the transition nucleus are due to collective excitations. A schematic diagram of some possible collective band structures is shown in Fig. II-17 for an even-even transition nucleus with a stable quadrupole deformation. The ground state band of the highly deformed transition nucleus has $K = 0$ with positive parity and spins of $0, 2, 4, 6, \ldots$. Several types of collective bands at low excitation energy are also shown with assignments. Particle excitations are absent in the low energy spectrum and become possible only for excitation energies exceeding the pairing energy of the transition nucleus $(2\Delta_f)$.

As already mentioned, dipole γ-ray absorption on an even-even target leads to a compound nucleus with spin and parity $1-$. The lowest lying $1-$ level in the intermediate even-even transition nucleus is expected to have $K = 0$. The angular distribution for this transition state is given by

$$\begin{aligned} W^{J=1}_{M=\pm 1, K=0}(\theta) &= \tfrac{1}{2}(2J+1)\{P(J=1, M=+1)\,|d^1_{1,0}(\theta)|^2 \\ &\quad + P(J=1, M=-1)\,|d^1_{-1,0}(\theta)|^2\} \\ &= \tfrac{3}{2}\{\tfrac{1}{2}|d^1_{1,0}(\theta)|^2 + \tfrac{1}{2}|d^1_{-1,0}(\theta)|^2\} \end{aligned} \qquad \text{(V-5)}$$

The probabilities $P(J=1, M=+1)$ and $P(J=1, M=-1)$ of forming the compound states $(J=1, M=+1)$ and $(J=1, M=-1)$, respectively, are the same and equal to $\frac{1}{2}$. Evaluation of the $d^J_{M,K}(\theta)$ functions defined by Eq. (V-2) gives

$$W^{J=1}_{M=\pm 1, K=0}(\theta) = \tfrac{3}{4}\sin^2\theta \qquad \text{(V-6)}$$

Equation (V-6) is normalized, i.e., $\int_0^\pi \frac{3}{4}\sin^2\theta \cdot \sin\theta\, d\theta = 1$.

In a similar type derivation, the angular distribution resulting from a transition state with $K = 1$ is

$$W^{J=1}_{M=\pm 1, K=\pm 1}(\theta) = \frac{3}{2}\left\{\frac{|d^1_{1,1}(\theta)|^2}{4} + \frac{|d^1_{-1,1}(\theta)|^2}{4} + \frac{|d^1_{1,-1}(\theta)|^2}{4} + \frac{|d^1_{-1,-1}(\theta)|^2}{4}\right\} \qquad \text{(V-7)}$$

The factors of $\frac{1}{4}$ arise in Eq. (V-7) because the probabilities of the two M states are equal and the two projections of $K(\pm 1)$ are to be given equal weight for each of the M values. All of the terms in Eq. (V-7) are not different and the total number of terms can be reduced to two, since $|d^1_{1,1}(\theta)|^2 = |d^1_{-1,-1}(\theta)|^2$ and $|d^1_{-1,1}(\theta)|^2 = |d^1_{1,-1}(\theta)|^2$. Evaluation of the $d^J_{M,K}(\theta)$ functions in Eq. (V-7) gives

$$W^{J=1}_{M=\pm 1, K=\pm 1}(\theta) = \tfrac{3}{4} - \tfrac{3}{8}\sin^2\theta \qquad \text{(V-8)}$$

Again, $\int_0^\pi (\frac{3}{4} - \frac{3}{8}\sin^2\theta)\sin\theta\, d\theta = 1$.

The angular distribution for a level in the transition nucleus with quantum numbers $J = 1$, $M = \pm 1$, and $K = 0$ which is given by Eq. (V-6) is plotted in Fig. V-2. This transition level gives no yield at 0° and peaks at 90°. The angular distribution resulting from a transition level with $J = 1$, $M = \pm 1$, and $K = 1$ [see Eq. (V-8)] is also given in Fig. V-2. This distribution peaks at 0° and decreases gradually to 90° where the intensity has fallen to $\frac{1}{2}$ the 0° intensity.

Dipole fission of an even-even target may occur through a transition level with either $K = 0$ [Eq. (V-6)] or $K = 1$ [Eq. (V-8)] or some combination of these two K levels. If we assume the fractional number of open $K = 0$ levels through which fission occurs is x and the fractional number of open $K = 1$ levels is y, then the angular distribution for a mixture of levels is given by

$$W(\theta)_{\mathrm{di(e-e)}} = [x\tfrac{3}{4}(\sin^2\theta) + 2y\{\tfrac{3}{4} - \tfrac{3}{8}(\sin^2\theta)\}]/(1+y) \qquad \text{(V-9)}$$

The factor of 2 in front of the second term on the right-hand side of the equals sign in Eq. (V-9) arises from the fact that all levels with a nonzero value of K must be counted twice in order to account for their double degeneracy in K, namely, positive and negative values. The factor $1/(1+y)$ is a normalization factor.

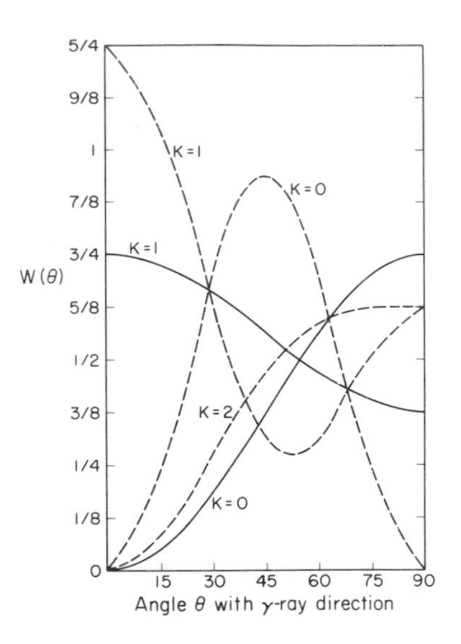

Fig. V-2. Photofission angular distributions for dipole and quadrupole absorption on an even-even target nucleus. [See Eqs. (V-6), (V-8), (V-12), (V-13, and (V-14).] Dipole (—) and quadrupole (---) absorption on even-even nucleus.

Simplification of Eq. (V-9) leads to

$$W(\theta)_{\mathrm{di(e-e)}} = [\tfrac{3}{2}y + \tfrac{3}{4}(x-y)\sin^2\theta]/(1+y) = a_{\mathrm{d}} + b_{\mathrm{d}}\sin^2\theta \quad \text{(V-10)}$$

where $a_{\mathrm{d}} = \frac{3}{2}y/(1+y)$ and $b_{\mathrm{d}} = \frac{3}{4}(x-y)/(1+y)$. Since $x = 1-y$, $b_{\mathrm{d}} = \frac{3}{4}(1-2y)/(1+y)$ and $b_{\mathrm{d}}/a_{\mathrm{d}} = (1-2y)/2y$. The ratio $W(\theta)/W(0°)$ is equal, therefore, to

$$[W(\theta)/W(0°)]_{\mathrm{di(e-e)}} = \frac{a_{\mathrm{d}}+b_{\mathrm{d}}\sin^2\theta}{a_{\mathrm{d}}} = 1 + \frac{b_{\mathrm{d}}}{a_{\mathrm{d}}}\sin^2\theta = 1 + \left(\frac{1-2y}{2y}\right)\sin^2\theta \quad \text{(V-11)}$$

Evaluation of the coefficients a_{d} and b_{d} from photofission angular distributions leads directly to the $K = 0$ and $K = 1$ contributions for dipole fission of an even-even target, as can be seen from Eqs. (V-10) and (V-11). A few MeV above the barrier where statistical theory is valid, the numbers of $K = 0$ and $K = 1$ levels are expected to be approximately equal (i.e., $y \approx x = \frac{1}{2}$). When this is true, the value of b/a is zero and isotropy is predicted [Eq. (V-11)] for dipole fission at moderate excitation energies. Experimentally we observe that b/a does approach zero as the energy increases. The prediction of statistical theory for the K distribution is given by Eq. (VI-5). Before discussing experimental results, we wish to present the theoretical angular distributions for quadrupole fission of even-even nuclei.

The quadrupole absorption of a photon by an even-even nucleus gives a compound nucleus with spin and parity of 2+ and $M = \pm 1$. The angular distributions for transition states with $K = 0$, 1, and 2 are, respectively,

$$W^{J=2}_{M=\pm 1, K=0}(\theta) = \frac{5}{2}\left\{\frac{|d^2_{1,0}|^2}{2} + \frac{|d^2_{-1,0}|^2}{2}\right\} = \frac{15}{16}\sin^2 2\theta \tag{V-12}$$

$$W^{J=2}_{M=\pm 1, K=\pm 1}(\theta) = \frac{5}{2}\left\{\frac{|d^2_{1,1}|^2}{4} + \frac{|d^2_{1,-1}|^2}{4} + \frac{|d^2_{-1,1}|^2}{4} + \frac{|d^2_{-1,-1}|^2}{4}\right\}$$
$$= \tfrac{5}{8}(2 - \sin^2\theta - \sin^2 2\theta) \tag{V-13}$$

$$W^{J=2}_{M=\pm 1, K=\pm 2}(\theta) = \frac{5}{2}\left\{\frac{|d^2_{1,2}|^2}{4} + \frac{|d^2_{1,-2}|^2}{4} + \frac{|d^2_{-1,2}|^2}{4} + \frac{|d^2_{-1,-2}|^2}{4}\right\}$$
$$= \tfrac{5}{8}[\sin^2\theta + \tfrac{1}{4}\sin^2 2\theta] \tag{V-14}$$

The results presented in Eqs. (V-12), (V-13), and (V-14) are normalized so that

$$\int_0^\pi \tfrac{15}{16}\sin^2 2\theta \sin\theta\, d\theta = \int_0^\pi \tfrac{5}{8}(2 - \sin^2\theta - \sin^2 2\theta)\sin\theta\, d\theta$$
$$= \int_0^\pi \tfrac{5}{8}[\sin^2\theta + \tfrac{1}{4}\sin^2 2\theta]\sin\theta\, d\theta = 1$$

The dashed curves in Fig. V-2 show plots of the angular distributions for quadrupole fission of an even-even target. The $K = 0$ curve is calculated with Eq. (V-12) and has a peak at 45°. The $K = 1$ curve calculated with Eq. (V-13) has a maximum at 0° and a minimum at approximately 52°. The $K = 2$ curve has zero intensity at 0° and increases monotonically with angle up to 90°.

If again we assume that the fractional numbers of $K = 0$, $K = 1$, and $K = 2$ levels through which quadrupole fission of an even-even target occurs are u, v, and w, respectively, then the angular distribution is given by

$$W(\theta)_{\text{quad(e-e)}} = u\tfrac{15}{16}\sin^2 2\theta + v\tfrac{5}{4}(2 - \sin^2\theta - \sin^2 2\theta)$$
$$+ w\tfrac{5}{4}[\sin^2\theta + \tfrac{1}{4}\sin^2 2\theta]$$
$$= \tfrac{5}{2}v + [\tfrac{5}{4}w - \tfrac{5}{4}v]\sin^2\theta$$
$$+ [\tfrac{15}{16}u - \tfrac{5}{4}v + \tfrac{5}{16}w]\sin^2 2\theta \tag{V-15}$$

If the relative fractions $u = v = w = \tfrac{1}{3}$, then Eq. (V-15) gives an isotropic angular distribution. In general, the angular distribution for quadrupole fission of an even-even target is described by the equation

$$W(\theta)_{\text{quad(e-e)}} = a_q + b_q \sin^2\theta + c_q \sin^2 2\theta \tag{V-16}$$

where

$$a_q = \tfrac{5}{2}v/(u+2v+2w)$$

$$b_q = \tfrac{5}{4}(w-v)/(u+2v+2w)$$

and

$$c_q = [\tfrac{15}{16}u - \tfrac{5}{4}v + \tfrac{5}{16}u]/(u+2v+2w)$$

Examination of Eqs. (V-10) and (V-16) reveals that the photofission angular distributions associated with dipole and quadrupole absorption are different. It is possible, therefore, to distinguish the contributions to fission of each of these two types of photon absorption. If d is the fraction of dipole fission and q the fraction of quadrupole fission, then by combining Eqs. (V-10) and (V-16) we obtain

$$W(\theta) = d(a_d + b_d \sin^2\theta) + q(a_q + b_q \sin^2\theta + c_q \sin^2 2\theta) = (da_d + qa_q) + (db_d + qb_q)\sin^2\theta + qc_q \sin^2 2\theta = a + b\sin^2\theta + c\sin^2 2\theta \tag{V-17}$$

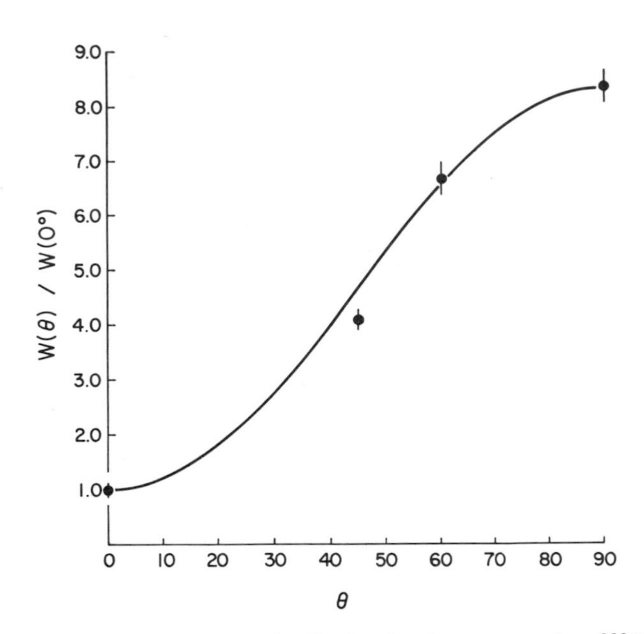

Fig. V-3. Fission-fragment angular distribution for the target nucleus ^{232}Th irradiated with bremsstrahlung of maximum energy of 7.0 MeV. [Data reproduced by permission of the National Research Council of Canada from Baerg, A. P., Bartholomew, R. M., Brown, F., Katz, L., and Kowalski, S. B., the *Canadian Journal of Physics*, *37*, 1418–1437 (1959).] The solid line is calculated with Eq. (V-11) where $(b_d/a_d) = 7.4$ and $y = 0.06$.

where a and b are mixed coefficients (arising from both dipole and quadrupole fission) and c comes entirely from quadrupole fission.

A major fraction of the existing experimental data on photofission has been obtained with a photon spectrum known as bremsstrahlung. Such a spectrum of γ rays is produced when a monoenergetic beam of electrons (accelerated to a fixed energy by, e.g., an electron linear accelerator) strikes a converter foil. The maximum energy of the photon spectrum E_{max} is the electron energy. In Fig. V-3, the fission-fragment angular distribution is plotted for the target nucleus ^{232}Th irradiated with bremsstrahlung of maximum energy 7.0 MeV. The points are the experimental ratios of $W(\theta)/W(0°)$ and the solid line is obtained with the function $1+7.4\sin^2\theta$. No evidence of any quadrupole fission is present in the data plotted in Fig. V-3. The angular distribution of fission fragments of ^{238}U irradiated with bremsstrahlung of maximum energy of 5.2 MeV is shown in Fig. V-4. This distribution cannot be fitted with a

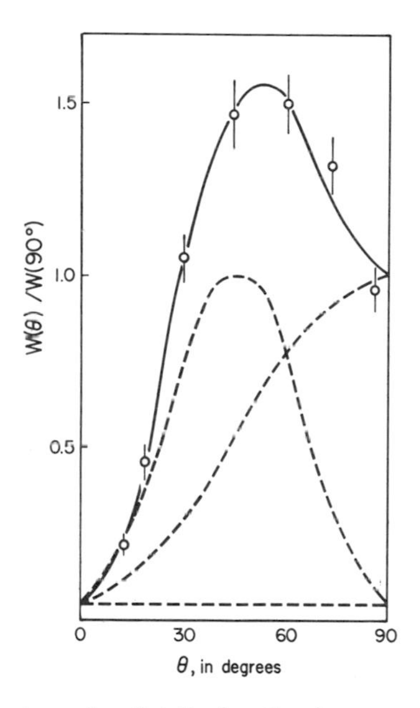

Fig. V-4. Fission-fragment angular distribution for the target nucleus ^{238}U irradiated with bremsstrahlung of maximum energy of 5.2 MeV. ○ Angular dependence of the experimental photofission cross section relative to the cross section at 90°. The dashed curve peaking at 45° is a theoretical curve for quadrupole fission of an even-even nucleus through a $K=0$ state. The dashed curve peaking at 90° is a theoretical curve for dipole fission of an even-even nucleus through a $K=0$ state (see Fig. V-3). — Composite of the dashed curves. [Data of Soldatov *et al.* (1965).]

theoretical expression containing only a constant and $\sin^2\theta$ terms. A sizeable contribution of a $\sin^2 2\theta$ term is evident in this distribution.

The energy dependence of b/a for ^{238}U fission induced by bremsstrahlung of various maximum energies E_{max} is shown in Fig. V-5. At low energies the $(J=1-,\ K=0)$ transition state dominates completely over the $(J=1-,\ K=1)$ transition state, as evidenced by the large values of b/a and, hence, the small values of y [see Eq. (V-11)]. This is a general feature of photofission of even-even nuclei. The $(J=1-,\ K=0)$ transition state is thought to be the lowest member of the mass asymmetry octupole band (see Fig. II-17). At very low energies, quadrupole fission becomes important for ^{238}U, as can be seen in Fig. V-5 by the rapid rise in the ratio c/b.

If the γ-ray energy is sufficient to excite only the lowest fission levels $J=1-$ and $2+$ each with $K=0$, then a of Eq. (V-17) is zero. In this case, c/b determines the quadrupole to dipole fission cross section ratio

$$c/b = \tfrac{5}{4}\{\sigma_{\gamma,\mathrm{f}}(J=2+,\ K=0)/\sigma_{\gamma,\mathrm{f}}(J=1-,\ K=0)\} \qquad \text{(V-18)}$$

On the other hand, if only dipole fission occurs, $c=0$, and the ratio of cross sections of dipole fission through two $1-$ states, one with $K=0$ and the other

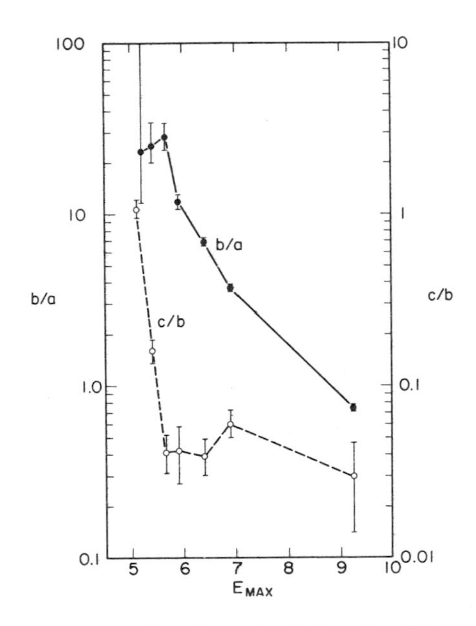

Fig. V-5. Dependence of the ratios b/a and c/b [see Eq. (V-17)] for photofission of ^{238}U as a function of maximum bremsstrahlung energy E_{max}. [Data of Rabotnov *et al.* (1965.)]

with $K = 1$, is related to b/a by

$$b/a = \{\sigma_{\gamma,\mathrm{f}}(J=1-, K=0)/\sigma_{\gamma,\mathrm{f}}(J=1-, K=1)\} - \tfrac{1}{2} \qquad \text{(V-19)}$$

In the comparison of Eqs. (V-11) and (V-19), we must recall that y in Eq. (V-11) is equivalent to

$$\sigma_{\gamma,\mathrm{f}}(J=1-, K=1)/\{2\sigma_{\gamma,\mathrm{f}}(J=1-, K=0) + \sigma_{\gamma,\mathrm{f}}(J=1-, K=1)\}$$

Some information on fission-fragment angular distributions has been obtained with monoenergetic (or essentially monoenergetic) γ-ray beams. These γ rays are produced by the $^{19}\mathrm{F}(\mathrm{p}, \alpha\gamma)$ reaction, various (n, γ) reactions, and positron annihilation. If a monoenergetic γ-ray line has a very narrow energy width, individual levels in the compound nucleus may be on or off resonance and caution must be used in the interpretation of such cross section data. The 6.14-MeV level in $^{16}\mathrm{O}$ which is excited by the $^{19}\mathrm{F}(\mathrm{p}, \alpha)$ reaction has a lifetime which is long compared to the stopping time of the recoil nucleus. Hence, the Doppler broadened 6.14-MeV γ-ray line has a very narrow width of approximately 10 eV which may be comparable to the spacing between $1-$ levels in an even-even nucleus. γ-ray lines produced by (n, γ) reactions may have very narrow energy spreads also.

The bulk of photofission studies have been performed with bremsstrahlung. With such beams, additional factors must be evaluated in order to interpret the fission cross section, namely, the energy dependence of both the photon absorption cross section and the photon spectrum. The photofission cross section is related to the photon cross section by

$$\sigma_{\gamma,\mathrm{f}}(E) = [\Gamma_{\mathrm{f}}/(\Gamma_{\mathrm{f}} + \Gamma_{\mathrm{n}} + \Gamma_{\gamma})]_E\, \sigma_\gamma(E) \qquad \text{(V-20)}$$

where Γ_{i} represents the width or probability for each specific deexcitation process and $\sigma_\gamma(E)$ is the total photonuclear absorption cross section. Deexcitation by charged particle emission is neglected in Eq. (V-20) because of its large inhibition by Coulomb forces. The ratio of the average fission width to the average total width of the compound nucleus for a two-humped barrier is given by Eq. (IV-18).

The giant dipole resonance in the total photonuclear absorption cross section can be approximately represented by a Lorentz line (Axel, 1962). For a limited range of energies near 7 MeV (well below the peak of the resonance) the following energy dependence of the cross section has been suggested by Axel (1962).

$$\sigma_{\gamma\mathrm{d}}(E) = 5.2(E/7\ \mathrm{MeV})^3(0.01A)^{8/3} \quad \mathrm{mb} \qquad \text{(V-21)}$$

where E is in units of MeV. In the derivation of this equation it is assumed that (1) the total integrated cross section in the giant resonance is given by the sum rule, (2) the peak of the giant resonance occurs at an energy equal to $80A^{-1/3}$, and (3) the full width of the giant resonance at half-maximum intensity is 5 MeV. For $A = 240$, Eq. (V-21) gives a predicted total cross section of 52 mb at $E_\gamma = 7$ MeV. This value is in good agreement with an experimentally determined value of $\sigma_\gamma(E_\gamma = 7\text{ MeV})$ of 60 ± 20 mb (Huizenga *et al.*, 1962). In an energy interval ΔE containing n levels, it is possible to relate the average cross section for dipole radiation to the γ-ray strength function $\bar{\Gamma}_0/D$ by the relation

$$\sigma_{\gamma d}(E) = \pi^2 \lambda\!\!\!^{-2} [(2I_e + 1)/(2I_g + 1)](\bar{\Gamma}_0/D) \tag{V-22}$$

where I_e and I_g are the spins of the excited and ground states, $\bar{\Gamma}_0$ is the average width of the elastic γ ray, and $D = \Delta E/n$. For $E_\gamma = 7$ MeV, we have $\pi^2 \lambda\!\!\!^{-2} = 7.84 \times 10^4$ mb. For spin zero target nuclei, it is therefore possible to deduce unambiguously the γ-ray strength function from the measured 7.0-MeV photon cross sections. The derived value of $(2.5 \pm 0.8) \times 10^{-4}$ for $\bar{\Gamma}_0/D$ is in good agreement with estimates of this quantity from neutron capture experiments.

The dipole photofission cross section of very heavy elements as a function of E is estimated from Eqs. (V-20) and (V-21) to be

$$\sigma_{\gamma,\mathrm{f}}(E) = 52(E/7\text{ MeV})^3 (\Gamma_\mathrm{f}/\Gamma_\mathrm{T})_\mathrm{E} \quad \text{mb} \tag{V-23}$$

where $\Gamma_\mathrm{T} = \Gamma_\mathrm{f} + \Gamma_\mathrm{n} + \Gamma_\gamma$ and these quantities are associated with the decay of compound nuclei formed by dipole absorption. At energies below the neutron threshold $\Gamma_\mathrm{n} = 0$ and the decay competition is between fission and radiation.

The average fission width $\langle \Gamma_{\lambda\mathrm{f}}(J, \pi, K, E) \rangle$ is defined by

$$\langle \Gamma_{\lambda\mathrm{f}}(J, \pi, K, E) \rangle = \frac{\langle D_{\lambda J\pi} \rangle}{2\pi} T_{\lambda\mathrm{f}}(J, \pi, K, E) \tag{V-24}$$

where $\langle D_{\lambda J\pi} \rangle$ is the average spacing between levels of a given total angular momentum and parity, $T_{\lambda\mathrm{f}}(J, \pi, K, E)$ is a transmission coefficient for fission through a specific channel. The average fission width defined in Eq. (V-24) is for a particular projection of K. All K values have two projections $(\pm K)$ except $K = 0$, which has only a single projection. We define the number of projections as $h(K)$. The penetrability through a single fission barrier† with the shape of an inverted parabola is (Hill and Wheeler, 1953)

$$T_\mathrm{f}(J, \pi, K, E) = (1 + \exp\{2\pi[B_\mathrm{f}(J, \pi, K) - E]/\hbar\omega(J, \pi, K)\})^{-1} \tag{V-25}$$

where $B_\mathrm{f}(J, \pi, K)$ is the fission barrier of state (J, π, K), E is the compound nucleus excitation energy, and $\hbar\omega(J, \pi, K)$ is a characteristic energy that defines the curvature of the barrier. The magnitude $\hbar\omega$ varies inversely to the thickness of the fission barrier. A small value of $\hbar\omega$ corresponds to a thick barrier and a large value of $\hbar\omega$ corresponds to a thin barrier. When the barrier is thick, the penetration is very small until one is near the top of the barrier.

† For a double-humped barrier, a more complicated expression, such as Eq. (V-76), is needed for barrier penetration.

If the total excitation energy E is less than the neutron binding energy B_n, the combination of Eqs. (V-23) and (V-24) leads to the following energy-dependent dipole photofission cross section.

$$\sigma_{\gamma,f}(J,\pi,E) = \frac{\text{const}\, E^3 \sum_K T_{\lambda f}(J,\pi,K,E)\, h(K)}{T_{\lambda\gamma}(J,\pi,E) + \sum_K T_{\lambda f}(J,\pi,K,E)\, h(K)} \tag{V-26}$$

where $\sigma_{\gamma f}(J,\pi,E)$ is in units of millibarns, E is in units of MeV, and the constant according to Eq. (V-23) is 0.15. However, this constant may vary considerably if the position and width of the Lorentz line are changed. The quantity $T_{\lambda\gamma}(J,\pi,E)$ in Eq. (V-26) is given by

$$T_{\lambda\gamma}(J,\pi,E) = \frac{2\pi \langle \Gamma_{\lambda\gamma}(E)\rangle}{\langle D(J,\pi,E)\rangle} \tag{V-27}$$

The quantity $\langle \Gamma_{\lambda\gamma}(E)\rangle$ is the average partial width for γ-ray decay of a compound state λ with total angular momentum J, parity π, and excitation energy E and $\langle D(J,\pi,E)\rangle$ is the average spacing of levels with total angular momentum J, parity π, and excitation energy E and $\langle D(J,\pi,E)\rangle$ is the average spacing of levels with total angular momentum J, parity π, and excitation energy E.

The energy dependence of the average radiation width for dipole γ-ray emission can be calculated from the relation

$$\langle \Gamma_{\lambda\gamma}(E)\rangle = C_1 \int_0^E \frac{\{\rho(J-1,E-\varepsilon) + \rho(J,E-\varepsilon) + \rho(J+1,E-\varepsilon)\}\,\varepsilon^3\, d\varepsilon}{\rho(J,E)} \tag{V-28}$$

where E is the initial excitation energy and ε is the γ-ray energy. The spin- and energy-dependent level density is

$$\rho(J,E) = C_2(2J+1)(E-\Delta)^{-2} \exp[2a^{1/2}(E-\Delta)^{1/2} - (J+\tfrac{1}{2})^2/2\sigma^2] \tag{V-29}$$

The angular momentum dependence of the level densities in the numerator and denominator of Eq. (V-28) cancels approximately to give a radiation width which is independent of angular momentum, namely

$$\langle \Gamma_{\lambda\gamma}(E)\rangle = C_1 \int_0^E \left\{\frac{(E-\Delta-\varepsilon)^{-2} \exp[2a^{1/2}(E-\Delta-\varepsilon)^{1/2}]}{(E-\Delta)^{-2} \exp[2a^{1/2}(E-\Delta)^{1/2}]}\right\} \varepsilon^3\, d\varepsilon \tag{V-30}$$

If the energy factor preceding the exponential in the numerator is neglected, the right-hand side of Eq. (V-30) may be integrated (Moldauer *et al.*, 1964) to give

$$\langle \Gamma_{\lambda\gamma}(E)\rangle = C_3\, X(E,a) = C_3\{x^4 - 10x^3 + 45x^2 - 105x + 105\} \tag{V-31}$$

where $x = 2a^{1/2}(E-\Delta)^{1/2}$. In Eqs. (V-28) to (V-31), C_1, C_2, C_3, Δ, and a are constants. The constant a is the level density parameter defined by $a = \pi^2 g_0/6$, where g_0 is the average number of single particle states per MeV, and Δ is an energy correction which accounts for pairing and shell corrections and defines the energy of a fictitious ground state with respect to the actual ground state.

With the result of Eqs. (V-28) to (V-31), the spin and energy dependence of $T_{\lambda\gamma}(J, \pi, E)$ may be written in the useful form

$$T_{\lambda\gamma}(J, \pi, E) = \frac{2\pi\langle\Gamma_{\lambda\gamma}(E')\rangle}{\langle D(J', \pi, E')\rangle} \frac{X(E, a)}{X(E', a)} \frac{\rho(J, E)}{\rho(J', E')} \tag{V-32}$$

where $\langle\Gamma_{\lambda\gamma}(E')\rangle/\langle D(J', \pi, E')\rangle$ is the average radiation width divided by the average level spacing for levels of spin J' at excitation energy E'. Resonance neutron capture in ^{239}Pu, for example, gives $\langle\Gamma_{\lambda\gamma}(6.47)\rangle = 0.040$ eV and $\langle D(1, +, 6.47)\rangle = 1.7 \pm 0.5$ eV. The resulting value of $T_{\lambda\gamma}(1, +, 6.47) =$

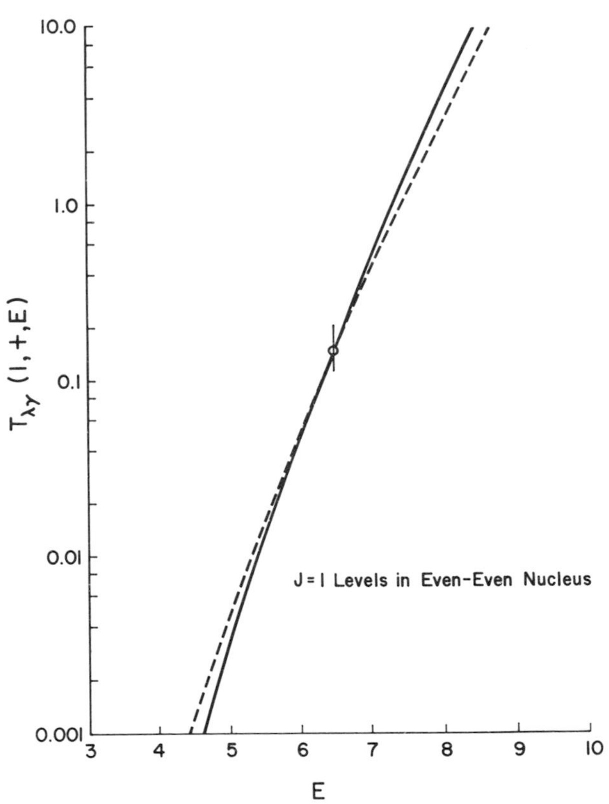

Fig. V-6. Dependence of $T_{\lambda\gamma}(1, +, E)$ on excitation energy E for an even-even heavy nucleus. The normalization is done with the experimental value of $2\pi\langle\Gamma_{\lambda\gamma}(6.47)\rangle$ divided by $\langle D(1, +, 6.47)\rangle$ for resonance capture on ^{239}Pu. The experimental value of $\langle\Gamma_{\lambda\gamma}\langle 6.47)\rangle$ is 0.040 eV and the value of $\langle D(1, +, 6.47)\rangle = 1.70 \pm 0.5$ eV [Farrell (1968)]. (—) Calculated with Eq. (V-32) for $a = 30$ MeV^{-1} and $\Delta = 1.1$ MeV, (---) results for either $a = 24$ MeV^{-1} and $\Delta = 1.1$ MeV or $a = 30$ MeV^{-1} and $\Delta = 0$. It is assumed that $T_{\lambda\gamma}(1, +, E) = T_{\lambda\gamma}(1, -, E)$.

$0.147^{+0.062}_{-0.033}$ is plotted as an open circle in Fig. V-6. The solid line in Fig. V-6 is calculated with Eq. (V-32) for $a = 30\ \text{MeV}^{-1}$ and $\Delta = 1.1$ MeV. The energy dependence of $T_{\lambda\gamma}(J, \pi, E)$ is rather insensitive to $X(E, A)$ and the neglect of the preexponential energy factor in the level density in the derivation of $X(E, A)$ gives a negligible error. Values $T_{\lambda\gamma}(J, \pi, E)$ may be calculated from the spin dependence of the level density and, for example, $T_{\lambda\gamma}(2, \pi, E)$ and $T_{\lambda\gamma}(0, \pi, E)$, and are equal to $1.58 T_{\lambda\gamma}(1, \pi, E)$ and $0.69 T_{\lambda\gamma}(1, \pi, E)$, respectively (for $\sigma = 6$).

At photon energies below the neutron binding energy the dipole photofission cross section of ^{232}Th for transition states $K = 0$ and $K = 1$ is predicted by Eq. (V-26) to be

$$\sigma_{\gamma,\mathrm{f}}(1, -, E) = \frac{\text{const}\, E^3 \{T_{\lambda\mathrm{f}}(1, -, 0, E) + 2T_{\lambda\mathrm{f}}(1, -, 1, E)\}}{T_{\lambda\gamma}(1, -, E) + T_{\lambda\mathrm{f}}(1, -, 0, E) + 2T_{\lambda\mathrm{f}}(1, -, 1, E)} \tag{V-33}$$

The theoretical photofission cross sections calculated with Eq. (V-33) are compared with experimental photofission cross sections of ^{232}Th measured by Knowles and Khan (1968) in Fig. V-7. The different lines are calculated with different values of the fission barriers $B_\mathrm{f}(J, \pi, K)$ and curvature energies $\hbar\omega(J, \pi, K)$. The values of $T_{\lambda\gamma}(1, -, E)$ are taken from the solid line of Fig. V-6. The theoretical photofission cross sections are normalized to 5.4 mb at $E_\gamma = 6.5$ MeV.

Information on the total photonuclear quadrupole absorption cross section of heavy elements at excitation energies near the fission barrier is sparse. Since the γ-ray wavelength at these energies is approximately an order of magnitude larger than the nuclear diameter, the dipole absorption cross section $\sigma_{\gamma\mathrm{d}}(E)$ is expected to be considerably larger than the quadrupole absorption cross section $\sigma_{\gamma\mathrm{q}}(E)$. An estimate of $(4/125)(R/\lambda\!\!\!^{-})^2$ for the electric radiation absorption ratio $\sigma_{\gamma\mathrm{q}}(E)/\sigma_{\gamma\mathrm{d}}(E)$ is obtained from Eq. (7.28) of Blatt and Weisskopf (1952). However, this equation probably overestimates electric dipole absorption for low energy photons (Blatt and Weisskopf, 1952). The electric quadrupole absorption cross section is approximately equal to (Blatt and Weisskopf, 1952)

$$\sigma_{\gamma\mathrm{q}}(E) \approx 1.2 E^3 (R/6 \times 10^{-13})^4 10^{-3} \quad \text{mb}$$

where E is in MeV and R is in cm. If we use this equation in conjunction with the electric dipole absorption cross section given by Eq. (V-21), we can calculate a $\sigma_{\gamma\mathrm{q}}(E)/\sigma_{\gamma\mathrm{d}}(E)$ ratio equal to 0.02 for low-energy γ rays.

Although the detection of such a small admixture of quadrupole absorption is difficult in general, Griffin (1959) pointed out that under particular conditions, the study of fission-fragment angular distributions is a sensitive technique for establishment of quadrupole absorption. If the fission threshold for a $(J = 2+, K = 0)$ level is lower than a $(J = 1-, K = 0)$ level, then for sub-

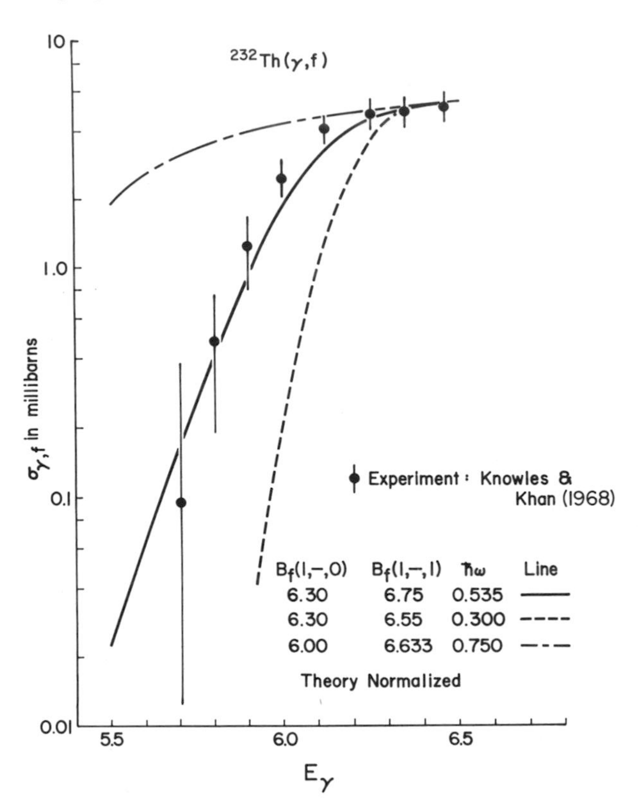

Fig. V-7. Comparison of theoretical and experimental photofission cross sections of ^{232}Th as a function of photon energy at energies below the neutron binding energy. [Experimental data are from Knowles and Khan (1968); see also Kahn and Knowles (1972).]

barrier fission the high penetrability [see Eq. (V-25)] of the barrier for quadrupole fission relative to dipole fission will considerably increase the ratio $\sigma_{\gamma,\mathrm{f}}(J=2+,\, K=0)/\sigma_{\gamma,\mathrm{f}}(J=1-,\, K=0)$. The increase of this ratio with decreasing energy for the photofission of ^{238}U is illustrated in Fig. V-5. The ratio c/b is directly related to $\sigma_{\gamma,\mathrm{f}}(J=2+,\, K=0)/\sigma_{\gamma,\mathrm{f}}(J=1-,\, K=0)$ by Eq. (V-18) (note the assumption that only those two states are excited) and increases sharply for the lowest two energies, as shown in Fig. V-5. The relation between the quadrupole and dipole photofission cross sections and absorption cross sections at energies below the neutron threshold is

$$\frac{\sigma_{\gamma,\mathrm{f}}(J=2+,\, K=0)}{\sigma_{\gamma,\mathrm{f}}(J=1-,\, K=0)} = \frac{\left\{\dfrac{T_{\lambda\mathrm{f}}(2,\, +,\, 0,\, E)}{T_{\lambda\gamma}(2,\, +,\, E)+T_{\lambda\mathrm{f}}(2,\, +,\, 0,\, E)}\right\}\sigma_{\gamma\mathrm{q}}(E)}{\left\{\dfrac{T_{\lambda\mathrm{f}}(1,\, -,\, 0,\, E)}{T_{\lambda\gamma}(1,\, -,\, E)+T_{\lambda\mathrm{f}}(1,\, -,\, 0,\, E)}\right\}\sigma_{\gamma\mathrm{d}}(E)} \tag{V-34}$$

The photon absorption cross section ratio $\sigma_{\gamma q}(E)/\sigma_{\gamma d}(E)$ may be written in the following way, making use of Eqs. (V-18) and (V-34).

$$\frac{\sigma_{\gamma q}(E)}{\sigma_{\gamma d}(E)} = \left(\frac{4}{5}\right)\left(\frac{c}{b}\right)\frac{\dfrac{T_{\lambda f}(1, -, 0, E)}{T_{\lambda\gamma}(1, -, E) + T_{\lambda f}(1, -, 0, E)}}{\dfrac{T_{\lambda f}(2, +, 0, E)}{T_{\lambda\gamma}(2, +, E) + T_{\lambda f}(2, +, 0, E)}} \tag{V-35}$$

Actual evaluation of this absorption cross section ratio from experimental values of c/b requires a knowledge of the transmission coefficients on the right-hand side of Eq. (V-35). Values of the γ-ray transmission coefficients $T_{\lambda\gamma}(1, -, E)$ and $T_{\lambda\gamma}(2, +, E)$ for even-even nuclei may be obtained from Fig. V-6. However, values of the fission barrier parameters are required for the evaluation of the fission transmission coefficients. We should remember that Eq. (V-35) is valid for the case where only $K = 0$ transition levels contribute to fission. At γ-ray energies where $T_{\lambda f}(1, -, 0, E)$ and $T_{\lambda f}(2, +, 0, E)$ are essentially unity (and $K > 0$ states are making a negligible contribution), Eq. (V-35) reduces to $\sigma_{\gamma q}(E)/\sigma_{\gamma d}(E) \approx c/b$. The values of c/b for ^{232}Th remain small even at the lowest photon energies. This is related to the fact that the outer barrier for ^{232}Th is higher than the inner barrier. An estimate of the quadrupole to dipole photon absorption cross section ratio may be obtained by substituting into Eq. (V-35) the values of the fission to total widths for the $(J = 1-, K = 0)$ and $(J = 2+, K = 0)$ states from Eq. (IV-18).

At excitation energies where contributions from transition states with $K > 0$ contribute, namely, $(J = 1-, K = 1)$, $(J = 2+, K = 2)$, and $(J = 2+, K = 1)$ states, the parameter b is made up of contributions from both dipole and quadrupole fission [see Eq. (V-17)]. From Eqs. (V-10), (V-16), and (V-17), the general relation for c/b is derived

$$\frac{c}{b} = \frac{q[\frac{15}{16}u - \frac{5}{4}v + \frac{5}{16}w]/(u+2v+2w)}{\{d[\frac{3}{4}(x-y)]/(1+y)\} + \{q[\frac{5}{4}(w-v)]/(u+2v+2w)\}} \tag{V-36}$$

where all the various symbols were defined previously. When $v = w = y = 0$ and only $K = 0$ states contribute, Eq. (V-36) reduces to Eq. (V-18). However, when all the K states are open or partially open, the interpretation of c/b is more difficult. In analogy to Eq. (V-36), the general expression for b/a is

$$\frac{b}{a} = \frac{\{d[\frac{3}{4}(x-y)]/(1+y)\} + \{q[\frac{5}{4}(w-v)]/(u+2v+2w)\}}{\{d(\frac{3}{2}y)/(1+y)\} + \{q(\frac{5}{2}v)/(u+2v+2w)\}} \tag{V-37}$$

In the low energy transition state spectrum it appears reasonable to assume that the contribution of the $(J = 2+, K = 1)$ state is negligible, i.e., $v = 0$. This gives some simplification of Eqs. (V-36) and (V-37). From this more simplified version of Eq. (V-37), we see that b/a increases as y decreases (see Fig. V-5). The interpretation of Eq. (V-36) is more complex. When y is large and the ratio b/a is small, the experimental ratio c/b is not related to the quadrupole to dipole absorption ratio in a simple way. However, when b/a is large (y is small), then the limiting value of c/b expressed by Eq. (V-18) is approached when v and w are zero. Some experimental ratios of b/a and c/b for photofission of ^{232}Th, ^{238}U, and ^{240}Pu induced by bremsstrahlung are listed in Table V-1.

The analysis of photofission with a continuous bremsstrahlung spectrum of γ rays is more complex than the previous analysis, which assumed monoenergetic γ rays. The bremsstrahlung spectrum near the electron energy E_{max} for a thick stopping foil may be represented to good approximation by the function (Rabotnov *et al.*, 1965)

$$F(E_{max}, E_\gamma)\, dE_\gamma = k(E_{max} - E_\gamma)^2\, dE_\gamma \tag{V-37a}$$

TABLE V-1

Parameters a, b, and c as Defined by Eq. (V-17) from Photofission Angular Distribution Measurements[a]

E_{max} (MeV)	b/a	c/b
	^{232}Th	
5.4	$111^{+\infty}_{-57}$	$0.03^{+0.027}_{-0.025}$
5.65	99^{+101}_{-34}	0
5.9	99^{+101}_{-34}	$0.085^{+0.016}_{-0.015}$
6.4	44^{+14}_{-8}	$0.022^{+0.015}_{-0.014}$
6.9	30^{+10}_{-6}	$0.021^{+0.023}_{-0.021}$
	^{238}U	
5.2	23^{+121}_{-11}	$1.06^{+0.14}_{-0.12}$
5.4	25^{+9}_{-5}	$0.16^{+0.03}_{-0.02}$
5.65	28^{+6}_{-4}	$0.041^{+0.011}_{-0.010}$
5.9	12^{+1}_{-1}	$0.042^{+0.016}_{-0.015}$
6.4	$6.9^{+0.3}_{-0.3}$	$0.039^{+0.010}_{-0.010}$
6.9	$3.7^{+0.1}_{-0.1}$	$0.060^{+0.011}_{-0.011}$
9.25	$0.75^{+0.02}_{-0.02}$	$0.030^{+0.017}_{-0.016}$
	^{240}Pu	
5.4	$4.85^{+2.16}_{-1.28}$	$1.29^{+0.21}_{-0.17}$
5.65	$2.32^{+0.53}_{-0.41}$	$0.90^{+0.22}_{-0.15}$
5.9	$0.78^{+0.05}_{-0.04}$	$0.69^{+0.08}_{-0.07}$
6.4	$0.52^{+0.02}_{-0.03}$	$0.31^{+0.05}_{-0.04}$
6.9	$0.45^{+0.04}_{-0.04}$	$0.22^{+0.09}_{-0.09}$
7.9	$0.38^{+0.03}_{-0.03}$	$0.27^{+0.09}_{-0.08}$

[a] Measurements made with bremsstrahlung produced by stopping electrons of maximum energy E_{max} in a thick stopping foil. [Data of Rabotnov *et al.* (1965).]

Since the values of E_{max} used in most experiments are near the fission barrier, only the very highest energy part of the γ-ray spectrum is effective in producing fission. The effective cross section for dipole photofission with bremsstrahlung in the approximation of Eq. (V-26) is

$$\sigma_{\gamma,f}(J,\pi,E_{max}) = \frac{\displaystyle\int_0^{E_{max}} \text{const}\, E^3 k(E_{max}-E)^2 \frac{\sum_K T_{\lambda f}(J,\pi,K,E)\,h(K)}{T_{\lambda\gamma}(J,\pi,E)+\sum_K T_{\lambda f}(J,\pi,K,E)\,h(K)}\,dE}{\displaystyle\int_0^{E_{max}} k(E_{max}-E)^2\,dE} \tag{V-38}$$

For a bremsstrahlung spectrum, Eq. (V-19) may be rewritten

$$\frac{b}{a}+\frac{1}{2} = \frac{\displaystyle\int_0^{E_{max}} \left\{\frac{E^3(E_{max}-E)^2 T_{\lambda f}(1,-,0,E)}{T_{\lambda\gamma}(1,-,E)+T_{\lambda f}(1,-,0,E)+2T_{\lambda f}(1,-,1,E)}\right\} dE}{\displaystyle\int_0^{E_{max}} \left\{\frac{2E^3(E_{max}-E)^2 T_{\lambda f}(1,-,1,E)}{T_{\lambda\gamma}(1,-,E)+T_{\lambda f}(1,-,0,E)+2T_{\lambda f}(1,-,1,E)}\right\} dE} \tag{V-39}$$

where the γ-ray and fission transmission coefficients are defined by Eqs. (V-32) and (V-25), respectively. Theoretical values of b/a for the photofission of ^{232}Th are shown in Fig. V-8 for various values of the parameters of the fission barriers. The values of $T_{\lambda\gamma}(1, -, E)$ were taken from Fig. V-6. The integrals were performed by summing over energy bins of 0.1 MeV. The theoretical values of b/a are compared in Fig. V-8 with the photofission data of ^{232}Th listed in Table V-1.

When $E_{max} \lesssim B_f(1, -, 0)$, Eq. (V-39) reduces to a simple limiting form for bremsstrahlung

$$\frac{b}{a}+\frac{1}{2} = \tfrac{1}{2}\exp\{2\pi[B_f(1,-,1)-B_f(1,-,0)]/\hbar\omega\} \tag{V-40}$$

where it is assumed that $\hbar\omega(1, -, 1) = \hbar\omega(1, -, 0)$. Hence, with an experimental value of $b/a = 100$ at the smallest values of E_{max}, the quantity $[B_f(1, -, 1) - B_f(1, -, 0)]/\hbar\omega$ is equal to 0.84 for ^{232}Th. The various sets of barrier parameters in Figs. V-7 and V-8 obey this relation. More recently, Rabotnov *et al.* (1969) have concluded that the bremsstrahlung spectrum near E_{max} for a thick stopping foil is nearly linear, i.e., $F(E_{max}, E_\gamma)\,dE_\gamma \approx k(E_{max} - E_\gamma)\,dE_\gamma$, rather than quadratic as given in Eq. (V-37a). This slightly alters the results obtained with Eq. (V-39) but, of course, does not alter Eq. (V-40).

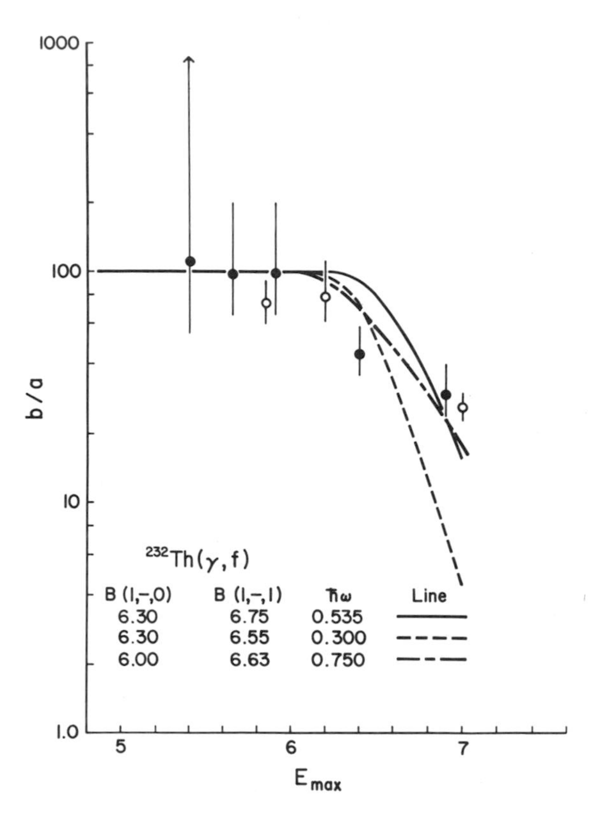

Fig. V-8. Comparison of theoretical and experimental values of b/a for photofission of ^{232}Th with bremsstrahlung. The theroretical values are derived with Eq. (V-39). [The experimental values are those of Rabotnov *et al.* (1965) (●), and Rabotnov *et al.* (1968) (○).]

The values of c/b for ^{240}Pu listed in Table V-1 are strikingly large. For example, at $E_{\max} = 6.4$ MeV, c/b has values of 0.02, 0.04, and 0.31 for ^{232}Th, ^{238}U, and ^{240}Pu, respectively. These results have prompted the suggestion that the quadrupole to dipole absorption ratio is much larger for ^{240}Pu than for the other two nuclei. However, alternatively, such results are explainable in terms of double-humped barriers of different heights and a constant quadrupole to dipole absorption ratio.

For even-even nuclei existing evidence suggests that $B_f(1, -, 0)$ is less than $B_f(1, -, 1)$. This condition leads to strong peaking of the fission fragments perpendicular to the photon direction for low energy photons and results in large values of b/a. The largest experimental values of b/a have been measured for the photofission of ^{232}Th (see Table V-1). The values of $B_f(2, +, 0)$ are predicted to be less than the values of $B_f(1, -, 0)$ for the inner barriers of heavy nuclei. However, these two barrier states are essentially degenerate at the second or outer barrier. In the case of ^{232}Th the outer barrier is the higher one and the very small relative yield of quadrupole fission for this nucleus is consistent with the expected small quadrupole to dipole absorption cross section ratio.

Both the trend for b/a to decrease and c/b to increase in going from ^{232}Th to ^{240}Pu can be qualitatively understood in terms of recent theoretical calculations of the potential energy surface. These calculations reveal a double-humped fission barrier. The inner barrier is stable with respect to mass asymmetric distortions, and for this barrier, $B_f(2,+,0)$ is expected to be considerably less than $B_f(1,-,0)$. However, the lowest energy second or outer barrier has an octupole deformation (Möller and Nilsson, 1970; Pauli *et al.*, 1971; Pashkevich, 1971) which can lead to a $K = 0$ ground state band with levels 0+, 1−, 2+, 3−, 4+,··· (Johansson, 1961). In this case, $B_f(1,-,0)$ is essentially degenerate with $B_f(2,+,0)$. The theoretical calculations also show a trend for the second barrier to decrease in energy relative to the first barrier in going from ^{232}Th to ^{240}Pu. The implications of these theoretical findings are indicated schematically in Fig. V-9. In the case of the lighter nucleus ^{232}Th where the outer barrier is higher and defines the transition state, the effective barrier heights for the 1− and 2+ states are comparable and the ratio c/b is determined primarily by the (small) ratio of quadrupole to dipole γ-ray absorption. At very low energy the reflection by the inner barrier, which is higher for the 1− than for the 2+ state, enhances

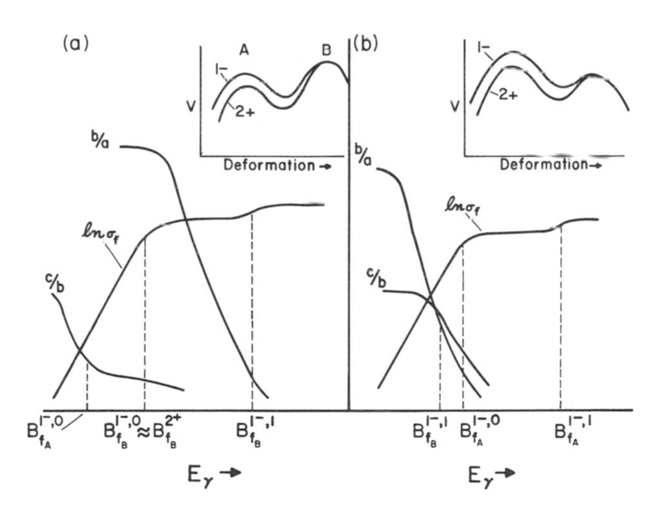

Fig. V-9. Schematic illustrations of the dependence of the angular distribution coefficient ratios and the total fission cross section on γ-ray energy E_γ for (a) the case where the inner barrier B_{fA} is lower than the outer barrier B_{fB}, and (b) the case where the inner barrier is higher in energy than the outer barrier.

the ratio c/b. The ratio b/a is large for fission through the $K = 0$ state and decreases when the energy becomes sufficiently high for the $K = 1$ states to contribute. This is at an energy well above the barrier for the $K = 0$ state where the fission cross section has reached an approximate plateau.

If in a heavier nucleus like ^{240}Pu the outer barrier is lower in energy than the inner barrier, the relative proportions of dipole and quadrupole fission will depend on the penetration of the inner barrier, even if the K value characterizing the angular distribution is determined by the outer barrier. (There is some evidence from neutron-induced fission that K is not preserved during the passage from the inner to the outer barrier.) Even though the $1-$ and $2+$ transition state channels at the second barrier are both open, the higher first barrier for nuclei with $J = 1-$ compared to nuclei with $J = 2+$ will inhibit dipole fission relative to quadrupole fission at energies lower than the height of the first barrier. Thus the predominance of dipole absorption may be largely compensated for at the smaller energies by the decreased penetration through the first barrier. The energy at which angular distributions characteristic of $J = 1-$, $K = 1$ states start appearing is modified also if the outer barrier is lower than the inner barrier. If, for example, the height of the inner barrier for the $J = 1-$, $K = 0$ state is higher than that for the $J = 1-$, $K = 1$ states at the outer barrier, the ratio b/a will decrease before the fission cross section has reached its plateau value. It has been emphasized by Rabotnov *et al.* (1970) that the experimentally observed decrease in b/a for ^{240}Pu at energies which are subbarrier with respect to the cross section is evidence that the inner barrier is appreciably higher in energy than the outer barrier.

2. Odd Mass Targets

Photofission of odd mass targets is expected to give an isotropic or near isotropic distribution of fission fragments at energies somewhat above the barrier. Preliminary measurements, some not at very low energy, made on ^{233}U, ^{235}U, ^{237}U, ^{239}Pu, and ^{241}Am indicate isotropic angular distributions (Baerg *et al.*, 1959). A more detailed theoretical discussion of the targets with different spins will be given later in this section. Since ^{239}Pu has the smallest nonzero spin ($I_0 = \frac{1}{2}$), it is expected to show an anisotropic distribution near the fission barrier. Precise measurements of the fission-fragment anisotropy for photofission of ^{239}Pu have shown anisotropic distributions at very low energies.

a. Dipole Photofission of ^{239}Pu

If only electric dipole absorption is assumed, compound states of ^{239}Pu are formed with $(J, M) = (\frac{3}{2}, \pm\frac{3}{2}), (\frac{3}{2}, \pm\frac{1}{2})$, and $(\frac{1}{2}, \pm\frac{1}{2})$ and probabilities $P(J, \pm M)$

of $\frac{1}{2}$, $\frac{1}{6}$, and $\frac{1}{3}$, respectively. The theoretical angular distributions for all the possible states formed by different combinations of J, M, and K are evaluated with Eq. (V-3) and the $d^{J}_{M,K}(\theta)$ functions defined by Eq. (V-2),

$$W^{\frac{3}{2}}_{\pm\frac{3}{2},\pm\frac{3}{2}}(\theta) = 2\left\{\frac{|d^{\frac{3}{2}}_{\frac{3}{2},\frac{3}{2}}(\theta)|^2}{4} + \frac{|d^{\frac{3}{2}}_{-\frac{3}{2},\frac{3}{2}}(\theta)|^2}{4} + \frac{|d^{\frac{3}{2}}_{\frac{3}{2},-\frac{3}{2}}(\theta)|^2}{4} + \frac{|d^{\frac{3}{2}}_{-\frac{3}{2},-\frac{3}{2}}(\theta)|^2}{4}\right\}$$

$$= 1 - \tfrac{3}{4}\sin^2\theta \tag{V-41}$$

$$W^{\frac{3}{2}}_{\pm\frac{1}{2},\pm\frac{3}{2}}(\theta) = 2\left\{\frac{|d^{\frac{3}{2}}_{\frac{1}{2},\frac{3}{2}}(\theta)|^2}{4} + \frac{|d^{\frac{3}{2}}_{-\frac{1}{2},\frac{3}{2}}(\theta)|^2}{4} + \frac{|d^{\frac{3}{2}}_{\frac{1}{2},-\frac{3}{2}}(\theta)|^2}{4} + \frac{|d^{\frac{3}{2}}_{-\frac{1}{2},-\frac{3}{2}}(\theta)|^2}{4}\right\}$$

$$= \tfrac{3}{4}\sin^2\theta \tag{V-42}$$

$$W^{\frac{3}{2}}_{\pm\frac{1}{2},\pm\frac{1}{2}}(\theta) = 2\left\{\frac{|d^{\frac{3}{2}}_{\frac{1}{2},\frac{1}{2}}(\theta)|^2}{4} + \frac{|d^{\frac{3}{2}}_{-\frac{1}{2},\frac{1}{2}}(\theta)|^2}{4} + \frac{|d^{\frac{3}{2}}_{\frac{1}{2},-\frac{1}{2}}(\theta)|^2}{4} + \frac{|d^{\frac{3}{2}}_{-\frac{1}{2},-\frac{1}{2}}(\theta)|^2}{4}\right\}$$

$$= 1 - \tfrac{3}{4}\sin^2\theta \tag{V-43}$$

$$W^{\frac{3}{2}}_{\pm\frac{3}{2},\pm\frac{1}{2}}(\theta) = 2\left\{\frac{|d^{\frac{3}{2}}_{\frac{3}{2},\frac{1}{2}}(\theta)|^2}{4} + \frac{|d^{\frac{3}{2}}_{-\frac{3}{2},\frac{1}{2}}(\theta)|^2}{4} + \frac{|d^{\frac{3}{2}}_{\frac{3}{2},-\frac{1}{2}}(\theta)|^2}{4} + \frac{|d^{\frac{3}{2}}_{-\frac{3}{2},-\frac{1}{2}}(\theta)|^2}{4}\right\}$$

$$= \tfrac{3}{4}\sin^2\theta \tag{V-44}$$

$$W^{\frac{1}{2}}_{\pm\frac{1}{2},\pm\frac{1}{2}}(\theta) = \left\{\frac{|d^{\frac{1}{2}}_{\frac{1}{2},\frac{1}{2}}(\theta)|^2}{4} + \frac{|d^{\frac{1}{2}}_{-\frac{1}{2},\frac{1}{2}}(\theta)|^2}{4} + \frac{|d^{\frac{1}{2}}_{\frac{1}{2},-\frac{1}{2}}(\theta)|^2}{4} + \frac{|d^{\frac{1}{2}}_{-\frac{1}{2},-\frac{1}{2}}(\theta)|^2}{4}\right\}$$

$$= \tfrac{1}{2} \tag{V-45}$$

[The origin of the factors of $\frac{1}{4}$ is given following Eq. (V-7).]

The $J = \frac{3}{2}-$ state may decay by fission through transition states with $K = \frac{1}{2}$ and $\frac{3}{2}$. If the probability for fission of the $J = \frac{3}{2}-$ state with $K = \frac{1}{2}$ is denoted by A and the probability with $K = \frac{3}{2}$ by $1 - A$, then the theoretical fission angular distribution for photofission of ^{239}Pu is

$$W^{239\text{Pu}}_{\gamma,\text{f}}(\theta) = \frac{X_{\frac{1}{2}-}}{3}\left(\frac{1}{2}\right) + \frac{X_{\frac{3}{2}-}}{2}\{A(\tfrac{3}{4}\sin^2\theta) + (1-A)(1-\tfrac{3}{4}\sin^2\theta)\}$$

$$+ \frac{X_{\frac{3}{2}-}}{6}\{A(1-\tfrac{3}{4}\sin^2\theta) + (1-A)\tfrac{3}{4}\sin^2\theta\} \tag{V-46}$$

where

$$X_{\frac{1}{2}-} = (\Gamma_{\text{f}}^{J=\frac{1}{2}-}/\Gamma_{\text{T}}^{J=\frac{1}{2}-}) \qquad \text{and} \qquad X_{\frac{3}{2}-} = (\Gamma_{\text{f}}^{J=\frac{3}{2}-}/\Gamma_{\text{T}}^{J=\frac{3}{2}-})$$

The total width Γ_{T} is the sum of all the partial widths. For example, if neutron emission is neglected,

$$X_{\frac{3}{2}-} = \frac{2T_{\lambda\text{f}}(\frac{3}{2}, -, \frac{3}{2}, E) + 2T_{\lambda\text{f}}(\frac{3}{2}, -, \frac{1}{2}, E)}{T_{\lambda\gamma}(\frac{3}{2}, -, E) + 2T_{\lambda\text{f}}(\frac{3}{2}, -, \frac{3}{2}, E) + 2T_{\lambda\text{f}}(\frac{3}{2}, -, \frac{1}{2}, E)} \tag{V-47}$$

Eq. (V-46) reduces to

$$W_{\gamma,\mathrm{f}}^{^{239}\mathrm{Pu}}(\theta) = \frac{X_{\frac{1}{2}-}}{6} + \frac{X_{\frac{3}{2}-}}{6}(3-2A) + \frac{X_{\frac{3}{2}-}}{4}(2A-1)\sin^2\theta \qquad \text{(V-48)}$$

and b/a is given by

$$\frac{b}{a} = \frac{X_{\frac{3}{2}-}(2A-1)/4}{(X_{\frac{1}{2}-}/6) + [X_{\frac{3}{2}-}(3-2A)/6]} \qquad \text{(V-49)}$$

When $X_{\frac{3}{2}-} = 0$, photofission is occurring only through the $J = \frac{1}{2}-$ state which gives isotropy ($b/a = 0$) according to Eq. (V-45). If $X_{\frac{3}{2}-} \neq 0$, then for $A = \frac{1}{2}$ the value of $b/a = 0$ and this resulting isotropy is independent of the values of $X_{\frac{1}{2}-}$ and $X_{\frac{3}{2}-}$. For positive values of b/a, A must be larger than $\frac{1}{2}$, and for negative values of b/a, A must be less than $\frac{1}{2}$. The absolute magnitude of b/a, however, depends on the values of $X_{\frac{1}{2}-}$ and $X_{\frac{3}{2}-}$.

Experimental values of b/a from a study of the photofission angular distribution of ^{239}Pu as a function of energy are plotted in Fig. V-10. The abscissa of Fig. V-10 represents the mean effective excitation energy which is computed

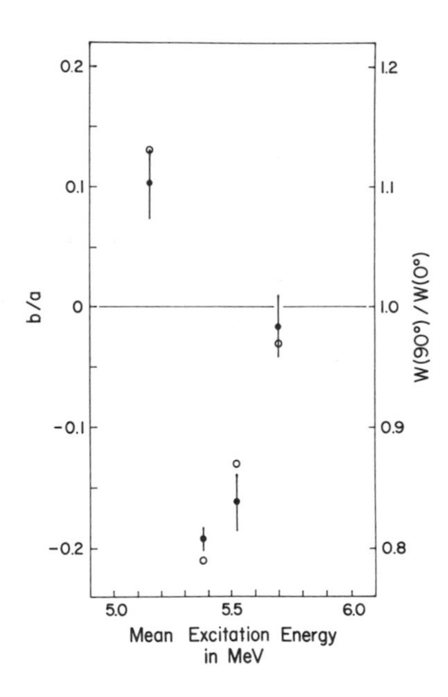

Fig. V-10. Dependence of b/a for photofission of ^{239}Pu as a function of mean excitation energy. ● Measurements of Rabotnov *et al.* (1966), ○ calculated values of b/a given in the last column of Table V-2 and described in the text. The mean excitation energy is defined by Eq. (V-50).

by the expression

$$\langle E_{\text{eff}} \rangle_{E_{\max}} = \frac{\int_0^{E_{\max}} E_\gamma \sigma_{\gamma,\text{f}}(E_\gamma) F(E_{\max}, E_\gamma)\, dE_\gamma}{\int_0^{E_{\max}} \sigma_{\gamma,\text{f}}(E_\gamma) F(E_{\max}, E_\gamma)\, dE_\gamma} \tag{V-50}$$

where $\sigma_{\gamma,\text{f}}(E_\gamma)$ values are calculated from Eq. (V-26) and $F(E_{\max}, E_\gamma)\, dE_\gamma$ is defined in Eq. (V-37a). The values of $\sigma_{\gamma,\text{f}}(E_\gamma)$ depend on the fission barrier parameters and the transmission coefficients of competing channels as shown by Eq. (V-26). The positive value of b/a at the lowest energy indicates that $A > \frac{1}{2}$ and the $K = \frac{1}{2}-$ state is more open than the $K = \frac{3}{2}-$ state. As the energy increases, b/a decreases. This decrease requires a $K = \frac{3}{2}-$ state contribution, and when $b/a = 0$, A must be equal to $\frac{1}{2}$. The value of A must be less than $\frac{1}{2}$ as b/a becomes negative at higher energies. If the transmission coefficients [see Eq. (V-25)] of a sequence of two states with $K = \frac{1}{2}-$ and $K = \frac{3}{2}-$ have the same energy dependence, b/a cannot become negative when the $K = \frac{1}{2}-$ state has the lower barrier energy. A second $K = \frac{3}{2}-$ state is required to make b/a negative. Alternatively, if the transmission coefficients of the $K = \frac{3}{2}-$ and $K = \frac{1}{2}-$ states have different energy dependences, it is possible to reproduce the energy dependence of b/a with only two states by adjusting their barrier heights and thicknesses appropriately (Vandenbosch, 1967). As an illustration, let us choose the two transition states with the following parameters; for the $K = \frac{3}{2}-$ state, $B_\text{f}(\frac{3}{2}, -, \frac{3}{2}) = 5.35$ MeV and $\hbar\omega(\frac{3}{2}, -, \frac{3}{2}) = 0.30$ MeV, and for the $K = \frac{1}{2}-$ state, $B_\text{f}(\frac{3}{2}, -, \frac{1}{2}) = B_\text{f}(\frac{1}{2}, -, \frac{1}{2}) = 5.50$ MeV and $\hbar\omega(\frac{3}{2}, -, \frac{1}{2}) = \hbar\omega(\frac{1}{2}, -, \frac{1}{2}) = 0.60$ MeV. The equalities assume that the two rotational states of the $K = \frac{1}{2}-$ band are degenerate.

Evaluation of A in column 2 of Table V-2 requires the transmission coefficients for particular fission channels defined by Eq. (V-25). The third column of Table V-2 is derived with Eq. (V-49) and requires only a knowledge of A. Evaluation of $X_{\frac{1}{2}-}/X_{\frac{3}{2}-}$ requires information about the appropriate radiation and neutron emission widths in addition to the fission widths, although for the

TABLE V-2

Calculated Parameters in Photofission of ^{239}Pu for Two Transition States $K = \frac{1}{2}-$ and $K = \frac{3}{2}-$ [a]

$\langle E_\text{f} \rangle$	A	b/a	$X_{\frac{1}{2}-}/X_{\frac{3}{2}-}$	b/a
5.15	0.62	$[4.89 + 2.78(X_{\frac{1}{2}-}/X_{\frac{3}{2}-})]^{-1}$	0.93	0.13
5.38	0.25	$[-3.33 - 1.33(X_{\frac{1}{2}-}/X_{\frac{3}{2}-})]^{-1}$	0.97	−0.21
5.52	0.36	$[-5.43 - 2.38(X_{\frac{1}{2}-}/X_{\frac{3}{2}-})]^{-1}$	0.99	−0.13
5.70	0.47	$[-22.8 - 11.1(X_{\frac{1}{2}-}/X_{\frac{3}{2}-})]^{-1}$	0.99	−0.03

[a] $B_\text{f}(\frac{3}{2}, -, \frac{3}{2}) = 5.35$ MeV and $\hbar\omega(\frac{3}{2}, -, \frac{3}{2}) = 0.30$ MeV; $B_\text{f}(\frac{3}{2}, -, \frac{1}{2}) = B_\text{f}(\frac{1}{2}, -, \frac{1}{2}) = 5.50$ MeV and $\hbar\omega(\frac{3}{2}, -, \frac{1}{2}) = \hbar\omega(\frac{1}{2}, -, \frac{1}{2}) = 0.60$ MeV.

values of $\langle E_\gamma \rangle$ in Table V-2 the neutron emission can be neglected, since $B_n = 5.62$ MeV.

In order to evaluate the ratio $X_{\frac{1}{2}-}/X_{\frac{3}{2}-}$ as a function of energy, we must have information on the absolute values and energy dependences of $T_{\lambda\gamma}(\frac{1}{2}, -, E)$ and $T_{\lambda\gamma}(\frac{3}{2}, -, E)$ [see Eq. (V-47]. Although the energy dependence of $T_{\lambda\gamma}(J, \pi, E)$ is given by Eq. (V-32), the normalization factor $2\pi\langle\Gamma_{\lambda\gamma}(E')\rangle/\langle D(J', \pi, E')\rangle$ must be reevaluated for the odd-A nucleus ^{239}Pu. This factor is determined for $\frac{1}{2}+$ states with data obtained from resonance neutron capture on even-even target nuclei. The resulting values of $T_{\lambda\gamma}(\frac{1}{2}, +, E)$ for odd-A nuclei are plotted as a solid line in Fig. V-11. The values of $T_{\lambda\gamma}(\frac{3}{2}, +, E)$ are 1.92 times larger than the corresponding $T_{\lambda\gamma}(\frac{1}{2}, +, E)$ values for a spin cutoff parameter $\sigma = 6$. We assume the $T_{\lambda\gamma}$ values are the same for the two parities.

The foregoing parameters give the ratios of $X_{\frac{1}{2}-}/X_{\frac{3}{2}-}$ listed in column 4 of

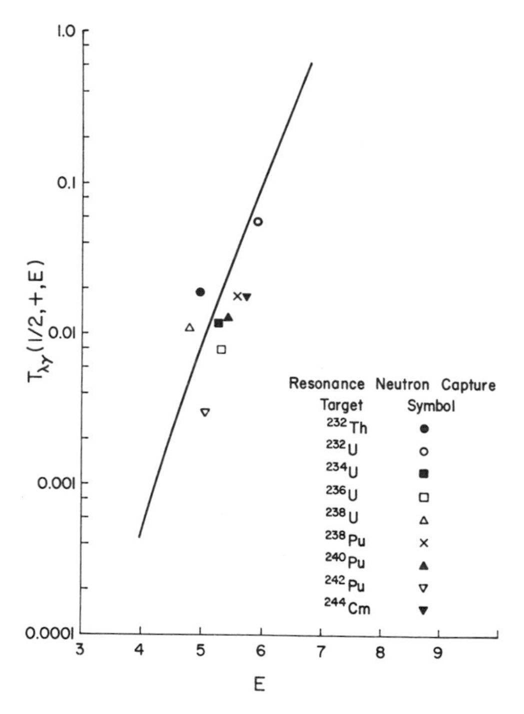

Fig. V-11. Dependence of $T_{\lambda\gamma}(\frac{1}{2}, +, E)$ on excitation energy E for an odd-A heavy nucleus. The normalization is performed with the experimental values of $2\pi\langle\Gamma_{\lambda\gamma}(B_n)\rangle$ divided by $\langle D(\frac{1}{2}, +, B_n)\rangle$ for resonance capture on the several nuclei listed on the figure. The data for resonance capture on ^{232}Th and ^{238}U are given greater weight because these isotopes have been studied more thoroughly. The solid line is calculated with Eq. (V−32) for $a = 30$ MeV^{-1} and $\Delta = 0.55$ MeV.

Table V-2. In the final column are listed the values of b/a calculated from Eq. (V-49) and the values of $X_{\frac{1}{2}-}/X_{\frac{3}{2}-}$ listed in column 4. These calculated values of b/a are plotted in Fig. V-10 along with the experimental values. The important point illustrated in Table V-2 and Fig. V-10 is that with two transition states we are able to reproduce the energy dependence of b/a up to a mean excitation energy of 5.7 MeV. The values of b/a are rather insensitive to changes in the radiation widths. If the values of $\hbar\omega$ for the $K = \frac{1}{2}-$ and $\frac{3}{2}-$ states are the same, several close lying states are required to give the observed change in b/a with energy. If the $K = \frac{1}{2}-$ state has the lowest energy, $A > \frac{1}{2}$ and b/a is positive. As a $K = \frac{3}{2}-$ channel opens up, A approaches $\frac{1}{2}$ and b/a approaches zero. However, for b/a to become negative a second $K = \frac{3}{2}-$ channel is required. These channels would have to be very close in energy to account for the experimental observations in this way. Recently it has been found that b/a becomes positive again at about 6 MeV (Soldatov *et al.*, 1970). This second change in the sign of b/a for a photon energy larger than the neutron binding energy has been confirmed by anisotropy measurements for neutron-induced fission of ^{238}Pu. Hence, a more reasonable explanation of the observed variation in b/a over the larger energy interval is in terms of the double-humped fission barrier. In particular, the structure in b/a is due to a resonance associated with a $K = \frac{3}{2}-$ state in the second minimum between the two barriers at an energy near the minimum value of b/a. The experimental cross section data are also consistent with such a resonance.

b. Dipole Photofission of Targets with Spin of $\frac{5}{2}$

As mentioned previously, photofission of odd-mass targets with sizeable spins leads to nearly isotropic fragment angular distributions. In the following discussion we will show in some detail the theoretical expectations for a target spin of $\frac{5}{2}$. These calculations then apply to targets like ^{233}U, ^{237}Np, ^{241}Am, and ^{243}Am which have ground state spins of $\frac{5}{2}$. Dipole absorption leads to compound states $(J, \pm M)$ of $(\frac{3}{2}, \pm\frac{1}{2})$, $(\frac{3}{2}, \pm\frac{3}{2})$, $(\frac{5}{2}, \pm\frac{1}{2})$, $(\frac{5}{2}, \pm\frac{3}{2})$, $(\frac{5}{2}, \pm\frac{5}{2})$, $(\frac{7}{2}, \pm\frac{1}{2})$, $(\frac{7}{2}, \pm\frac{3}{2})$, $(\frac{7}{2}, \pm\frac{5}{2})$, and $(\frac{7}{2}, \pm\frac{7}{2})$. The relative probability of formation of each of these compound states and the appropriate Clebsch–Gordan coefficients and their values are given in Table V-3. The corresponding fission-fragment angular distributions for a transition state $K = \frac{1}{2}$, $W^J_{\pm M, \pm\frac{1}{2}}(\theta)$, are also given in Table V-3. The theoretical dipole photofission-fragment angular distribution for a positive parity target is given by

$$W_{K=\frac{1}{2}}(\theta) = \frac{\displaystyle\sum_{J,M} P(J, \pm M)\left\{\frac{2T_{\lambda f}(J, -, \frac{1}{2}, E)}{T_{\lambda\gamma}(J, -, E) + 2T_{\lambda f}(J, -\frac{1}{2}, E)}\right\} W^J_{\pm M, \pm\frac{1}{2}}}{\displaystyle\sum_{J,M} P(J, \pm M)\left\{\frac{2T_{\lambda f}(J, -, \frac{1}{2}, E)}{T_{\lambda\gamma}(J, -, E) + 2T_{\lambda f}(J, -, \frac{1}{2}, E)}\right\}} \tag{V-51}$$

TABLE V-3

Dipole Photon Absorption by a Target with Spin $\frac{5}{2}$; Calculations of $P(J, \pm M)$ and $W^J_{\pm M, \pm\frac{1}{2}}(\theta)$

J	$C(I_0, l, J; \mu, l_m, M)$	$\lvert C^{I_0, l, J}_{\mu, l_m, M}\rvert^2$		$(J, \pm M)$	$P(J, \pm M)$	$W^J_{\pm M, \pm\frac{1}{2}}(\theta)$
1.5	C(2.5, 1.0, 1.5; 1.5, −1.0, 0.5)	0.400000				
	C(2.5, 1.0, 1.5; −0.5, 1.0, 0.5)	0.200000	1.200000	$(\frac{3}{2}, \pm\frac{1}{2})$	0.100000	$\frac{1}{4}(1+3\cos^2\theta)$
	C(2.5, 1.0, 1.5; 0.5, −1.0, −0.5)	0.200000				
	C(2.5, 1.0, 1.5; −1.5, 1.0, −0.5)	0.400000				
	C(2.5, 1.0, 1.5; 2.5, −1.0, 1.5)	0.666667				
	C(2.5, 1.0, 1.5; 0.5, 1.0, 1.5)	0.066667	1.466668	$(\frac{3}{2}, \pm\frac{3}{2})$	0.122222	$\frac{3}{4}(1-\cos^2\theta)$
	C(2.5, 1.0, 1.5; −0.5, −1.0, −1.5)	0.066667				
	C(2.5, 1.0, 1.5; −2.5, 1.0, −1.5)	0.666667				
2.5	C(2.5, 1.0, 2.5; 1.5, −1.0, 0.5)	0.457143				
	C(2.5, 1.0, 2.5; −0.5, 1.0, 0.5)	0.514286	1.942858	$(\frac{5}{2}, \pm\frac{1}{2})$	0.161905	$\frac{3}{8}(1-2\cos^2\theta+5\cos^4\theta)$
	C(2.5, 1.0, 2.5; 0.5, −1.0, −0.5)	0.514286				
	C(2.5, 1.0, 2.5; −1.5, 1.0, −0.5)	0.457143				
	C(2.5, 1.0, 2.5; 2.5, −1.0, 1.5)	0.285714				
	C(2.5, 1.0, 2.5; 0.5, 1.0, 1.5)	0.457143	1.485714	$(\frac{5}{2}, \pm\frac{3}{2})$	0.123810	$\frac{3}{16}(1+14\cos^2\theta-15\cos^4\theta)$
	C(2.5, 1.0, 2.5; −0.5, −1.0, −1.5)	0.457143				
	C(2.5, 1.0, 2.5; −2.5, 1.0, −1.5)	0.285714				
	C(2.5, 1.0, 2.5; 1.5, 1.0, 2.5)	0.285714	0.571428	$(\frac{5}{2}, \pm\frac{5}{2})$	0.047619	$\frac{15}{16}(1-2\cos^2\theta+\cos^4\theta)$
	C(2.5, 1.0, 2.5; −1.5, −1.0, −2.5)	0.285714				
3.5	C(2.5, 1.0, 3.5; 1.5, −1.0, 0.5)	0.142857				
	C(2.5, 1.0, 3.5; −0.5, 1.0, 0.5)	0.285714	0.857142	$(\frac{7}{2}, \pm\frac{1}{2})$	0.071429	$\frac{1}{32}(9+45\cos^2\theta-165\cos^4\theta+175\cos^6\theta)$
	C(2.5, 1.0, 3.5; 0.5, −1.0, −0.5)	0.285714				
	C(2.5, 1.0, 3.5; −1.5, 1.0, −0.5)	0.142857				
	C(2.5, 1.0, 3.5; 2.5, −1.0, 1.5)	0.047619				
	C(2.5, 1.0, 3.5; 0.5, 1.0, 1.5)	0.476190	1.047618	$(\frac{7}{2}, \pm\frac{3}{2})$	0.087302	$\frac{15}{32}(1-7\cos^2\theta+27\cos^4\theta-21\cos^6\theta)$
	C(2.5, 1.0, 3.5; −0.5, −1.0, −1.5)	0.476190				
	C(2.5, 1.0, 3.5; −2.5, 1.0, −1.5)	0.047619				
	C(2.5, 1.0, 3.5; 1.5, 1.0, 2.5)	0.714286	1.428572	$(\frac{7}{2}, \pm\frac{5}{2})$	0.119048	$\frac{5}{32}(1+33\cos^2\theta-69\cos^4\theta+35\cos^6\theta)$
	C(2.5, 1.0, 3.5; −1.5, −1.0, −2.5)	0.714286				
	C(2.5, 1.0, 3.5; 2.5, 1.0, 3.5)	1.000000	2.000000	$(\frac{7}{2}, \pm\frac{7}{2})$	0.166667	$\frac{35}{32}(1-3\cos^2\theta+3\cos^4\theta-\cos^6\theta)$
	C(2.5, 1.0, 3.5; −2.5, −1.0, −3.5)	1.000000				

where the two types of transmission coefficients are defined by Eqs. (V-25) and (V-32). Equation (V-51) is valid at energies where neutron emission is negligible, otherwise the $T_{\lambda n}$ values must be included also. Furthermore, it is assumed in Eq. (V-51) that a single transition level $K = \frac{1}{2}$ is contributing to fission.

In the limiting case where $T_{\lambda f}(J, -, \frac{1}{2}, E) \gg T_{\lambda\gamma}(J, -, E)$, and the $T_{\lambda f}(J, -, \frac{1}{2}, E)$ values are independent of J (i.e., the fission transmission coefficients of the rotational members of a particular band are equal), Eq. (V-51) reduces to

$$W_{K=\frac{1}{2}}(\theta) = \sum_{J,M} P(J, \pm M) W^{J}_{\pm M, \pm\frac{1}{2}}(\theta) = 0.4857 + 0.0214 \sin^2\theta \quad \text{(V-52)}$$

Equation (V-52) gives a ratio of $b/a = 0.044$ and $W(90°)/W(0°) = 1.044$. Hence, with the foregoing special assumption for a transition state with $K = \frac{1}{2}$, dipole photofission of a target with spin of $\frac{5}{2}$ is predicted to give a 4% anisotropy of the fission fragments. The individual values of $P(J, \pm M)$ and $W^{J}_{\pm M, \pm\frac{1}{2}}$ necessary for the evaluation of the coefficients a and b in Eq. (V-52) are listed in Table V-3.

Theoretical photofission-fragment angular distributions for a level with $K = \frac{3}{2}, \frac{5}{2}$, or $\frac{7}{2}$ may be written in a form analogous to Eq. (V-51). If again the γ ray transmission coefficients are assumed negligible, and the values of $T_{\lambda f}(J, -, K, E)$ in a particular K band are assumed to be independent of J, the following angular distributions result for $K = \frac{3}{2}, \frac{5}{2}$, and $\frac{7}{2}$ levels, respectively.

$$W_{K=\frac{3}{2}}(\theta) = 0.483 + 0.026 \sin^2\theta, \qquad b/a = 0.054 \quad \text{(V-53)}$$

$$W_{K=\frac{5}{2}}(\theta) = 0.398 + 0.153 \sin^2\theta, \qquad b/a = 0.385 \quad \text{(V-54)}$$

$$W_{K=\frac{7}{2}}(\theta) = 0.750 - 0.375 \sin^2\theta, \qquad b/a = -0.500 \quad \text{(V-55)}$$

Equations (V-54) and (V-55) are quite anisotropic distributions of the opposite sense. The parameters necessary for the calculation of Eqs. (V-53) to (V-55) are tabulated in Table V-4.

If an equal number of transition states with K values of $\frac{1}{2}, \frac{3}{2}, \frac{5}{2}$, and $\frac{7}{2}$ are degenerate in energy in the transition nucleus, then in the approximation of Eqs. (V-52) to (V-55) an isotropic fission-fragment angular distribution results. In this calculation we must include the appropriate weighting factors, which take into account that the $J = \frac{7}{2}$ compound states decay by fission through four transition states ($K = \frac{7}{2}, \frac{5}{2}, \frac{3}{2}$, and $\frac{1}{2}$), $J = \frac{5}{2}$ compound states through three transition states ($K = \frac{5}{2}, \frac{3}{2}$, and $\frac{1}{2}$), $J = \frac{3}{2}$ compound states through two transition states ($K = \frac{3}{2}$ and $\frac{1}{2}$), and $J = \frac{1}{2}$ compound states through one transition state ($K = \frac{1}{2}$).

At energies in excess of the fission barrier, statistical theory [see Eq. (VI-5)] predicts approximately equal numbers of states with $K = \frac{1}{2}, \frac{3}{2}, \frac{5}{2}$, and $\frac{7}{2}$. Hence, isotropy is predicted for the dipole photofission angular distributions at these energies. Experiments with bremsstrahlung produced with electrons of energies

TABLE V-4

Dipole Photon Absorption by a Target with Spin of $\frac{5}{2}$; Tabulation of $P(J, \pm M)$, $W^J_{\pm M, \pm \frac{3}{2}}(\theta)$, $W^J_{\pm M, \pm \frac{5}{2}}(\theta)$, and $W^J_{\pm M, \pm \frac{7}{2}}(\theta)$

$(J, \pm M)$	$P(J, \pm M)$	$W^J_{\pm M, \pm \frac{3}{2}}(\theta)$	$W^J_{\pm M, \pm \frac{5}{2}}(\theta)$	$W^J_{\pm M, \pm \frac{7}{2}}(\theta)$
$(\frac{3}{2}, \pm\frac{1}{2})$	0.100000	$\frac{3}{4}(1-\cos^2\theta)$		
$(\frac{3}{2}, \pm\frac{3}{2})$	0.122222	$\frac{1}{4}(1+3\cos^2\theta)$		
$(\frac{5}{2}, \pm\frac{1}{2})$	0.161905	$\frac{3}{16}(1+14\cos^2\theta-15\cos^4\theta)$	$\frac{15}{16}(1-2\cos^2\theta+\cos^4\theta)$	
$(\frac{5}{2}, \pm\frac{3}{2})$	0.123810	$\frac{3}{32}(9-38\cos^2\theta+45\cos^4\theta)$	$\frac{15}{32}(1+2\cos^2\theta-3\cos^4\theta)$	
$(\frac{5}{2}, \pm\frac{5}{2})$	0.047619	$\frac{15}{32}(1+2\cos^2\theta-3\cos^4\theta)$	$\frac{3}{32}(1+10\cos^2\theta+5\cos^4\theta)$	
$(\frac{7}{2}, \pm\frac{1}{2})$	0.071429	$\frac{15}{32}(1-7\cos^2\theta+27\cos^4\theta-21\cos^6\theta)$	$\frac{5}{32}(1+33\cos^2\theta-69\cos^4\theta+35\cos^6\theta)$	$\frac{35}{32}(1-3\cos^2\theta+3\cos^4\theta-\cos^6\theta)$
$(\frac{7}{2}, \pm\frac{3}{2})$	0.087302	$\frac{1}{32}(1+261\cos^2\theta-765\cos^4\theta+567\cos^6\theta)$	$\frac{3}{32}(9-59\cos^2\theta+155\cos^4\theta-105\cos^6\theta)$	$\frac{21}{32}(1+\cos^2\theta-5\cos^4\theta+3\cos^6\theta)$
$(\frac{7}{2}, \pm\frac{5}{2})$	0.119048	$\frac{3}{32}(9-59\cos^2\theta+155\cos^4\theta-105\cos^6\theta)$	$\frac{1}{32}(25-51\cos^2\theta-85\cos^4\theta+175\cos^6\theta)$	$\frac{7}{32}(1+9\cos^2\theta-5\cos^4\theta-5\cos^6\theta)$
$(\frac{7}{2}, \pm\frac{7}{2})$	0.166667	$\frac{21}{32}(1+\cos^2\theta-5\cos^4\theta+3\cos^6\theta)$	$\frac{7}{32}(1+9\cos^2\theta-5\cos^4\theta-5\cos^6\theta)$	$\frac{1}{32}(1+21\cos^2\theta+35\cos^4\theta+7\cos^6\theta)$

greater than 8 MeV have shown essentially isotropic distributions (Baerg *et al.*, 1959) within rather large experimental errors. No experiments with odd-mass targets of spin $\frac{5}{2}$ have been reported with very low energies. Theoretically, sizeable anisotropies may result if a single particular K state, e.g., $K = \frac{5}{2}$ or $K = \frac{7}{2}$, is responsible for a major fraction of the fission. Hence, low energy photofission of targets like ^{233}U, ^{237}Np, ^{241}Am, and ^{243}Am is a sensitive method of determining the quantum numbers of the lowest lying transition state of such nuclei. Photoabsorption strongly excites high spin states for target nuclei of large spin, $J = I_0 + 1, I_0$, and $I_0 - 1$; hence, all transition states with $K \leqslant J$ may in principle be excited. The parity of the transition states excited by dipole photoabsorption is always opposite to that of the target nucleus.

c. Dipole Photofission of Targets with Spin of $\frac{7}{2}$

The calculations of the theoretical angular distributions for dipole photo-fission of targets, such as ^{235}U, with spin $\frac{7}{2}$ are similar to those described in the previous section for targets with spin $\frac{5}{2}$. The theoretical angular distributions for the various K states under the same assumptions as Eq. (V-52) are:

$$W_{K=\frac{1}{2}}(\theta) = 0.492 + 0.012 \sin^2\theta \tag{V-56}$$

$$W_{K=\frac{3}{2}}(\theta) = 0.491 + 0.013 \sin^2\theta \tag{V-57}$$

$$W_{K=\frac{5}{2}}(\theta) = 0.489 + 0.016 \sin^2\theta \tag{V-58}$$

$$W_{K=\frac{7}{2}}(\theta) = 0.398 + 0.153 \sin^2\theta \tag{V-59}$$

$$W_{K=\frac{7}{2}}(\theta) = 0.750 - 0.375 \sin^2\theta \tag{V-60}$$

C. Neutron-Induced Fission

1. Odd-Mass Transition Nuclei (Formed by Neutron Fission of Even-Even Targets)

Neutron-induced fission of even-even nuclei provides a sensitive probe of the level structure of the intermediate odd-mass transition nuclei. To reach the fission thresholds of several even-even target nuclei, neutrons with energy of a few hundred keV are required. Hence, it is possible with monoenergetic neutron beams to investigate fission-fragment angular distributions and cross sections for odd-mass transition nuclei in the energy region of the fission barrier and below.

The first interpretation of a fission-fragment angular distribution in terms of a particular K state of the transition nucleus was made by Wilets and Chase

(1956). They proposed an explanation of the sideways peaked angular distribution of fission fragments from neutron-induced fission of ^{232}Th with 1.6 MeV neutrons. The more detailed early work of Lampherc (1962) must also be mentioned. He examined the fission-fragment anisotropy $\sigma_f(0°)/\sigma_f(90°)$ from the ^{234}U(n, f) reaction as a function of neutron energy and interpreted the observed variation in the anisotropy in terms of new fission channels which were composed of different K bands.

For neutron-induced fission of an even-even target nucleus with zero spin, $I_0 = 0$, Eq. (V-3) gives explicitly the fission-fragment angular distribution

$$W^J_{\pm\frac{1}{2}, \pm K}(\theta) = \frac{(2J+1)}{4}\{|d^J_{\frac{1}{2}, K}(\theta)|^2 + |d^J_{-\frac{1}{2}, K}(\theta)|^2\} \qquad \text{(V-61)}$$

Since the target spin is zero and the neutron spin is $\frac{1}{2}$, only two values of M are allowed, $M = \pm\frac{1}{2}$, and each has an equal probability of formation. Equation (V-61) has already been simplified by the symmetry relations $|d^J_{\frac{1}{2}, K}(\theta)|^2 = |d^J_{-\frac{1}{2}, -K}(\theta)|^2$ and $|d^J_{-\frac{1}{2}, K}(\theta)|^2 = |d^J_{\frac{1}{2}, -K}(\theta)|^2$, and is normalized so that

$$\int_0^\pi W^J_{\pm\frac{1}{2}, \pm K}(\theta) \sin\theta \, d\theta = 1$$

As already discussed, information about the low lying levels of the transition nucleus may be obtained from fission-fragment angular distributions. For neutron-induced fission of even-even targets the theoretical angular distributions of the fragments for transition states of different J and K as calculated by Eq. (V-61) are shown in Figs. V-12 and V-13. It is interesting to note that rotational members of the $K = \frac{1}{2}$ band give fragment peaking along the neutron beam direction, while all other K bands give no yield of fragments at 0° to the neutron beam direction. A casual examination of the curves in Figs. V-12 and V-13 shows that the different transition states give, in general, quite different fragment angular distributions. While the anisotropy $\sigma_f(0°)/\sigma_f(90°)$ may not give a sensitive characterization of the state, in many cases a full fragment angular distribution does give a sensitive identification of the transition state.

Experimentally the energy dependence of the fission-fragment cross section and angular distribution is measured near the barrier where one or only a few fission channels are contributing. A transition state is characterized by J, K, and π (parity) as well as by its barrier energy $B(J, \pi, K)$ and a characteristic energy $\hbar\omega(J, \pi, K)$ which defines the curvature of the barrier. The cross section gives a measure of the number of open or partially open fission channels. In the analysis of neutron fission to be presented here, we use both the cross sections and angular distributions to deduce information about the quantum numbers of the transitions states and their characteristic energies (Vandenbosch, 1967; Behkami *et al.*, 1968a). Whereas it is sometimes relatively easy to fit the energy dependence of the cross sections or angular distributions

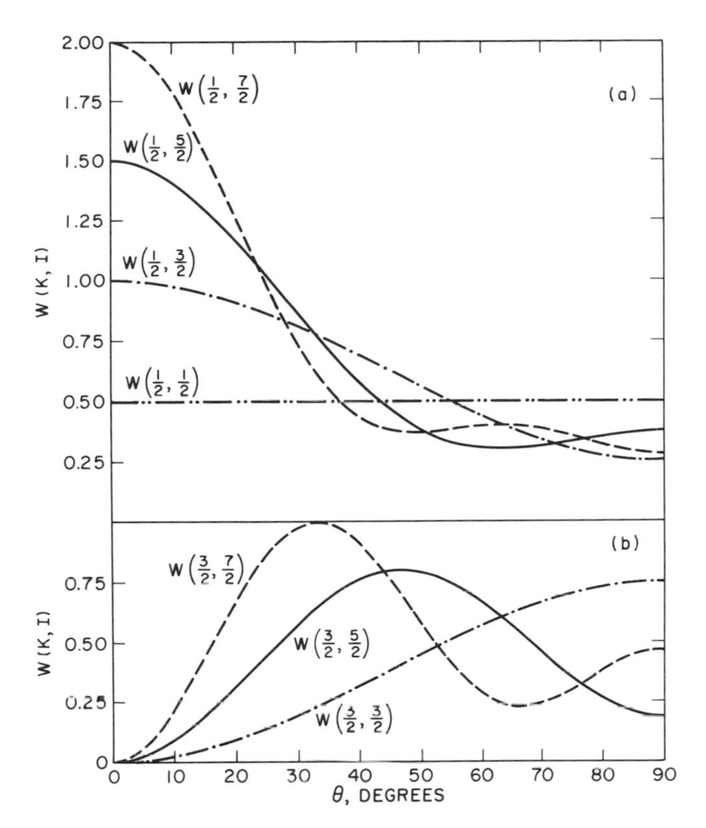

Fig. V-12. Theoretical fission-fragment angular distributions for neutron fission of even-even targets calculated with Eq. (V-61). The axis of quantization is along the beam direction and both values of M, $+\frac{1}{2}$ and $-\frac{1}{2}$, are included in the results. (a) Fission through states in a band with $K = \frac{1}{2}$ and J values of $\frac{1}{2}$, $\frac{3}{2}$, $\frac{5}{2}$, and $\frac{7}{2}$. Each curve is normalized such that

$$\int_0^{\pi} W^J_{+\frac{1}{2}, \pm K}(\theta) \sin\theta \, d\theta = 1$$

(b) Fission through states in a band with $K = \frac{3}{2}$ and J values of $\frac{3}{2}$, $\frac{5}{2}$, and $\frac{7}{2}$. Each curve is normalized as, in (a).

separately, fits to both types of data give a much better evaluation of the transition state parameters.

Capture of neutrons with energies of hundreds of keV gives compound nuclei with several J values. With a neutron spin s, orbital angular momentum l, and target spin $I_0 = 0$, J has values $|l \pm s|$. The relative probability of each J value is determined from optical model neutron transmission coefficients. In the calculations to be described we assume compound nucleus formation and decay; furthermore, we explicitly assume that the partial widths for different entrance and exit channels are not correlated. In the following paragraphs we

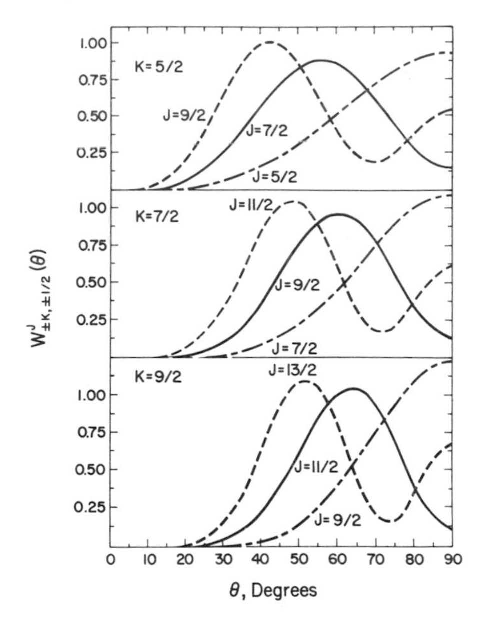

Fig. V-13. Continuation of Fig. V-12 for $K = \frac{5}{2}, \frac{7}{2}$, and $\frac{9}{2}$.

describe the calculation of the cross section for a particular exit channel. The total cross section for a particular process is obtained by summing the cross sections over the individual exit channels for that process. The total cross section results from summing the cross section for the different types of exit channels, namely, fission, neutron emission, and γ-ray emission.

For the neutron bombardment of an even-even nucleus, the Breit–Wigner formula for the cross section for partial wave l, entrance channel α, and exit channel α' near an isolated resonance λ of total angular momentum J gives

$$\sigma_{Jl}^{\alpha\alpha'} = \frac{(2J+1)}{2} \frac{\pi}{k^2} \frac{\Gamma_{\lambda Jl}^{(\alpha)} \Gamma_{\lambda Jl}^{(\alpha')}}{(E_\lambda - E)^2 + (\Gamma_{\lambda J}/2)^2} \tag{V-62}$$

where E_λ is the resonance energy, $k/2\pi$ is the wave number of the incident neutron, $\Gamma_{\lambda Jl}^{(\alpha)}$ is the partial width for entrance channel α, $\Gamma_{\lambda Jl}^{(\alpha')}$ is the partial width for exit channel α', and $\Gamma_{\lambda J}$ is the total width of the resonance. The factor of 2 in the denominator arises from the product $(2I_0+1)(2s+1)$ where the target spin $I_0 = 0$ and the neutron spin $s = \frac{1}{2}$.

The peak cross section of a resonance at $E = E_\lambda$ is

$$\sigma_{Jl}^{\alpha\alpha'} = \frac{2(2J+1)\pi}{k^2} \frac{\Gamma_{\lambda Jl}^{(\alpha)} \Gamma_{\lambda Jl}^{(\alpha')}}{\Gamma_{\lambda J}^2} \tag{V-63}$$

The integral of the cross section over the resonance is given by

$$\int_{-\infty}^{E-E_\lambda=\infty} \sigma_{Jl}^{\alpha\alpha'}\, dE = \frac{2(2J+1)\pi}{k^2} \frac{\Gamma_{\lambda Jl}^{(\alpha)}\Gamma_{\lambda Jl}^{(\alpha')}}{\Gamma_{\lambda J}^2} \int_{-\infty}^{E-E_\lambda=\infty} \frac{(\Gamma_{\lambda J}/2)^2}{(E_\lambda - E)^2 + (\Gamma_{\lambda J}/2)^2}\, dE \tag{V-64}$$

where the integral on the right-hand side is equal to $\pi\Gamma_{\lambda J}/2$. Therefore, the cross section integrated over one resonance is expressed by

$$\int_{-\infty}^{E-E_\lambda=\infty} \sigma_{Jl}^{\alpha\alpha'}\, dE = \frac{(2J+1)\pi^2}{k^2} \frac{\Gamma_{\lambda Jl}^{(\alpha)}\Gamma_{\lambda Jl}^{(\alpha')}}{\Gamma_{\lambda J}} \tag{V-65}$$

If there are n resonances of a given total angular momentum and parity in an energy interval ΔE, then the average spacing between levels in the compound nucleus $\langle D_{\lambda J}\rangle = \Delta E/n$ and the average cross section for n resonances in the energy interval ΔE is obtained by multiplying each side of Eq. (V-65) by $n/\Delta E$.

$$\langle \sigma_{Jl}^{\alpha\alpha'} \rangle = \frac{(2J+1)\pi^2}{\langle D_{\lambda J}\rangle k^2} \left\langle \frac{\Gamma_{\lambda Jl}^{(\alpha)}\Gamma_{\lambda Jl}^{(\alpha')}}{\Gamma_{\lambda J}} \right\rangle \tag{V-66}$$

Experimentally the energy resolution of the neutron beam is much larger than $\langle D_{\lambda J}\rangle$; hence, the measured cross section is to be compared with a theoretical cross section calculated by averaging over many resonances. Since the average of a ratio is not, in general, equal to the ratio of the averages of the individual quantities, we cannot substitute

$$\frac{\langle\Gamma_{\lambda Jl}^{(\alpha)}\rangle\langle\Gamma_{\lambda Jl}^{(\alpha')}\rangle}{\langle\Gamma_{\lambda J}\rangle} \quad \text{for} \quad \left\langle \frac{\Gamma_{\lambda Jl}^{(\alpha)}\Gamma_{\lambda Jl}^{(\alpha')}}{\Gamma_{\lambda J}} \right\rangle$$

A quantity $S_{\alpha\alpha'}$ (the level width fluctuation correction factor) is defined as (Moldauer, 1961)

$$S_{\alpha\alpha'} = \frac{\left\langle \dfrac{\Gamma_{\lambda Jl}^{(\alpha)}\Gamma_{\lambda Jl}^{(\alpha')}}{\Gamma_{\lambda J}} \right\rangle}{\dfrac{\langle\Gamma_{\lambda Jl}^{(\alpha)}\rangle\langle\Gamma_{\lambda Jl}^{(\alpha')}\rangle}{\langle\Gamma_{\lambda J}\rangle}} \tag{V-67}$$

and in substitution of Eq. (V-67) into Eq. (V-66), we obtain

$$\langle \sigma_{Jl}^{\alpha\alpha'} \rangle = \frac{\pi^2}{k^2}(2J+1)\frac{\langle\Gamma_{\lambda Jl}^{(\alpha)}\rangle\langle\Gamma_{\lambda Jl}^{(\alpha')}\rangle}{\langle D_{\lambda J}\rangle\langle\Gamma_{\lambda J}\rangle} S_{\alpha\alpha'} \tag{V-68}$$

Use of Eq. (V-68) along with the appropriate summations allows us to calculate partial fission cross sections as well as other partial cross sections, such as the neutron and γ-ray emission cross sections.

The partial width for fission through an exit channel of given (J, π, K) has been defined by Eq. (V-24). The transmission coefficients for fission through an inverted parabolic barrier are defined by Eq. (V-25). The compound nucleus excitation energy E is given by the sum of the neutron binding energy B_n and the neutron kinetic energy E_n. The fission barrier heights of the rotational members of a particular K band in the transition nucleus are calculated with the expression

$$B_f(J, \pi, K) = B_0(\pi, K) + \frac{\hbar^2}{2\mathscr{I}_\perp}\{J(J+1) - \alpha(-1)^{J+\frac{1}{2}}(J+\tfrac{1}{2})\,\delta_{K,\frac{1}{2}}\} \quad \text{(V-69)}$$

where $B_0(\pi, K)$ is a constant for a particular (π, K) rotational band, $\mathscr{I}_\perp$ is an effective moment of intertia, α is the decoupling constant for $K = \frac{1}{2}$ bands, and $\delta_{K,\frac{1}{2}}$ is the Kronecker delta. In most calculations, we assume $\hbar^2/2\mathscr{I}_\perp$ equal to 5 keV and $\alpha = 1$. The effect of the variation of these parameters on the partial fission width will be discussed later in this section.

The partial widths for neutron entrance and exit channels in Eq. (V-69) are replaced with optical model neutron transmission coefficients using a relation analogous to Eq. (V-24)

$$\langle \Gamma_{\lambda n}(J, \pi, l, E)\rangle = \frac{\langle D_{\lambda J \pi}\rangle}{2\pi}\, T_{\lambda n}(J, \pi, l, E) \quad \text{(V-70)}$$

Unless otherwise stated we use the neutron transmission coefficients from the Perey–Buck optical model (Auerbach and Perey, 1962).

The γ-emission (radiation) channels are treated assuming dipole γ-ray emission and a statistical model of the level density. The energy-dependent γ-ray emission transmission coefficients $T_{\lambda\gamma}(J, \pi, E)$ are defined by Eq. (V-32). The functions required for the calculation of $T_{\lambda\gamma}(J, \pi, E)$ are given in Eqs. (V-29) and (V-31). In actual calculations, we have used the parameter values $a = 30$ MeV^{-1}, $\Delta = 0.55$ MeV, $\sigma = 6$, and a value of $2\pi\langle \Gamma_{\lambda\gamma}(B_n)\rangle / \langle D(\frac{1}{2}, \pi, B_n)\rangle$ at the neutron binding energy from Fig. V-11 (neutron capture on even-even targets gives odd-mass compound nuclei).

The denominator of Eq. (V-68) contains the average total width $\langle \Gamma_{\lambda J}\rangle$ of all competing channels for compound spin J. Hence, the calculation of the cross section of a particular channel is similar to a standard Hauser–Feshbach (1952) calculation except that the level width fluctuations are taken into account. In this calculation, as in a Hauser–Feshbach calculation, we assume no interference between different entrance and exit channels; hence, each channel is a well-defined entity whose partial width is in no way correlated to the partial widths of other channels.

By substitution of the transmission coefficients for the fission, neutron, and γ-ray widths as defined by Eqs. (V-24), (V-70), and (V-32), respectively, into Eq. (V-68), we can write the neutron fission cross section of a zero spin target for a particular fission channel (J, π, K) in the following form (Huizenga *et al.*, 1969).

$$\sigma_f(J, \pi, K, E) = \tfrac{1}{2}\pi\lambda\!\!\!^{-2}(2J+1)\{T_{\lambda n}^{l-\frac{1}{2}}(J, \pi, l, E_n) + T_{\lambda n}^{l+\frac{1}{2}}(J, \pi, l, E_n)\} R \tag{V-71}$$

where

$$R = \frac{2T_{\lambda f}(J, \pi, K, E)\, S_{\alpha\alpha'}}{\sum_i 2T_{\lambda f}(J, \pi, K, E) + \sum_{I'} \sum_j T_{\lambda n}(J, \pi, l, E_n') + T_{\lambda\gamma}(J, \pi, E)}$$

The sum over i represents a sum over all fission transition states with parity π and $K \leqslant J$. The factor of 2 represents $h(K)$, which accounts for the double degeneracy $(\pm K)$ of the half-integral K values. Hence, the cross section $\sigma_f(J, \pi, K, E)$ as given by Eq. (V-71) is for both K projections. The sum over j (where $j = l \pm \frac{1}{2}$) represents a sum of all the neutron transmission coefficients which are allowed in the decay of the compound state J to the residual nucleus with spin I'. The lower limit on this sum is $j = |J - I'|$ and the upper limit $j = |J + I'|$. For a particular l value, there may be two, one, or zero values of the appropriate neutron transmission coefficients in the sum j, depending on whether the conditions on spin and parity are satisfied for both channel spins j, one channel spin, or neither. A sum is also required over all states I' in the residual nucleus which are reached by neutron emission.

The level width fluctuation correction factor $S_{\alpha\alpha'}$ defined by Eq. (V-67) is calculated with the additional assumption that the various partial widths are distributed according to a χ^2 distribution defined by Eq. (IV-3). We assume each of the neutron and fission partial widths has a χ^2 distribution with one degree of freedom (a Porter–Thomas distribution with $\nu = 1$). For $\nu = 1$, $S_{\alpha\alpha'}$ varies from 1 to $\frac{1}{2}$ for $\alpha \neq \alpha'$ and from 1 to 3 for $\alpha = \alpha'$ (i.e., an enhancement of compound elastic scattering). For the radiation channels, it is well known that ν is relatively large.

The initial step in an (n, f) reaction is the formation of a compound nucleus with a particular (J, π) distribution. This compound nucleus decays by fission, neutron emission, and γ-ray emission. Neutron emission leads to states of the target nucleus, γ-ray emission populates states of the compound nucleus, and fission occurs through states in the transition nucleus. Equation (V-71) is used to calculate the partial fission cross section for each member (J, π) of the rotational band built upon a particular state of the transition nucleus (π, K). If the fission cross section of a particular state (J, π, K, E) calculated by Eq. (V-71) is denoted by $\sigma_f(J, \pi, K, E)$, the total fission cross section is given by

$$\sigma_f(E) = \sum_{J, \pi, K} \sigma_f(J, \pi, K, E) \tag{V-72}$$

and the angular distribution of fragments by

$$W(\theta, E) = \sum_{J, \pi, K} \sigma_f(J, \pi, K, E)\, W^J_{\pm\frac{1}{2}, \pm K}(\theta) \tag{V-73}$$

where $W^J_{\pm\frac{1}{2},\pm K}(\theta)$ is defined by (V-61). The quantum number M does not appear in the summations of Eqs. (V-72) and (V-73) because for neutron fission of an even-even target with zero spin only two values of M are possible, $\pm\frac{1}{2}$. These two M values are equally probable and both projections are included already in $\sigma_f(J, \pi, K, E)$ and $W^J_{\pm\frac{1}{2},\pm K}(\theta)$. The summations over K in Eqs. (V-72) and (V-73) are to be performed over only one of the two possible projections of K ($\pm K$), since the equations contain this degeneracy in K.

By using the equations above, it is possible to calculate the fission cross section and angular distribution for any incident neutron energy by specifying the number of accessible states of the transition nucleus and the J, π, K, B_0, and $\hbar\omega$ values associated with each state, the energies and quantum numbers of the states reached by neutron emission, and values of the parameters for γ-ray decay discussed earlier. In actual practice, we search for the best fit to the energy variation of the experimental fission cross section and angular distribution data by varying the number of accessible states of the transition nucleus and the parameters J, π, K, B_0, and $\hbar\omega$ associated with each of these states. In the calculations it is assumed that $\hbar\omega(J, \pi, K, E)$ is a constant for a particular (π, K) band.

Because of the low fission cross sections near threshold, a highly efficient method of measuring fission-fragment angular distributions is usually required. One such method utilizes solid state nuclear track detectors. It is well known that radiation-damaged sites caused by fission fragments entering a number of insulating materials can be enlarged by chemical etching until they can be seen with an optical microscope. By choosing a suitable material, the number of fission events can be recorded uniquely in the presence of a high background of low mass particles.

One detector arrangement used to study fragment angular distributions from neutron-induced fission is illustrated in Fig. V-14. This detector is a polycarbonate resin (Makrofol) of 200-μm thickness. It is arranged in the form

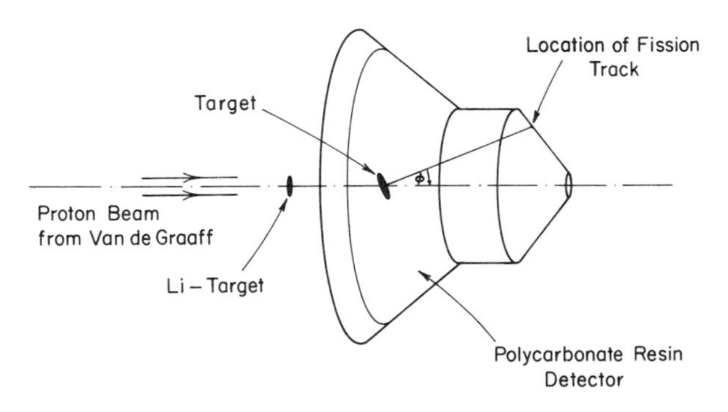

Fig. V-14. Schematic diagram of an experimental arrangement to measure fission-fragment angular distributions with nearly 2π geometry. [After Behkami *et al.* (1968a).]

of a truncated cone at the base, supporting a cylindrical section, which in turn supports a top cone. The top cone is sliced at an angle of 33° so that an elliptical section is at the top near the polar angle of 0°. Some reasons for choosing this geometry are

(a) it guarantees that all fission fragments from a source at the center of the base of the bottom cone will enter the detector material at angles between 20° and 70°, thus ensuring proper track registration;

(b) the symmetry of the detector ensures rapid reading of the data, since all the tracks along a circular ring perpendicular to the axis of symmetry (the beam axis) will correspond to the same angle θ; and

(c) it affords the 2π geometry necessary for measuring angular distributions in the low cross section region near threshold.

After irradiation, the detector material is chemically etched (6 N NaOH for 40 min at 70°C or 50 hr at room temperature) and the resulting fragment "tracks" are viewed with an optical microscope.

In order to read the data we merely count the number of fragment tracks for a given circular ring perpendicular to the symmetry axis (the beam direction) on a conical or cylindrical portion of the detector. The corresponding angle θ which the fragment made with the beam direction is calculated from distance measurements on the known geometrical configuration.

One reaction which may be used to produce monoenergetic neutrons is the ^{7}Li(p, n)^{7}Be reaction. The neutron energy spread from this reaction is mainly determined by the lithium target thickness and the size of the angle subtended by the fissionable target.

Neutron intensities attainable by this reaction depend on several factors and are approximately 2×10^{5} neutrons per second per microampere of protons current per keV of Li target thickness per steradian. The Q value for the ^{7}Li(p, n)^{7}Be reaction is 1.645 MeV, and neutrons up to 2.0 MeV are readily produced with a 4-MeV Van de Graaff. One disadvantage of this reaction is the admixture of a small intensity group of lower energy neutrons produced from the ^{7}Li(p, n)^{7*}Be reaction which leaves ^{7}Be in its first excited state at 477 keV.

Experimental fission-fragment angular distributions for monoenergetic neutron-induced fission of ^{232}Th, ^{234}U, ^{236}U, and ^{242}Pu are shown in Figs. V-15 to V-18.

With the theoretical framework presented in this section we are now in a position to utilize neutron fission cross section and angular distribution data to determine the K quantum number and energies of levels in the transition nucleus. The actual calculations are done in the following manner. One or more states of particular K and π are chosen (the rotational members of the chosen K bands are always included). In addition, a range of values of $B_0(\pi, K)$ and $\hbar\omega(J, \pi, K)$ are chosen with appropriately selected increments in each

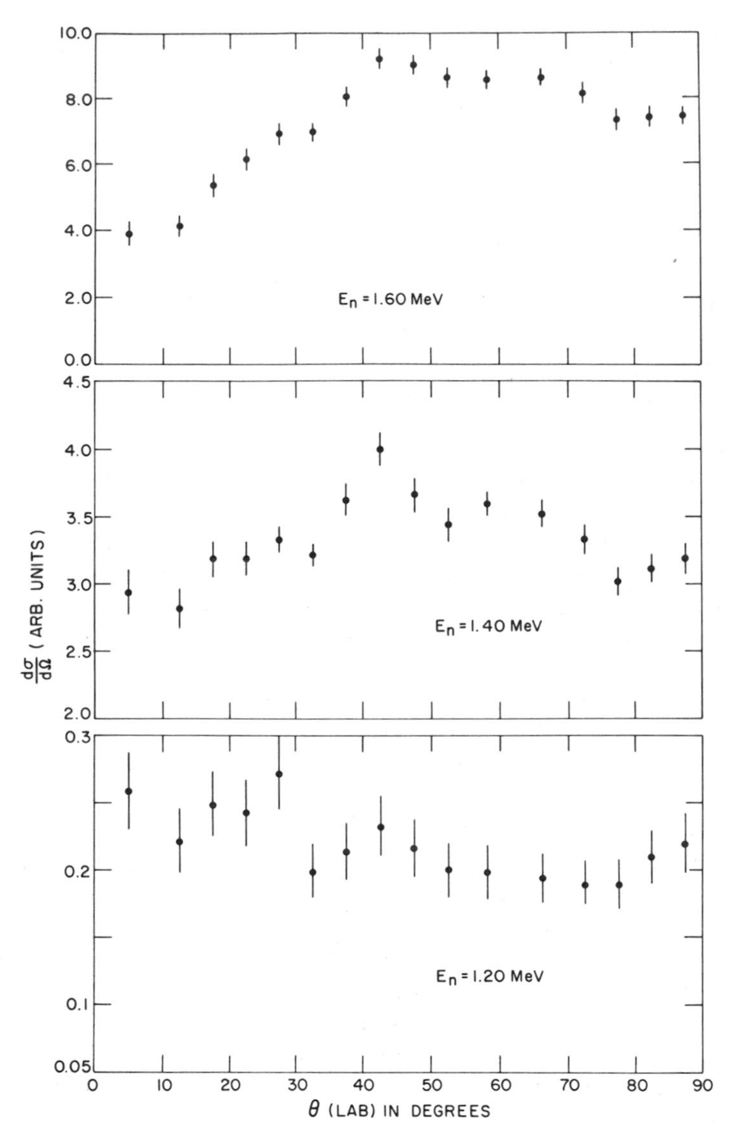

Fig. V-15. Fission-fragment angular distributions for the ^{232}Th(n,f) reaction at three incident neutron energies. The neutron energy resolution is ± 20 keV. [After Behkami *et al.* (1968b).]

quantity. The computer program calculates in succession the theoretical values of $\sigma_f(E)$ and $W(\theta, E)$ for each combination of sets of the parameters $B_0(\pi, K)$ and $\hbar\omega(J, \pi, K)$ and compares these values as a function of energy with the appropriate experimental values. Such a search can involve hundreds of calculations when two or more channels are included. As a guide in the quantitative evaluation of the agreement between the theoretical and experimental fission cross sections and angular distributions, a χ^2 criterion is used. Hence, for a chosen set of (π, K) states the values of $B_0(\pi, K)$ and $\hbar\omega(J, \pi, K)$ which give the best agreement with experiments are determined. Then another set of (π, K) states are chosen and the calculations repeated.

The best fits for the angular distributions from the $^{234}U(n, f)$ reaction at two

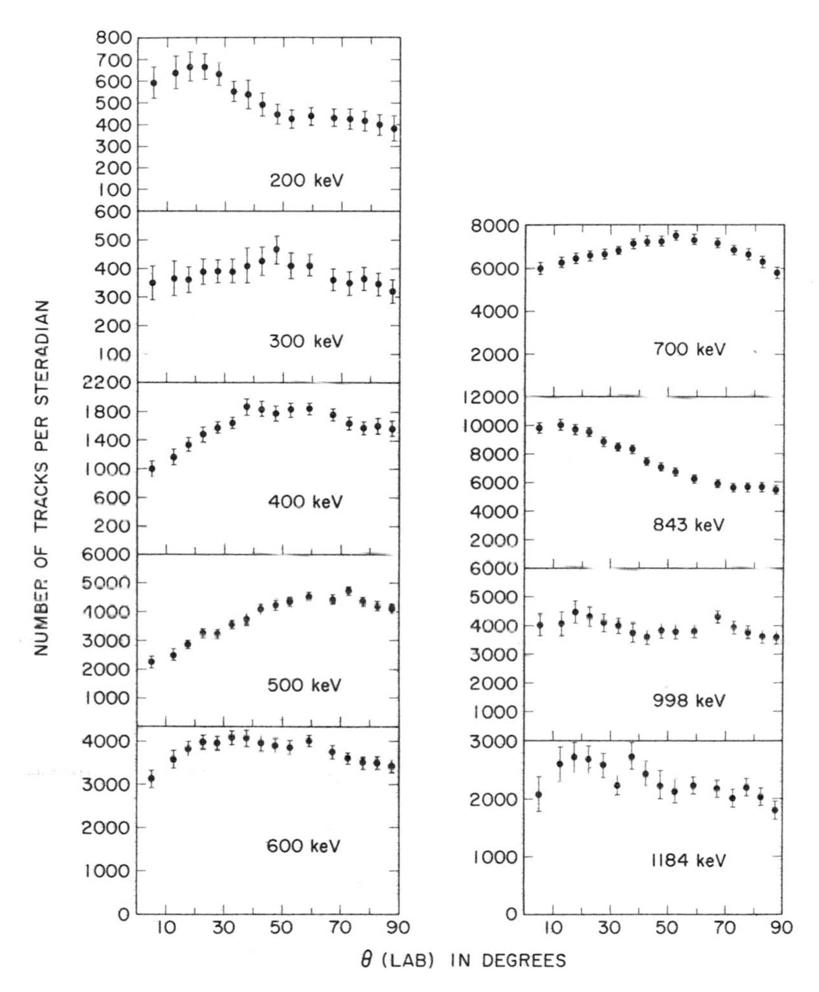

Fig. V-16. Fission-fragment angular distributions for the $^{234}U(n,f)$ reaction at several incident neutron energies. The neutron energy resolution is ± 15 keV. [After Behkami *et al.* (1968a).]

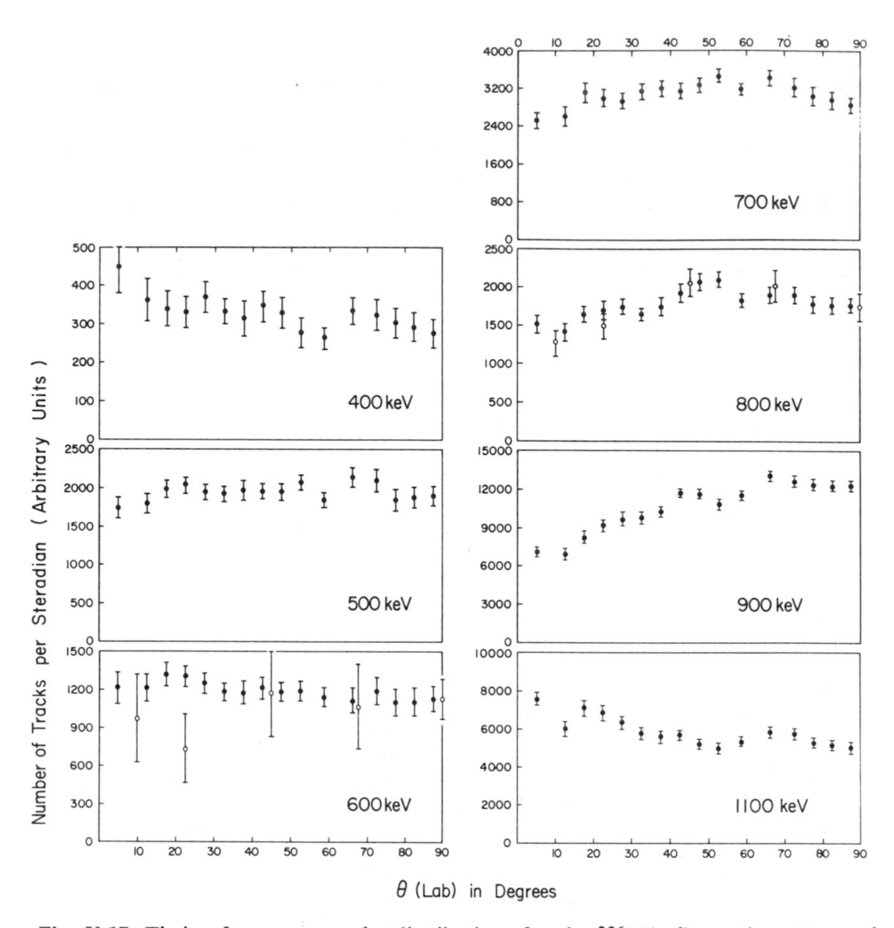

Fig. V-17. Fission-fragment angular distributions for the $^{236}U(n,f)$ reaction at several incident neutron energies. The neutron energy resolution is ± 20 keV. [After Huizenga *et al.* (1969).]

neutron energies, namely, 200 and 300 keV, are illustrated in Fig. V-19. The only acceptable fit to both the cross section and angular distribution data at these energies gives two transition states $(K, \pi) = \frac{1}{2}+$ and $\frac{3}{2}+$. The best values of $B_0(\frac{3}{2}+)$, $B_0(\frac{1}{2}+)$, $\hbar\omega(\frac{3}{2}+)$, and $\hbar\omega(\frac{1}{2}+)$ are 375, 600 (or 550), 275, and 625 (or 550) keV, respectively. Variations of values of $B_0(\frac{1}{2}+)$ and $\hbar\omega(\frac{1}{2}+)$ by 50 and 75 keV do not change χ^2 to any extent. In making this search it is assumed that we should use the minimum number of accessible states of the transition nucleus at any given energy. This assumption, made for simplicity and precision in the determination of the free parameters, means that there may be many other hypotheses involving weakly excited states which will fit the data. The calculation includes the rotational members of the specified K bands. The fission thresholds of the various members of a particular rotational band are related by Eq. (V-69). Hence, values of the rotational constant $\hbar^2/2\mathscr{I}_\perp$ as well as the decoupling constant α for $K = \frac{1}{2}$ bands must be supplied within reasonable limits. The results are rather insensitive to different values of these constants. The low lying levels in the target nucleus ^{234}U reached by neutron emission are well known. The numerical values of the constants used in the foregoing calculations for the γ-ray emission channels were mentioned previously. The results of the calculation are quite insensitive to variation of these parameters by as much as 100%.

As the neutron energy is increased, acceptable fits to all the ^{234}U(n, f) data require additional transition states (Behkami *et al.*, 1968a). For example, the data from the 400- and 500-keV bombardments require a third transition state. The parameters of the first two states ($\frac{3}{2}+$) and ($\frac{1}{2}+$) are kept fixed and the

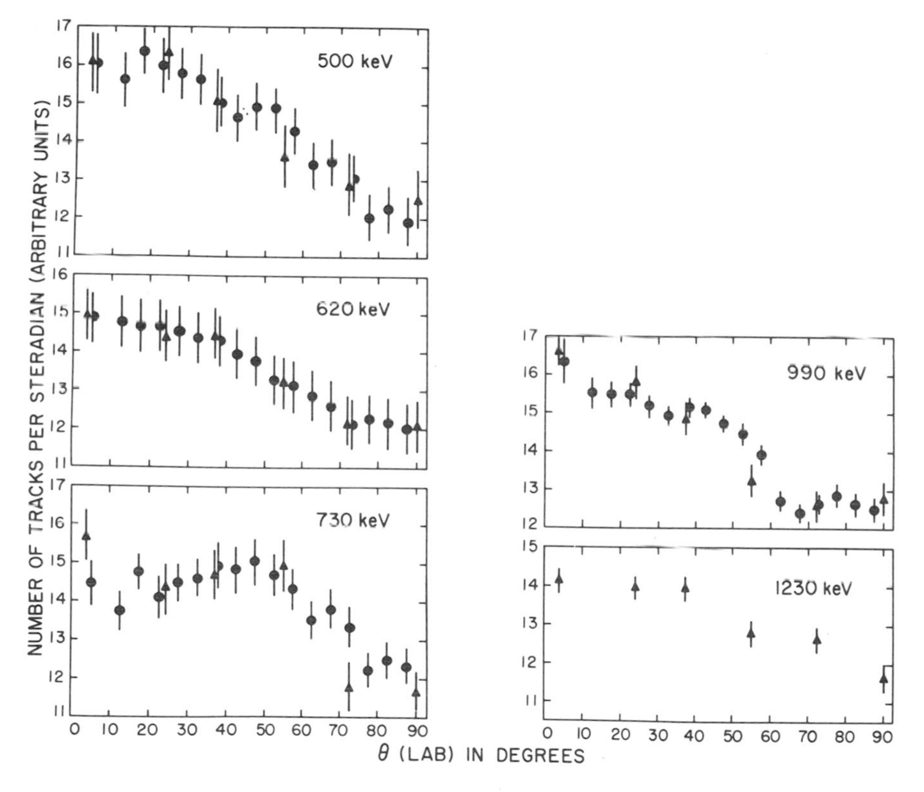

Fig. V-18. Fission-fragment angular distributions for the ^{242}Pu(n,f) reaction at several incident neutron energies. The neutron energy resolution is ± 20 keV. [After Otozai *et al.* (1970).]

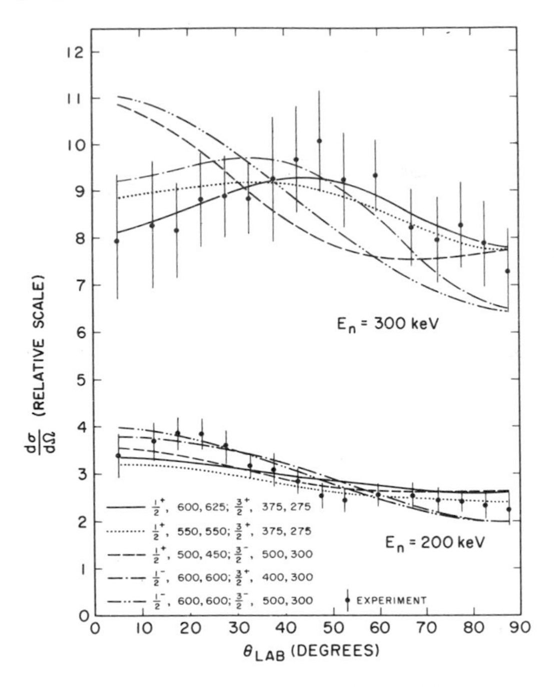

Fig. V-19. Fission-fragment angular distributions for the ^{234}U(n,f) reaction at $E_n = 200$ and 300 keV. The points are the experimental data and the curves represent best fits to the angular distributions and cross sections with two accessible states of the transition nucleus. The transition state parameters for these best fit curves are given in the figure. The notation $\frac{1}{2}+$, 600, 625 means $(K, \pi) = \frac{1}{2}+$, $B_0 = 600$ keV, and $\hbar\omega = 625$ keV. [Adapted from Behkami *et al.* (1968a).]

values of K, π, $B_0(\pi, K)$, and $\hbar\omega(J, \pi, K)$ for the third state are varied. Best fits to the cross sections at 200, 300, 400, and 500 keV and to the angular distributions at 400 and 500 keV are shown in Figs. V-20 and V-21 for four different choices of (K, π) for the third state. Only the $K = \frac{3}{2}-$ state is acceptable.

2. Odd-Mass Transition Nuclei (Two-Humped Barrier Penetration)

The previous calculations have employed a fission barrier with the shape of an inverted parabola. The penetrability through such a barrier can be solved exactly, and this result is given in Eq. (V-25). In this section we present calculations which make use of transmission coefficients $T_{\lambda f}$ for fission through a double-humped potential energy barrier.

The penetrability has been computed exactly for a fission barrier defined in terms of two parabolic peaks connected smoothly with a third parabola forming the intermediate well (Cramer and Nix, 1970). Such a potential is usually specified by the two peak energies and the minimum energy of the intermediate well along with the individual curvature energies $\hbar\omega$ of the three parabolas. For energies below the barrier tops, the exact penetrabilities are very well reproduced by the WKB approximation. In the calculations to be described, we use this approximate (WKB) penetrability $P(E)$ of a two-peaked barrier with maxima A and B as given by (Cramer and Nix, 1970),

$$P(E) = P_A P_B \{[(P_A + P_B)^2/4] \sin^2 \phi(E) + [(P_A P_B + 16)/8]^2 \cos^2 \phi(E)\}^{-1} \tag{V-74}$$

The quantities P_A and P_B are the quasi-classical penetrabilities of the separate parabolic barriers A and B, respectively, at incident energy E, defined previously by Eq. (V-25). The energy-dependent phase $\phi(E)$ is given by the integral

$$\phi(E) = \int_\alpha^\beta \{(2\mu/\hbar^2)[E - V(\varepsilon)]\}^{1/2} \, d\varepsilon \tag{V-75}$$

The limits of integration α and β are the values of the deformation ε on the

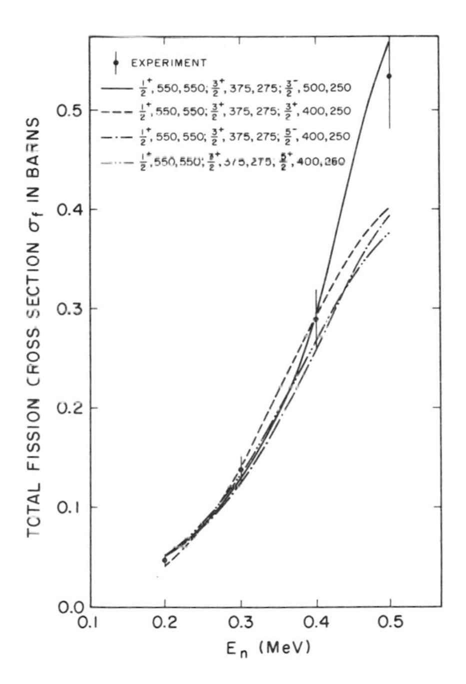

Fig. V-20. Total fission cross section for the $^{234}U(n,f)$ reaction as a function of incident neutron energy; $E_n = 200$, 300, 400, and 500 keV. The points are the experimental data of R. W. Lamphere ["Physics and Chemistry of Fission," Vol. 1, IAEA, Vienna, 1965, p. 63] with a $\pm 10\%$ uncertainty as indicated by the work of W. G. Davey [*Nucl. Sci. Eng.* **26**, 149 (1966)]. The curves show the best fit calculations which include level width fluctuation corrections. The notation is the same as in Fig. V-19. [Adapted from Behkami *et al.* (1968a).]

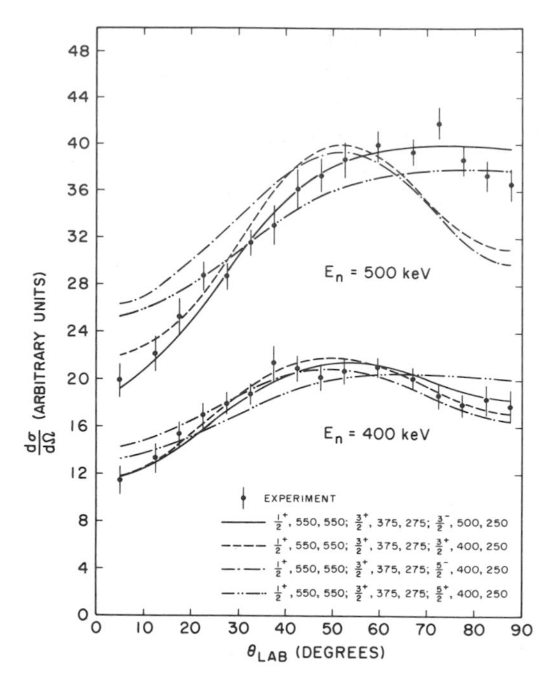

Fig. V-21. Fission-fragment angular distributions for the $^{234}U(n,f)$ reaction at $E_n = 400$ and 500 keV. The points are experimental data and the curves represent best fits to the angular distributions and cross sections with three accessible states of the transition nucleus (notation as in Fig. V-19). [Adapted from Behkami *et al.* (1968a).]

inner sides of the two peaks at which the integrand vanishes. If the potential energy $V(\varepsilon)$ in the second well is assumed to be given by $V(\varepsilon) = E_2 + \frac{1}{2}\mu\omega^2\varepsilon^2$, then the differential of the phase $\phi(E)$ with respect to energy is $(d\phi/dE) = \pi/\hbar\omega$.

The barrier penetrability $P(E)$ oscillates with variation in energy E and reaches maximum values at the energies corresponding to quasi-bound levels $E_n(J, \pi, K)$ in the second potential well. At a resonance, $\phi[E_n(J, \pi, K)] = \pi(n + \frac{1}{2})$. For an energy displacement of ΔE from a resonance the phase angle is given by $\phi + \Delta\phi$, where $\Delta\phi = \pi\,\Delta E/\hbar\omega$ and $\hbar\omega$ is the vibrational level spacing in the second potential well. For small values of $\Delta\phi$, we can with good accuracy make the substitutions of $\sin(\phi + \Delta\phi) = 1$ and $\cos(\phi + \Delta\phi) = \Delta\phi = \pi\,\Delta E/\hbar\omega$ into Eq. (V-74). With the additional assumption that $P_A P_B \ll 16$, the penetrability in the vicinity of a quasi-stable level has a Lorentzian dependence on energy (Lynn, 1969; Ignatyuk *et al.*, 1969)

$$P[E_n(J, \pi, K) + \Delta E] = \frac{\Gamma_A \Gamma_B}{[E - E_n(J, \pi, K)]^2 + \frac{1}{4}(\Gamma_A + \Gamma_B)^2} \qquad \text{(V-76)}$$

where $\Delta E = E - E_n(J, \pi, K)$, $\Gamma_A = (\hbar\omega/2\pi)\,P_A$ and $\Gamma_B = (\hbar\omega/2\pi)\,P_B$.

The neutron fission cross section of a zero spin target for a particular fission channel (J, π, K) is calculated in this section with Eq. (V-71) and the fission transmission coefficients of Eq. (V-76). The total fission cross section has been defined by Eq. (V-72). In this section we wish to discuss the structure in the cross section of the ^{230}Th(n, f) reaction. This reaction exhibits the most pronounced anomaly in the fission cross section of any of the heavy nuclei. A very strong peak in the neutron fission cross section occurs with neutrons of approximately 720 keV. Detailed calculations have shown that no reasonable assumptions about competing decay modes, such as inelastic neutron scattering, can satisfactorily explain the observed maximum in the fission cross section at the foregoing energy where the fission cross section is expected to increase exponentially.

If the second minimum in the potential energy surface is shallow, structure in the fission yield as a function of energy may be ascribed to a vibrational level in the second well which is only partially damped. The unusual peak in the excitation function of the ^{230}Th(n, f) reaction is interpreted in terms of such a vibrational mode resonance state in a potential well of a two-humped fission barrier.

The procedure followed in these calculations is identical to that described earlier except that we make use of the fission transmission coefficients given by Eq. (V-76). The total fission cross section $\sigma_f(E)$ and angular distribution $W(\theta, E)$ are computed in the vicinity of the resonance for positive and negative parity $K = \frac{1}{2}$ bands. The parameters Γ_A, Γ_B, $E_0(\pi, K)$, $\hbar^2/2\mathscr{I}_\perp$, and α are varied and the theoretical results are compared with experimental fission cross sections and angular distributions. Examples of these comparisons for a $K = \frac{1}{2}+$ state are shown in Figs. V-22 and V-23. For a positive parity vibrational level $K = \frac{1}{2}+$, orbital angular momenta $l = 0$ and 2 contribute at these energies, leading to a rotational band with $J = \frac{1}{2}+, \frac{3}{2}+$, and $\frac{5}{2}+$.

With a two-humped barrier it is possible to obtain a reasonable fit of the cross section data with either a $K = \frac{1}{2}+$ or $K = \frac{1}{2}-$ vibrational state. Variation of the parameters in Eq. (V-69) enables us to shift the energy positions of the rotational members of the $K = \frac{1}{2}$ bands without seriously disturbing the theoretical total fission cross sections. This shift in the energies of the members of a rotational band causes considerable change in the energy dependence of the theoretical fission-fragment angular distribution. For example, the experimental angular distribution for the neutron energy of 730 keV exhibits a shape which is characteristic of angular distributions having a dominant contribution from a $J \geqslant \frac{3}{2}$ member of a $K = \frac{1}{2}$ band. A $J = \frac{5}{2}$ contribution is possible for a $K = \frac{1}{2}+$ band at these neutron energies since the $l = 2$ transmission coefficients are sizeable, giving the theoretical partial fission cross sections shown in Fig. V-22. Contributions from $J = \frac{5}{2}$ and $\frac{7}{2}$ states are possible for a $K = \frac{1}{2}-$ band in that the $l = 3$ neutron transmission coefficients are still of importance at

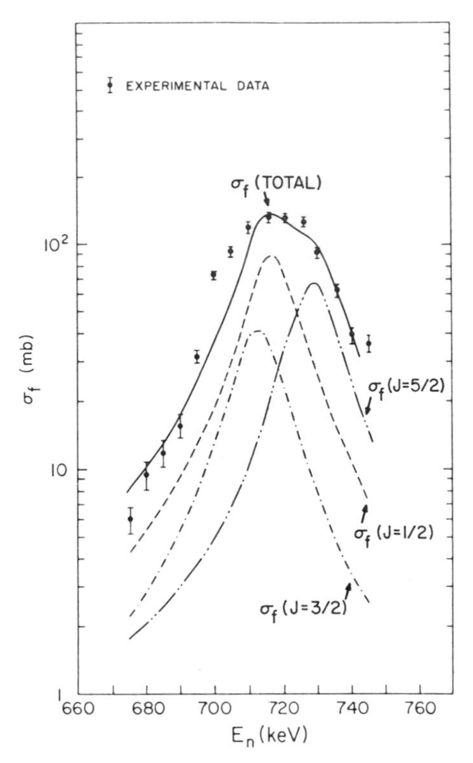

Fig. V-22. Comparison of theoretical and experimental fission cross sections for the ^{230}Th(n, f) reaction at neutron energies corresponding to the postulated vibrational state resonance. The experimental data are from James *et al.* (1972) and are in units of millibarns. The theoretical calculation exhibited was performed for a $K=\frac{1}{2}+$ state with the parameters Γ_A, Γ_B, $E_0(\pi, K)$, $\hbar^2/2\mathscr{I}_\perp$, and α equal to 12.0, 1.9, 714.9, and 0.87 keV, and 2.9, respectively. Various dashed lines: theoretical partial fission cross sections for individual J states; solid line: sum of the partial cross sections or the total fission cross section. [From Yuen *et al.* (1971).]

these neutron energies. On the basis of present fits, it is not possible to make a firm assignment of parity for the $K = \frac{1}{2}$ vibrational state. Although the angular distribution for the neutron energy of 730 keV is well fitted with a $K = \frac{1}{2}+$ assignment, the value of $\hbar^2/2\mathscr{I}$ used in this calculation is unrealistically small. Calculations by James *et al.* (1972) in which the widths of the individual spin states are allowed to be different favor a negative parity assignment for the resonance. The latter calculations give a more reasonable value of 2.5 keV for the rotational parameter in the second well. The value of $\hbar^2/2\mathscr{I}$ in the second well is of considerable theoretical interest.

The binding energy of the last neutron in ^{231}Th is 5.09 MeV. Hence, the

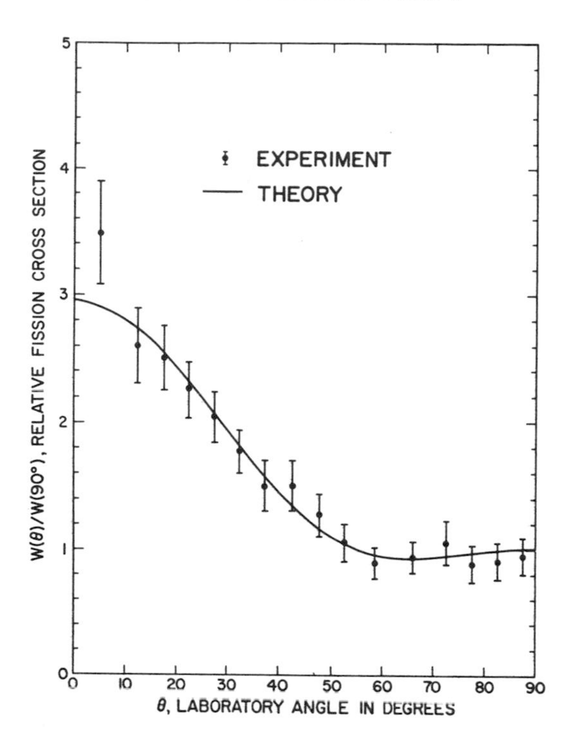

Fig. V-23. Comparison of theoretical and experimental fission-fragment angular distributions for the ^{230}Th(n,f) reaction at a neutron energy of 730 keV. This neutron energy is in the vicinity of the postulated vibrational state resonance. The experiment was performed with 5-keV neutron energy resolution. The parameters are the same as in Fig. V-22. [Adapted from Yuen *et al.* (1971).]

vibrational resonance in the second potential well is displaced 5.8 MeV above the ground state energy. If the vibrational spacing $\hbar\omega$ in the second potential well is estimated to be 1 MeV, then the penetrabilities P_A and P_B for the parameters used in Figs. V-22 and V-23 are 0.08 and 0.01, respectively. In this mass region, theoretical calculations indicate that the second barrier is the higher one. Computation of the individual barrier heights requires knowledge about their widths and inertial parameters. If the simple expression given by Eq. (V-25) is used along with a barrier curvature of 600 keV for each barrier, the heights of barriers A and B are 6.03 and 6.24 MeV, respectively. If, on the other hand, it is assumed that the curvature for the barriers is 1 MeV, then the heights of barriers A and B are increased by 0.16 and 0.29 MeV, respectively.

Bjørnholm and Strutinsky (1969) have noted a systematic change in the character of the angular distributions when going from the lighter to the

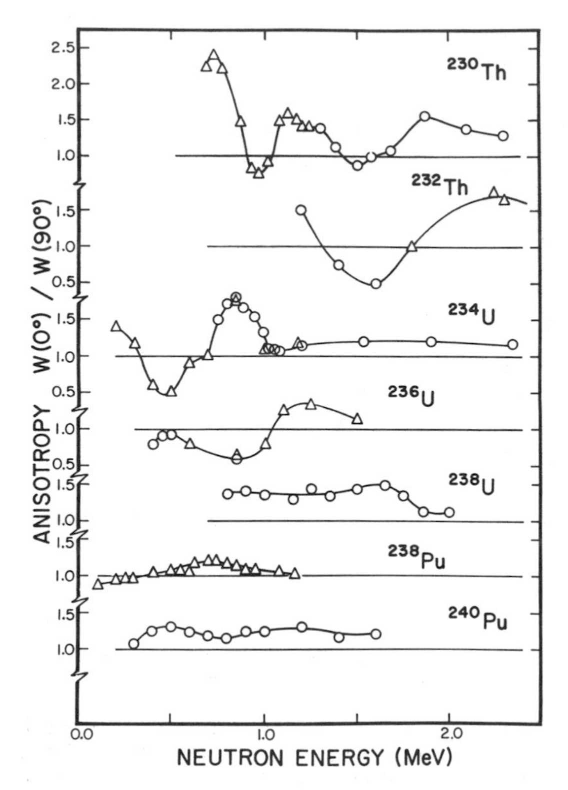

Fig. V-24. Fission-fragment anisotropies for neutron-induced fission in several heavy target nuclides.

heavier nuclei, as illustrated in Fig. V-24. The dependence of the anisotropy varies strongly with energy for the lighter elements, but becomes rather structureless for the heavier elements. This behavior can be qualitatively understood as a consequence of the systematic trend for the outer barrier *B* to change from an energy higher than barrier *A* for the lighter elements to an energy lower than barrier *A* for the heavier elements. The fission-fragment anisotropies of the heavier elements can be understood on the basis of the assumption that the *K* values are not conserved during the passage through the second minimum. The time for passage through the region of the second minimum is sufficiently long so that the *K* values of the first barrier are forgotten and reestablished at the second barrier. When the second barrier is lower than the first barrier by 1 MeV or more, a near statistical distribution of *K* values may prevail and the resulting anisotropy is rather smooth with energy.

3. Even-Even Transition Nuclei (Formed by Neutron Fission of Odd-*N* Targets)

The capture of a neutron in an odd-*N* target produces an even-even fissioning nucleus. One of the basic differences in the capture of a neutron in an odd-*N* target compared to an even-even target is in the larger neutron binding energy. Hence, even the capture of a thermal neutron leads to an excitation energy in the compound nucleus which is considerably in excess of the lowest fission barrier for the even-even transition nucleus. Hence, most of the analyses of angular distribution data for fission induced by neutron capture of odd-*N* targets have been performed with the assumption of a Gaussian *K* distribution [see Eq. (VI-5)].

The relative fission-fragment angular cross section may be written in the following form if we assume that the compound nucleus spin distribution is not fractionated during decay.

$$W(\theta) \propto \sum_j \sum_J \sum_M P(j; J, M) W(J, M; \theta) \tag{V-77}$$

The first term is a partial probability of formation of a state of the compound nucleus from a particular j value

$$P(j; J, M) = \frac{(2j+1)\{T_{l=j-\frac{1}{2}}(E) + T_{l=j+\frac{1}{2}}(E)\} \sum_\mu \sum_\sigma |C^{j;\, I_0,\, J}_{\sigma,\, \mu,\, M}|^2}{(2s+1)(2I_0+1) \sum_j (2j+1)\{T_{l=j-\frac{1}{2}}(E) + T_{l=j+\frac{1}{2}}(E)\}} \tag{V-78}$$

The transmission coefficients $T_l^{\,j}(E)$ are derived from the optical model with spin-orbit interaction, where $\mathbf{j} = \mathbf{l} + \mathbf{s}$ and s is the neutron spin with projection σ on the space-fixed axis. The target spin and its projection on the space-fixed axis are denoted by I_0 and μ, respectively. The total angular momentum J of the compound nucleus is given by $\mathbf{J} = \mathbf{j} + \mathbf{I}_0$ and the projection of J on the space-fixed axis is given by M. The quantity $C^{j,\, I_0,\, J}_{\sigma,\, \mu,\, M}$ is a Clebsch–Gordan coefficient. The second term $W(J, M; \theta)$ gives the angular distribution of the fission fragments emitted from the compound state (J, M)

$$W(J, M; \theta) = \sum_{K=-J}^{+J} \frac{\exp(-K^2/2K_0^{\,2})(2J+1)|d^J_{M,K}(\theta)|^2}{\sum_{K=-J}^{+J} \exp(-K^2/2K_0^{\,2})} \tag{V-79}$$

where the $d^J_{M,K}(\theta)$ function is defined by Eq. (V-2). The relative cross section at angle θ is obtained by summing the product $P(j, J, M)W(J, M; \theta)$ as given by Eq. (V-77) over all values of M, J, and j.

Calculation of the values of $P(j; J, M)$ for the reaction ^{239}Pu + 400 keV neutrons is illustrated in Tables V-5 to V-7. First of all, the transmission coefficients for the various channel spins j are given in Table V-5. The compound nucleus spin J is formed by the coupling of the target spin I_0 and the

TABLE V-5

Transmission Coefficients for the Interaction of 400 keV Neutrons with ^{239}Pu Assuming a Spin-Orbit Interaction

l	j, π	$T_{l=j\pm\frac{1}{2}}$
0	$\frac{1}{2}+$	0.3194
1	$\frac{1}{2}-$	0.6933
	$\frac{3}{2}-$	0.7766
2	$\frac{3}{2}+$	0.0485
	$\frac{5}{2}+$	0.0446
3	$\frac{5}{2}-$	0.0105
	$\frac{7}{2}-$	0.0170
4	$\frac{7}{2}+$	0.0001
	$\frac{9}{2}+$	0.0001

TABLE V-6

Calculation of the Partial Probability of Formation of State (J, M) of a Compound Nucleus from a Particular j Value[a]

I_0, π	μ	l	j, π	σ	$T_{l=j\pm\frac{1}{2}}$	R	J, π	M	$\lvert C^{j;I_0;J}_{\sigma,\mu,M}\rvert^2$	j: J, M	$P(j; J, M)$	
$\frac{1}{2}+$	$+\frac{1}{2}$	0	$\frac{1}{2}+$	$-\frac{1}{2}$	0.3194	0.086293	0+	0	$\frac{1}{2}$	$\frac{1}{2}$; 0, 0	0.013781	0.08739
$\frac{1}{2}+$	$-\frac{1}{2}$	0	$\frac{1}{2}+$	$+\frac{1}{2}$	0.3194	0.086293	0+	0	$\frac{1}{2}$	$\frac{1}{2}$; 0, 0	0.013781	
$\frac{1}{2}+$	$+\frac{1}{2}$	1	$\frac{1}{2}-$	$-\frac{1}{2}$	0.6933	0.086293	0−	0	$\frac{1}{2}$	$\frac{1}{2}$; 0, 0	0.029914	
$\frac{1}{2}+$	$-\frac{1}{2}$	1	$\frac{1}{2}-$	$+\frac{1}{2}$	0.6933	0.086293	0−	0	$\frac{1}{2}$	$\frac{1}{2}$; 0, 0	0.029914	

$$R = \frac{(2j+1)}{(2s+1)(2I_0+1)\sum_j(2j+1)\{T_{l=j-\frac{1}{2}}(E)+T_{l=j+\frac{1}{2}}(E)\}}$$

[a] I_0 and μ are the target spin and its projection on the space-fixed axis, respectively. The neutron spin is denoted by s and the channel spin $j = \underline{l} \pm s$. The projection of j on the space-fixed axis is σ. The transmission coefficients $T_{l=j\pm\frac{1}{2}}$ are from Table V-5. This illustration is for the reaction ^{239}Pu + 400-keV neutrons. The spin and parity of ^{239}Pu is $\frac{1}{2}+$. The quantities $C^{j,I_0,J}_{\sigma,\mu,M}$ are Clebsch–Gordan coefficients.

channel spin j. As shown in Table V-6, there are four ways to produce the compound state ($J = 0$, $M = 0$) by combining a target spin of $\frac{1}{2}$ and a channel spin of $\frac{1}{2}$ when parity is included. In the present calculation these four probabilities may be summed to give $P(\frac{1}{2}; 0, 0)$, since the value of $W(J, M; \theta)$, as defined in Eq. (V-79), depends only on J and M and is independent of parity. Equation (V-78) is formulated to do the sum illustrated in Table V-6, and the parity enters only in an indirect manner through the transmission coefficients $T_{l=j-\frac{1}{2}}$ and $T_{l=j+\frac{1}{2}}$. The total probability $P(\frac{1}{2}; 0, 0)$ given in Table V-6 is shown as the first entry in Table V-7. The remaining values of $P(j; J, M)$ given in Table V-7 are computed in an identical way.

TABLE V-7

Partial Probabilities of Formation of States (J, M) of the Compound Nucleus from a Particular j[a]

$(j; J, M)$	$P(j; J, M)$	$(j; J, M)$	$P(j; J, M)$	$(j; J, M)$	$P(j; J, M)$
$\frac{1}{2}; 0, 0$	0.087389	$\frac{3}{2}; 2, 1$	0.106800	$\frac{7}{2}; 4, -1$	0.003689
$\frac{1}{2}; 1, -1$	0.087389	$\frac{5}{2}; 2, -1$	0.004755	$\frac{7}{2}; 4, 0$	0.005902
$\frac{1}{2}; 1, 0$	0.087389	$\frac{5}{2}; 2, 0$	0.014264	$\frac{7}{2}; 4, 1$	0.003689
$\frac{1}{2}; 1, 1$	0.087389	$\frac{5}{2}; 2, 1$	0.004755	$\frac{9}{2}; 4, -1$	0.000017
$\frac{3}{2}; 1, -1$	0.035600	$\frac{5}{2}; 3, -1$	0.009510	$\frac{9}{2}; 4, 0$	0.000043
$\frac{3}{2}; 1, 0$	0.142400	$\frac{5}{2}; 3, 0$	0.014264	$\frac{9}{2}; 4, 1$	0.000017
$\frac{3}{2}; 1, 1$	0.035600	$\frac{5}{2}; 3, 1$	0.009510	$\frac{9}{2}; 5, -1$	0.000026
$\frac{3}{2}; 2, -1$	0.106800	$\frac{7}{2}; 3, -1$	0.002213	$\frac{9}{2}; 5, 0$	0.000043
$\frac{3}{2}; 2, 0$	0.142400	$\frac{7}{2}; 3, 0$	0.005902	$\frac{9}{2}; 5, 1$	0.000026
		$\frac{7}{2}; 3, 1$	0.002213		

[a] A more detailed calculation of the first entry is given in Table V-6. These values are calculated for the reaction ^{239}Pu + 400-keV neutrons.

The Gaussian weighting of K values in Eq. (V-79) is an assumption which is not expected to be valid for low energy neutron fission of odd-N targets. However, this assumption allows us to parameterize the theoretical angular distribution in a simple and straightforward way, so that we can search for discontinuities or rapid changes in the K distribution.

The experimental fission-fragment angular distributions for neutron-induced fission of ^{239}Pu have been studied extensively as a function of neutron energy. Examples of the experimental angular distributions are shown for two neutron energies, 400 ± 25 and 1300 ± 25 keV, in Figs. V-25 and V-26, respectively. With a Gaussian K distribution and Perey–Buck neutron transmission coefficients (Auerbach and Perey, 1962), the theoretical fragment angular distributions as calculated with Eqs. (V-77) to (V-79) are compared with the experimental values for various values of K_0^2 in Figs. V-25 and V-26. As seen in these figures, the experimental angular distributions are well fitted for selected values of K_0^2.

The general behavior of K_0^2 as a function of neutron energy for the ^{239}Pu + n reaction is shown in Fig. V-27. At the lower neutron energies the determined values of K_0^2 are of the order of 5 to 6. These small values of K_0^2 are interpreted to mean that only collective levels are excited in the fissioning transition nucleus. The ground state band of an even-even transition nucleus has $K = 0$, and for a few hundred keV neutrons a number of collective excitations and their rotational bands are expected to contribute. For example, if five collective K states are open with $K = 0, 1, 2, 2$, and 4, the resulting value of K_0^2 is 5.0. A

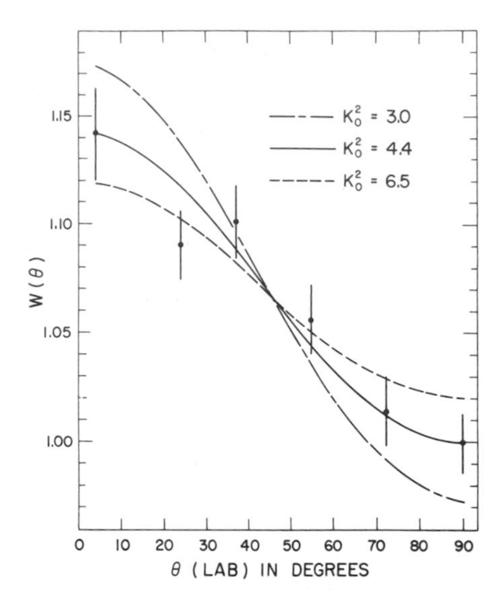

Fig. V-25. Experimental and theoretical fission-fragment angular distributions for fission of ^{239}Pu induced with 400 ± 25-keV neutrons. ● Experimental points; the lines correspond to theoretical results with different values of $K_0{}^2$. Perey–Buck neutron transmission coefficients are used in these theoretical calculations. [After Huizenga *et al.* (1968).]

similar value of $K_0{}^2$ has been obtained in the same excitation energy region from a study of the fragment angular distribution for the $^{239}Pu(d, pf)$ reaction.

The value of $K_0{}^2$ increases rapidly for neutrons of energies from 600 to

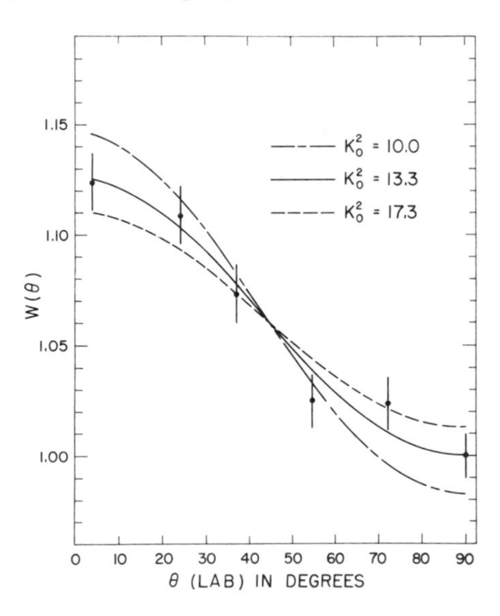

Fig. V-26. Same as Fig. V-25 except 1350 ± 25 keV-neutrons.

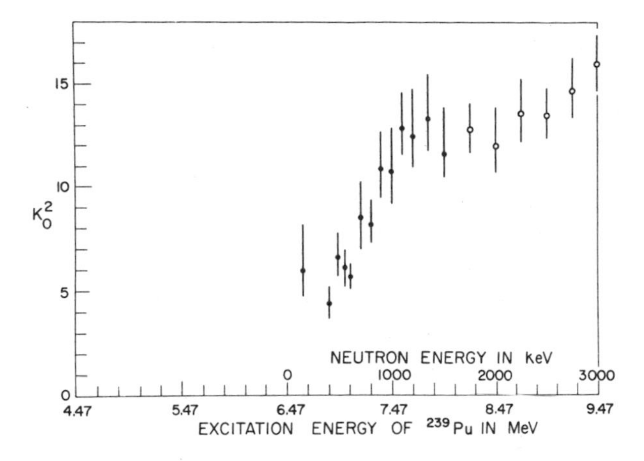

Fig. V-27. The derived values of $K_0{}^2$ are plotted as a function both of neutron energy in keV and of the excitation energy of ^{240}Pu in MeV. The solid circles are from Huizenga *et al.* (1968) and the open circles are from normalized data of J. E. Simmons [reported by Griffin (1963).]

1000 keV and reaches a plateau value of about 13 at 1100 keV. Such a value of $K_0{}^2$ is consistent with the expectation for two-quasi-particle states in a heavy nucleus. If the excitation energy is sufficient to break either a neutron or a proton pair, $\langle K^2 \rangle$ is a weighted average over two neutron and two proton quasi particles.

The rapid increase in $K_0{}^2$ for the even-even transition nucleus ^{240}Pu is interpreted as the threshold for exciting two quasi particles. This threshold occurs at an excitation energy of about 7.27 MeV.

D. Fission of Nuclei Excited in Direct Reactions

1. Introduction

A convenient technique for investigating the properties of transition states is to observe the decay by fission of nuclei excited in a direct reaction. The properties of transition states over a continuous region of excitation energies can be studied by measurements of the energies of the outgoing direct particles in coincidence with fission fragments. Excitation of a nucleus by direct reaction to induce fission is an especially important process for studying the transition states in an even-even nucleus where it is possible to make measurements at excitation energies corresponding to negative neutron energies for the appropriate neutron capture reaction. Northrup *et al.* (1959) used the (d, pf) reaction technique to make the first measurements of the fission barrier heights of ^{234}U, ^{236}U, and ^{240}Pu.

As already discussed in Sections A and B of this chapter, information on the nuclear states at the deformed saddle configuration can be derived from fission-fragment angular distributions. The applications of direct reactions to excite even-even nuclei and study the fission-fragment angular distributions were first reported for the (d, pf) reaction by Britt *et al.* (1965) and the (α, α'f) reaction by Wilkins *et al.* (1964). Similar correlations for odd-A transition nuclei were first studied by Vandenbosch *et al.* (1965).

In the analysis of fission of nuclei excited in direct reactions, it is assumed that the overall reaction proceeds in two independent steps. In the first step, isolated nuclear levels with the target deformation are excited by the direct process. In the second step, these levels decay by fission, neutron, and γ-ray emission via the compound nucleus mechanism. We need to know the direct reaction cross section as a function of energy for each angular momentum and parity in order to interpret the angular distribution data. The excitation energy region of interest is usually 5–7 MeV, a region near the fission barrier. One of the most uncertain aspects of this type of fission reaction is the initial angular momentum distribution.

One approach to the direct reaction mechanism has been to assume a plane wave description for the process. This approximation results in a uniform distribution, in a plane perpendicular to the recoil axis of the compound nucleus, of the vectors describing the orbital angular momentum transferred by the direct reaction. For a fixed charged particle coincidence angle this gives a fission-fragment angular distribution analogous to that produced by a beam of neutrons incident along the compound nucleus recoil direction. For a (d, pf) reaction, the in-plane proton fission-fragment angular correlations, as illustrated in Fig. V-28, give information similar to that obtained by measuring the fission-fragment angular distributions from the (n, f) reaction. Also, the angular correlation would be isotropic for fission-fragment directions in the plane perpendicular to the recoil axis.

Some of the direct reactions, such as the (d, pf) experiments, have been performed using projectile energies relatively near in energy to the Coulomb barrier. It is, therefore, likely that Coulomb distortion effects occur in the direct reaction. Distorted wave born approximation (DWBA) theory predicts that the angular correlation will be more complex than that calculated by the use of plane waves. The angular momentum transfer from the direct reaction no longer necessarily results in orbital angular momentum vectors lying in a plane perpendicular to the recoil axis. In general, the angular momentum vectors have a smeared directional distribution, with an average direction perpendicular to a symmetry axis differing from the recoil direction. The angular momentum vectors are also not expected to be uniformly distributed

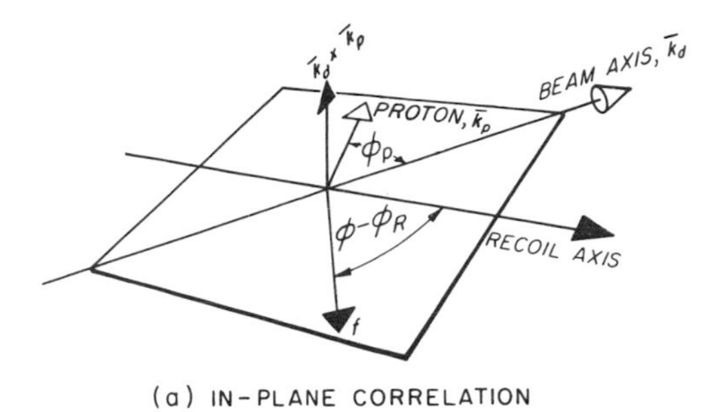

(a) IN-PLANE CORRELATION

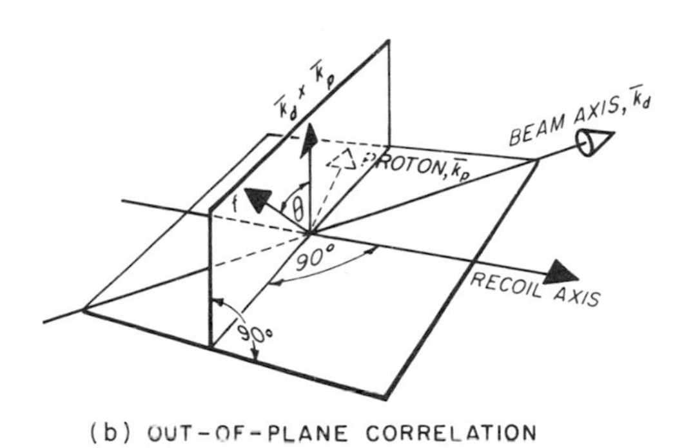

(b) OUT-OF-PLANE CORRELATION

Fig. V-28. (a) The geometry for measurements taken in the reaction plane (in-plane measurements). The angle of the fission fragment with respect to the recoil direction is designated by $\phi-\phi_R$. (b) A plane perpendicular to the recoil axis. The out-of-plane measurements correspond to observation of fission fragments in this plane as a function of the angle θ with respect to the direction of $\mathbf{k}_d \times \mathbf{k}_p$, which is perpendicular to the reaction plane. ϕ_p is the angle of the proton measured relative to the deuteron beam direction. [After Wolf *et al.* (1968).]

in the plane perpendicular to the symmetry axis, resulting in a nonisotropic out-of-plane angular correlation for different fragment directions in the plane perpendicular to the recoil axis.

Angular correlations, both in and out of the (d, p) reaction plane, have been measured for the ^{239}Pu(d, pf) reaction (Wolf *et al.*, 1968). The observed correlations are strongly dependent on the angle of emission of the protons, as shown in Fig. V-29 for an excitation energy corresponding to 0.5–1.5 MeV above the fission threshold. Figure V-29a shows the out-of-plane angular correlations for three different proton angles ϕ_p, where θ is the out-of-plane angle describing the fission-fragment direction with respect to the vector $\mathbf{k}_d \times \mathbf{k}_p$. Hence, the in-plane correlation is given by $\theta = 90°$. When the proton counter is moved to a backward angle of 140°, the out-of-plane correlation is negligible, and the plane-wave approximation provides an adequate

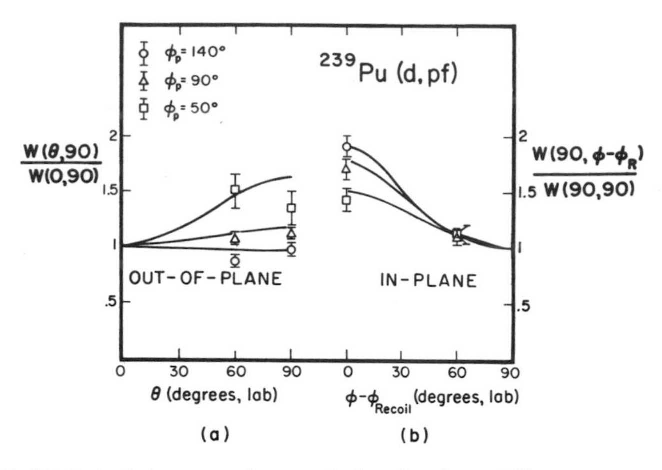

Fig. V-29. (a) Out-of-plane angular correlation for three different proton angles ϕ_p, with θ the angle describing the fission-fragment direction with respect to the vector $\mathbf{k}_d \times \mathbf{k}_p$; (b) in-plane correlations as a function of the difference between the fragment direction ϕ and the recoil direction ϕ_R. The full curves are from the DWBA calculations. The in-plane anisotropies for $\phi_p = 140°$ are taken from the work of Britt *et al.* (1965). Data are for an excitation energy region corresponding to $E^* - E_f$ values of 0.5–1.5 MeV. [After Wolf *et al.* (1968).]

description of the observed correlations. From Fig. V-29b we observe that the in-plane correlations are largest when the proton counter is at the largest angle with respect to the beam direction.

Although some direct reactions have the property of forming states of moderately high angular momentum at excitation energies in the fissioning nuclei near the barrier, a quantitative calculation of the initial angular momentum intensity distribution following the direct reaction is not presently possible. This is one of the most serious shortcomings of the technique which utilizes direct reactions for producing the excitation energy necessary to induce fission. Andersen *et al.* (1970) have calculated the spin-dependent formation cross sections by a DWBA calculation using single particle wave functions in a deformed Woods–Saxon potential. In the case of particle transfer to highly excited states, coupling of states with the same Ω and parity but with N differing by 2 is important and is included in the foregoing calculations. Hence, after evaluation of the spectroscopic factors, the cross section for stripping into a definite state of the final nucleus with specific quantum numbers is calculated in the standard way. Up to this point, the final state is assumed to have a structure similar to that known from low excitation energies. However, at higher excitation energies, the coupling to all the other degrees of freedom will distribute the strength of the simple states over a rather large energy interval. The assumption is made that the cross section to a particular single particle state is distributed over the nearby compound states by a strength function factor. Compound states of the residual nucleus of definite angular momentum may be formed by contributions from several single particle states. The cross sections for each angular momentum and parity are calculated as a function of excitation energy. The results of such a calculation are shown in Fig. V-30 for the ^{239}Pu(d, p) reaction. The distance between neighboring curves gives the probability of finding the spin and parity indicated in the upper curve. Hence, the distance between the 0− and 1− states gives the probability of having states with spin and parity of 1−.

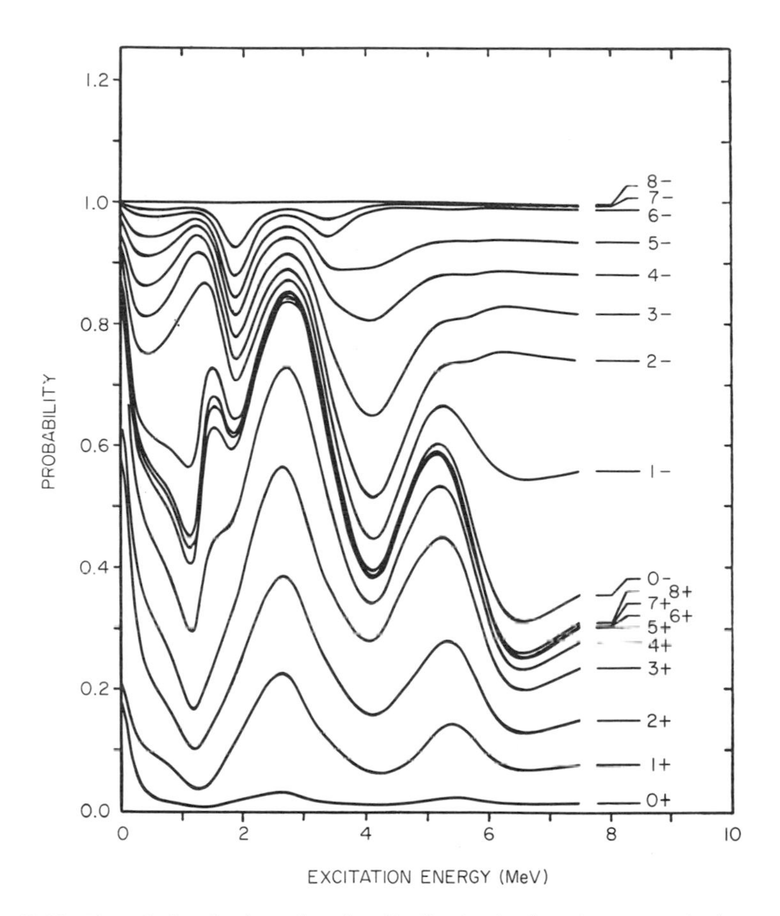

Fig. V-30. The calculated spin and parity distribution is plotted versus excitation energy in ^{239}Pu(d,p) ($\theta = 140°$; $E_d = 13$ MeV), so that the distance between neighboring curves gives the probability of finding the spin and parity indicated at the upper curve. Hence, the distance between the 0− and 1− states gives the probability of having states with spin and parity of 1−. [After Andersen *et al.* (1970).]

A summary of transition nuclei which have been investigated by the technique of excitation by direct reaction is given in Table V-8.

TABLE V-8

Transition nucleus	Reaction	Projectile energy	Year	Ref.
^{232}Th	$^{232}Th(\alpha, \alpha'f)$	42	1966	*a*
^{234}U	$^{233}U(d, pf)$	14.0	1959	*b*
	$^{233}U(d, pf)$	14.9	1965	*c*
	$^{233}U(d, pf)$	18.0	1968	*d*
	$^{233}U(d, pf)$	12.0	1969	*e*
	$^{233}U(d, pf)$	13.0	1969, 1970	*f, g*
	$^{233}U(t, df)$	18.0	1969	*h*
^{235}U	$^{234}U(d, pf)$	13.0	1965	*i*
	$^{234}U(d, pf)$	18.0	1970	*j*
	$^{233}U(t, pf)$	18.0	1970	*j*
^{236}U	$^{235}U(d, pf)$	14.0	1959	*b*
	$^{235}U(d, pf)$	12.5	1966	*k*
	$^{235}U(d, pf)$	10, 13, 15, 21	1967	*l*
	$^{235}U(d, pf)$	18.0	1968	*d*
	$^{235}U(d, pf)$	13.0	1969	*e*
	$^{235}U(d, pf)$	13.0	1969, 1970	*f, g*
	$^{235}U(t, df)$	18.0	1969	*h*
	$^{234}U(t, pf)$	13.0	1965	*m*
	$^{234}U(t, pf)$	18.0	1968	*d*
^{237}U	$^{236}U(d, pf)$	18.0	1970	*j*
	$^{236}U(t, df)$	18.0	1970	*j*
	$^{235}U(t, pf)$	18.0	1970	*j*
^{238}U	$^{238}U(\alpha, \alpha'f)$	43.0	1964, 1965	*n, o*
	$^{238}U(\alpha, \alpha'f)$	40.0	1966	*p*
^{239}U	$^{238}U(d, pf)$	14.0	1959	*b*
	$^{238}U(d, pf)$	18.0	1970	*j*
^{238}Np	$^{237}Np(d, pf)$	13.0	1971	*g*
^{239}Pu	$^{238}Pu(d, pf)$	13.0	1971	*g*
^{240}Pu	$^{239}Pu(d, pf)$	14.0	1959	*b*
	$^{239}Pu(d, pf)$	14.9	1965	*c*
	$^{239}Pu(d, pf)$	12.5	1966	*k*
	$^{239}Pu(d, pf)$	15.0	1968	*q*
	$^{239}Pu(d, pf)$	15.0	1968	*d*
	$^{239}Pu(d, pf)$	11.5	1969	*r*
	$^{239}Pu(d, pf)$	13.0	1969	*e*
	$^{239}Pu(d, pf)$	13.0	1969, 1970	*f, g*
	$^{239}Pu(t, df)$	18.0	1969	*h*
	$^{240}Pu(\alpha, \alpha'f)$	38.1	1966	*p*
	$^{240}Pu(\alpha, \alpha'f)$	42.0		*a*
	$^{240}Pu(p, p'f)$	20.0	1969	*s*

TABLE V-8 (continued)

Transition nucleus	Reaction	Projectile energy	Year	Ref.
^{241}Pu	^{240}Pu(d, pf)	18.0	1970	*j*
	^{240}Pu(d, pf)	13.0	1971	*g*
	^{240}Pu(t, df)	18.0	1970	*j*
	^{239}Pu(t, pf)	18.0	1970	*j*
^{242}Pu	^{241}Pu(d, pf)	13.0	1969	*e*
	^{241}Pu(d, pf)	13.0	1969, 1970	*f, g*
^{243}Pu	^{242}Pu(d, pf)	18.0	1970	*j*
	^{242}Pu(t, df)	18.0	1970	*j*
^{242}Am	^{241}Am(d, pf)	13.0	1971	*g*
^{244}Am	^{243}Am(d, pf)	13.0	1971	*g*

[a] Unik (1966)
[b] Northrop *et al.* (1959)
[c] Britt *et al.* (1965)
[d] Britt *et al.* (1968)
[e] Pedersen and Kuzminov (1969)
[f] Back *et al.* (1969)
[g] Back *et al.* (1971)
[h] Britt and Cramer (1969)
[i] Vandenbosch *et al.* (1965)
[j] Britt and Cramer (1970)
[k] Specht *et al.* (1966)
[l] Vandenbosch *et al.* (1967)
[m] Eccleshall and Yates (1965)
[n] Wilkens *et al.* (1964)
[o] Huizenga *et al.* (1965)
[p] Britt and Plasil (1966)
[q] Wolf *et al.* (1968)
[r] Specht *et al.* (1969)
[s] Britt *et al.* (1969)

2. Fission of Even-Even Transition Nuclei

The introduction of excitation energy by a direct reaction allows us to study fission at energies below the barrier. In contrast to this, neutron-induced fission can be used only at excitation energies exceeding the neutron binding energy. Since even-even final nuclei are produced by neutron capture on nuclei with odd A and odd N, the lowest possible excitation energy usually exceeds to some extent the fission barrier. Hence, it is not possible to study the near-barrier transition states of even-even nuclei by the neutron capture reaction.

The $(\alpha, \alpha'\mathrm{f})$ reaction on even-even targets gives very large fission-fragment anisotropies near threshold excitation. In such a reaction, all the relevant spins are zero; furthermore, the angular momentum transfers are larger than those in a (d, p) reaction. The angular distribution of fission fragments observed in the $^{238}\mathrm{U}(\alpha, \alpha'\mathrm{f})$ experiment for excitation energies between 5.6 and 6.2 MeV (Wilkins *et al.*, 1964) is shown in Fig. V-31. The angles on the abscissa of this figure are measured relative to the angular momentum symmetry axis which has been measured to be within the experimental error ($\pm 3°$) identical to the recoil axis. For α' angles of at least 75°, measurements have shown that the angular correlations are independent of the azimuthal angle and the reaction may be interpreted in terms of plane wave theory. In addition, DWBA

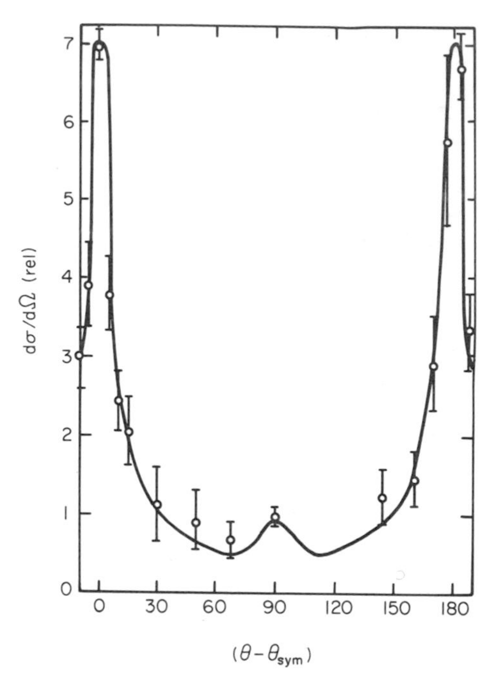

Fig. V-31. Angular distribution of fission fragments observed in the $^{238}U(\alpha, \alpha' f)$ experiment for excitation energies between 5.6 and 6.2 MeV. The angles are measured relative to the angular momentum symmetry axis θ_{sym}, which in this case was measured to be within experimental error ($\pm 3°$) identical to the recoil axis. [After Wilkins *et al.* (1964).]

calculations have confirmed the validity of the plane wave approximation (Britt and Plasil, 1966).

The angular distribution of fragments for the $(\alpha, \alpha' f)$ reaction can be expressed in terms of symmetric top wave functions as

$$W^J_{M,K}(\theta) = \frac{2J+1}{2}\left|\sum_M a_M\, D^J_{M,K}(\theta, \phi, \psi)\right|^2 \qquad \text{(V-80)}$$

where M is the projection of J on some space-fixed axis and the coefficients a_M represent the relative amplitudes of the possible M substates. When the plane wave approximation is valid and the space-fixed axis is taken in the direction of the kinematic recoil of the residual nucleus considerable simplification occurs for Eq. (V-80). For the foregoing case, only $M = 0$ substates contribute (i.e., $a_M = 0$ for all $M \neq 0$) and the angular distribution in the reaction plane for a given J, K is

$$W^J_{0,K}(\theta) = \frac{(2J+1)}{2}\,\frac{(J-K)!}{(J+K)!}\,|P_J{}^K(\cos\theta)|^2 \qquad \text{(V-81)}$$

where $P_J^K(\cos\theta)$ are Legendre polynomials. The total observed angular distributions may be written as

$$W(\theta) = \sum_K A_K \sum_J b_J W_{0,K}^J(\theta) \tag{V-82}$$

In Eq. (V-82), it is assumed that the transition state spectrum is made up of rotational bands built on vibrational states of a specific K projection. The quantity A_K represents the relative probability of fissioning through all of the available transition states of a given K. This form assumes that the relative probability of fission through a given vibrational band is independent of the total angular momentum J for the transition state involved except that only natural parity states are allowed. The quantity b_J represents the relative probability of fissioning through a specific J state. As already indicated, the (α, α') reaction can excite only natural parity states, so the sum over J in Eq. (V-82) is taken over only even J values for positive parity transition states and over only odd J values for negative parity transition states.

The measured angular distribution for an excitation energy bin very near the fission barrier for the $^{238}U(\alpha, \alpha'f)$ reaction is strong evidence for a $K = 0+$ vibration state in the intermediate transition nucleus with rotational members $J = 0, 2, 4, 6, 8, 10, \ldots$. For such a rotational band, Eq. (V-82) reduces to

$$W(\theta) = \sum_J \frac{(2J+1)}{2} b_J |P_J(\cos\theta)|^2 \tag{V-83}$$

The solid curve in Fig. V-31 shows an empirical fit to the experimental data for excitation energies between 5.6 and 6.2 MeV with $b_J = \exp -(J/7)^2$ and summing over only even values of J.

As the excitation energy is increased, the angular correlation measurements for the $(\alpha, \alpha'f)$ reaction are markedly changed. These results are interpreted to mean that additional vibrational states with $K > 0$ are making significant contributions at the higher energies. Negative parity $K = 0-$ and $K = 1-$ bands are known to exist in even-even transition nuclei from photofission experiments. The $K = 0-$ band is expected to give an angular correlation similar to the $K = 0+$ band. In principle, it is possible to make a definite determination of whether a band is $K = 0+$ or $0-$ by a careful investigation of the $(\alpha, \alpha'f)$ angular correlation near 90°. This is illustrated in Fig. V-32, where the Legendre polynomials for odd and even J are plotted for $K = 0$. The yields for the various even J states reinforce each other at 90°, whereas the odd-J states all have zero yield.

Comparison of angular correlation from the $^{239}Pu(d, pf)$ reaction (Britt *et al.*, 1965) and the $^{240}Pu(\alpha, \alpha'f)$ reaction (Britt and Plasil, 1966) has led to some assignments of vibrational transition states in addition to the $K = 0+$ band.

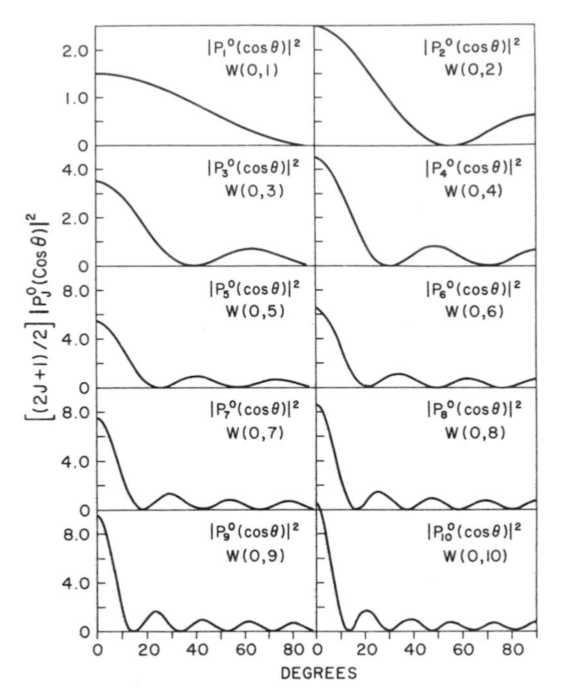

Fig. V-32. Square of Legendre polynomials normalized such that

$$\int_{-1}^{1} \frac{2J+1}{2} |P_J{}^0(\cos\theta)|^2 = 1$$

If the angular momentum J lies with equal probability in all directions in the plane perpendicular to the axis of quantization ($M=0$), and the fission is occurring through states with $K=0$ only, the curves give the fission-fragment angular distributions measured relative to the axis of quantization for various values of J. [After Huizenga *et al.* (1965).]

However, these assignments must be considered very tentative due to the large number of assumptions in the data analyses.

The fission probabilities of even-even nuclei have been measured by the (d, pf) reaction at very low excitation energy. Structure in the fission probability functions has been observed for the nuclei (see Fig. V-33) and interpreted in terms of vibrational states in the second potential well. In order to illustrate the evidence for these vibrational states in the second minimum, we will discuss the results from the ^{239}Pu(d, pf) reaction (Back *et al.*, 1971). The fission probability for ^{240}Pu shows a pronounced resonance structure at 5.0 MeV and a weaker structure at about 4.4 MeV (see Fig. V-33). Anisotropy measurements of the fission fragments indicate that the resonance at 5.0 MeV is due to a $K = 0$ band (Britt *et al.*, 1965). The resonance at 5.0 MeV has also been

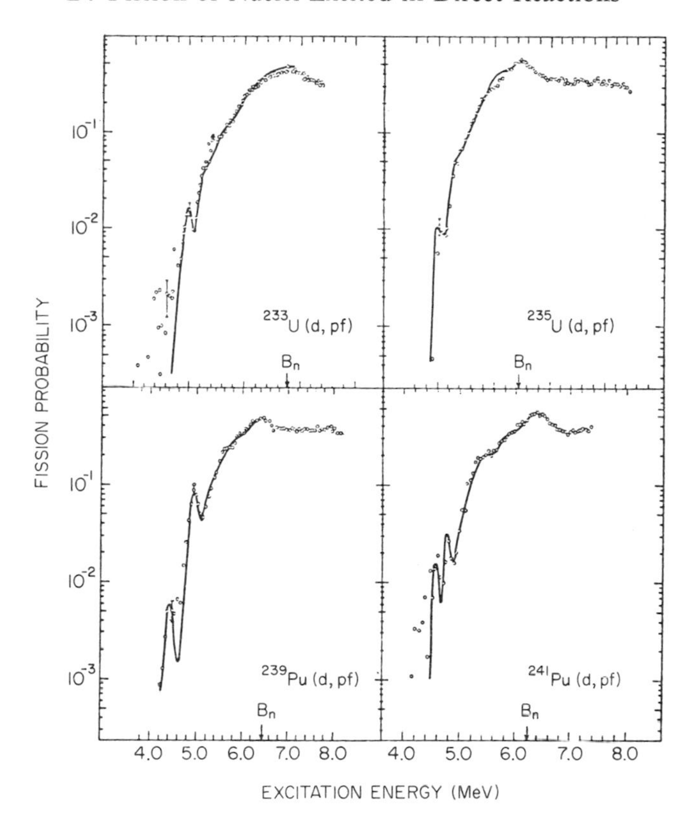

Fig. V-33. Experimental and calculated fission probabilities for the even-even fissioning nuclei ^{234}U, ^{236}U, ^{240}Pu, and ^{242}Pu ($E_d = 13$ MeV, $\theta_p = 140°$, $\theta_f = 90°$). o Experimental values; — fits to the sets of experimental points; B_n indicates the neutron binding energy up to which energy the calculations have been performed. The parameters of the calculation are given in the original paper. [After Back *et al.* (1971).]

studied with 20-keV energy resolution (Specht *et al.*, 1969). The overall resonance with a total width of approximately 250 keV was resolved into eight to ten peaks, each with a width that may be smaller than or equal to the experimental resolution and with an average spacing of about 35 keV (see Fig. V-34). If we assume that this energy corresponds to the average spacing between 2+ states in the transition nucleus, the corresponding excitation energy in the second potential well is 2.5–3.0 MeV. From the ^{239}Pu(n, f) reaction, the average level spacing at the neutron binding energy for 1+ states in the first and second potential wells is 3 and 460 eV, respectively. These values give an energy difference between the two wells of 2.3 MeV. This result is in reasonable agreement with a recent determination of this quantity of 2.6 ± 0.3 MeV from a threshold measurement (Britt *et al.*, 1971). Hence, for

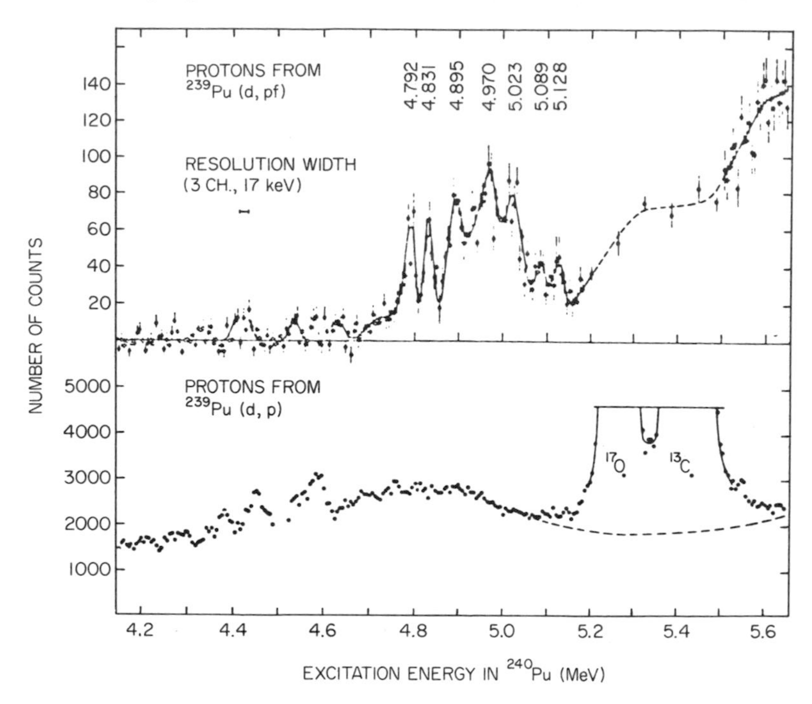

Fig. V-34. Direct experimental results showing singles and coincidence proton spectra for the ^{239}Pu(d,p) and ^{239}Pu(d,pf) reactions. Error bars give statistical deviations. [After Specht *et al.* (1969).]

5-MeV excitation in the first well, we expect an excitation energy of about 2.5 MeV in the second well, a value in good agreement with the spacing of the assumed 2+ states in the resonance.

The theoretical calculations of the fission probabilities displayed in Fig. V-33 depend on several assumptions. One of the enhancements to the fission probability comes from the introduction of damping in the second well. This enhancement arises from the fact that at the peak of the resonance the fission transmission coefficient can exceed the radiation transmission coefficient of Eq. (V-27) by several orders of magnitude. Although damping decreases the peak value of the fission transmission coefficient, the resulting larger width can increase greatly the energy region over which the fission transmission coefficient exceeds the corresponding radiation coefficient, resulting in an increase in the area of the fission resonance. The magnitude and energy dependence of the damping depend on an energy-dependent imaginary potential which is adjusted to fit the resonance width at 5 MeV and to be consistent with no resonances at higher energy. The question of the initial population of states by a direct reaction has been discussed earlier. It is interesting to note that the area of the resonance is smaller in the ^{240}Pu(p; p′f) reaction than in the ^{239}Pu(d, pf) reaction. This is consistent with an enhancement in the population of positive parity states in the $^{239}Pu(d, pf)$ reaction (see Fig. V-30). At energies below the neutron binding energy the only exit channels competing with fission are the γ-decay channels. Calculation of the fission cross section or probability is performed by Eq. (V-71) with fission transmission coefficients determined for a two-humped fission barrier.

3. Fission of Odd-*A* Transition Nuclei

The first measurements of fission-fragment angular distributions for odd-*A* transition nuclei excited via a direct reaction fission process was for the $^{234}U(d, pf)$ reaction (Vandenbosch *et al.*, 1965). Although the principal motivation for this experiment was to search for a possible correlation between angular anisotropy and mass symmetry, the measured anisotropies at low excitation energies were found to be very similar to the $^{234}U(n, f)$ data, as shown in Fig. V-35.

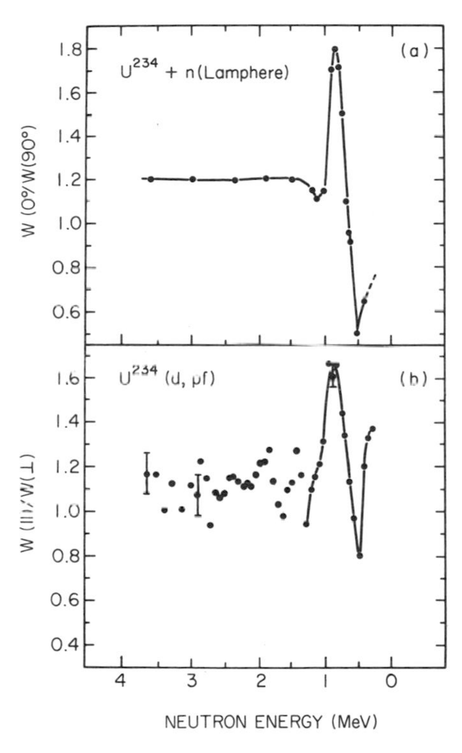

Fig. V-35. (a) Angular anisotropy as a function of neutron energy obtained by Lamphere (1962), (b) anisotropy obtained in the $^{234}U(d,pf)$ reaction. [After Vandenbosch *et al.* (1965).]

More recently, fission-fragment angular correlations and fission probabilities have been measured for a series of odd-A uranium and plutonium isotopes excited by (d, p), (t, d), and (t, p) reactions (Britt and Cramer, 1970). Because of the complexity of both the fission and direct reaction processes, it has not been possible to determine details of the fission barrier shapes or the transition state spectra from direct reaction fission results. The fundamental difference between the (n, f) and direct reaction fission processes is in the distribution of angular momentum values which can be excited in each case. The calculated average angular momentum transfers $\langle l \rangle$ in the (n, f) reaction are 0.6, 1.2, 1.7, and 2.3 for E_n values of 0.1, 0.5, 1.0, and 2.0 MeV, respectively. These values are to be compared to $\langle l \rangle$ equal to 3.1 and 4.3 for the (d, pf) and (t, pf) reactions, respectively. The result of these differences is that the different reactions are expected to excite particular transition states with quite different intensities when the fission barrier is less than a few tenths of a MeV above the neutron binding energy. When the fission barrier corresponds to high energy neutrons, the two types of reactions should give very similar results.

Another feature of the odd-A nuclei studied by the direct reaction fission process is the lack of evidence for vibrational states in these nuclei. The slight decrease in the fission probability around the neutron binding energy for ^{239}Pu is explained in terms of neutron competition. In order to reproduce the lack of resonance structure, the theory requires a stronger damping in the second potential well (Back *et al.*, 1971). Such a result is expected on the basis of the higher level density. It would be interesting to measure the fission probabilities of odd uranium and thorium isotopes in some detail by direct reaction fission processes, especially for ^{231}Th.

The doubly odd fissioning nuclei are extremely complex to interpret due to the expected large density of K bands near the barrier. Several such cases have been studied (Back *et al.*, 1971). For these nuclei it was necessary to assume complete damping in the second well to reproduce the absence of any resonance structure below the neutron binding energy. Since the second barrier for these nuclei is of the order of 1 MeV lower than the first one, the complete damping in the second well gives transparencies which are essentially those of a single barrier.

References

Andersen, B. L., Back, B. B., and Bang, J. M. (1970). *Nucl. Phys. A* **147**, 33.

Auerbach, E. H., and Perey, F. G. J. (1962). Rep. BNL 765. Brookhaven Nat. Lab. Upton, New York, unpublished.

Axel, P. (1962). *Phys. Rev.* **126**, 671.

Back, B. B., Bondorf, J. P., Otroshenko, G. A., Pedersen, J., and Rasmussen, B. (1969). *Proc. IAEA Symp. Phys. Chem. Fission, 2nd, Vienna, 1969*, p. 351. IAEA, Vienna.

Back, B. B., Bondorf, J. P., Otroshenko, G. A. Pedersen, J., and Rasmussen, B. (1971). *Nucl. Phys. A* **165**, 449.
Baerg, A. P., Bartholomew, R. M., Brown, F., Katz, L., and Kowalski, S. B. (1959). *Can. J. Phys.* **37**, 1418.
Behkami, A. N. (1971). *Nucl. Data Tables* **10**, 1 (Academic Press, New York).
Behkami, A. N., Roberts, J. H., Loveland, W. D., and Huizenga, J. R. (1968a). *Phys. Rev.* **171**, 1267.
Behkami, A. N., Huizenga, J. R., and Roberts, J. H. (1968b). *Nucl. Phys. A* **118**, 65.
Bjørnholm, S., and Strutinsky, V. M. (1969). *Nucl. Phys. A* **136**, 1.
Blatt, J. M., and Weisskopf, V. F. (1952). "Theoretical Nuclear Physics", p. 654. Wiley, New York.
Bohr, A. (1956). *Proc. U.N. Int. Conf. Peaceful Uses At. Energy*, Vol. **2**, p. 151. New York.
Britt, H. C., and Cramer, J. D. (1969). *Phys. Rev.* **181**, 1634.
Britt, H. C., and Cramer, J. D. (1970). *Phys. Rev. C* **2**, 1758.
Britt, H. C., and Plasil, F. (1966). *Phys. Rev.* **144**, 1046.
Britt, H. C., Gibbs, W. R., Griffin, J. J., and Stokes, R. H. (1965). *Phys. Rev.* **139**, B 354.
Britt, H. C., Rickey, F. A., Jr., and Hall, W. S. (1968). *Phys. Rev.* **175**, 1525.
Britt, H. C., Burnett, S. C., and Cramer, J. D. (1969). *Proc. IAEA Symp. Phys. Chem. Fission, 2nd, Vienna, 1969*, p. 375. IAEA, Vienna.
Britt, H. C., Burnett, S. C., Erkkila, B. H., Lynn, J. E., and Stein, W. E. (1971). *Phys. Rev. C* **4**, 1444.
Cramer, J. D., and Nix, J. R. (1970). *Phys. Rev. C* **2**, 1048 and references therein.
Eccleshall, D., and Yates, M. J. L. (1965). *Proc. IAEA Symp. Phys. Chem. Fission, Salzburg, 1965*, **1**, p. 77. IAEA, Vienna.
Farrell, J. A. (1968). *Phys. Rev.* **165**, 1371.
Griffin, J. J. (1959). *Phys. Rev.* **116**, 107.
Griffin, J. J. (1963). *Phys. Rev.* **132**, 2204.
Hauser, W., and Feshbach, H. (1952). *Phys. Rev.* **87**, 366.
Hill, D. L., and Wheeler, J. A. (1953). *Phys. Rev.* **89**, 1102.
Huizenga, J. R., Clarke, K. M., Gindler, J. E., and Vandenbosch, R. (1962). *Nucl. Phys.* **34**, 439.
Huizenga, J. R., Unik, J. P., and Wilkens, B. D. (1965). *Proc. Int. Conf. Phys. Chem. Fission, Salzburg, 1965*, **1**, p. 11. IAEA, Vienna.
Huizenga, J. R., Behkami, A. N., Meadows, J. W., and Klema, E. D. (1968). *Phys. Rev.* **174**, 1539.
Huizenga, J. R., Behkami, A. N., and Roberts, J. H. (1969). *Proc. IAEA Symp. Phys. Chem. Fission, 2nd, Vienna, 1969*, p. 403. IAEA, Vienna.
Ignatyuk, A. V., Rabotnov, N. S., and Smirenkin, G. N. (1969). *Phys. Lett. B* **29**, 209.
James, G. D., Lynn, J. E., and Earwaker, L. G. (1972). *Nucl. Phys.* **A 189**, 225.
Johansson, S. A. E. (1961). *Nucl. Phys.* **22**, 529.
Kahn, A. M., and Knowles, J. W. (1972). *Nucl. Phys. A* **179**, 333.
Knowles, J. W. (1970). Private communication.
Knowles, J. W., and Khan, A. M. (1968). *Can. Phys. Soc. Meeting, 1968*.
Lamphere, R. W. (1962). *Nucl. Phys.* **38**, 561.
Lynn, J. E. (1969). *Proc. IAEA Symp. Phys. Chem. Fission, 2nd, Vienna, 1969*, p. 249. IAEA, Vienna.
Moldauer, P. A. (1961). *Phys. Rev.* **123**, 968.
Moldauer, P. A., Engelbrecht, C. A., and Duffy, G. J. (1964). Rep. No. ANL-6978. Argonne Nat. Lab. Argonne, Illinois.
Möller, P., and Nilsson, S. G. (1970). *Phys. Lett.* **B 31**, 283.

Northrop, J. A., Stokes, R. H., and Boyer, K. (1959). *Phys. Rev.* **115**, 1277.
Otozai, K., Meadows, J. W., Behkami, A. N., and Huizenga, J. R. (1970). *Nucl. Phys. A* **144**, 502.
Pashkevich, V. V. (1971). *Nucl. Phys.* **A 169**, 275.
Pauli, H. C., Ledergerber, T., and Brack, M. (1971). *Phys. Lett.* **B 34**, 264.
Pedersen, J., and Kuzminov, B. D. (1969). *Phys. Lett. B* **29**, 176.
Preston, M. A. (1962). "Physics of the Nucleus." Addison-Wesley, Reading, Massachusetts.
Rabotnov, N. S., Smirenkin, G. N., Soldatov, A. S., Usachev, L. N., Kapitza, S. P., and Tsipenyuk, Yu. M. (1965). *Proc. IAEA Symp. Phys. Chem. Fission, Salzburg, 1965*, **1**, p. 135. IAEA, Vienna.
Rabotnov, N. S., Smirenkin, G. N., Soldatov, A. S., Usachev, L. N., Kapitza, S. P., and Tsipenyuk, Yu. M. (1966). *Nucl. Phys.* **77**, 92.
Rabotnov, N. S., Smirenkin, G. N., Soldatov, A. S., Usachev, L. N., Kapitza, S. P., and Tsipenyuk, Yu. M. (1968). *Phys. Lett. B* **26**, 218.
Rabotnov, N. S., Smirenkin, G. N., Soldatov, A. S., Usachev, L. N., Kapitza, S. P., and Tsipenyuk, Yu. M. (1969). Institute of Physics and Energetics, Report FEI-170 [Translation in Los Alamos Sc. Lab. Rep. LA-4385-tr (1970)].
Rabotnov, N. S., Smirenkin, G. N., Soldatov, A. S., Usachev, L. N., Kapitza, S. P., and Tsipenyuk, Yu. M. (1970). *Sov. J. Nucl. Phys.* **11**, 285.
Rose, M. E. (1957). "Elementary Theory of Angular Momentum." Wiley, New York.
Specht, H. J., Fraser, J. S., and Milton, J. C. D. (1966). *Phys. Rev. Lett.* **17**, 1187.
Specht, H. J., Fraser, J. S., Milton, J. C. D., and Davies, W. G. (1969). *Proc. IAEA Symp. Phys. Chem. Fission, 2nd Vienna, 1969*, p. 363. IAEA, Vienna.
Soldatov, A. S., *et al.* (1965). *Phys. Lett.* **14**, 217.
Soldatov, A. S., Tsipenyuk, Yu. M., and Smirenkin, G. N. (1970). *Sov. J. Nucl. Phys.* **11**, 552.
Unik, J. P. (1966). Private communication.
Vandenbosch, R. (1967). *Nucl. Phys. A* **101**, 460.
Vandenbosch, R., Unik, J. P., and Huizenga, J. R. (1965). *Proc. IAEA Symp. Phys. Chem. Fission, Salzburg, 1965*, **1**, p. 547. IAEA, Vienna.
Vandenbosch, R., Wolf, K. L., Unik, J. P., Stephan, C., and Huizenga, J. R. (1967). *Phys. Rev. Lett.* **19**, 1138.
Wheeler, J. A. (1963). *In* "Fast Neutron Physics" (J. B. Marion and J. L. Fowler, eds.), Pt. II. Wiley (Interscience), New York.
Wilets, L., and Chase, D. M. (1956). *Phys. Rev.* **103**, 1296.
Wilkins, B. D., Unik, J. P., and Huizenga, J. R. (1964). *Phys. Lett.* **12**, 243.
Wolf, K. L., Vandenbosch, R., and Loveland, W. D. (1968). *Phys. Rev.* **170**, 1059.
Yuen, G., Rizzo, G. T., Behkami, A. N., and Huizenga, J. R. (1971). *Nucl. Phys. A* **171**, 614

CHAPTER

VI

Fission-Fragment Angular Distributions at Moderate to High Excitation Energies

A. Introduction

In the last chapter a theory of fission-fragment angular distributions was developed for isolated transition states with specific quantum numbers. In the present chapter this theory will be extended to higher excitation energies where the transition levels can be described by statistical methods.

Angular distributions have been studied in the excitation region up to several tens of MeV for many targets and for a variety of projectiles, including neutrons, protons, deuterons, α particles, and heavy ions. Some of the more striking features of the experimental observations are summarized in the following statements.

(a) The fission fragments have the largest differential cross sections in the directions forward and backward along the beam.

(b) The anisotropies are largest for the heaviest projectiles and smallest for neutron and proton bombardments.

(c) The anisotropy increases whenever a threshold is reached where it becomes energetically possible for fission to occur in the residual nucleus which is left behind after the evaporation of some definite number of neutrons (see Figs. VI-13, VI-14, and VI-12).

(d) The anisotropies are approximately the same for odd-A targets as for even-even targets (see Fig. VI-20).

(e) The anisotropy decreases as the value of Z^2/A of the target increases (see Figs. VI-19 and VI-20).

The angular distribution of fission fragments depends on two quantities: the angular momentum brought in by the projectile and the fraction of this angular momentum which is converted into orbital angular momentum between the fragments. This fraction is characterized through a parameter K, where K is defined as the projection of the angular momentum J on the nuclear symmetry axis (see Fig. V-1). Again, the two assumptions already discussed in Chapter V are necessary to give a quantitative description of fission-fragment angular distributions. These assumptions are that (a) the fragments separate along the nuclear symmetry axis and (b) although K is not a good quantum number between the original nucleus and the saddle point, it is a good quantum number beyond the saddle point.

The distribution of angular momenta of the fissioning system may be controlled by the experimenter through his choice of the projectile and its energy. The distribution of K, however, is determined by the nucleus itself. From measurements of fission-fragment angular distributions, it is possible, therefore, to deduce the K spectrum of states at the saddle point deformation.

B. Theory of Angular Distributions for Transition States Described by a Statistical Model

1. Exact Theoretical Expression

The density of levels in the transition nucleus with spin J and projection of J on the nuclear symmetry axis equal to K is given by the approximate relation

$$\rho(J, K) \propto \exp[(E - E_{\text{rot}}^{J,K})/t] \qquad \text{(VI-1)}$$

where E is the total energy, $E_{\text{rot}}^{J,K}$ is the energy tied up in rotation for transition state (J, K), and t is the thermodynamic temperature. The derivation of Eq. (VI-1) requires the constancy of t for small changes in excitation energy around E. The thermodynamic energy available to the nucleus is the quantity $(E - E_{\text{rot}}^{J,K})$. The rotational energy of a nucleus in its saddle point deformation is

$$E_{\text{rot}}^{J,K} = (\hbar^2/2\mathcal{I}_\perp)(J^2 - K^2) + (\hbar^2/2\mathcal{I}_\parallel)K^2 \qquad \text{(VI-2)}$$

where $\mathcal{I}_\perp$ and $\mathcal{I}_\parallel$ are nuclear moments of inertia about axes perpendicular and parallel to the symmetry axis, respectively. Substitution of Eq. (VI-2) into Eq. (VI-1) gives

$$\rho(J, K) \propto \exp\{(E/t) - (\hbar^2 J^2/2\mathcal{I}_\perp t) - (\hbar^2 K^2/2t)[(1/\mathcal{I}_\parallel) - (1/\mathcal{I}_\perp)]\} \qquad \text{(VI-3)}$$

For fixed values of E and t the number of transition levels $\rho(J, K)$ depends on two quantities, $(\hbar^2 J^2/2\mathscr{I}_\perp t)$ and $(\hbar^2 K^2/2t)[(1/\mathscr{I}_\parallel)-(1/\mathscr{I}_\perp)]$. If, in addition, J is fixed, then the distribution in K becomes

$$\rho(K) \propto \exp\{-(\hbar^2 K^2/2t)[(1/\mathscr{I}_\parallel) - (1/\mathscr{I}_\perp)]\} \tag{VI-4}$$

This equation is equivalent to a Gaussian K distribution (Halpern and Strutinsky, 1958)

$$\begin{aligned} \rho(K) &\propto \exp(-K^2/2K_0^2), \qquad K \leqslant J \\ &= 0, \qquad K > J \end{aligned} \tag{VI-5}$$

where $K_0^2 = t/\hbar^2[(1/\mathscr{I}_\parallel)-(1/\mathscr{I}_\perp)]$. If the quantity $[(1/\mathscr{I}_\parallel)-(1/\mathscr{I}_\perp)]$ is replaced by $1/\mathscr{I}_{\text{eff}}$, then

$$K_0^2 = t\mathscr{I}_{\text{eff}}/\hbar^2 \tag{VI-6}$$

For a Gaussian K distribution an exact expression for the fission-fragment angular distribution may be derived from Eq. (V-3) by proper weightings of J, M, and K. If the target and projectile spins are included, the exact equation for the angular distribution is

$$W(\theta) \propto \sum_{J=0}^{\infty} \sum_{M=-(I_0+s)}^{+I_0+s} \left\{ \sum_{l=0}^{\infty} \sum_{S=|I_0-s|}^{I_0+s} \sum_{\mu=-I_0}^{+I_0} \frac{(2l+1)\, T_l\, |C_{M,0,M}^{S,l,J}|^2\, |C_{\mu,M-\mu,M}^{I_0,s,S}|^2}{\sum_{l=0}^{\infty}(2l+1)\, T_l} \right\} \times \sum_{K=-J}^{J} \frac{(2J+1)|d_{M,K}^{J}(\theta)|^2 \exp(-K^2/2K_0^2)}{\sum_{K=-J}^{J} \exp(-K^2/2K_0^2)} \tag{VI-7}$$

The quantities I_0, s, and S are the target spin, projectile spin, and channel spin, respectively. The channel spin S is defined by the relation $\mathbf{S} = \mathbf{I}_0 + \mathbf{s}$. The total angular momentum J is given by the sum of the channel spin and orbital angular momentum, $\mathbf{J} = \mathbf{S} + \boldsymbol{l}$. The projection of I_0 on the space-fixed axis is given by μ, whereas the projection of S (and J) on the space-fixed axis is M. The quantity in the bracket gives the weighting factor for a particular (J, M) combination. This value multiplies the angular dependent term for the allowable K states (K distribution is weighted also) of a particular J. This product is summed first over all M values for a particular J and finally over all J values. The $d_{M,K}^{J}(\theta)$ function is defined in Eq. (V-2), and the quantities $C_{M,0,M}^{S,l,J}$ and $C_{\mu,M-\mu,M}^{I_0,s,S}$ are Clebsch–Gordan coefficients.

Equation (VI-7) is an exact theoretical expression for the fission-fragment angular distribution when all the excited compound nuclei decay by fission. It is valid also for cases where the compound nucleus decays by other channels if no fractionation occurs in the compound nucleus spin distribution. The latter restriction is necessary because Eq. (VI-7) contains a weighting over the J distribution which results from the formation of the compound nucleus. If spin fractionation occurs in the decay of the compound nucleus through the various exit channels, a change must be made in the weighting factors of J in Eq. (VI-7) in order to obtain the appropriate J distribution for nuclei which decay through the fission channels.

The relative number of levels $\rho(J, K)$ in the transition nucleus at a particular energy is given by Eq. (VI-3). However, we will observe that the normalization of Eq. (VI-7) is done separately for each value of J with the Gaussian K distribution defined by Eq. (VI-5). This means that the relative number of transition states with different values of J does not enter into the normalization. It is implicitly assumed that the relative J distribution is fixed in the initial stage of compound nucleus formation and remains fixed (no spin fractionation) throughout the fission process; hence, it is independent of the relative numbers of levels with different values of J in the transition nucleus. The total intensity of a particular J value distributes itself over the available levels in the transition nucleus. If more levels are available, the fraction of the intensity to each transition level is reduced accordingly.

It is perhaps instructive to understand the physical origin of the angular distribution given by Eq. (VI-7). Neutron- and charged particle-induced fission give distributions which typically have more fragments emitted forward and backward along the beam direction than perpendicular to the beam direction, as illustrated in Figs. VI-4, VI-6, VI-7, and VI-8. This result is most easily understood for the case where the target and projectile spins are zero. In this case, each value of the compound nuclear angular momentum arises from an orbital angular momentum associated with a particular impact parameter of the collision. These compound nuclear angular momentum vectors are uniformly distributed in the plane perpendicular to the beam direction. If all K values were equally probable, an isotropic distribution would result. The density of transition states with particular values of J and K is given by Eq. (VI-1), where the familiar Boltzmann factor contains the rotational energy for the particular state of angular momentum J and projection K on the nuclear symmetry axis. For a particular value of J, the rotational energy is smallest for $K = 0$ (rotation about an axis perpendicular to the nuclear symmetry axis) due to the fact that $\mathscr{I}_{\perp}$ is larger than $\mathscr{I}_{\parallel}$ [see Eq. (VI-2)]. This leads to the result of Eq. (VI-5), where the density $\rho(K)$ is largest for $K = 0$.

Fragments originating from fission through the $K = 0$ states will be aligned perpendicular to the angular momentum vectors J, which in turn lie in the plane perpendicular to the beam. Integration over all the allowed directions in space leads to a concentration of fragments in the fore-aft direction relative to the beam. At the other extreme, fission through states with $K = J$ results in fragments which are concentrated in a direction perpendicular to the beam. These events with large K are less likely due to their smaller state density, which results from their larger rotational energy [Eq. (VI-2)] and less favorable Boltzmann factor. This qualitatively explains the fore-aft peaking for a K distribution given by Eq. (VI-5).

The result of nonzero target or projectile spin (assumed to be randomly oriented in space) is to introduce a component of the total compound nuclear angular momentum which need not lie in the plane perpendicular to the beam. This results in a smearing of the fore-aft peaking and a decrease in the anisotropy, as illustrated in Fig. VI-5 for increasing target spin.

2. Approximate Theoretical Expressions

For energetic projectiles many angular momentum states are excited in the compound nucleus and calculations of angular distributions become very time consuming, even with large computers. Some inherent simplicity results, however, for high energy projectiles. For neutron and charged particle projectiles, for example, the orbital angular momenta are in a plane perpendicular to the respective beam directions. In addition, when the orbital angular momenta are very large relative to the target and projectile spins, the projections of the total angular momenta M on the space-fixed z axis (beam direction) are very small relative to the total angular momenta J.

In the limit when the target and projectile spins are zero and no particle emission from the initial compound nucleus occurs before fission (i.e., $M = 0$), the angular distribution for a particular J with the assumption of a Gaussian K distribution is

$$W^J_{M=0}(\theta) \propto \sum_{K=-J}^{J} \frac{(2J+1)|d^J_{M=0,K}(\theta)|^2 \exp(-K^2/2K_0^2)}{\sum_{K=-J}^{J} \exp(-K^2/2K_0^2)} \tag{VI-8}$$

When many J values of the compound nucleus contribute and the transmission coefficients T_l are known, the overall angular distribution for $M = 0$ becomes

$$W(\theta) \propto \sum_{J=0}^{\infty} (2J+1) T_J \sum_{K=-J}^{J} \frac{(2J+1)|d^J_{M=0,K}(\theta)|^2 \exp(-K^2/2K_0^2)}{\sum_{K=-J}^{J} \exp(-K^2/2K_0^2)} \tag{VI-9}$$

where the transmission coefficients are written as T_J since $l = J$ when $M = 0$.

Equation (VI-9) is an exact theoretical expression for computation of fission-fragment angular distributions when both the target and projectile spins are zero. Examples of such reactions are α particle-induced fission of even-even target nuclei (limited to reactions where only first-chance fission occurs, since M is no longer zero if fission follows initial deexcitation by neutron emission). The use of Eqs. (VI-7) and (VI-9) requires the evaluation of many $d^J_{M,K}(\theta)$ functions; hence, both of these equations have been used at moderate excitation energies for the first time only recently (Huizenga *et al.*, 1969).

Since the evaluation of many $d^J_{M,K}(\theta)$ functions in either Eq. (VI-7) or (VI-9) takes considerable computer time, a theoretical expression for fission-fragment angular distributions will be developed in which the $d^J_{M,K}(\theta)$ functions are approximated with the following relation (Wheeler, 1963).

$$|d^J_{M,K}(\theta)|^2 \cong \pi^{-1}[(J+\tfrac{1}{2})^2 \sin^2\theta - M^2 - K^2 + 2MK\cos\theta]^{-1/2} \tag{VI-10}$$

The substitution of this expression into Eq. (V-3) for the case where M = 0 gives

$$W^J_{M=0,K}(\theta) = \frac{2J+1}{2\pi}[(J+\tfrac{1}{2})^2 \sin^2\theta - K^2]^{-1/2} \tag{VI-11}$$

With a Gaussian K distribution, the angular distribution for a particular J (for $M=0$) is

$$W^J_{M=0}(\theta) = \frac{(2J+1)\int_0^{(J+1/2)\sin\theta}[(J+\tfrac{1}{2})^2\sin^2\theta - K^2]^{-1/2}\exp(-K^2/2K_0{}^2)\,dK}{2\pi\int_0^{(J+1/2)}\exp(-K^2/2K_0{}^2)\,dK} \tag{VI-12}$$

The integrals in both the numerator and the denominator can be evaluated and written in the following explicit form (Huizenga *et al.*, 1969).

$$W^J_{M=0}(\theta) = \frac{(\pi/2)(2J+1)\exp[-(J+\tfrac{1}{2})^2\sin^2\theta/4K_0{}^2]J_0[i(J+\tfrac{1}{2})^2\sin^2\theta/4K_0{}^2]}{2\pi(\pi^{1/2}/2)(2K_0{}^2)^{1/2}\,\mathrm{erf}[(J+\tfrac{1}{2})/(2K_0{}^2)^{1/2}]} \tag{VI-13}$$

where J_0 is the zero order Bessel function with an imaginary argument, and $\mathrm{erf}[(J+\tfrac{1}{2})/(2K_0{}^2)^{1/2}]$ is the error function defined by

$$\mathrm{erf}(x) = (2/\pi^{1/2})\int_0^x \exp(-t^2)\,dt$$

If each J value is weighted by $(2J+1)T_J$, the overall fission-fragment angular distribution in the approximation ($M=0$) is

$$W(\theta) \propto \sum_{J=0}^{\infty} \frac{(2J+1)^2 T_J \exp[-(J+\tfrac{1}{2})^2\sin^2\theta/4K_0{}^2]J_0[i(J+\tfrac{1}{2})^2\sin^2\theta/4K_0{}^2]}{\mathrm{erf}[(J+\tfrac{1}{2})/(2K_0{}^2)^{1/2}]} \tag{VI-14}$$

Equations (VI-9) and (VI-14) are theoretical expressions for the fragment angular distributions from particle-induced fission for reactions in which the target and projectile have zero spins ($M=0$). Although Eq. (VI-9) is an exact expression, Eq. (VI-14) is an inexact expression, since the $d^J_{M,K}(\theta)$ functions are approximated by Eq. (VI-10) in its derivation.

3. Comparisons of the Various Theoretical Expressions

The validity of the approximate relation given in Eq. (VI-10) for moderate energy particle-induced fission is illustrated in Fig. VI-1. The theoretical anisotropy $W(171.7°)/W(93.8°)$ of fission fragments from the $^{206}Pb(\alpha,f)$ reaction with 38-MeV α particles is plotted as a function of $K_0{}^2$ for various theoretical equations. The experimental anisotropy of 2.56 is also shown. Curves (b) and (c) are computed with Eqs. (VI-14) and (VI-9), respectively. The agreement between these two curves is very good and illustrates the degree of validity of Eq. (VI-10) for such a reaction. Essentially equally good agreement between Eqs. (VI-14) and (VI-9) is obtained for a wide range of projectile energies (remembering, of course, that M must be zero).

Curve (a) is computed with an equation similar to Eq. (VI-14) except that

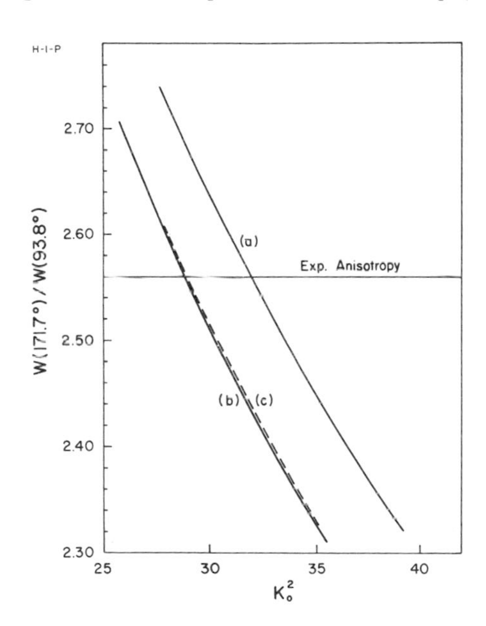

Fig. VI-1. Anisotropy $W(171.7°)/W(93.8°)$ of fission fragments for the $^{206}Pb(\alpha,f)$ reaction with 38-MeV α particles. The experimental anisotropy is 2.56. The values of $K_0{}^2$ calculated with different theoretical expressions described in the text are designated as follows. Curve (c) is calculated with Eq. (VI-9), the exact expression for $M = 0$; curve (b) with Eq. (VI-14), in which the $d^J_{M,K}(\theta)$ function is approximated by Eq. (VI-10); and curve (a) with an equation similar to Eq. (VI-14) except that the error function in the denominator is neglected.

the error function in the denominator is omitted. The value of K_0^2 computed by neglecting the error function is too large by 12%. The magnitude of the error resulting from the omission of the error function in the denominator of Eq. (VI-14) depends upon the projectile energy; however, the error is significant for most energies.

When either the target or the projectile (or both) has a finite spin, an exact calculation of the fission-fragment angular distribution may be performed with Eq. (VI-7). Comparisons of Eqs. (VI-9) and (VI-14) with the exact Eq. (VI-7) for the target ^{209}Bi with spin $\frac{9}{2}$ for the (d,f) reaction with 21-MeV deuterons and the (α,f) reaction with 43-MeV α particles are shown in Figs. VI-2 and VI-3, respectively. Again, the theoretical anisotropies computed with Eqs. (VI-9) and (VI-14) (assuming $M = 0$) and shown by curves (c) and (b), respectively, are almost identical. The results of exact theory with Eq. (VI-7) shown as curves (d) in Figs. VI-2 and VI-3 lead to values of K_0^2 which are several percent smaller for each reaction. Neglect of the error function gives the curve (a) in each of the two figures.

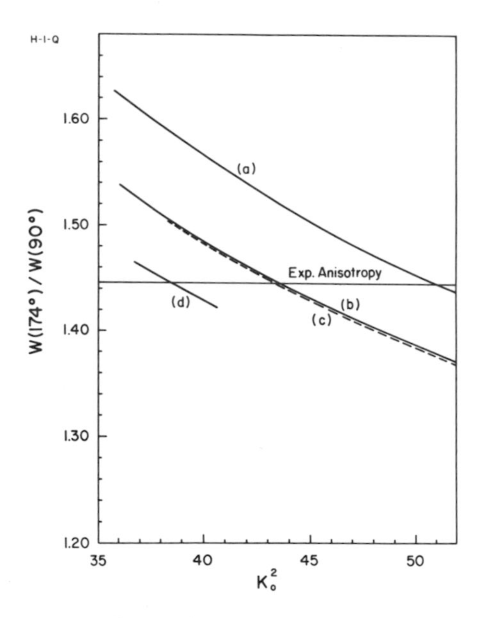

Fig. VI-2. Anisotropy $W(174°)/W(90°)$ of fission fragments for the ^{209}Bi(d, f) reaction with 21-MeV deuterons. The experimental anisotropy is 1.446. The values of K_0^2 calculated with the different theoretical expressions described in the text are designated as follows. Curve (d) is calculated with Eq. (VI-7), the exact expression including target and projectile spin; curve (c) with Eq. (VI-9), which assumes zero target and projectile spins ($M = 0$); curve (b) with Eq. (VI-14), which assumes zero target and projectile spins ($M = 0$) [in addition, the $d^J_{M,K}(\theta)$ functions are approximated by Eq. (VI-10)]; curve (a) with an equation similar to Eq. (VI-14) except that the error function in the denominator is neglected.

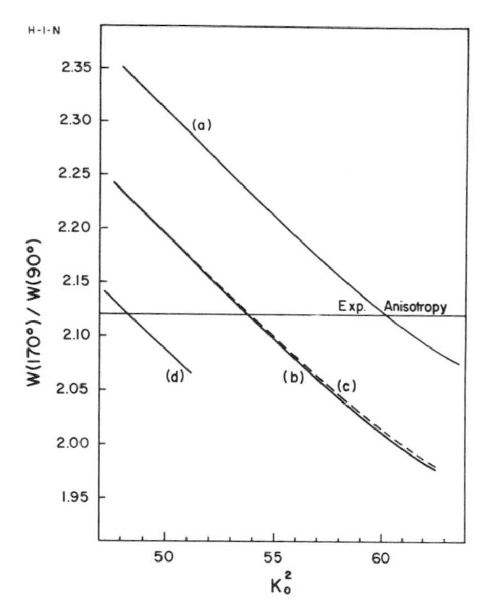

Fig. VI-3. Anisotropy $W(170°)/W(90°)$ of fission fragments for the $^{209}Bi(\alpha, f)$ reaction with 43-MeV α particles. The experimental anisotropy is 2.12. The values of $K_0{}^2$ calculated with the different theoretical expressions described in the text have the same designations as in Fig. VI-2.

Comparison of experimental and theoretical fission-fragment angular distributions in the past have been done with theoretical expressions which neglect the error function in the denominator [see Eq. (VI-14)]. The various approximate theoretical expressions used earlier have also differed in other ways; e.g., $(J+\frac{1}{2})$ has been replaced with J. A comparison of the results from several other approximate equations has been published (Huizenga *et al.*, 1969).

In Fig. VI-4 a comparison is made between the theoretical fragment angular distributions computed for the $^{206}Pb(\alpha, f)$ reaction at $E_\alpha = 30$ MeV with Eq. (VI-14) and with an equation similar to Eq. (VI-14) except that the error function in the denominator is omitted. The two theoretical expressions give an equally good fit to the experimental data. However, with Eq. (VI-14) the solid-line fit is obtained with a value of 8.83 for $K_0{}^2$, whereas the dashed-line fit obtained without the error function included in Eq. (VI-14) is obtained with a value of $K_0{}^2$ equal to 9.47.

The effect of target spin on anisotropy is illustrated in Fig. VI-5 for 43-MeV helium-ion bombardment of elemental bismuth. In the theoretical calculation of the anisotropy $W(170°)/W(90°)$, it is assumed that $K_0{}^2 = 51$ (see Fig. VI-3) and that the target has different values of spin. The theoretical anisotropy

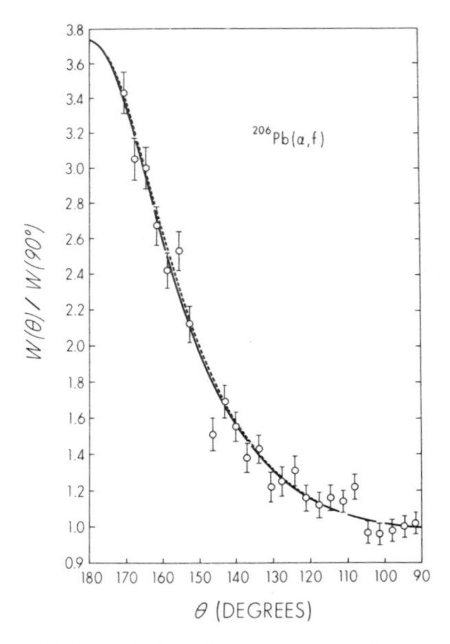

Fig. VI-4. Comparison of theoretical and experimental fragment angular distributions from 30-MeV α particle-induced fission of ^{206}Pb. The solid line is calculated with Eq. (VI-14) and gives $K_0^2 = 8.83$. The dashed line is calculated with an equation similar to Eq. (VI-14) except that the error function is omitted and gives $K_0^2 = 9.47$. [The experimental data are from Moretto *et al.* (1969).]

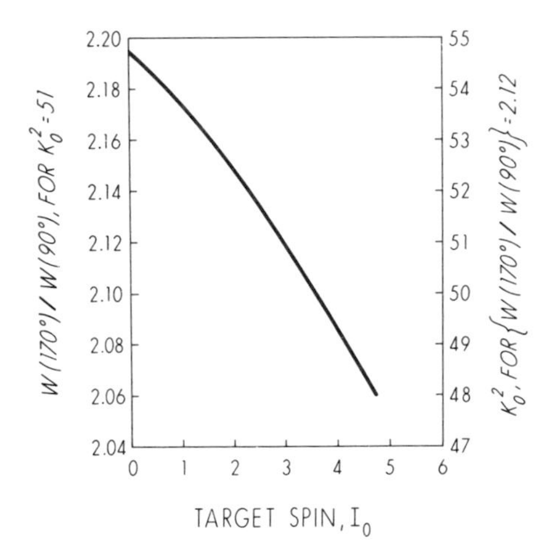

Fig. VI-5. Calculation of fission-fragment anisotropy $W(170°)/W(90°)$ and K_0^2 as a function of target spin employing Eq. (VI-7). The transmission coefficients T_l are computed for 42-MeV helium-ion-induced fission of bismuth. The left-hand ordinate $W(170°)/W(90°)$ is calculated as a function of target spin for constant $K_0^2 = 51$. The right-hand ordinate K_0^2 is calculated as a function of target spin for constant value of $W(170°)/W(90°)$ equal to 2.12 (the experimental anisotropy).

decreases approximately 1.3% for an increase in target spin of one unit. Or looked at in a different way, if the anisotropy $W(170°)/W(90°)$ is fixed at the experimental value of 2.12, then the computed value of K_0^2 decreases with increasing target spin. The magnitude of K_0^2 decreases about 3% per unit increase in target spin. The value of K_0^2 decreases with increasing target spin because the coupling of the target spin with the orbital angular momentum serves to decrease the alignment of the total angular momentum J.

C. Experimental Fission-Fragment Angular Distributions

In Figs. VI-6 to VI-8 a number of experimental fragment angular distributions are plotted for different fissioning systems. In general, the angular

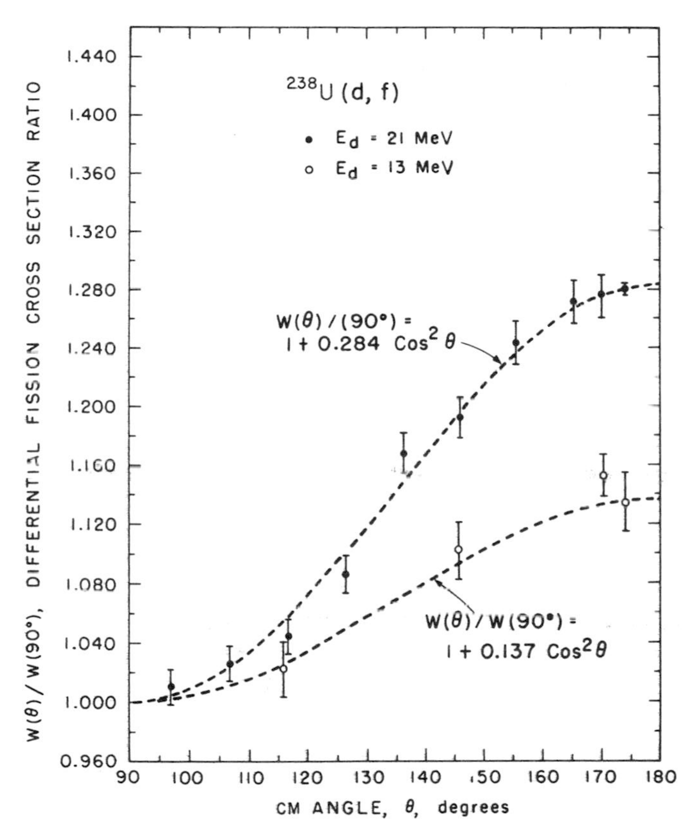

Fig. VI-6. Fragment angular distributions $W(\theta)/W(90°)$ as a function of the center-of-mass angle in degrees for 13-MeV and 21-MeV deuteron-induced fission of ^{238}U. The plotted points are experimental values with statistical uncertainties. The dashed curves are analytical fits for the expression $[W(\theta)/W(90°)] = 1 + C \cos^2 \theta$. [From Bate *et al.* (1963).]

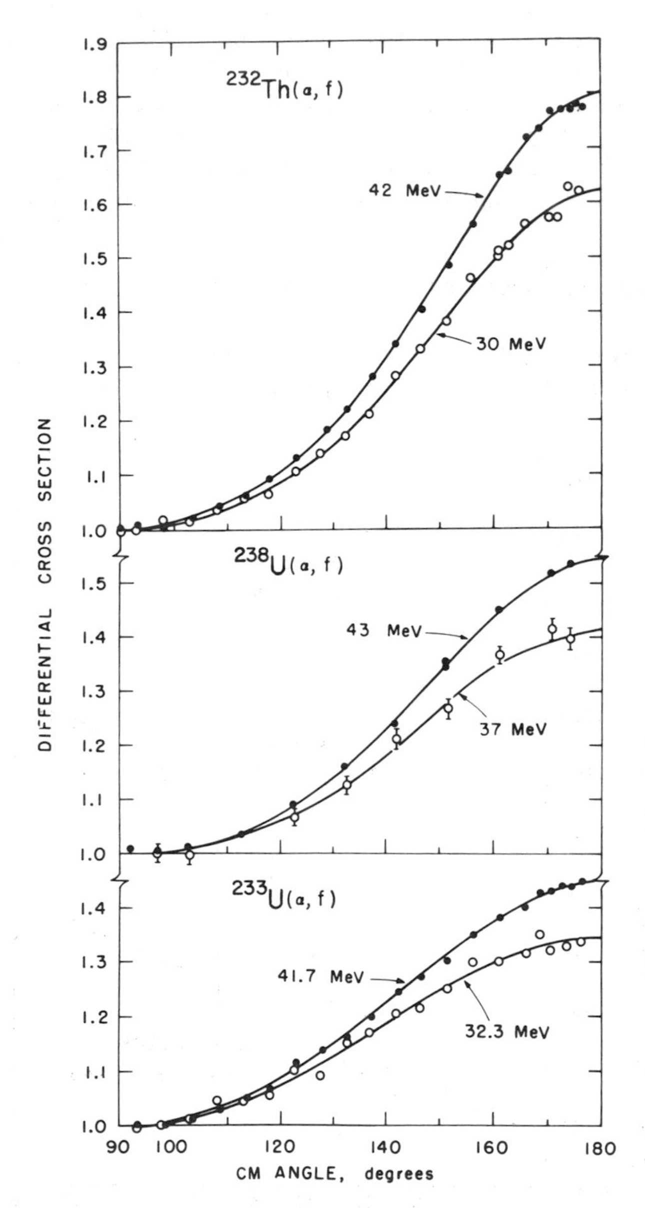

Fig. VI-7. Fragment angular distributions as a function of the center-of-mass angle in degrees for helium-ion-induced fission of ^{232}Th, ^{238}U, and ^{233}U. The projectile energies are shown on each section of the figure. The standard deviations are approximately the size of the symbols except for ^{238}U, where the errors are somewhat larger. [From Vandenbosch *et al.* (1961).]

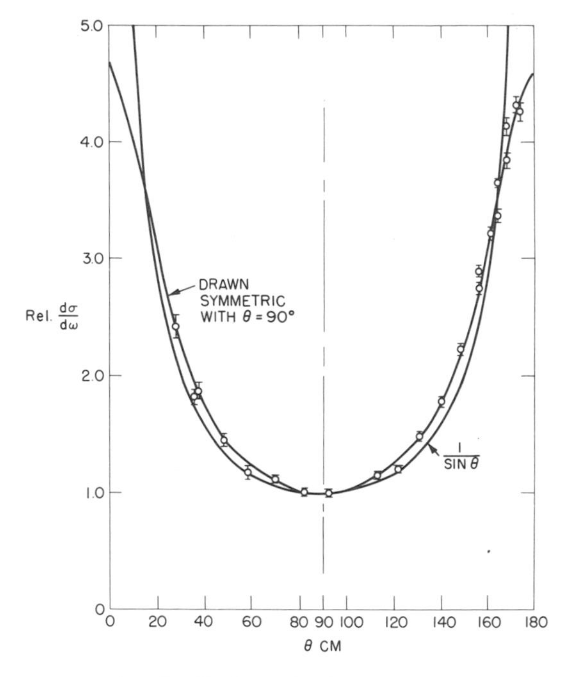

Fig. VI-8. Fragment angular distribution as a function of the center-of-mass angle in degrees for carbon-ion-induced fission of ^{197}Au. The experimental angular distribution is compared to a $1/\sin\theta$ distribution. [From Gordon *et al.* (1960).]

distributions are more strongly peaked along the beam direction as either the beam energy or the projectile mass is increased.

The degree of fission-fragment anisotropy for particle reactions depends on both the J and K distributions. Increasing the weighting of high J states serves to increase the anisotropy. This may be accomplished by the experimenter by increasing the projectile energy or mass. On the other hand, the K distribution is a nuclear property characterized by K_0^2, which depends on the quantity $\mathscr{I}_{\text{eff}}\, t/\hbar^2$, as shown earlier by Eq. (VI-6). As the excitation energy increases, the value of the temperature t increases, leading to larger values of K_0^2 (assuming $\mathscr{I}_{\text{eff}}$ to be independent of excitation energy). The anisotropy decreases with increasing K_0^2. Hence the excitation energy dependence of the anisotropy depends on the J and K distributions, which for increasing values of the two quantities change the anisotropy in opposite directions.

In a more quantitative discussion of the excitation energy dependence of the fission-fragment angular distribution it is convenient to show a ratio of differential fission cross sections rather than the full angular distribution. Actually this has already been done in several figures in this chapter, since the same information on K_0^2 is contained in this ratio as in the full angular

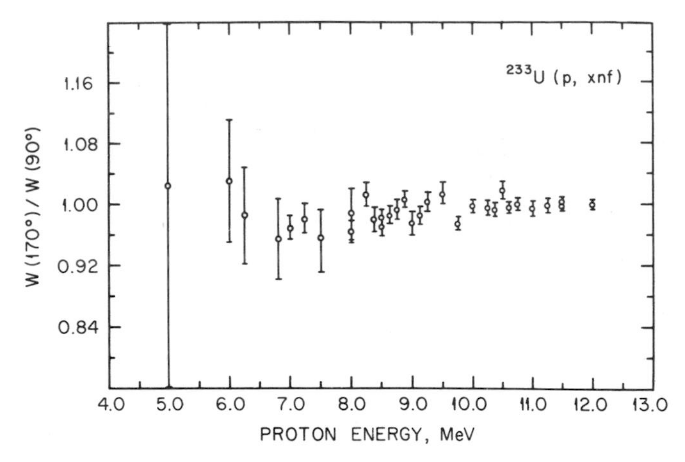

Fig. VI-9. Fission-fragment anisotropy ratio for proton-induced fission of ^{233}U as a function of proton energy. [After Bate and Huizenga (1964).]

distribution. Examples of the dependence of the fission-fragment anisotropy on excitation energy are given in Figs. VI-9 to VI-12. These data include proton-induced fission of ^{233}U, deuteron-induced fission of ^{232}Th, and helium-ion-induced fission of ^{209}Bi and ^{226}Ra.

For the proton energies included in Fig. VI-9, the angular momenta are quite small and the fission fragments are emitted nearly isotropically at all energies. The anisotropy of the deuteron-induced fission of ^{232}Th given in Fig. VI-10 increases rather smoothly with increasing energy. This is qualitatively a result of the greater importance of the increasing angular momenta relative to the increasing values of $K_0{}^2$ with excitation energy. The anisotropy of helium-ion-induced fission of ^{209}Bi shown in Fig. VI-11 is rather constant with excitation energy. This is qualitatively understood in terms of the equal importance of the two opposing factors in question. The anisotropy of helium-ion-induced fission of ^{226}Ra shown in Fig. VI-12 has a general upward trend, although considerable structure is also present as a function of excitation energy.

The periodic structure in the anisotropy shown in Fig. VI-12 is a result of multichance fission. The concept of multichance fission is illustrated in Fig. VI-13. The initial compound nuclei (Z, A) at excitation $E^*(Z, A)$ decay partially by fission and partially by neutron emission, forming nuclei $(Z, A-1)$ with average excitation energy $\langle E^*(Z, A-1)\rangle$. The average excitation energy of the nuclei formed by single neutron emission is reduced from the initial excitation energy by the neutron binding energy $B_n(Z, A)$ and the average neutron kinetic energy $\langle E_n(Z, A)\rangle$. The nuclei $(Z, A-1)$ are subject to a similar competition between fission and neutron emission and the branching continues until the excitation energy is smaller than the neutron binding energy and fission threshold.

Similar structure in the anisotropy is observed for neutron-induced fission of several targets, as shown in Fig. VI-14. In addition to the low energy structure discussed in Chapter V, a structure occurs at an excitation energy where second chance-fission (n, n′f) becomes energetically possible. The fragment anisotropy of this second-chance fission will be unusually large due to the small value of K_0^2 associated with the low excitation energy and nuclear temperature near the fission barrier. This structure near 6- or 7-MeV neutron energy is especially pronounced for even-even target nuclei where the pairing energy gap plays an important role in the threshold energy region of second-chance fission.

The foregoing structure in the anisotropy correlates rather closely with a

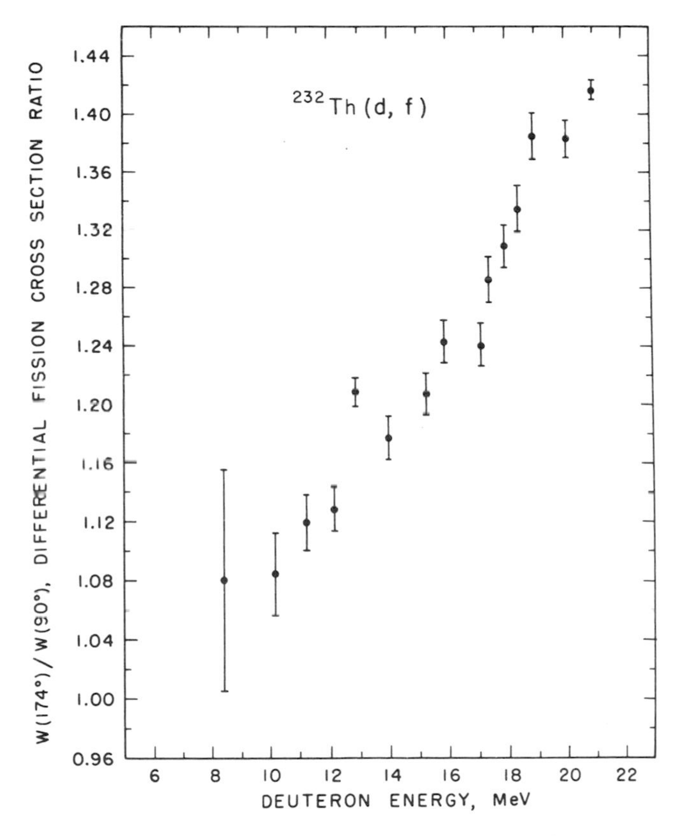

Fig. VI-10. Differential fission cross section ratio $W(174°)/W(90°)$ for deuteron-induced fission of ^{232}Th as a function of deuteron energy in the laboratory system. [After Bate *et al.* (1963).]

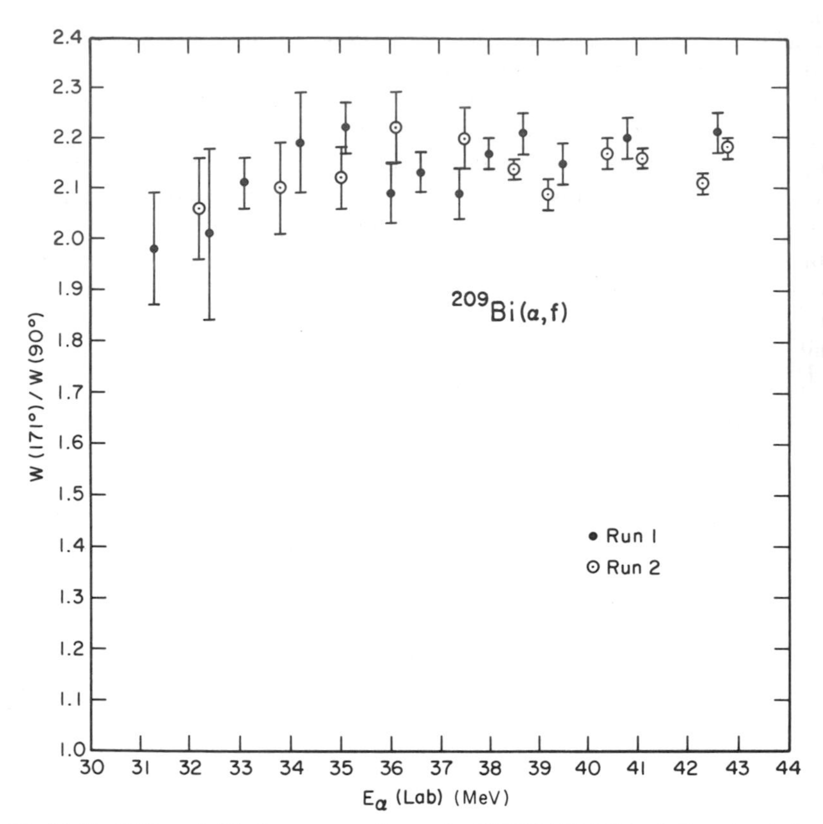

Fig. VI-11. Differential fission cross section ratio $W(171°)/W(90°)$ for helium-ion-induced fission of ^{209}Bi as a function of helium-ion energy in the laboratory system. [After Huizenga (1965).]

similar structure in the total fission cross section as a function of projectile energy, a subject discussed in Chapter VII. As second-, third-, etc., chance fission becomes energetically possible, the total fission cross section exhibits a sudden increase with each of these additional reactions. This increase is followed by a leveling off of the total fission cross section until another, later-chance fission reaction becomes energetically possible. Hence, it is well established that multichance fission occurs at various excitation energies according to the schematic diagram shown in Fig. VI-13.

If the amount of angular momentum carried away by evaporated neutrons is very small compared to the angular momentum of the compound system, then the angular momentum distribution may be considered approximately constant for all fissioning species. However, the excitation energy and, as a result, $K_0{}^2$ of the fissioning nucleus change greatly with the number of neutrons emitted prior to fission. Therefore, at an excitation energy that is low, i.e., one slightly above the threshold energy of ith-chance fission, the angular anisotropy for that particular ith-chance fission should be large. This large anisotropy is then superimposed upon the anisotropies resulting from $(i-1)$-, $(i-2)$-, ..., first-chance fission. If the contributions of the first-, ..., $(i-2)$-, and $(i-1)$-chance fission are small compared to that of the ith-chance fission, then a large increase in anisotropy is expected near the threshold of the ith-chance fission. On the other hand, if the contributions from earlier-chance fission are large, then a smaller change in the anisotropy is expected near the threshold of the ith-chance fission.

Since the magnitude of the anisotropy is largely dependent upon the value of K_0^2, it becomes important to know how this quantity varies as a function of the excitation energy. Statistical theory predicts the relationship between K_0^2 and t given by Eq. (VI-6). When statistical theory is applicable, the interpretation of the energy dependence of the anisotropy is straightforward if only first-chance fission is occurring. The values of K_0^2 calculated from the anisotropies measured for the $^{209}Bi(\alpha, f)$ reaction (see Fig. VI-11) are plotted in Fig. VI-15 for the assumption of single-chance fission.

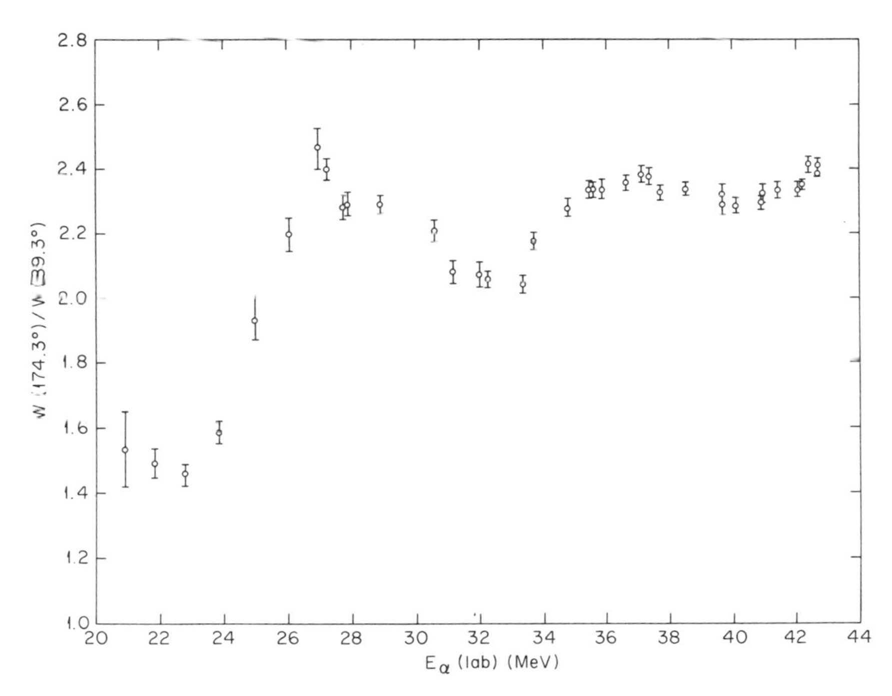

Fig. VI-12. Helium-ion-induced fission-fragment anisotropy from ^{226}Ra as a function of projectile energy. [After Gindler *et al.* (1964).]

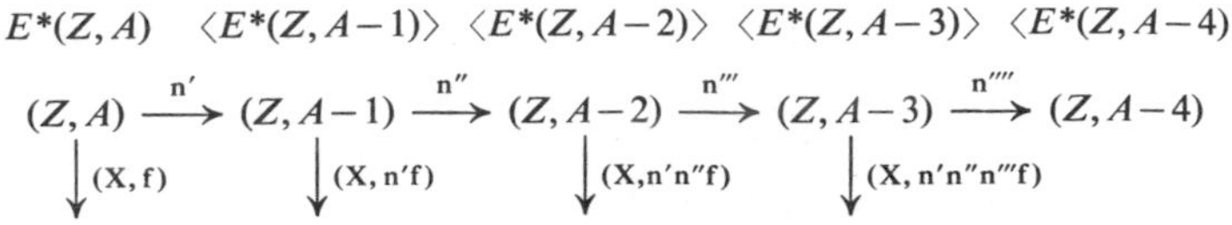

$$
\begin{array}{ccccccccc}
E^*(Z,A) & & \langle E^*(Z,A-1)\rangle & & \langle E^*(Z,A-2)\rangle & & \langle E^*(Z,A-3)\rangle & & \langle E^*(Z,A-4)\rangle \\
(Z,A) & \xrightarrow{\mathrm{n}'} & (Z,A-1) & \xrightarrow{\mathrm{n}''} & (Z,A-2) & \xrightarrow{\mathrm{n}'''} & (Z,A-3) & \xrightarrow{\mathrm{n}''''} & (Z,A-4) \\
\downarrow \mathrm{(X,f)} & & \downarrow \mathrm{(X,n'f)} & & \downarrow \mathrm{(X,n'n''f)} & & \downarrow \mathrm{(X,n'n''n'''f)} & &
\end{array}
$$

Initial compound nucleus excitation energy $E^*(Z, A)$

$$\langle E^*(Z,A-1)\rangle = E^*(Z,A) - B_{\mathrm{n}}(Z,A) - \langle E_{\mathrm{n}}(Z,A)\rangle$$

Fig. VI-13. Neutron emission and fission competition in compound nuclei (Z, A) formed by projectile X at excitation energy $E^*(Z, A)$.

The calculation of $K_0{}^2$ from the anisotropy by a relation such as Eq. (VI-7) requires projectile transmission coefficients T_l which are computed with optical model theory. The solid line in Fig. VI-15 is a theoretical fit of the $K_0{}^2$ points. The behavior of $K_0{}^2$ is approximately accounted for by the equation of state $E = a_{\mathrm{f}} t^2 - t$ and a constant moment of inertia at all excitation energies. However, the agreement between experiment and theory over the excitation energy region shown in Fig. VI-15 is not a critical test of statistical theory.

The interpretation of the energy dependence of the anisotropy for heavy

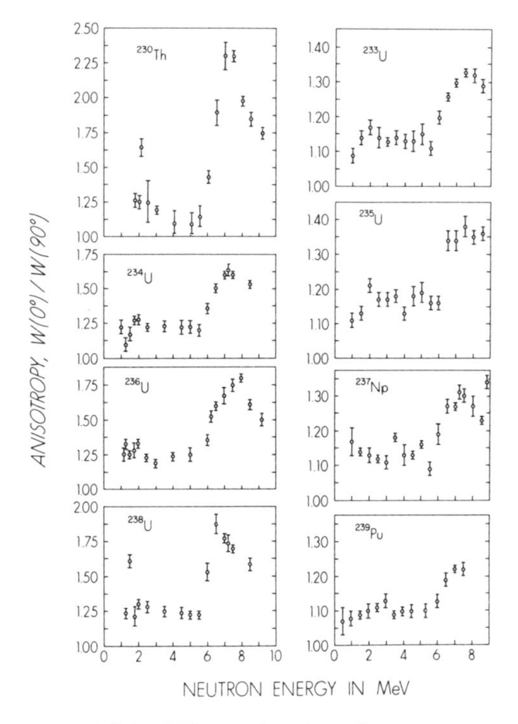

Fig. VI-14. Anisotropy $W(0°)/W(90°)$ as a function of neutron energy for ^{230}Th, ^{234}U, ^{236}U, ^{238}U, ^{233}U, ^{235}U, ^{237}Np, and ^{239}Pu target nuclei. [Data of Simmons and Henkel (1960).]

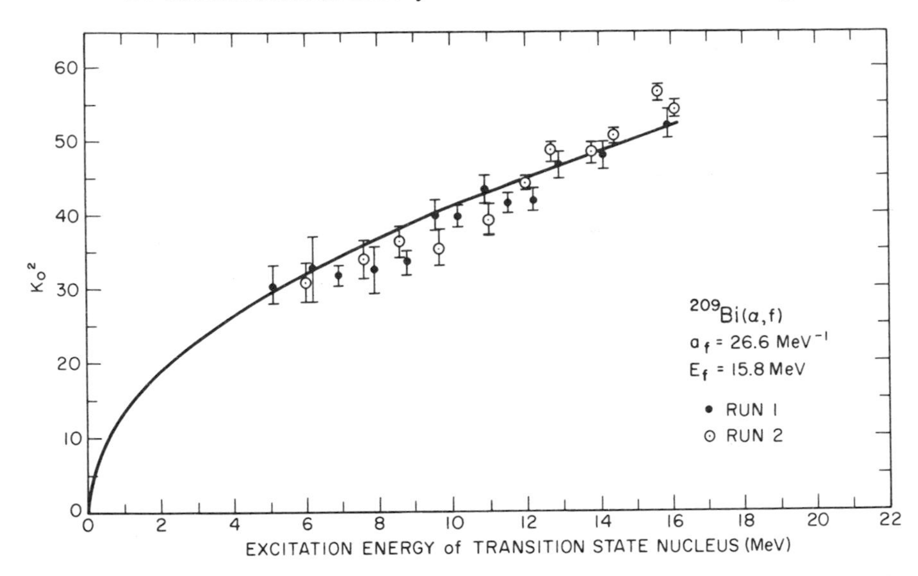

Fig. VI-15. Excitation energy dependence of K_0^2 deduced from helium-ion-induced fission of ^{209}Bi for the assumption of single-chance fission. The abscissa is the excitation energy above the fission barrier where $E_f = 15.8$ MeV. (—) Theoretical energy dependence predicted by statistical theory given in Eq. (VI-6). [After Huizenga (1965).]

elements, in general, is further complicated by multichance fission. The experimental anisotropy results from several-chance fission, each chance fission having its own angular momentum distribution (neutron emission alters both the J and M distributions), K_0^2, and temperature. Attempts to fit such multichance fission anisotropies as those shown in Figs. VI-12 and VI-14 with statistical theory have not been very successful.

D. Modification in Theory to Account for Nuclear Pairing

Griffin (1963) has described the derivation of K_0^2 in a superfluid system which includes nuclear pairing interactions. In this section, we give a brief description of his derivation of the energy dependence of K_0^2. In the superconductor theory of the nucleus it is assumed that the same type of residual interaction which takes place in a superconducting metal between pairs of electrons of equal and opposite momentum also takes place between pairs of nucleons within the shell being filled. The strength of the interaction is reflected in the correlation function θ_0. An energy gap of approximately $2\theta_0$ is produced in an even-even system. The Bardeen, Cooper, Shrieffer (1957) theory (BCS) gives the excitation energy E and entropy S as a function of the thermodynamic temperature t of the superconducting system, and gives an expression for the occupation probability of the "quasi-particle states" of the system as a function of t.

From these quantities we can derive the level density, the various nuclear temperatures, and the moment of inertia. The main differences between the superconducting model and the Fermi gas independent particle model are the following. The ground state of the superconductor lies below the Fermi gas ground state by a condensation energy C_0. The superconductor has an energy gap of $2\theta_0$ in the excitation energy spectrum. At a certain critical temperature t_c and corresponding critical energy E_c of the order of 15 MeV, the pairing energy disappears. Above the critical energy the superconductor is predicted to behave like a normal Fermi gas whose ground state is shifted upward by an amount equal to the condenstation energy C_0. At excitation energies exceeding the critical energy there is no reduction of the moment of inertia from the rigid-body value. Below E_c the theory results in approximately constant nuclear temperature and a moment of inertia which strongly decreases with decreasing excitation energy.

For the ground state the theory gives a condensation energy C_0 of

$$C_0 = \tfrac{1}{4} g \theta_0^2 = 0.47 a t_c^2 \tag{VI-15}$$

where

$$a = \tfrac{1}{6}\pi^2 g \tag{VI-16}$$

$$t_c = \tfrac{4}{7}\theta_0 \tag{VI-17}$$

g is the number of single particle states (neutrons and protons) per MeV of a nucleus without residual interactions, and $2\theta_0$ is the energy gap for $t = 0$. The transition or critical energy above which the superconductor nucleus behaves like a normal Fermi gas whose ground state is shifted upward by the condensation energy C_0 is given by

$$E_c = a t_c^2 + C_0 = 1.473 a t_c^2 \tag{VI-18}$$

The modifications of the independent particle model by the BCS pairing interaction are especially relevant for fission-fragment angular distributions, because they depend directly on the effective moment of inertia. This relationship is parameterized through K_0^2 (the mean square value of the projection of angular momentum on the nuclear symmetry axis). Because of the pairing interaction of single particle states, the particle excitations which contribute to the total projection on the symmetry axis do not extend continuously to the ground state. For a given excitation energy a superfluid nucleus has a significantly reduced K_0^2. Hence, the dependence of K_0^2 upon excitation energy is a rather sensitive method of distinguishing between the two models.

The value of K_0^2 was derived earlier and given in Eq. (VI-6), which with the definition $\mathscr{I}_{\text{eff}} = \mathscr{I}_{\|}\mathscr{I}_{\perp}/(\mathscr{I}_{\perp} - \mathscr{I}_{\|})$ becomes

$$K_0^2 = \frac{t\mathscr{I}_{\|}}{\hbar^2}\left[\frac{\mathscr{I}_{\perp}}{\mathscr{I}_{\perp} - \mathscr{I}_{\|}}\right] \qquad \text{(VI-19)}$$

The Fermi gas model gives an equation of state relating energy and temperature; furthermore, it predicts rigid moments of inertia. Hence, this model gives a unique prediction of the dependence of K_0^2 on excitation energy. In evaluating K_0^2 for the superconductor model we have to use the equation of state of the superconducting system and calculate the moments of inertia as a function of the superconductor temperature. For determination of fission-fragment anisotropies in the superconductor model, the most important parameter to be calculated is $\mathscr{I}_{\|}\, t/\hbar^2$. This quantity is directly related to the average of K^2 over the particle spectrum and is given by

$$\frac{\mathscr{I}_{\|}\, t}{\hbar^2} = \sum_k \frac{m_k^2 \exp[(\varepsilon_k^2 + \theta^2)^{1/2}/t]}{\{1 + \exp[(\varepsilon_k^2 + \theta^2)^{1/2}/t]\}^2} \qquad \text{(VI-20)}$$

where m_k^2 is the square of the projection of the angular momentum on the symmetry axis for the quasi-particle excitation with energy $(\varepsilon_k^2 + \theta^2)^{1/2}$ and the remainder of the expression is the probability that *one* and *only one* of the degenerate pair $\pm k$ is excited. If m_k^2 is replaced by $\langle m^2 \rangle$, which may be factored out, and $x = \varepsilon_k/t = \beta\varepsilon_k$, then by replacing the summation with an integral, we obtain

$$\frac{\mathscr{I}_{\|}\, t}{\hbar^2} = tg\langle m^2 \rangle \int_{-\infty}^{\infty} \frac{\exp(x^2 + \beta^2\theta^2)^{1/2}\, dx}{[1 + \exp(x^2 + \beta^2\theta^2)^{1/2}]^2} \qquad \text{(VI-21)}$$

The quantity $g\langle m^2 \rangle$ is defined by $\mathscr{I}_{\|}^{\text{rigid}}/\hbar^2$. Values of the temperature-dependent integral are calculated numerically and appear in Table VI-1. This integral will be denoted by $[A(t/t_c)]$. Multiplying the numerator and denominator of Eq. (VI-21) by the critical temperature t_c and making the substitutions above gives

$$\frac{\mathscr{I}_{\|}\, t}{\hbar^2} = \frac{t_c \mathscr{I}_{\|}^{\text{rigid}}}{\hbar^2}\left(\frac{t}{t_c}\right)[A(t/t_c)] \qquad \text{(VI-22)}$$

where $\mathscr{I}_{\|} = \mathscr{I}_{\|}^{\text{rigid}}[A(t/t_c)]$. At the critical energy (or temperature) $[A(t/t_c)] = 1$ and $\mathscr{I}_{\|}$ reverts back to the rigid moment. However, at energies less than the critical energy, $\mathscr{I}_{\|}$ is always less than the rigid body value of the moment of inertia and may be considerably less.

In analogy to Eq. (VI-19), the quantity K_0^2 for the superconducting system is given by

$$K_0^2 = \{\mathscr{I}_{\perp}/[\mathscr{I}_{\perp} - A(t/t_c)\mathscr{I}_{\|}^{\text{rigid}}]\}(t_c \mathscr{I}_{\|}^{\text{rigid}}/\hbar^2)(t/t_c)[A(t/t_c)] \qquad \text{(VI-23)}$$

TABLE VI-1

Thermodynamic Properties for a Superconductor

t/t_c	θ/θ_o	$S/2at_c$	$E/2at_c^2$	$\int_{-\infty}^{\infty} \frac{\exp(x^2+\theta^2\beta^2)^{1/2}\,dx}{[1+\exp(x^2+\theta^2\beta^2)^{1/2}]^2}$
1.00	0.0000	1.0000	0.7364	1.000
0.98	0.2436	0.9519	0.6888	0.961
0.96	0.3416	0.9048	0.6431	0.921
0.94	0.4148	0.8587	0.5993	0.883
0.92	0.4749	0.8136	0.5573	0.845
0.90	0.5263	0.7694	0.5172	0.807
0.88	0.5715	0.7263	0.4787	0.769
0.86	0.6117	0.6842	0.4421	0.732
0.84	0.6480	0.6432	0.4073	0.695
0.82	0.6810	0.6032	0.3741	0.659
0.80	0.7110	0.5643	0.3425	0.623
0.78	0.7386	0.5266	0.3127	0.587
0.76	0.7640	0.4900	0.2846	0.552
0.74	0.7874	0.4546	0.2580	0.518
0.72	0.8089	0.4203	0.2330	0.484
0.70	0.8288	0.3873	0.2096	0.451
0.68	0.8471	0.3554	0.1876	0.419
0.66	0.8640	0.3249	0.1671	0.387
0.64	0.8796	0.2956	0.1481	0.357
0.62	0.8939	0.2676	0.1304	0.327
0.60	0.9070	0.2410	0.1142	0.298
0.58	0.9190	0.2157	0.0993	0.270
0.56	0.9299	0.1918	0.0856	0.245
0.54	0.9399	0.1693	0.0733	0.218
0.52	0.9488	0.1482	0.0621	0.193
0.50	0.9569	0.1285	0.0521	0.170
0.48	0.9641	0.1103	0.0431	0.148
0.46	0.9704	0.0937	0.0353	0.127
0.44	0.9760	0.0784	0.0284	0.108
0.42	0.9809	0.0646	0.0225	0.090
0.40	0.9850	0.0524	0.0175	0.074
0.38	0.9885	0.0416	0.0133	0.060
0.36	0.9915	0.0322	0.0098	0.047
0.34	0.9938	0.0243	0.0071	0.036
0.32	0.9957	0.0177	0.0049	0.027
0.30	0.9971	0.0124	0.0032	0.019
0.28	0.9982	0.0082	0.0020	0.0129
0.26	0.9989	0.0051	0.0011	0.0082
0.24	0.9994	0.0030	0.0006	0.0049
0.22	0.9997	0.0016	0.00025	0.0026
0.20	0.9999	0.0007	0.00014	0.0012
0.18	1.0000	0.0003	0.00005	0.00049
0.16	1.0000	0.0001	0.00002	0.00015
0.14	1.0000	0.0000	0.00000	0.00003

The moment of inertia $\mathscr{I}_\perp$ perpendicular to the symmetry axis is estimated from a model. Migdal (1959) has shown that

$$\mathscr{I}_\perp^{\text{irrot}} < \mathscr{I}_\perp < \mathscr{I}_\perp^{\text{rigid}} \tag{VI-24}$$

and that, as the deformation becomes larger, $\mathscr{I}_\perp$ approaches $\mathscr{I}_\perp^{\text{rigid}}$. For saddle shapes of heavy nuclei, $\mathscr{I}_\perp^{\text{irrot}}$ is already of the order of $\frac{3}{4}$ the rigid moment. The approximation $\mathscr{I}_\perp \equiv \mathscr{I}_\perp^{\text{rigid}}$ has very little influence (Griffin, 1963) at all temperatures on the derived values of K_0^2. For a particular transition state nucleus, $(t_c \mathscr{I}_\parallel^{\text{rigid}}/\hbar^2)$ is constant and Eq. (VI-23) reduces to (Huizenga, 1965)

$$K_0^2 = \text{const}[D(t/t_c)](t/t_c)[A(t/t_c)] \tag{VI-25}$$

where $D(t/t_c) = \mathscr{I}_\perp/[\mathscr{I}_\perp - A(t/t_c)\mathscr{I}_\parallel^{\text{rigid}}]$. Relative values of K_0^2 are plotted as a function of excitation energy $(E/2at_c^2)$ in Fig. VI-16 for the following cases:

(1) $D(t/t_c) = \text{const}$ and $A(t/t_c) = 1$, Fermi gas model;
(2) $D(t/t_c) = 1$;

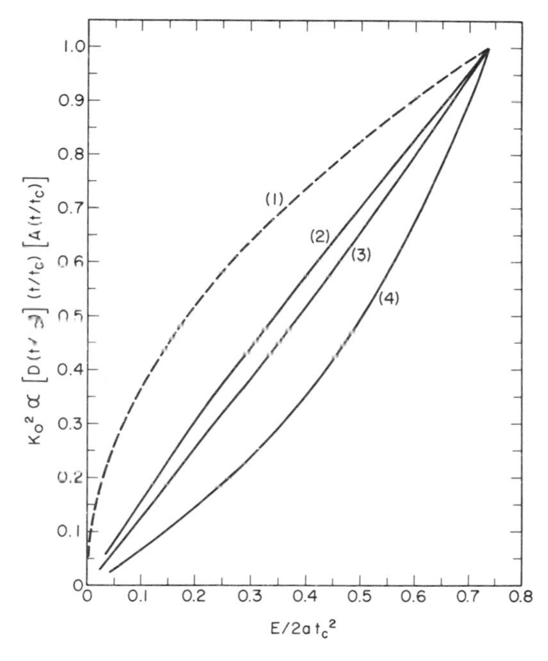

Fig. VI-16. Dependence of K_0^2 on excitation energy predicted by the Fermi gas statistical theory (curve 1) and a superconductor theory (curves 2, 3, and 4). Curves 2, 3, and 4 result from different approximations of $[D(t/t_c)]$ which are discussed in the text. Qualitatively, nuclei with saddle configurations which are very deformed approach curve 2, whereas nuclei with less deformed saddle configurations are more nearly represented by curve 4. [After Huizenga (1965).]

(3) $D(t/t_c) = \frac{3}{4}\{4/[4 - A(t/t_c)]\}$, which is the condition for $(\mathscr{I}_{\perp}^{\text{rigid}}/\mathscr{I}_{\parallel}^{\text{rigid}}) = 4$ and $(\mathscr{I}_{\perp}/\mathscr{I}_{\perp}^{\text{rigid}}) = 1$ for all t;

(4) $D(t/t_c) = \frac{1}{3}\{1.5/[1.5 - A(t/t_c)]\}$, which is the condition for $(\mathscr{I}_{\perp}^{\text{rigid}}/\mathscr{I}_{\parallel}^{\text{rigid}}) = 2$ and $(\mathscr{I}_{\perp}/\mathscr{I}_{\perp}^{\text{rigid}}) = \frac{3}{4}$ for all t.

Each of the four curves giving K_0^2 as a function of energy in Fig. VI-16 is normalized at the critical energy. For the Fermi gas model (condition 1) the ordinate of Fig. VI-16 reduces to a constant times the temperature, and the dependence of K_0^2 on energy is given by curve 1. This is the same relationship as given by Eq. (VI-6) for a specified transition state nucleus. The solid curves in Fig. VI-16 give the K_0^2 dependence on excitation energy for the superconductor model with different assumptions about $D(t/t_c)$ (see conditions 2, 3, and 4 above). Condition 2 represents an extreme in the sense that this is the least reduction in K_0^2 predicted by the superconductor model. More realistic evaluations of $D(t/t_c)$ give an even larger reduction in K_0^2. It is interesting to

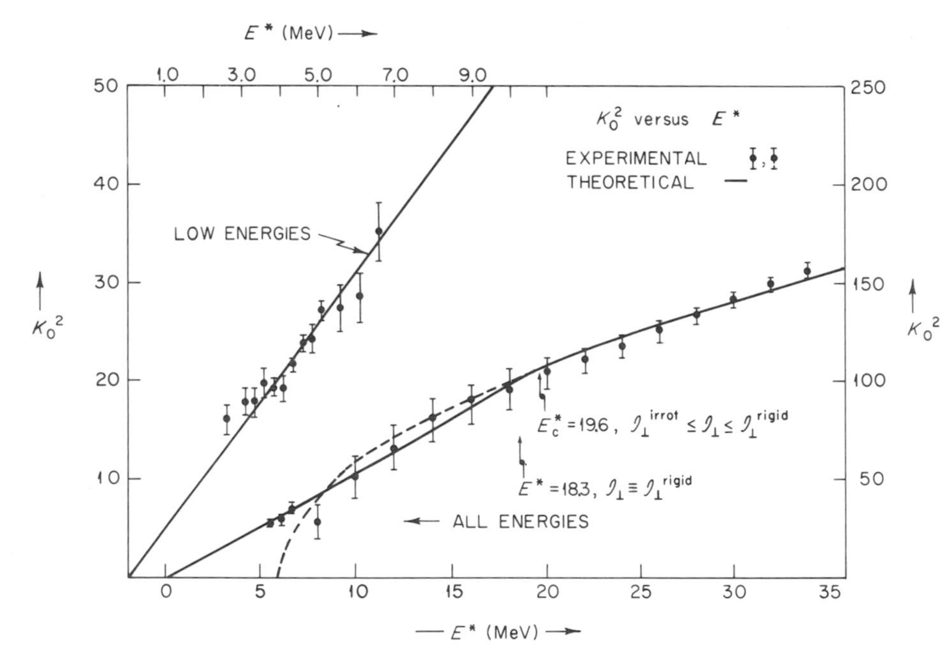

Fig. VI-17. Comparison of experimental and theoretical energy dependence of K_0^2. At low energies the experimental values of K_0^2 are obtained from the fragment angular distributions for the ^{239}Pu(n, f) reaction. At high energies the experimental values are from helium-ion-induced fission (Vandenbosch *et al.*, 1961). The Fermi gas statistical theory is shown by the broken-line parabolic curve. The superconductor model gives a linear curve at low excitation energy and a parabolic curve above the critical energy E_c. The difference between zero excitation energy and the energy where the parabola goes to zero is the energy gain due to nuclear pairing. [After Griffin (1965).]

note that the least deformed transition state nuclei give the largest reduction in K_0^2 (compare conditions 3 and 4).

The analyses of several fissioning systems indicate that a reduction in K_0^2 from the Fermi gas values at low excitation energy is required to fit experimental angular distributions. As discussed previously in this chapter, if the nucleus were composed of independent particles in a potential well exhibiting a uniform single particle spacing (Fermi gas statistical model), $\mathscr{I}_{\text{eff}}$ could be estimated from the appropriate rigid body moments of the nucleus, and t determined from the equation of state $E = at^2 - t$. The predicted excitation energy dependence of K_0^2 given by Eq. (VI-6) is proportional to the temperature t of the transition nucleus. The Fermi gas temperature varies essentially as the square root of the excitation energy E where $E = E^* - B_f$. Fission-fragment angular distribution experiments at low excitation energy have indicated that K_0^2 varies directly with E and not as the square root of E. The BCS superconductor model can give an approximately linear relationship between K_0^2 and E, as shown in Fig. VI-16. However, this explanation of the experimental data is not unique. It is well known that the fission barrier of heavy elements has a structure consisting of two peaks with a potential minimum between them. Hence, the value of $\mathscr{I}_{\text{eff}}$ may vary initially with excitation energy due to a change in the saddle point shape itself (Ramamurthy *et al.*, 1970). At higher excitation energies the value of $\mathscr{I}_{\text{eff}}$ is expected to approach a constant value characteristic of the liquid drop theory of the transition nucleus. This subject will be discussed further in the next section.

Examples of fitting experimental angular distributions with this superconductor model are shown in Figs. VI-17 and VI-18 (see p. 204). In general, improvement in the agreement with experiment is obtained with the superconductor theory.

E. Shape of the Transition Nucleus

At reasonably large excitation energies the Fermi gas statistical model is expected to describe reasonably accurately the K distribution of the levels in the intermediate transition nucleus. As already discussed, such a model predicts the K distribution to be Gaussian. The square of the standard deviation of the Gaussian, K_0^2, is given by Eq. (VI-6) in terms of $\mathscr{I}_{\text{eff}}/\hbar^2$ and the temperature t.

In the statistical model the quantity $\mathscr{I}_{\text{eff}}$ is defined in terms of the rigid body nuclear moments of inertia $\mathscr{I}_{\perp}$ and $\mathscr{I}_{\parallel}$ where the subscripts refer to axes perpendicular and parallel to the symmetry axis, respectively. The temperature is defined in terms of the excitation energy of the transition nucleus

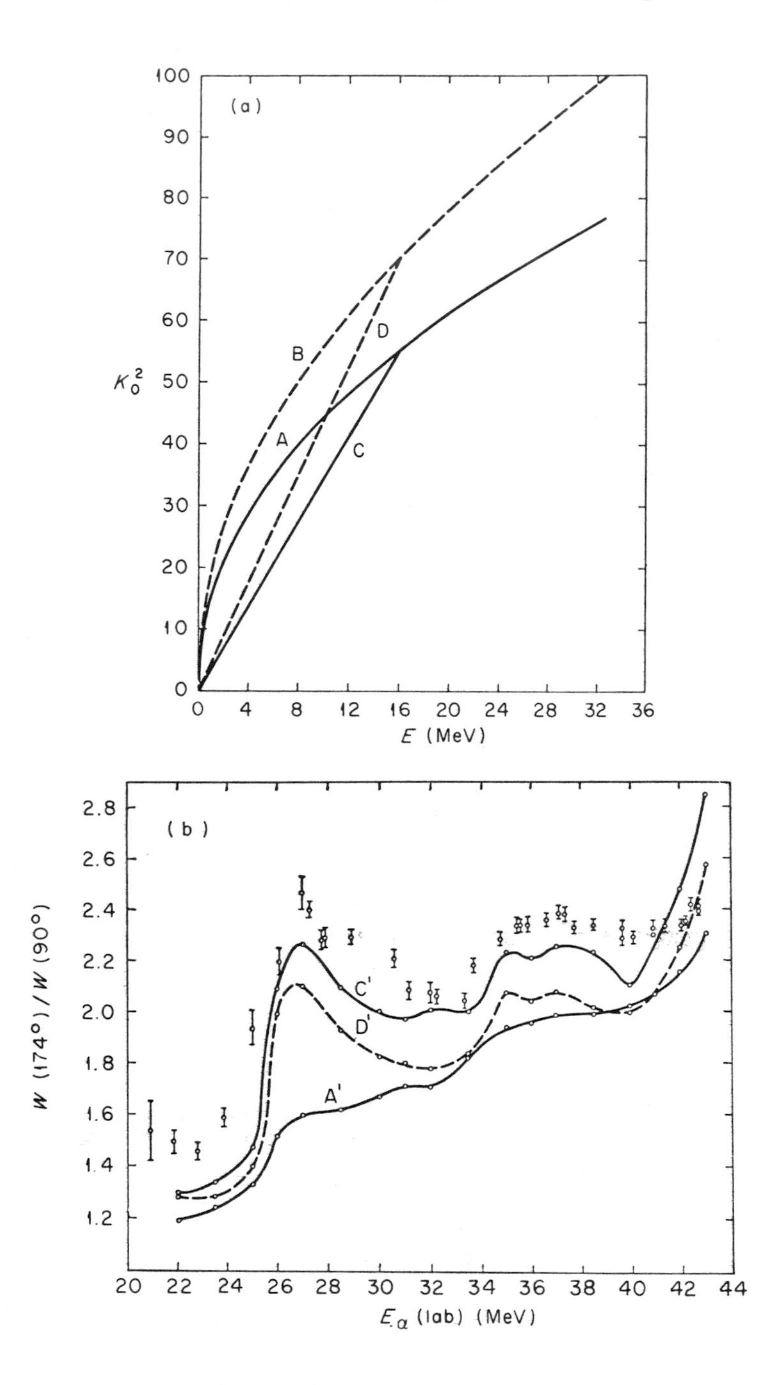

Fig. VI-18.

through the equation of state $E = E^* - B_f = at^2 - t$, where a is a level density parameter. Hence, if a is known, an experimental determination of K_0^2 gives a measurement of $\mathscr{I}_{eff}$ and the shape of the transition nucleus.

The experimental anisotropies for 21-MeV deuteron-induced fission (Bate *et al.*, 1963) and 42.8-MeV helium-ion-induced fission (Reising *et al.*, 1966) are plotted in Figs. VI-19 and VI-20. Values of K_0^2 for the various fissioning nuclei are deduced from these anisotropies by utilization of transmission coefficients which are calculated with an optical model potential. Although the excitation energies of the initial compound nuclei are large enough for the

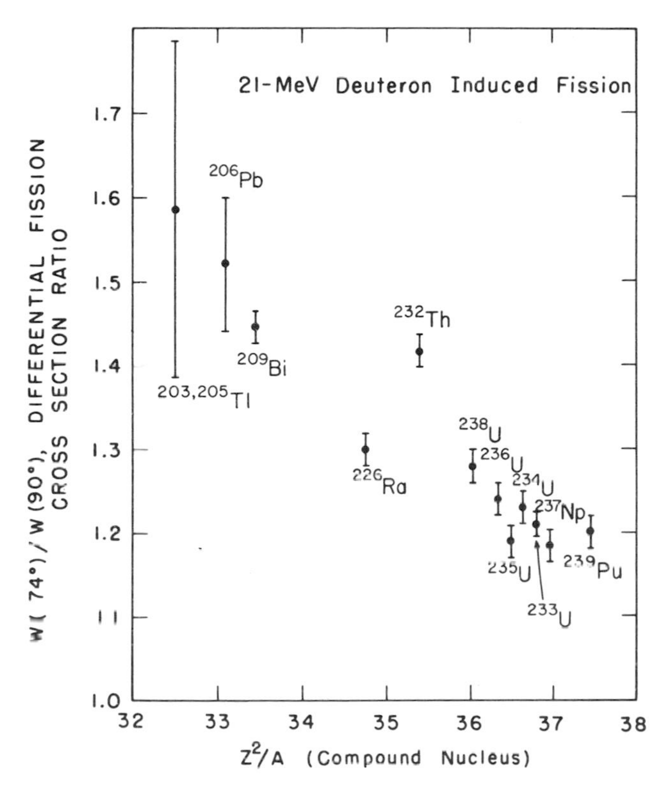

Fig. VI-19. Variation of fission-fragment anisotropies for deuteron-induced fission in various target nuclides (as labeled) as a function of Z^2/A for the compound nucleus. [After Bate *et al.* (1963).]

Fig. VI-18. (a) Variation of K_0^2 with excitation energy. These values of K_0^2 are used to calculate the energy dependence of helium-ion-induced fission-fragment anisotropy of radium shown in (b). The experimental anisotropies are given as points with error bars. The curves are labeled with primes to indicate that a small correction in the theoretical unprimed curves shown in (a) has been made for the odd-A nuclei (for odd-A nuclei, K_0^2 is larger). [After Gindler *et al.* (1964).]

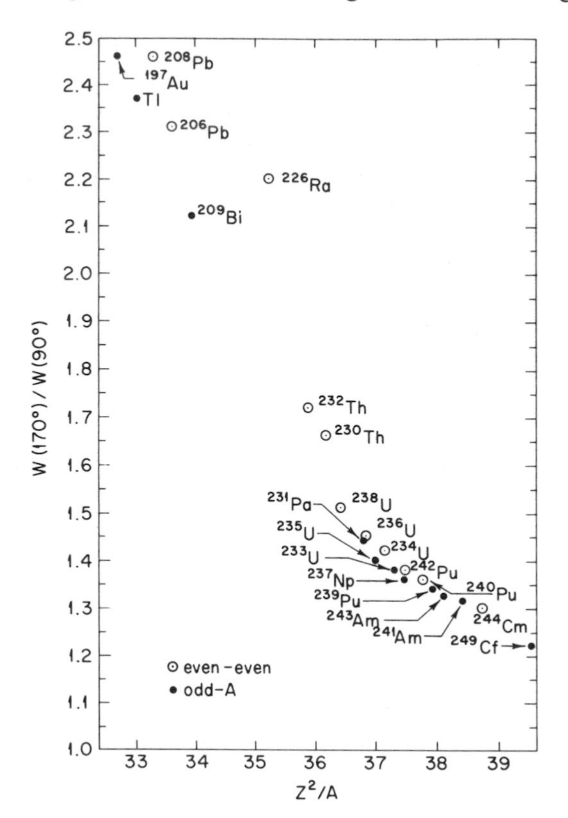

Fig. VI-20. Variation of fission-fragment anisotropies for 42.8-MeV helium-ion-induced fission in various target nuclides (as labeled) as a function of Z^2/A for the compound nucleus. [After Reising *et al.* (1966).]

statistical model to be applicable to the transition nuclei, multichance fission complicates a straightforward calculation of $\mathscr{I}_{\text{eff}}$ with Eq. (VI-6). When multichance fission occurs, the experimental value of K_0^2 is an average value associated with the various-chance fissions, each chance fission having a different transition nucleus and associated temperature.

If α_i represents the fraction of ith-chance fission, Eq. (VI-6) can be used directly when α_1 is the dominant term. For target nuclei with $Z \leqslant 83$, α_1 is expected to dominate, since the fission-to-neutron width Γ_f/Γ_n decreases sharply with decreasing excitation energy for these nuclei (Chapter VII). At the other extreme of the abscissas of Figs. VI-19 and VI-20, where Z^2/A has the largest values, α_1 again dominates. For the intermediate values of Z^2/A, however, second- and higher-chance fission contribute to the experimental anisotropy.

First we will discuss the gross features of the anisotropy as a function of

Z^2/A of the fissioning nucleus. For both the 21-MeV deuteron-induced fission and the 42.8-MeV helium-ion-induced fission, the anisotropy increases sharply as the value of Z^2/A decreases. This is a direct result of the fact that $\mathscr{I}_{\text{eff}}$ decreases sharply with decreasing Z^2/A.

Some irregularity exists in the anisotropies of nuclei with intermediate values of Z^2/A. This is due to a contribution of second- and higher-chance fission which occurs at lower excitation energies, where $K_0{}^2$ is appreciably smaller. For deuteron-induced fission the anisotropy for ^{232}Th is larger than that for ^{226}Ra. Whereas deuteron-induced fission of ^{232}Th has an appreciable contribution of higher-chance fission, deuteron-induced fission of ^{226}Ra is mostly first-chance fission. Likewise, for helium-ion-induced fission of ^{226}Ra, ^{232}Th, and ^{230}Th various contributions of higher-chance fission are present.

By application of fission-to-neutron level width ratios Γ_f/Γ_n (see Chapter VII), it is possible to correct the observed anisotropies for second- and higher-chance fission. The inclusion of this small correction permits us to compare the first-chance anisotropies of all the targets at a nearly uniform and rather high excitation energy. The resulting values of $K_0{}^2$ associated with first-chance fission are used to calculate the nuclear shape parameter $\mathscr{I}_{\text{eff}}$ as a function of Z^2/A. As mentioned previously, we assume rigid body moments of inertia and a temperature defined by $E^* - B_f = at^2 - t$, where $E^* - B_f$ represents the excitation energy of the transition nucleus.

The values of $\mathscr{I}_{\text{eff}}$ as a function of Z^2/A are plotted in Fig. VI-21. The

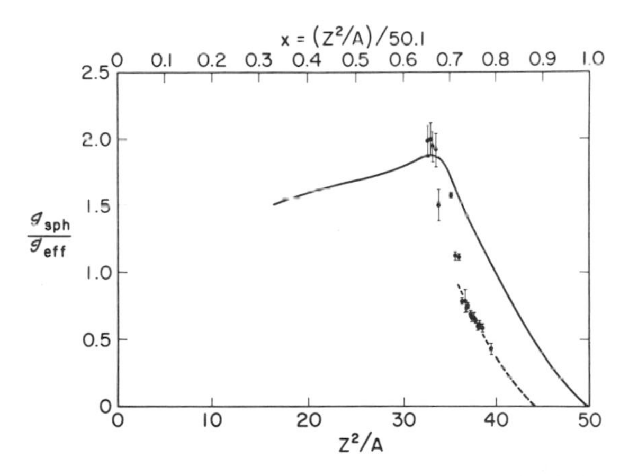

Fig. VI-21. Saddle or transition state deformation computed from first-chance fission anisotropies as a function of Z^2/A for the compound nucleus. (—) Theoretical predictions based on charged liquid drop model calculations using $(Z^2/A)_{\text{crit}} = 50.1$ (Cohen and Swiatecki, 1963; Strutinsky *et al.* 1963). (---) Extrapolation from ^{231}Pa and heavier elements to zero saddle deformation for revised estimate of $(Z^2/A)_{\text{crit}} = 45$. The extrapolated curve is based on $a = A/8$ for the level density parameter of the transition nucleus. [From Reising *et al.* (1966).]

ordinate is actually the ratio $\mathscr{I}_{sph}/\mathscr{I}_{eff}$, where $\mathscr{I}_{sph}$ is a normalizing factor and equal to the rigid-body moment of inertia of a sphere for a nucleus of equal volume. The radius parameter used for the calculation of $\mathscr{I}_{sph} = \frac{2}{5}MR^2$ is $1.216 \times 10^{-13} A^{1/3}$ cm. The experimentally determined values of $\mathscr{I}_{sph}/\mathscr{I}_{eff}$ increase rapidly with decreasing values of Z^2/A in accordance with theoretical calculations. The solid curve in Fig. VI-21 represents the theoretical dependence of $\mathscr{I}_{sph}/\mathscr{I}_{eff}$ on Z^2/A, based on charged liquid drop model calculations by Cohen and Swiatecki (1963) and by Strutinsky *et al.* (1963) using a value of $(Z^2/A)_{crit} = 50.1$.

The discrepancy between experiment and theory as illustrated in Fig. VI-21 appears real. Strutinsky has improved the theoretical fit by a modification of the liquid drop model of the nucleus which allows for the surface tension to vary with the curvature of the nuclear surface. A further modification of the liquid drop model at low excitation energies is suggested by the discovery of fission isomers which are a result of double-humped fission barriers.

Since the actual fission barrier is considerably different from the liquid drop fission barrier, we can rightly ask which barrier shape controls the fission anisotropy. At low excitation energies, Strutinsky has postulated that the second barrier (see Fig. V-24 and the related discussion) does so. As the excitation energy is incresaed, however, the shape of the intermediate transition nucleus is expected to change from that of the second barrier to the liquid drop shape (Ramamurthy *et al.*, 1970). This is so because for high excitation energies, the minimum number of levels on the path to fission corresponds to the liquid drop barrier shape.

State density calculations have been performed for heavy nuclei as a function of both excitation energy and deformation with different sets of single particle levels. If a nucleus is excited to a particular excitation energy in its ground state deformation, we are interested in determining the deformation which gives the minimum state density or saddle point shape for the given energy. The excitation energy as a function of deformation is given by the difference between the initial excitation energy and the potential energy of deformation relative to the ground state deformation. The potential energy surface is calculated by the Strutinsky method. For fission at low compound nucleus excitation energies the minimum state density at the saddle point occurs at a deformation where the shell correction gives a local maximum in the potential energy (i.e., at either the first or second barrier). This is so because the deformation dependence of the available excitation energy, which is lowest at barrier deformations, is more important in the determination of the state density than is the deformation dependence of the single particle level density parameter, which is largest at barrier deformations (see Section VII, B, 3). As the compound nucleus excitation energy increases, the fractional variation in the available excitation energy with deformation decreases and the deformation dependence of the single particle level density parameter plays a more important role. At very high energies the interplay between these two effects results in a minimum state density at a deformation near the liquid drop saddle. This high energy limit can be described in an alternative manner by noting that at high energies shell effects disappear, so that if the available excitation energy is obtained by using the liquid drop potential energy surface, and if we use a deformation-independent single particle level density parameter, the minimum in the state density will occur at the liquid drop saddle point deformation.

The excitation energy at which the fission saddle deformation coincides with the liquid drop saddle shape is not well defined by the present theory. This is mainly due to uncertainties in the potential energy surface at larger deformations. For example, in the actinide element region the introduction of P_3 and P_5 degrees of freedom reduces the second barrier by a substantial amount. The rate at which the potential energy decreases at large deformations influences the deformation at which we calculate the minimum number of exit channels. The discrepancy between the experimental saddle shapes and liquid drop theory mentioned earlier and displayed in Fig. VI-21 may in part be due to the saddle deformation's being larger than the liquid drop saddle essentially to the scission shape. Therefore, some correlation between the deformation at medium excitation energies. A more detailed understanding of the medium energy anisotropy results in terms of saddle point shapes depends on more accurate estimates of the potential energy surfaces and state densities of these elements.

F. Angular Anisotropy of Fission Fragments of Particular Mass

In some of the very early studies of fission-fragment anisotropies it was observed that asymmetric fission fragments exhibit larger anisotropies than symmetric fragments. However, the interpretation of this phenomenon is obscured for the fissioning systems studied earlier because of multichance fission. Hence, it has long been suspected that this effect reflects the variation of symmetric to asymmetric fission as a function of excitation energy, and not necessarily a more fundamental relationship between angular anisotropy and mass asymmetry. As discussed earlier in this chapter for multichance fission, those fission events occurring after the emission of one or more neutrons have a larger anisotropy because of the resulting smaller temperature t and reduced values of K_0^2. At the same time, these lower energy fissioning events are richer in asymmetric fragments accounting for their larger anisotropy.

In order to investigate the dependence of fission-fragment anisotropy on mass, fissioning systems must be chosen where single-chance (α_1) fission is dominant. This is the case for fissioning nuclei in the vicinity of lead, even at rather high excitation energies. For the heavier elements the excitation energies must be kept low enough so that only single-chance fission is energetically possible.

The fissioning systems investigated in the lead region were 42-MeV helium-ion-induced fission of ^{206}Pb and ^{209}Bi. Fission fragments were collected in thin aluminum catcher foils mounted in a scattering chamber at center-of-mass angles of 90° and 165° to the beam direction. Several elements were chemically separated from the foils and purified by standard radiochemical procedures. The differential cross section ratios $W(165°)/W(90°)$ for ruthenium (^{105}Ru), palladium ($^{109,112}Pd$), silver ($^{111,112,113}Ag$), strontium ($^{91,92}Sr$), molybdenum (^{99}Mo), and bromine (^{83}Br) were measured. The results are shown in Fig. VI-22. The average values of $W(165°)/W(90°)$ determined in these experiments are in excellent agreement with results obtained with solid state counters. Furthermore, it can be seen from Fig. VI-22 that, within the experimental errors, the anisotropy is independent of mass from a mass ratio of 1.0 to 1.3. All of these mass ratios arise from a configuration with an essentially common shape (ignoring changes in the nuclear temperature). These nuclei differ from the very heavy nuclei in that their saddle configurations are deformed essentially to the scission shape. Therefore, some correlation between the fission-fragment angular anisotropy and mass would not be surprising for these nuclei. The anisotropy of ^{83}Br ($M_1/M_2 = 1.53$) from the fission of ^{209}Bi was 10% less than that for other fragments. This may be associated with an increase in $\mathscr{I}_{\text{eff}}$ for a configuration with a $50N$ shell.

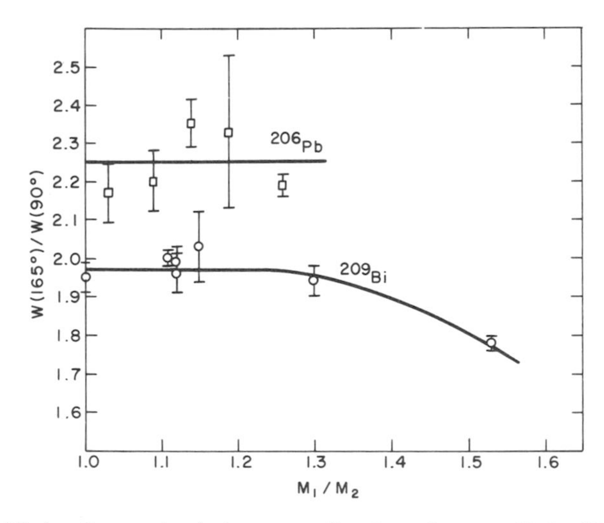

Fig. VI-22. Fission-fragment anisotropy as a function of mass split for 42-MeV helium-ion-induced fission of ^{206}Pb and ^{209}Bi. [After Flynn *et al.* (1964).]

Measurements of the fission-fragment anisotropy as a function of mass asymmetry for heavy elements must be done at low energy, where only single-chance fission occurs. One obvious method of performing such experiments is by neutron-induced fission with MeV neutrons. Such neutrons give detectable anisotropies, and if their energy is kept below approximately 5 MeV, the (n, n′f) process is energetically forbidden. One of the difficulties, however, in studying the relationship between anisotropy and mass asymmetry with MeV-neutron-induced fission is the low intensity of such neutron beams which are available. Although a change in anisotropy with fission-fragment mass for 4-MeV neutron-induced fission of ^{235}U has been reported, this result must be considered inconclusive at this time.

The (d, pf) reaction is another method for studying fission-fragment anisotropy as a function of mass asymmetry at a fixed excitation energy which

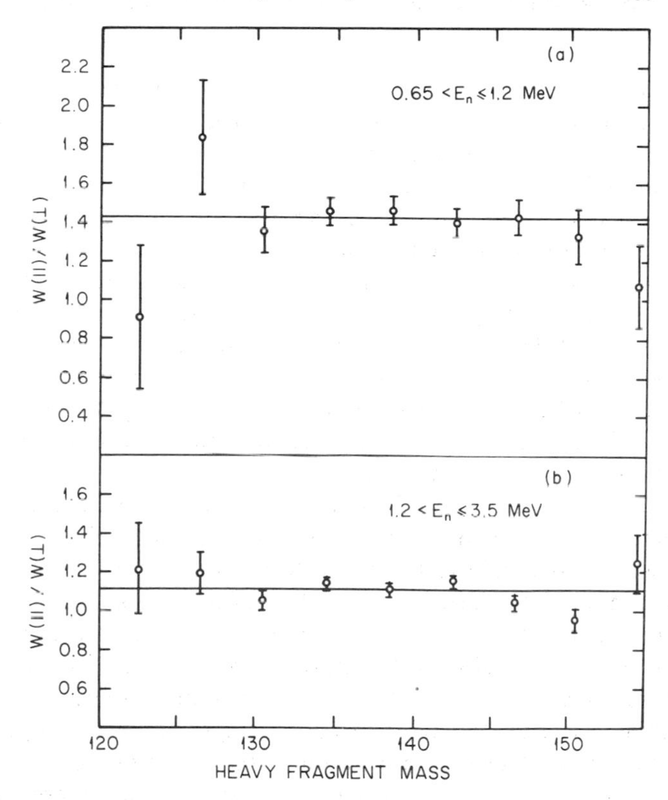

Fig. VI-23. Fission-fragment anisotropy as a function of fission-fragment mass for two different energy bins represented in terms of the kinetic energy of the captured neutrons. At these excitation energies only single-chance fission is energetically possible. The ordinate is in terms of the ratio of yields of fragments parallel and perpendicular to the heavy nucleus recoil direction. [After Vandenbosch *et al.* (1965).]

is low enough to eliminate the possibility of second-chance fission. A coincidence is required between fission and the emitted proton. Selection of desired events is accomplished by choosing a particular proton energy bin. The fragment mass distribution is obtained by two fission-fragment energy detectors in coincidence and the requirement that linear momentum be conserved. The results from the $^{234}U(d, pf)$ experiment are shown in Fig. VI-23. The fission-fragment anisotropy is plotted as a function of fission-fragment mass for two different energy bins. At each of these excitation energies, only single-chance fission is energetically possible. Within the accuracy of this experiment, no detectable dependence of the anisotropy on mass asymmetry is observed.

G. Angular Correlation between Fission Fragments

In order to compare fission-fragment angular distributions with statistical theory as a function of excitation energy, it is necessary to know the initial angular momentum distribution as a function of bombarding energy. At medium excitation energies the assumption is usually made that the entire inelastic cross section goes into compound nucleus formation. The optical model transmission coefficients are then employed to calculate the initial angular momentum distribution of the compound nucleus. For light projectiles of higher energies (Kapoor *et al.*, 1966; Viola *et al.*, 1971) and for heavy ion (Sikkeland *et al.*, 1962) projectiles, the foregoing assumption may not always be valid. Hence, if we are to compare experiment with theory, some estimate of the fraction of direct reactions must be made. In particular, we must estimate the change in the initial angular momentum distribution due to a contribution from direct reactions.

The first direct estimate of the amount of fission occurring after a direct interaction (as compared to that after compound nucleus formation) was made by Nicholson and Halpern (1959). A "direct interaction" is defined here as a reaction in which the forward momentum of the struck nucleus at the time of fission is different from what it would have been if the incident projectile were initially completely absorbed. Information about the momentum of the fissioning nucleus may be obtained from a measurement of the angular correlation of coincident pairs of fission fragments. If a compound nucleus is formed, the total forward momentum of the absorbed projectile must appear in the fission fragments. Schematic vector diagrams representing binary fission reactions following compound nucleus formation and direct interaction are shown in Fig. VI-24.

The in-plane angular correlation between coincident pairs of fission fragments for a 63.5-MeV α-particle bombardment of ^{238}U is shown in Fig.

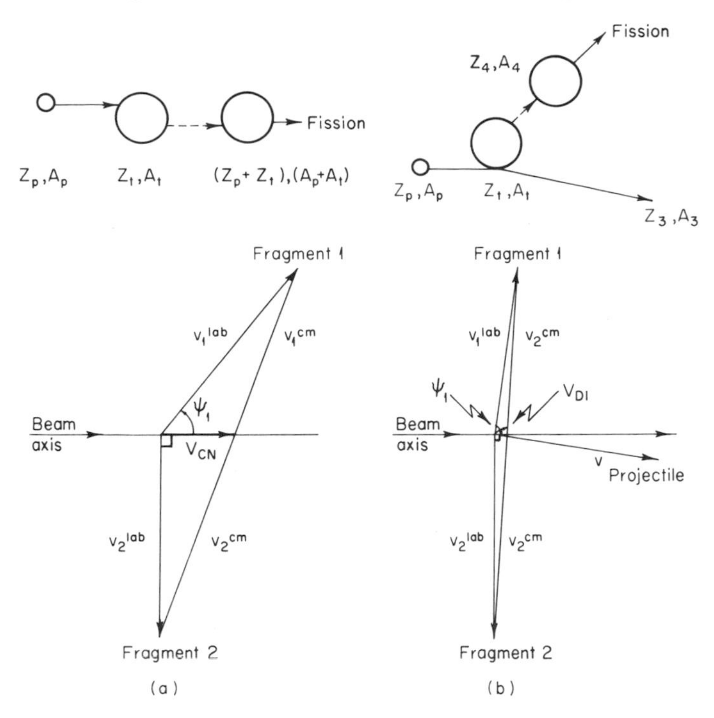

Fig. VI-24. Schematic vector diagrams represent binary fission reactions following (a) compound nucleus formation and (b) direct interaction. In the vector diagram V_{CN} and V_{DI} are the velocities of the recoiling nucleus after compound nucleus formation and direct interaction, respectively. The quantities v^{lab} and v^{cm} are the laboratory and center-of-mass velocities of the fission fragments, respectively. The laboratory angle for fragment 2 is fixed at 270° and ψ_1 represents the correlation angle of the complementary fragment 1. [After Viola *et al.* (1971).]

VI-25. The most probable correlation angle is $80.5 \pm 0.2°$. Within the experimental error, this value is in good agreement with the calculated value based on compound nucleus formation. However, the angular correlation is not symmetric about the angle 80.5° but skewed towards angles near 90°. This effect is due to fission following direct interaction in which less than complete linear momentum is transferred from projectile to target.

In a quantitative determination of the fraction of fission events following direct interaction, we assumed that the angular correlation function for compound nucleus fission events is symmetric about the most probable correlation angle. Reflection of the low angle portion of the angular correlation about the most probable correlation angle gives the dot-dashed curve in Fig. VI-25. Those events under this curve are assumed to represent fission following compound nucleus formation. The remaining events (shaded area)

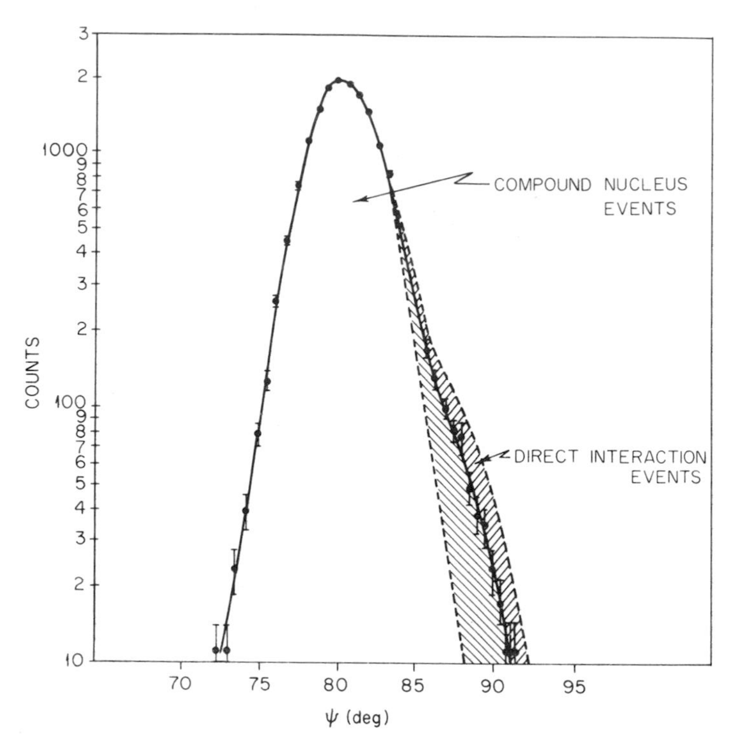

Fig. VI-25. In-plane angular correlation of fission fragments produced by 63.5-MeV α-particle bombardment of ^{238}U. The solid curve represents the gross experimental angular correlation. The inner (dot-dashed) curve represents fission events following compound nucleus formation obtained by reflection of the smaller angle results about the most probable correlation angle of 80.5°. The broken-line curve gives the added contribution due to the broader out-of-plane angular correlation for the fission events following direct interaction. The shaded area represents the total fraction of fission events following direct interaction. [After Viola *et al.* (1971).]

are attributed to fission following a direct interaction. In order to integrate the total three-dimensional angular correlation function and obtain the total fraction of fission events following direct interaction, it is necessary to measure both in-plane and out-of-plane correlations. A value of $7.5 \pm 1.0\%$ was obtained for the percentage of fission events following direct interaction for the foregoing reaction. After correcting for these events, Viola *et al.* (1971) concluded that the value of $\mathscr{I}_{\text{eff}}$ for ^{242}Pu is independent of excitation energy and angular momentum for bombarding energies up to 63.5 MeV.

References

Bardeen, J., Cooper, L. N., and Schrieffer, J. R. (1957). *Phys. Rev.* **108**, 1175.

Bate, G. L., and Huizenga, J. R. (1964). *Phys. Rev.* **133**, *B* 1471.

Bate, G. L., Chaudhry, R., and Huizenga, J. R. (1963). *Phys. Rev.* **131**, 722.

Cohen, S., and Swiatecki, W. J. (1963). *Ann. Phys.* (*New York*) **22**, 406.

Flynn, K. F., Glendenin, L. E., and Huizenga, J. R. (1964). *Nucl. Phys.* **58**, 321.

Gindler, J. E., Bate, G. L., and Huizenga, J. R. (1964). *Phys. Rev.* **136**, *B* 1333.

Gordon, G. E., Larsh, A. E., Sikkeland, T., and Seaborg, G. T. (1960). *Phys. Rev.* **120**, 1341.

Griffin, J. J. (1963). *Phys. Rev.* **132**, 2204.

Griffin, J. J. (1965). *Proc. IAEA Symp. Phys. Chem. Fission, Salzburg, 1965*, **1**, p. 23. IAEA, Vienna.

Halpern, I., and Strutinsky, V. M. (1958). *Proc. U.N. Int. Conf. Peaceful Uses At. Energy, 2nd*, **15**, p. 408, P/1513. United Nations, New York.

Huizenga, J. R. (1965). *In* "Nuclear Structure and Electromagnetic Interactions" (N. MacDonald, ed.), p. 319. Oliver & Boyd, Edinburgh.

Huizenga, J. R., Behkami, A. N., and Moretto, L. G. (1969). *Phys. Rev.* **177**, 1826.

Kapoor, S. S., Baba, H., and Thompson, S. G. (1966). *Phys. Rev.* **149**, 965.

Migdal, A. B. (1959). *Nucl. Phys.* **13**, 655.

Moretto, L. G., Gatti, R. C., Thompson, S. G., Huizenga, J. R., and Rasmussen, J. O. (1969). *Phys. Rev.* **178**, 1845.

Nicholson, W. J., and Halpern, I. (1959). *Phys. Rev.* **116**, 175.

Ramamurthy, V. S., Kapoor, S. S., and Kataria, S. K. (1970). *Phys. Rev. Lett.* **25**, 386.

Reising, R. F., Bate, G. L., and Huizenga, J. R. (1966). *Phys. Rev.* **141**, 1161.

Sikkeland, T., Haines, E. L., and Viola, V. E., Jr. (1962). *Phys. Rev.* **125**, 1350.

Simmons, J. E., and Henkel, R. L. (1960). *Phys. Rev.* **120**, 198.

Strutinsky, V. M., Lyashchenko. N. Ya., and Popov, N. A. (1963). *Nucl. Phys.* **46**, 639.

Vandenbosch, R., Warhanek, H., and Huizenga, J. R. (1961). *Phys. Rev.* **124**, 846.

Vandenbosch, R., Unik, J. P., and Huizenga, J. R. (1965). *Proc. IAEA Symp. Phys. Chem. Fission, Salzburg, 1965*, **1**, p. 547. IAEA, Vienna.

Viola, V. E., Jr., Minor, M. M., Salwin, A. E., Bondelid, R. O., and Theus, R. B. (1971). *Nucl. Phys. A* **174**, 321.

Wheeler, J. A. (1963). *In* "Fast Neutron Physics" (J. B. Marion and J. L. Fowler, eds.), Pt. II, p. 2051. Wiley (Interscience), New York.

CHAPTER

Competition between Fission and Neutron Emission

A. Experimental Determination of Γ_n/Γ_f Values at Low and Moderate Excitation Energy

There are a number of different types of data available from which we can deduce neutron emission to fission width ratios. The range of excitation energy which can be examined depends on both the type of data being considered and the magnitude and excitation energy dependence of Γ_n/Γ_f.

1. Fission Cross Sections

For the lighter elements with fission barriers which are much larger than neutron binding energies, only a small fraction of the reaction cross section goes into fission. The relative probability for fission compared to neutron emission for these nuclei also is a strongly increasing function of excitation energy, so that fission following neutron emission can be neglected. In these circumstances, the ratio of the total fission cross section to the reaction cross section σ_f/σ_R is equal to Γ_f/Γ_n for the initial nucleus and excitation energy. (For historical reasons we will characterize the competition by Γ_f/Γ_n for the less fissionable nuclei and by Γ_n/Γ_f for the more fissionable elements.) At very high excitation energy Γ_f/Γ_n increases more slowly and contributions from second-chance fission (fission following neutron emission from the original compound nucleus) begin, and appropriate corrections must be made. Charged particle emission may also start to compete with neutron emission and fission. Determinations of Γ_f/Γ_n from particle-induced fission cross sections and calculated

TABLE VII-1

Survey of σ_f/σ_R for the Less Fissionable Elements

Target	Projectile	Compound nucleus	Lowest and highest excitation energy	$\frac{\sigma_f}{\sigma_R} \approx \Gamma_f/\Gamma_n$ for these energies	Ref.
^{169}Tm	α	^{173}Lu	40.5	3.2×10^{-9}	*a*
			75.7	1.2×10^{-5}	
^{175}Lu	α	^{179}Ta	36.2	1.7×10^{-9}	*a*
			75	4.2×10^{-5}	
^{185,187}Re	α	^{189,191}Ir	29.5	3.5×10^{-9}	*a*
			75	8.2×10^{-4}	
^{197}Au	p	^{198}Hg	22.5	7.8×10^{-10}	*b*
			56.8	1.7×10^{-3}	
^{197}Au	α	^{201}Tl	23	3.8×10^{-11}	*b,c,d*
			58.8	4×10^{-3}	
^{203,205}Tl	α	^{207,209}Bi	32	2.3×10^{-5}	*d*
			38.2	3.5×10^{-4}	
^{206}Pb	p	^{207}Bi	21.5	2.4×10^{-11}	*b*
			53	6.4×10^{-3}	
^{208}Pb	p	^{209}Bi	23	7.1×10^{-12}	*b*
			53	3.1×10^{-3}	
^{206}Pb	α	^{210}Po	19.7	7.8×10^{-12}	*b,d,e*
			120	$2.45 \times 10^{-1\,f}$	
^{209}Bi	p	^{210}Po	20.2	3.8×10^{-11}	*b*
			54.4	$1.8 \times 10^{-2\,f}$	
^{207}Pb	α	^{211}Po	28.1	5.3×10^{-5}	*b,e*
			100	$1.2 \times 10^{-1\,f}$	
^{208}Pb	α	^{212}Po	15.6	1.4×10^{-11}	*b,e*
			116.5	$1.7 \times 10^{-1\,f}$	
^{209}Bi	α	^{213}At	15.4	3.2×10^{-11}	*b,d,e*
			41.4	$2 \times 10^{-2\,f}$	

[a] Raisbeck and Cobble (1967). [b] Khodai-Joopari (1966). [c] Burnett *et al.* (1964).
[d] Huizenga *et al.* (1962). [e] Halpern and Nicholson (1960).
[f] These values of σ_f/σ_R are sufficiently large that the approximation $\Gamma_f/\Gamma_n \approx \sigma_f/\sigma_R$ is no longer valid.

reaction cross sections have been made for a considerable number of nuclides, including the heavier rare earths and elements in the lead–bismuth region. A summary of the available data is given in Table VII-1. We have omitted from this table data obtained with heavy ion projectiles due to possible complications associated with angular momentum effects.

If we attempt to extend this type of analysis to charged particle-induced fission of the heavier elements, we find that as soon as we have sufficient energy to get through the Coulomb barrier, the excitation energy is sufficiently high and the fissionability so large that second-chance fission contributes significantly. We can, however, extract information about Γ_n/Γ_f from fast neutron fission cross sections for neutron energies below the threshold for second-chance fission. Neglecting proton emission and γ-ray deexcitation, the neutron fission cross section is expressed by

$$\sigma_f = \sigma_R \Gamma_f/(\Gamma_f + \Gamma_n)$$

When σ_f and σ_R are known, Γ_n/Γ_f is given by

$$\Gamma_n/\Gamma_f = (\sigma_R/\sigma_f) - 1$$

Table VII-2 summarizes the available data. For the most fissionable nuclei the analysis is rather sensitive to the reaction cross section, which has been assumed to be 3.1 barns on the basis of the analysis of Howerton (1958). A closely related type of measurement is that of the ratio of the number of protons of a given energy in coincidence with fission to the total number of protons for (d, p) and (t, p) reactions. This ratio is equal to $\Gamma_f/(\Gamma_f + \Gamma_n)$, assuming again that neutron emission and fission are the only important modes of decay. The (d, pf) reaction is not of particular interest here, as the same compound nucleus can be studied more easily for a given target by fast neutrons. The (t, pf) reaction, however, can make nuclides not accessible by neutron-induced fission. Table VII-3 lists Γ_n/Γ_f values deduced by this method at an excitation energy 2 MeV above the neutron binding energy. These values tend to be somewhat higher than those obtained with fast neutrons for those cases where comparisons can be made. The reason for this discrepancy is not known.

2. Photofission and Photoneutron Cross Sections

If the absorption cross section for photons were known, we could determine Γ_n/Γ_f from photofission measurements in the same way as for fast neutron fission. Unfortunately photoabsorption cross sections are poorly known, so a measurement of the photoneutron cross section is also desirable. Photofission and photoneutron cross sections have been measured for ^{232}Th and ^{238}U and the resulting values of Γ_n/Γ_f appear to be independent of excitation energy in the energy range 8–12 MeV within experimental error. On the other hand,

TABLE VII-2

Γ_n/Γ_f Values Deduced from Fast Neutron Fission Cross Sections at 2- and 5-MeV Neutron Energy Assuming a Reaction Cross Section of 3.1 Barns[a]

Compound nucleus	$(\Gamma_n/\Gamma_f)_2$ MeV	$(\Gamma_n/\Gamma_f)_5$ MeV
^{227}Ra		1.2×10^5 [g]
^{231}Th	6.6[b]	
^{233}Th	27	20
^{232}Pa	1.2[c]	
^{234}U	0.62	1.17
^{235}U	1.05	1.48
^{236}U	1.37	2.0
^{237}U	2.7	2.7
^{239}U	4.9	5.2
^{238}Np	0.82	1.2
^{239}Pu	0.3[c,e]	
^{240}Pu	0.56	0.9
^{241}Pu	0.94	1.14
^{242}Pu	0.74	1.17
^{243}Pu	1.4[e]	
^{245}Pu	2.1[h]	2.73[h]
^{242}Am	0.8[e]	
^{244}Am	1.21	
^{245}Cm	0.55[d]	
^{246}Cm	0.41[i]	
^{247}Cm	0.77[i]	
^{249}Cm	0.79[i]	
^{250}Bk	1.6[f]	
^{253}Cf	0.17[j]	

[a] The fission cross sections have been taken from the evaluation of Davey (1968) unless otherwise indicated.
[b] *Sov. J. At. Energy* (1961).
[c] BNL 325 (1965).
[d] *Nat. Bur. Stand.* (*U.S.*) (1968). Spec. Publ. 299.
[e] Fomushkin and Gutnikova (1970).
[f] Vorotnikov *et al.* (1970); Fomushkin *et al.* (1972).
[g] Babenko *et al.* (1969).
[h] Auchampaugh *et al.* (1971).
[i] Moore and Keyworth (1971).
[j] Moore *et al.* (1971).

TABLE VII-3

Γ_n/Γ_f Values Deduced from (t, pf) Experiments at an Excitation Energy 2 MeV Above the Neutron Binding Energy[a]

Compound nucleus	Γ_n/Γ_f
^{232}Th	16
^{234}Th	30
^{235}U	1.5
^{236}U	1.7
^{237}U	3.3
^{238}U	4.6
^{240}U	6.1
^{241}Pu	1.4
^{242}Pu	1.3
^{244}Pu	1.8

[a] From Cramer and Britt (1970).

measurements by a different technique (Bowman *et al.*, 1964) indicate that Γ_n/Γ_f is increasing considerably between 7 and 10 MeV in the case of ^{235}U. Most measurements of photofission and photoneutron emission have been performed with bremstrahlung radiation; hence, they represent an average over all energies less than the maximum energy. To the extent that Γ_n/Γ_f is independent of excitation energy and the difference between the neutron emission threshold and fission threshold is small, the bremsstrahlung results should be representative of Γ_n/Γ_f at energies between threshold and the maximum bremsstrahlung energy. Table VII-4 summarizes the available data. These values of Γ_n/Γ_f are considerably more uncertain than those obtained from fast neutron fission cross sections.

3. Mean Values of Γ_n/Γ_f Deduced from Higher Energy Neutron Fission Cross Sections

The methods previously described enable us to obtain Γ_n/Γ_f at a unique or near unique (in the case of photofission) excitation energy. If we examine a fast neutron fission excitation function, we find that a plateau is usually achieved within 1 MeV above the fission threshold. Above 5- or 6-MeV neutron energy the fission cross section is observed to start to rise again,

TABLE VII-4

Γ_n/Γ_f Values Deduced from Photofission and Photoneutron Emission Cross Sections for 12-MeV Bremsstrahlung

	Γ_n/Γ_f	
Compound nucleus	Huizenga (1958)	Katz *et al.* (1958)
^{230}Th	4.9	
^{232}Th	12	10
^{233}U	1.0	1.6
^{234}U	1.6	
^{235}U	1.6	
^{236}U	2.1	
^{238}U	4.0	4.9
^{237}Np	1.0	
^{239}Pu	0.4	0.3

reaching a new plateau by about 8 MeV. A typical example is illustrated in Fig. VII-1. This additional contribution to the fission cross section arises from fission following neutron emission. If we want to investigate higher excitation energies or more fissionable nuclei, we have to consider information on Γ_n/Γ_f which represents an average for two or more excitation energies corresponding

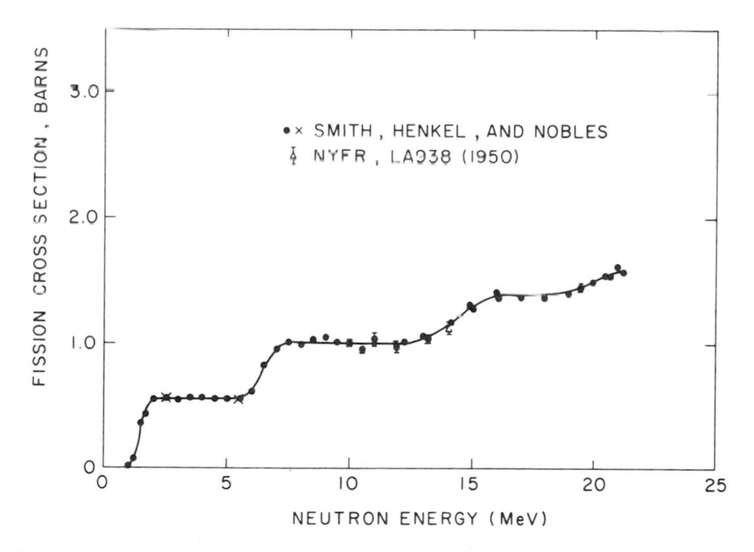

Fig. VII-1. The neutron-induced fission cross section of ^{238}U is plotted as a function of neutron energy. [From Huizenga and Vandenbosch (1962).]

to competition both before and after neutron emission from the initial compound nucleus. Consider, for example, 10-MeV neutron fission cross sections. The fraction of the initial nuclei which emit a neutron is given by the ratio of the neutron emission width to the neutron plus fission width for the compound nucleus with mass number A and excitation energy $E^* = 10 + B_n$, $[\Gamma_n/(\Gamma_n + \Gamma_f)]_{A, E^* = 10 + B_n}$. Similarly, the residual nucleus following neutron emission will have sufficient excitation energy either to emit a second neutron or to fission, with the probability of emitting a neutron given by $[\Gamma_n/(\Gamma_n + \Gamma_f)]_{A-1, E^* = 10 - 2T}$ where the competition is occurring in the nucleus with mass number $A-1$ at an average excitation energy which is about $(B_n + 2T) \approx 8$ MeV lower than that of the initial compound nucleus. The neutron is assumed to have an average kinetic energy of twice the nuclear temperature, with typical nuclear temperatures of 0.5–1 MeV. The fraction of nuclei which survive fission both before and after emission of a neutron, $(\sigma_c - \sigma_f)/\sigma_c$, will be given by

$$(\sigma_c - \sigma_f)/\sigma_c = [\Gamma_n/(\Gamma_n + \Gamma_f)]_{A, E^* = 10 + B_n} [\Gamma_n/(\Gamma_n + \Gamma_f)]_{A-1, E^* = 10 - 2T}$$

A mean value of Γ_n/Γ_f can then be obtained by the relation

$$\Gamma_n/\Gamma_f = [(\sigma_c - \sigma_f)/\sigma_c]^{1/2} / \{1 - [(\sigma_c - \sigma_f)/\sigma_c]^{1/2}\}$$

The mean fissioning nucleus is symbolically characterized by a mass number $A - 0.5$. Mean values of Γ_n/Γ_f deduced from fission cross sections at 10-MeV neutron energy are given in Table VII-5.

In principle we might hope to be able to determine Γ_n/Γ_f uniquely for the first step of the competition if Γ_n/Γ_f were known for the second, low excitation energy step from other data (such as neutron fission cross sections at lower energy for a target with mass number one less than the target whose second-plateau fission cross section is being considered). An attempt (Vandenbosch and Huizenga, 1958) to extract such information has been made, but unfortunately the results are quite sensitive to the poorly known reaction cross sections, particularly for the more fissionable nuclei.

4. Mean Values of Γ_n/Γ_f Derived from Charged Particle-Induced Spallation Cross Sections

When heavy element nuclides are bombarded with charged particles, the resulting excitation energy is usually high enough so that fission has a chance to compete with neutron emission along an evaporation chain composed of the parent compound nucleus and the various successive daughter nuclei formed by neutron evaporation. This effect is illustrated in Fig. VII-2, where we compare some excitation functions for the $(\alpha, 2n)$, $(\alpha, 3n)$, and $(\alpha, 4n)$ reactions for the slightly fissionable nucleus ^{209}Bi with the easily fissionable

TABLE VII-5

Mean Value of Γ_n/Γ_f from Neutron-Induced Fission Cross Sections at 10 MeV[a]

Average fissioning nucleus	Γ_n/Γ_f
$^{232.5}Th$	19
$^{233.5}U$	1.3
$^{234.5}U$	1.47
$^{235.5}U$	1.94
$^{236.5}U$	1.94
$^{238.5}U$	4.5
$^{237.5}Np$	0.89
$^{239.5}Pu$	1.05
$^{240.5}Pu$	1.94
$^{241.5}Pu$	1.05

[a] Data from evaluation of Davey (1968). The total reaction cross section has been assumed to be 2.9 barns.

nucleus ^{235}U. In the latter case the spallation yields are greatly decreased due to fission competition. For each additional neutron emitted, fission has another chance to compete, so that the excitation functions are increasingly attenuated as the number of neutrons emitted increases. A measure of the competition between neutron emission and fission can therefore be obtained by comparing spallation yields of fissionable isotopes with corresponding hypothetical yields had fission not occurred.

In order to deduce the extent to which fission competition has reduced the yield of a particular evaporation product, we need to know what yield would have been expected had fission not taken place. Jackson (1956) has described a simple model for neutron evaporation incorporating the following assumptions.

(1) The neutron energy spectrum is given by $\varepsilon \exp(-\varepsilon/T)$ where ε is the kinetic energy of the neutron and T is the nuclear temperature;
(2) neutron emission occurs whenever it is energetically possible;
(3) proton evaporation is neglected; and
(4) the nuclear temperature T is independent of excitation energy.

This last assumption is a rather crude approximation, but probably does not

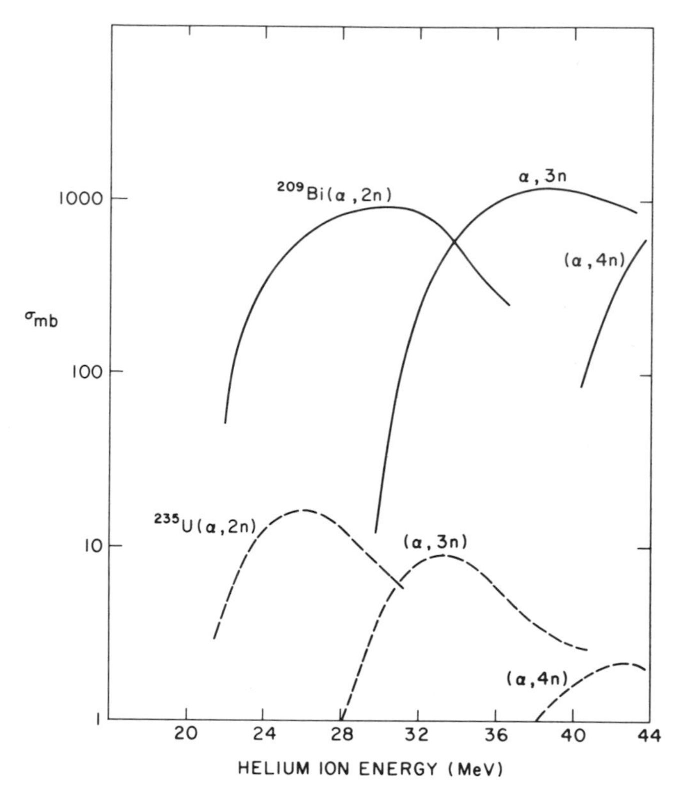

Fig. VII-2. Spallation excitation functions for the slightly fissionable target ^{209}Bi and for the fissionable target ^{235}U. [^{209}Bi data from Ramler *et al.* (1959), ^{235}U data from Vandenbosch *et al.* (1958).]

introduce a large uncertainty in the calculation. If $P(E,x)$ is the calculated probability that a nucleus with initial excitation energy E will evaporate exactly x neutrons, then the cross section for a particular spallation reaction at excitation energy E is

$$\sigma_{x\text{n}} = P(E,x)\sigma_{\text{c}}$$

Jackson has shown that $P(E,x)$ is given by

$$I(\Delta_x, 2x-3) - I(\Delta_{x+1}, 2x-1)$$

where I is Pearson's incomplete gamma function and

$$\Delta_x = \left(E - \sum_{i=1}^{x} B_i\right)\Big/ T$$

where B_i is the binding energy of the ith neutron and T is the nuclear temperature. The first incomplete gamma function gives the probability that the original compound nucleus will emit at least x neutrons; the second, the probability that the residual nucleus will have an excitation energy greater than the binding energy of the last neutron.

To include fission competition we must make two modifications. For each step i of the deexcitation where fission can compete with neutron emission we must include a multiplicative factor $[\Gamma_n/(\Gamma_n+\Gamma_f)]_i$, hereafter designated G_{ni}. We must also take into account the possibility that the fission threshold E_f of the final residual nucleus corresponding to the emission of i neutrons may be less than the neutron binding energy of that nucleus. Residual nuclei with excitation energy greater than the fission threshold but insufficient for neutron evaporation are assumed to fission. This effect is incorporated in the formalism above by replacing B_{x+1} by E_f in the expression for Δ_{x+1}. Assuming further that $\Gamma_n/(\Gamma_f+\Gamma_n)$ is independent of excitation energy over the excitation energy range corresponding to the possible kinetic energies of the emitted neutrons, we have

$$\sigma_{xn} = \sigma_c G_{n_1} G_{n_2} \cdots G_{n_x} [P(E,x)]$$

By matching the magnitude of the observed excitation function to that predicted by the expression above, the normalizing constant

$$G_{n_1} G_{n_2} \cdots G_{n_x} = \prod_{i=1}^{x} G_{n_i}$$

can be determined. Mean values of Γ_n/Γ_f for the nuclei and excitation energies sampled can be determined from the geometric mean value of G_{n_i}

$$\bar{G}_n = \left(\prod_{i=1}^{x} G_{n_i} \right)^{1/x}$$

An analysis of data in which 2–4 neutrons were emitted has been given previously (Vandenbosch and Huizenga, 1958). These values together with newer data are summarized in Table VII-6.

5. Summary of Experimental Values of Γ_n/Γ_f

The values of Γ_n/Γ_f for the heavier nuclei which are not strongly dependent on excitation energy are illustrated in Fig. VII-3. The data are seen to define fairly distinct parallel lines for different atomic numbers when plotted versus mass number. The strong dependence of Γ_n/Γ_f on mass number for a given Z is primarily due to the essentially monotonic decrease of neutron binding energies with increasing mass number, and to a much lesser extent to the increase in barrier heights with mass number. One of the most interesting features of this presentation is the fairly definite break in the variation of

TABLE VII-6

Geometric Mean Values of Neutron Emission to Fission Width Ratios Deduced from Charged Particle-Induced Spallation Reactions[a]

Target nucleus	Reaction	Cross section reference	Average fissioning nucleus	$\overline{(\Gamma_n/\Gamma_f)}$	Target nucleus	Reaction	Cross section reference	Average fissioning nucleus	$\overline{(\Gamma_n/\Gamma_f)}$
^{226}Ra	α, 4n	10	$^{228.5}$Th	4.2	^{239}Pu	d, 2n	21	$^{240.5}$Am	0.44
^{232}Th	p, 3n	19	^{232}Pa	2.2	^{239}Pu	d, 3n	21	^{240}Am	0.35
^{230}Th	α, 4n	10	$^{232.5}$U	0.50	^{240}Pu	d, 2n	24	$^{241.5}$Am	0.46
^{232}Th	α, 4n	20	$^{234.5}$U	0.92	^{240}Pu	d, 3n	24	^{241}Am	0.36
^{233}U	d, 2n	21	$^{234.5}$Np	0.25	^{238}Pu	α, 2n	9	$^{241.5}$Cm	0.33
^{233}U	d, 3n	21	^{234}Np	0.28	^{238}Pu	α, 4n	9	$^{240.5}$Cm	0.15
^{235}U	d, 2n	13	$^{236.5}$Np	0.43	^{239}Pu	α, 2n	9	$^{242.5}$Cm	0.28
^{235}U	d, 3n	13	^{236}Np	0.37	^{239}Pu	α, 3n	9	^{242}Cm	0.22
^{238}U	d, 2n	13	$^{239.5}$Np	0.89	^{239}Pu	α, 4n	9	$^{241.5}$Cm	0.22
^{233}U	α, 2n	11	$^{236.5}$Pu	0.16	^{240}Pu	α, 4n	25	$^{242.5}$Cm	0.20
^{233}U	α, 3n	11	^{236}Pu	0.12	^{242}Pu	α, 2n	9	$^{245.5}$Cm	2.8
^{234}U	α, 4n	22	$^{236.5}$Pu	0.21	^{242}Pu	α, 4n	9	$^{244.5}$Cm	0.46
^{235}U	α, 4n	11	$^{237.5}$Pu	0.28	^{243}Am	α, 2n	26	$^{246.5}$Bk	0.81
^{235}U	α, 3n	11	^{238}Pu	0.29	^{243}Am	α, 4n	26	$^{245.5}$Bk	0.57
^{235}U	α, 2n	11	$^{238.5}$Pu	0.38	^{244}Cm	α, 2n	27	$^{247.5}$Cf	0.34
^{236}U	α, 4n	10	$^{238.5}$Pu	0.34	^{244}Cm	α, 3n	27	^{247}Cf	0.23
^{238}U	α, 4n	23	$^{240.5}$Pu	0.61	^{244}Cm	α, 4n	27	$^{246.5}$Cf	0.18
^{237}Np	α, 2n	21	$^{240.5}$Am	0.34	^{249}Bk	α, 2n	28	$^{252.5}$E	0.53
^{237}Np	α, 3n	21	^{240}Am	0.33	^{249}Bk	α, 4n	28	$^{251.5}$E	0.41
^{238}Pu	d, 2n	24	$^{239.5}$Am	0.24	^{252}Cf	α, 4n	29	$^{254.5}$Fm	0.24[b]
^{238}Pu	d, 3n	24	^{239}Am	0.26					
		Addendum							
^{238}U	α, 2n	*c*	$^{241.5}$Pu	1.15					
^{238}U	α, 3n	*c*	^{241}Pu	0.84					
^{238}U	α, 4n	*c*	$^{240.5}$Pu	0.61					
^{235}U	α, 3n	*c*	^{237}Pu	0.33					
^{238}U	d, 4n	*d*	$^{238.5}$Np	0.62					
^{236}U	d, 3n	*d*	^{237}Np	0.53					
^{234}U	d, 3n	*d*	^{235}Np	0.31					

[a] From Vandenbosch and Huizenga (1958); the reference numbers refer to references in this paper.
[b] Lower limit.
[c] Wing *et al.* (1959).
[d] Lessler (1958).

Γ_n/Γ_f with atomic number at Z of approximately 93. For higher atomic numbers Γ_n/Γ_f does not depend as strongly on the atomic number as it does for lower atomic numbers. This is a consequence of the tendency for fission barriers in this region not to decrease as fast with Z^2/A as would be expected on the basis of the liquid drop model.

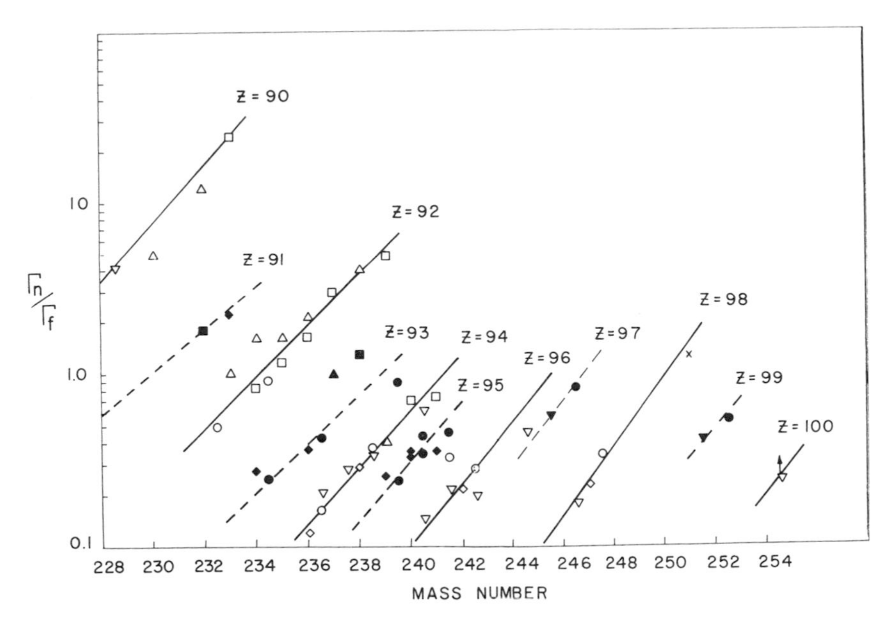

Fig. VII-3. Neutron emission to fission width ratios are plotted as a function of mass number. Open and closed symbols refer to fissioning nuclei with even and odd atomic number, respectively. (△) Data obtained from photoneutron and photofission experiments and corresponding to an excitation energy of 8 to 12 MeV; (□) data derived from 3-MeV neutron fission cross sections and corresponding to an excitation energy of 8 to 10 MeV; (○, ◇, ▽) mean values of Γ_n/Γ_f obtained from spallation excitation functions and corresponding to average excitation energies of approximately 13, 18, and 23 MeV, respectively. [From Vandenbosch and Huizenga (1958).]

B. Theoretical Expectations for Γ_n/Γ_f

1. The Neutron and Fission Widths at Low Energies

In order to appreciate the assumptions inherent in the usual expressions for Γ_n/Γ_f, let us start with the expressions for the neutron width and the fission width used in our previous discussion of low energy resonance fission. For a single neutron decay channel the neutron width is $\Gamma_n = (D/2\pi)T$, where T is a transmission coefficient. If more than one final state can be reached, a sum over the transmission coefficients T leading to various final states appears,

$$\Gamma_n = (D/2\pi) \sum T$$

where D is the level spacing of levels of the appropriate spin and parity of the compound nucleus. A similar expression has been given for the fission width. The expression above for the neutron width is valid at low energies where only

s-wave ($l = 0$) emission occurs. At higher energy, where other partial waves contribute, the sum over angular momentum (but not over different final energy levels) becomes $\sum_l (2l+1) T_l$, which at high energies can be replaced by an integral $\int (2l+1) T_l \, dl$. If we assume that T_l is unity up to the maximum possible angular momentum the neutron can carry away, l_m, and equal to zero for $l > l_\mathrm{m}$, the integral becomes

$$\int_0^{l_\mathrm{m}} (2l+1) T_l \, dl \simeq l_\mathrm{m}^{\,2} = (Rp/\hbar)^2 = 2mR^2\varepsilon/\hbar^2$$

where l_m has been evaluated by considering emission of a neutron of energy ε and mass m from the nuclear surface at radius R in a direction tangent to the surface. If we now replace the sum over final levels by an integral over the energy levels of the residual nucleus, $\rho(E - B_\mathrm{n} - \varepsilon)$, and take into account the neutron intrinsic spin degeneracy g, we obtain

$$\Gamma_\mathrm{n} = (D/2\pi)(2mR^2 g/\hbar^2) \int_0^{E-B_\mathrm{n}} \varepsilon\rho(E - B_\mathrm{n} - \varepsilon) \, d\varepsilon \qquad \text{(VII-1)}$$

For the fission width we usually simply replace the sum of transmission (barrier penetration) coefficients in the expression

$$\Gamma_\mathrm{f} = (D/2\pi) \sum T$$

by an integral over the energy levels at the saddle, assuming $T = 1$ for energies above the barrier E and $T = 0$ for energies below the barrier, to obtain

$$\Gamma_\mathrm{f} = (D/2\pi) \int_0^{E-E_\mathrm{f}} \rho(E - E_\mathrm{f} - K) \, dK \qquad \text{(VII-2)}$$

where K is the kinetic energy in the fission degree of freedom. If barrier penetration is included, Eq. (VII-2) becomes

$$\Gamma_\mathrm{f} = \frac{D}{2\pi} \int_{-\infty}^{E-E_\mathrm{f}} \left[\frac{\rho(E - E_\mathrm{f} - K)}{1 + \exp(-2\pi K/\hbar\omega)} \right] dK$$

Negative values of K are associated with tunneling through barriers corresponding to transition states at an energy of $E_\mathrm{f} - K$ with respect to the ground state. In practice the only significant contribution to the integral comes from K values close to zero.

It is now apparent that we have introduced additional degrees of freedom for neutron emission relative to fission by considering neutron emission as a three-dimensional problem, whereas we have treated fission as a one-dimensional problem in the fission coordinate.

The three-dimensional aspect of neutron emission arises because neutron emission exploits all the possible phase space of a free particle. This is not the case for fission, since the nucleus becomes committed to fission as it passes over the saddle, long before the final fragments achieve their identities. There is, however, little direct evidence to support the assumption that passage over the saddle is strictly a one-dimensional process, and it would be of considerable interest to identify other relevant simple degrees of freedom at the saddle points and to include them explicitly in the evaluation of the fission width. If, however, the frequencies associated with these degrees of freedom at the saddle point are fairly high, contribution at higher energies to the phase space will be overshadowed by the exponential increase of the nuclear level density with excitation energy and their effect will be difficult to isolate.

Expressions (VII-1) and (VII-2) can be combined to yield an expression for Γ_n/Γ_f

$$\frac{\Gamma_n}{\Gamma_f} = \frac{2mr_0^2 g}{\hbar^2} \frac{A^{2/3} \int_0^{E-B_n} \varepsilon\rho(E-B_n-\varepsilon)\, d\varepsilon}{\int_0^{E-E_f} \rho(E-E_f-K)\, dK} \qquad \text{(VII-3)}$$

For suitable functional forms of the nuclear level density, Eq. (VII-3) can be integrated to yield closed function expressions for Γ_n/Γ_f. The simplest such form for the level density is a constant temperature level density

$$\rho(E) = \text{const} \exp(E/T) \qquad \text{(VII-4)}$$

This yields, dropping some small terms,

$$\frac{\Gamma_n}{\Gamma_f} = \frac{2TA^{2/3}}{K_0} \exp[(E_f - B_n)/T] \qquad \text{(VII-5)}$$

where $K_0 = \hbar^2/2mr_0^2 \approx 10$ MeV. This expression is expected to be valid only over a rather limited range of excitation energy and where E_f and B_n are of comparable magnitude. The existence of a fairly flat plateau for neutron-induced fission cross sections of heavy elements is qualitatively consistent with the assumption of constant temperature over an energy range of a few MeV. The functional form of the expression above suggests that a plot of the logarithm of Γ_n/Γ_f should be correlated with the difference between the fission threshold and the neutron binding energy. Figure VII-4 is such a plot for the Γ_n/Γ_f values obtained for heavy elements from 2-MeV neutron fission cross sections. The fission thresholds used are experimental values given in Table VIII-1. The experimental data are seen to fall on one of several lines, depending on the even-odd character of the fissioning nucleus. For nuclei of a given type a very strong correlation of Γ_n/Γ_f with the difference $E_f - B_n$ is observed. The slope of the line yields a nuclear temperature of 0.33 MeV. This is not unreasonable in view of the fact that the average excitation energy of the residual nuclei is only about 2 MeV. The displacement of the data points for different nuclear types can be readily understood in terms of odd-even effects on nuclear level densities. If we take odd-A nuclei as a reference surface, then we expect even-even nuclei to have an effective excitation energy less than the nominal

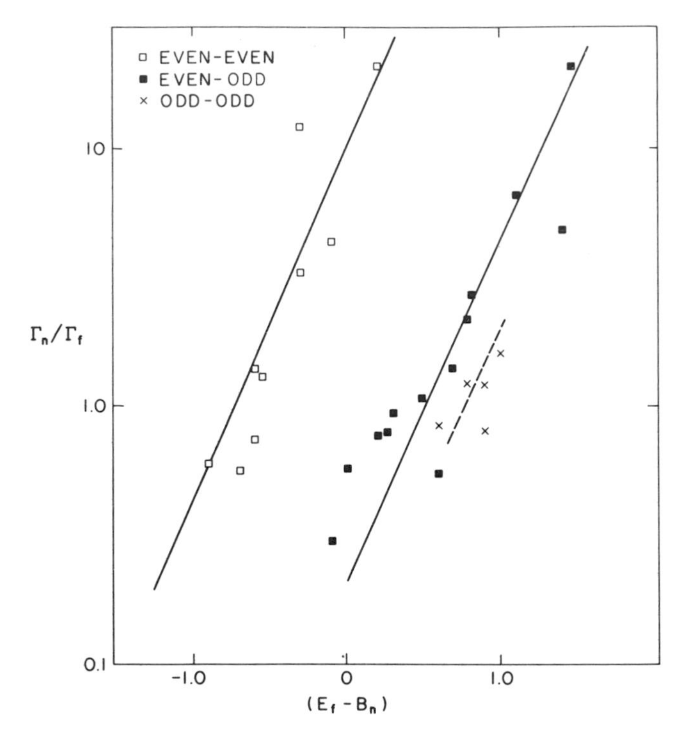

Fig. VII-4. Γ_n/Γ_f values from 2-MeV neutron-induced fission cross sections or from (t, pf) experiments (the latter values have been decreased by a factor of 0.7 in keeping with the observation that when data obtained by both methods are available, the values obtained from (t, p) experiments are consistently high) are plotted against the difference between the experimental fission thresholds and neutron binding energies.

excitation energy by an amount Δ, and odd-odd nuclei to have an effective excitation energy increased by an amount Δ. If we allow for the fact that Δ may be different at the saddle point than at the equilibrium deformation appropriate to neutron emission, the difference between the effective fission threshold E_f' and the effective neutron binding energy B_n' becomes

$$E_f' - B_n' = E_f - B_n + \Delta_f \qquad \text{(even-even fissioning nuclei)}$$

$$E_f' - B_n' = E_f - B_n - \Delta_n \qquad \text{(even-odd fissioning nuclei)}$$

These corrections are illustrated graphically for an even-even compound nucleus and an odd-A compound nucleus in Figs. VII-5 and VII-6.

In addition to pairing effects on level densities, there are also shell effects which are often incorporated in the Δ corrections. These shell effects are excitation energy dependent and when large (in the vicinity of closed shells) must be treated explicitly and in addition to the pairing effects discussed above.

The displacement between the Γ_n/Γ_f values for even-even and even-odd fissioning nuclei indicates that $\Delta_n+\Delta_f = 1.25$ MeV. The few points for odd-odd nuclei exhibit more scatter but suggest that $\Delta_f-\Delta_n = 0.25$. This would give $\Delta_n \simeq 0.5$ and $\Delta_f \simeq 0.7$, although the uncertainties are such that Δ_f could be essentially equal to Δ_n, especially if the barriers for odd-odd nuclei are a bit high due to specialization energy effects. We might expect a systematic effect of this sort, as the only determination of barriers in these nuclei has been from neutron fission excitation functions. The value of Δ_n deduced from this analysis is in good agreement with the odd-even differences in the ground state masses which correspond to $\Delta_n \simeq 0.6$ MeV.

If we adopt estimates of 0.5 for Δ_n and 0.7 for Δ_f, we can plot mean values of Γ_n/Γ_f for the heavy elements versus the average value of $E_f'-B_n'$. Such a plot is shown in Fig. VII-7. There is considerably more scatter in the data, partly because the data now represent a considerably larger range of excitation energy than that for which the nuclear temperature can be reasonably expected to be constant, and partly due to inaccuracies in the data. The slope of the line drawn corresponds to $T = 0.4$ MeV, a value only slightly higher than for the

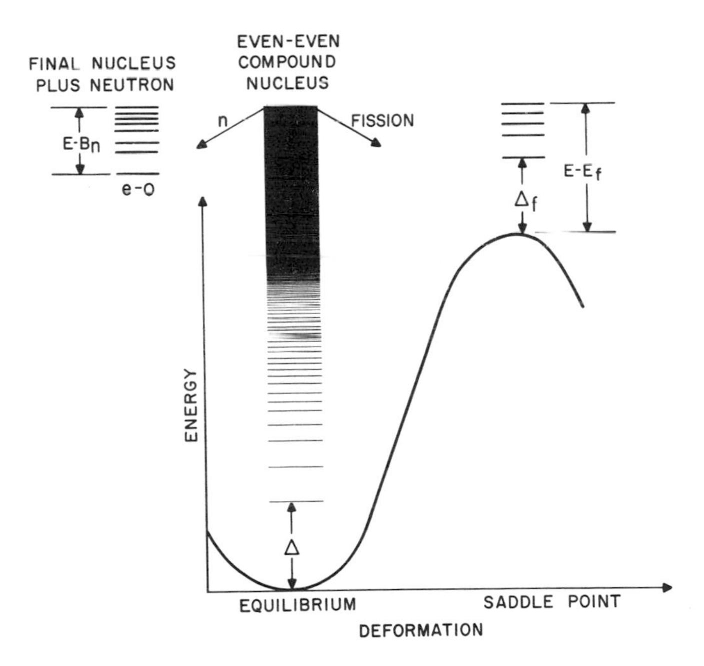

Fig. VII-5. The spectrum of intrinsic nucleonic states at the equilibrium and at the fission saddle point deformations for an even-even nucleus, and in the residual (even-odd) nucleus following neutron emission. The reference surface is assumed to be that of an odd-A nucleus. [From Huizenga and Vandenbosch (1962).]

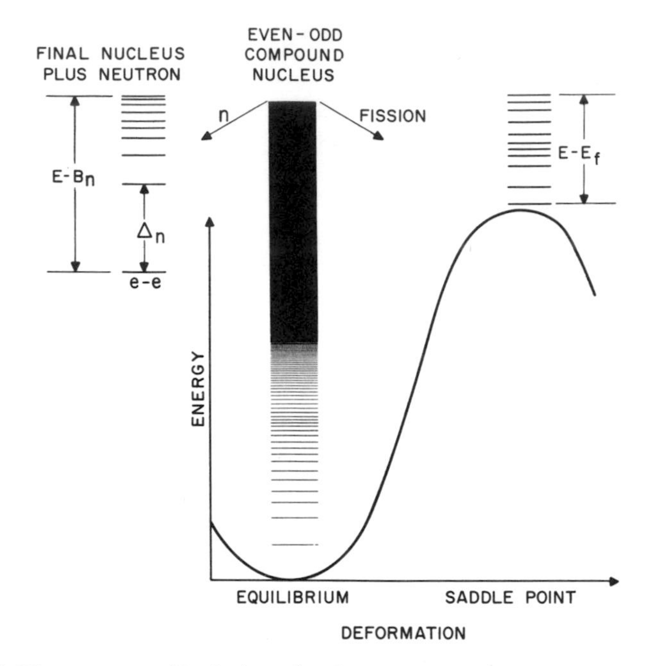

Fig. VII-6. The spectrum of intrinsic nucleonic states at the equilibrium and at the fission saddle point deformations for an even-odd nucleus, and in the residual nucleus following neutron emission. The reference surface is assumed to be that of an odd-*A* nucleus. [From Huizenga and Vandenbosch (1962).]

low energy neutron-induced fission data, but at least in the right direction considering the higher average excitation energy at which the competition is being sampled. The absolute value of Γ_n/Γ_f at $E_f' - B_n' = 0$ is less than one half of the value expected from Eq. (VII-5).

Detailed examination of Fig. VII-7 reveals that there is a systematic tendency for the Γ_n/Γ_f values for the points for Cm (the highest Z included in this graph) nuclei to fall to the right of the line. This may reflect the fact that the Cm fission barriers are very high, apparently due to strong shell effects. Some of these shell effects may have been weakened at the higher excitation energies sampled in these reactions. This effect may also account for the steepness of the plot of (Γ_n/Γ_f) versus $E_f' - B_n'$ and the resulting low value of the nuclear temperature.

2. Excitation Energy Dependence of Γ_n/Γ_f

A discussion of the excitation energy dependence of Γ_n/Γ_f requires a more realistic level density expression than the constant temperature expression

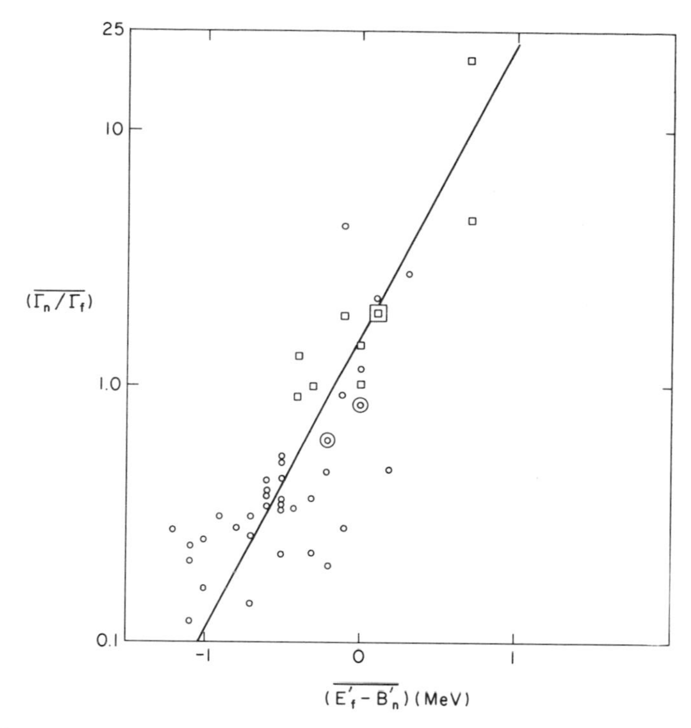

Fig. VII-7. Mean values of Γ_n/Γ_f from spallation cross section analyses (open circles) and from fast neutron second-plateau fission cross sections (open squares) plotted as a function of the average value of the difference between the effective fission threshold and the effective neutron binding energy. Data for the trans-curium elements for which no experimental fission thresholds are known and for which extrapolations are not reliable have been omitted.

given previously. One expression is the so called Fermi gas expression

$$\rho(E) \propto \exp[2(aE)^{1/2}] \tag{VII-6}$$

where a is the nuclear level density parameter. A more rigorous form of the level density expression contains a preexponential energy factor, but the resulting expressions for the neutron width and the fission width cannot be integrated in closed form. Substitution of (VII-6) into (VII-3) yields, after dropping some small terms,

$$\frac{\Gamma_n}{\Gamma_f} = \frac{4A^{2/3}a_f(E-B_n)}{K_0\, a_n[2a_f^{1/2}(E-E_f)^{1/2}-1]} \exp[2a_n^{1/2}(E-B_n)^{1/2} - 2a_f^{1/2}(E-E_f)^{1/2}] \tag{VII-7}$$

where we have anticipated the possibility that the level density parameter appropriate to the saddle point, a_f, may be different from that appropriate to the equilibrium deformation, a_n. The magnitude of the variation of Γ_n/Γ_f with excitation energy and the dependence of this variation on $(E_f - B_n)$ is illustrated in Fig. VII-8. Numerical integration of the expressions obtained with inclusion of the preexponential energy dependence of the level density results in curves very similar to those obtained with Eq. (VII-7) if the a values are increased approximately 20%. The dependence on excitation energy is clearly strongest when the fission barrier and neutron binding energy are very different. There are no experimental data available for the case where $E_f \ll B_n$, but there do exist considerable data for $E_f \gg B_n$, as reviewed in Section VII, A, 1. Halpern and Nicholson (1960) observed soon after such data became available that it was impossible to fit both the absolute magnitude and the energy dependence

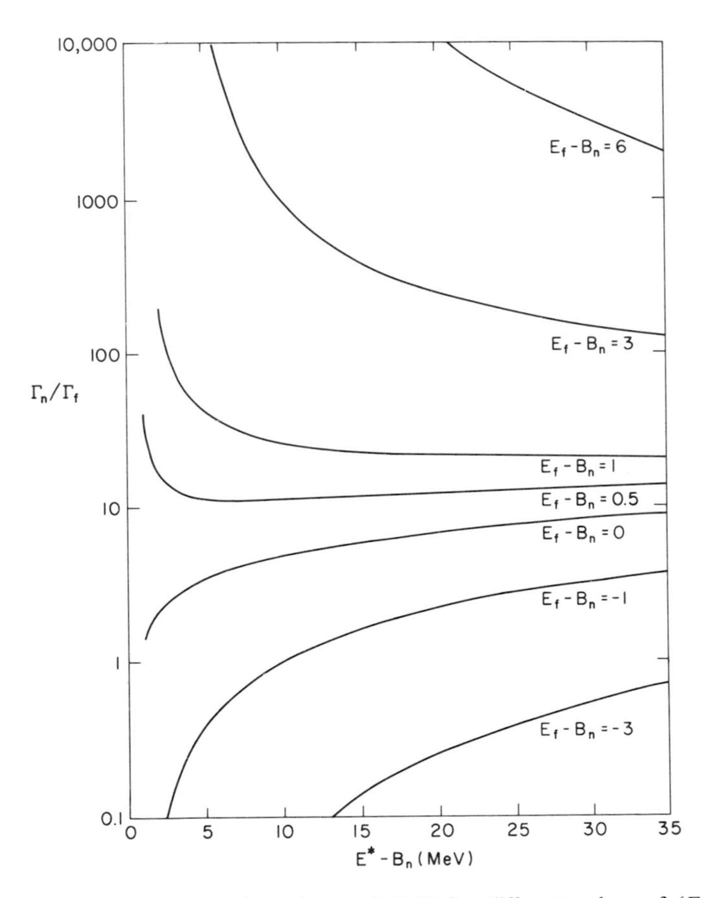

Fig. VII-8. Excitation energy dependence of Γ_n/Γ_f for different values of $(E_f - B_n)$ as given by Eq. (VII-7). The level density parameters a_f and a_n were assumed equal to 25 MeV^{-1} and B_n to 6 MeV in this example.

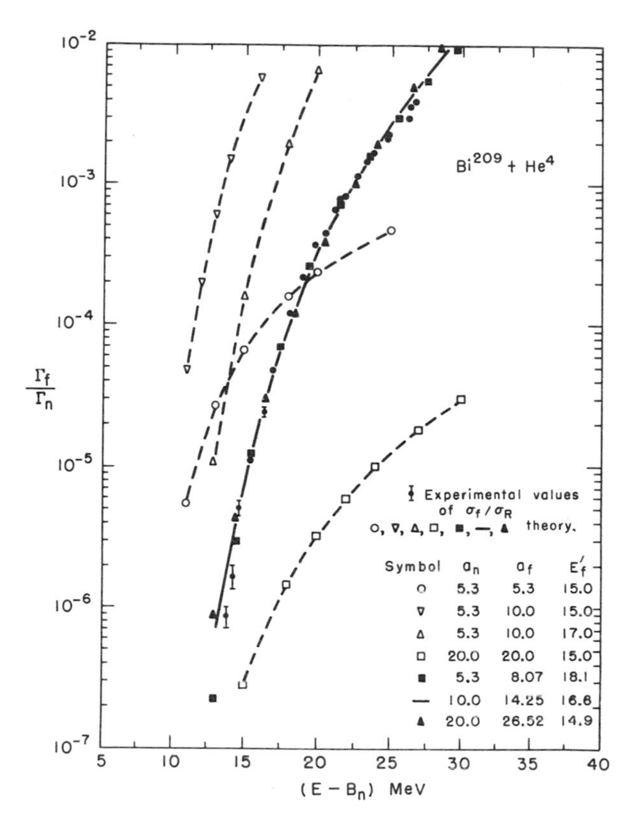

Fig. VII-9. Illustration of the necessity for a_f to be larger than a_n to fit Γ_f/Γ_n as a function of excitation energy. The closed circles are the experimental data. [From Huizenga *et al.* (1962).]

of Γ_f/Γ_n if $a_n = a_f$. Invariably the ratio of a_f/a_n had to be greater than unity, usually by 20–30%. An illustration of this effect is shown in Fig. VII-9. For some time this was believed to be due to the fact that the nuclei considered were close to the $Z = 82$, $N = 126$ shell, and it was felt that a_n would be anomalously small due to shell effects, whereas these shell effects would have disappeared at the nuclear deformation corresponding to the top of the fission barrier. As still lighter nuclei away from closed shells were investigated, it became clear that a_f/a_n was still larger than unity, although somewhat less than in the lead–bismuth region. A possible origin of this effect is discussed in a later section.

When experimental data for Γ_f/Γ_n as a function of excitation energy are fit by an expression such as Eq. (VII-7), the ratio a_f/a_n and the difference $E_f - B_n$ can be determined fairly uniquely. The fits are relatively insensitive to the absolute value of a_f (or a_n) and somewhat more sensitive to the absolute value of B_n (or E_f). One aim of fitting to such data is to determine the value of the fission threshold. In Table VII-7 values of E_f and a_f/a_n are listed for analyses where the experimental neutron binding energies were used. This simple form of analysis is inadequate in the region around the $Z = 82$, $N = 126$ shell. An approximate procedure is to absorb the shell correction into an effective neutron binding energy. The magnitude of this correction (or alternatively, the value of a_n) is in fact expected to be a function of excitation energy. One way of estimating a shell correction Δ_n^s to the excitation energy is to add to B_n the energy equivalent of some fraction of the difference between a smooth liquid drop mass and the true experimental shell-affected mass. Inclusion of shell effects through an effective B_n' tends to increase the fission thresholds and decrease the ratio a_f/a_n. A more realistic treatment of shell effects is described in Section B, 3.

TABLE VII-7

Values of E_f and a_f/a_n from Analyses of Fission Excitation Functions

Compound nucleus	E_f' (MeV)	a_f/a_n	Ref.
^{173}Lu	28.7 ± 2.5	1.08	Raisbeck and Cobble (1967)
^{178}W	23.0 ± 3.5	1.25	Sikkeland *et al.* (1971)
^{179}W	25.2 ± 3.5	1.24	Sikkeland *et al.* (1971)
^{179}Ta	27.5 ± 2.5	1.11	Raisbeck and Cobble (1967)
^{180}W	28.7 ± 3.5	1.19	Sikkeland *et al.* (1971)
^{181}Re	25.0 ± 3.5	1.21	Sikkeland *et al.* (1971)
^{186}Os	25.8 ± 3.5	1.20	Sikkeland *et al.* (1971)
^{187}Ir	21.8 ± 3.5	1.20	Sikkeland *et al.* (1971)
^{189}Ir	21.7 ± 2.5	1.10	Raisbeck and Cobble (1967)
^{191}Ir	22.8 ± 2.5	1.10	Raisbeck and Cobble (1967)
^{194}Hg	19.4 ± 3.5	1.20	Sikkeland *et al.* (1971)
^{198}Hg	21.8 ± 1.5	1.19	Khodai-Joopari (1966)
^{201}Tl	22.5 ± 1.5	1.35	Burnett *et al.* (1964)
^{207}Bi	21.2 ± 1.5	1.41	Khodai-Joopari (1966)
^{209}Bi	22.6 ± 1.5	1.44	Khodai-Joopari (1966)
^{210}Po	20.4 ± 1.5	1.46	Khodai-Joopari (1966)
^{212}Po	18.6 ± 1.5	1.40	Khodai-Joopari (1966)
^{213}At	16.8 ± 1.5	1.51	Khodai-Joopari (1966)

The excitation energy dependence of Γ_n/Γ_f becomes very difficult to investigate when the fission threshold and the neutron binding energy are approximately equal, both because of the weak dependence expected and the nature of the experimental data available. Since any excitation energy dependence is expected to be strongest at low excitation energy, the slope of the first "plateau" for neutron-induced fission might be expected to reflect this dependence. To qualitatively predict the type of dependence expected, we will assume that $(\Gamma_n/\Gamma_f)_2$ for 2-MeV neutrons is proportional to $T_2 \exp[(E_f' - B_n')/T_2]$ and for 5-MeV neutrons $(\Gamma_n/\Gamma_f)_5 \propto T_5 \exp[(E_f' - B_n')/T_5]$. This leads to the expression

$$\frac{(\Gamma_n/\Gamma_f)_5}{(\Gamma_n/\Gamma_f)_2} = \frac{T_5}{T_2} \exp\left[(E_f' - B_n')\Big/\left(\frac{1}{T_5} - \frac{1}{T_2}\right)\right].$$

In Fig. VII-10 we plot the ratio of Γ_n/Γ_f at two energies versus $E_f' - B_n'$, the difference between the experimental thresholds each corrected for odd-even effects. The decrease in the ratio $(\Gamma_n/\Gamma_f)_5/(\Gamma_n/\Gamma_f)_2$ with increasing values of $E_f' - B_n'$ may be understood in the following way. As the quantity $E_f' - B_n'$ becomes larger, the corresponding nuclei become less fissionable, and for these nuclei it is known that Γ_n/Γ_f decreases with excitation energy. The ratio $(\Gamma_n/\Gamma_f)_5/(\Gamma_n/\Gamma_f)_2$ is unity for a slightly positive value of $E_f' - B_n'$, as expected from the preexponential temperature dependence of Γ_n/Γ_f. Assuming $T_2 = 0.33$ from Fig. VII-4, the slope and the crossover of the line drawn through the data in Fig. VII-10 give independent but identical estimates of $T_5 = 0.5$ MeV.

Although a strong excitation energy dependence of Γ_n/Γ_f for the less fissionable elements has been observed, and there is evidence for the heavier nuclei

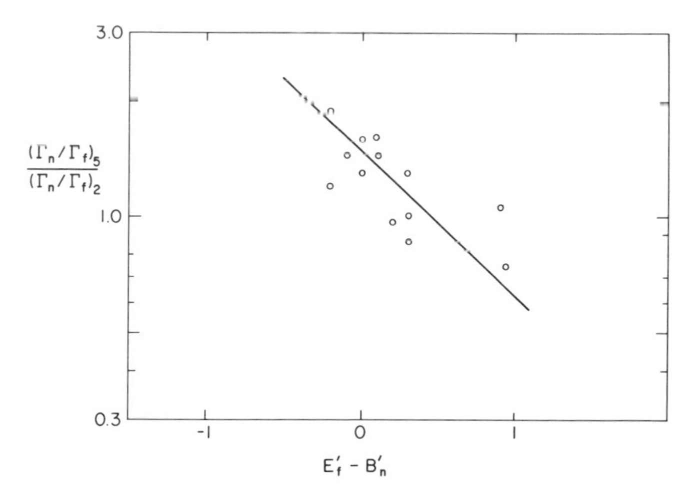

Fig. VII-10. The ratio of (Γ_n/Γ_f) from 5-MeV neutron-induced fission to that from 2-MeV neutron-induced fission is plotted as a function of the difference between the effective thresholds for fission and neutron emission.

that there are smaller variations at low energies which correlate with the difference $(E_f' - B_n')$, the excitation energy dependence of Γ_n/Γ_f for the heavier elements at higher excitation energy is rather unclear. There are a number of results which indicate that Γ_n/Γ_f for the heavier elements does not vary much with excitation energy.

In a few cases it is possible, by comparison of spallation cross sections for adjacent isotopes leading to the same final nucleus, to determine Γ_n/Γ_f at a fairly high excitation energy. If we consider the spallation cross section ratio for the $^{236}U(\alpha, 4n)$ reaction and the $^{235}U(\alpha, 3n)$ reaction at bombarding energies such that the average excitation energy of the identical final product is the same in both cases, we find that

$$\frac{\sigma_{\alpha,4n}}{\sigma_{\alpha,3n}} = \frac{G_{240}G_{239}G_{238}G_{237}P_4\sigma_R}{G_{239}G_{238}G_{237}P_3\sigma_R}$$

The value of Γ_n/Γ_f for ^{240}Pu at an excitation energy of about 36 MeV can be deduced. This calculation is possible since the last three competing steps of the $(\alpha, 4n)$ reaction occur in the same nuclei and at the same excitation energies as the corresponding competing steps of the $(\alpha, 3n)$ reaction and therefore cancel. The resulting value for Γ_n/Γ_f of 0.62 for ^{240}Pu at an excitation energy of 36 MeV is quite similar to the value of 0.74 for the same nucleus at an excitation energy of 9 MeV, the latter value being obtained from the 3-MeV neutron fission cross section of ^{239}Pu. A similar comparison of the $^{236}U(d, 3n)$ and $^{235}U(d, 2n)$ reactions yields a value of $\Gamma_n/\Gamma_f = 0.85$ for ^{238}Np at an excitation energy of about 24 MeV. This may be compared with a valuc of 0.9 at an excitation energy of 9 MeV as obtained from the 3-MeV neutron fission cross section of ^{237}Np.

TABLE VII-8

Comparison of Γ_n/Γ_f Values for the Same Average Fissioning Nucleus but Corresponding to Different Average Excitation Energies[a]

Reaction	Average intermediate nucleus	Γ_n/Γ_f	Initial excitation energy	Average excitation energy
$^{232}Th + p$	^{232}Pa	2.2	27	18
$^{231}Pa + n$	^{232}Pa	1.85	9	9
$^{232}Th + \alpha$	$^{234.5}U$	0.92	36	22.5
$^{233}U, ^{234}U + n$	$^{234.5}U$	1.0	9	9
$^{238}U + \alpha$	$^{240.5}Pu$	0.61	36	22.5
$^{239}Pu, ^{240}Pu + n$	$^{240.5}Pu$	0.72	9	9

[a] From Vandenbosch and Huizenga (1958).

There are a number of other examples where mean values of Γ_n/Γ_f are available for the same average fissioning nucleus but from data corresponding to different excitation energies. For example, the (Γ_n/Γ_f) value obtained from the $(\alpha, 4n)$ excitation function of ^{232}Th represents fission competition in the average fissioning nucleus $^{234.5}$U (the halfway point along the evaporation chain). If we interpolate between the values of Γ_n/Γ_f obtained from 3-MeV neutrons on ^{233}U and ^{234}U, we can obtain a value of Γ_n/Γ_f for $^{234.5}$U at lower excitation energy. In Table VII-8 are listed those cases for which a comparison of this type can be made. Again Γ_n/Γ_f is seen to vary relatively slowly with excitation energy. In all these cases, both those in Table VII-8 and the two more unique determinations mentioned in the preceding paragraph, the difference $E_f' - B_n'$ is very close to zero. From Fig. VII-8 we can see that little excitation energy dependence is expected, although for $a_f = a_n$ we would expect an increase in Γ_n/Γ_f with excitation energy somewhat larger than indicated experimentally.

There is some evidence that the approximate excitation energy independence of Γ_n/Γ_f extends to appreciably higher energies than thus far considered. Attempts have been made to calculate spallation cross sections for reactions induced by protons with energies in excess of 300 MeV. The initial stages of the reaction are treated by following nucleon–nucleon collisions by Monte Carlo techniques. The residual nuclei following the fast cascade can have fairly high excitation energy, and the effect of fission competition with each of the ten or so nucleons evaporated is investigated. The experimental cross sections can be reproduced fairly well with the assumption that Γ_n/Γ_f values deduced for low excitations are valid at high excitation energy as well. Calculations in which Γ_n/Γ_f was allowed to vary with excitation energy as given by Eq. (VII-7) were less successful, although in at least one such calculation the theoretical fission thresholds used may not have been very realistic.

In addition to the experimental methods mentioned previously, there is another type of experimental measurement from which we might hope to deduce information about Γ_n/Γ_f and in particular its excitation energy dependence. This is the determination of the relative numbers of prefission and postfission neutrons from the angular correlation of the neutrons with respect to the fragment direction. The prefission neutrons are essentially isotropic with respect to the fragment direction, while the postfission neutrons are strongly correlated with the fragment direction. This technique has been applied to the proton-induced fission of ^{238}U at two rather different energies. In one experiment with 155-MeV protons Cheifetz *et al.* (1970) concluded that the nearly equal fraction of prefission and postfission neutrons is inconsistent with the assumption that Γ_n/Γ_f is independent of excitation energy, but is qualitatively consistent with the energy dependence given by Eq. (VII-7). In another study (Bishop *et al.*, 1972) the excitation energy dependence was investigated in greater detail at lower energies by varying the proton energy from 11.5 to 22 MeV. The observed excitation energy dependence could not be accounted for unless a_f/a_n was between 1.1 and 1.15.

3. Deformation Dependence of the Level Density Parameter *a*

It has been known for some time that a_f/a_n is appreciably larger than unity in the lead–bismuth region. As one moves away from the lead–bismuth region to lighter nuclei, a_f/a_n decreases from the values obtained close to the closed shells $Z = 82$, $N = 126$. The large values of a_f/a_n in this region have been generally attributed to shell effects on the level density for spherical or near spherical distortions appropriate to neutron emission. There is experimental evidence that nuclei exhibiting a_f/a_n values different from unity are not restricted to the vicinity of a closed shell. In addition to the finding mentioned in the previous section that a_f is greater than a_n for proton-induced fission of uranium, there is also evidence that a_f/a_n has not returned to unity for nuclei much lighter than the $Z = 82$, $N = 126$ closed shell, with midshell heavy rare earths exhibiting a_f/a_n values of the order of 1.1 to 1.2. Thus the deviation of a_f/a_n from unity cannot be wholly due to the shell effects for spherical magic nuclei.

There is a systematic feature of the deformation dependence of the single particle level density which explains why a_f/a_n tends to be greater than unity for nuclei far away from closed shells as well as for nuclei close to closed shells. Myers and Swiatecki (1967) have shown the close connection between the local single particle level density at the Fermi surface and the nuclear mass. (See Chapter II.) A low single particle level density for a given nucleus and deformation will result in a lower mass than that given by a smooth liquid drop mass formula, whereas a high single particle level density will give the opposite effect. For a given nucleus, the single particle level density may vary strongly enough with deformation so that nonspherical shapes may be stabilized. Strutinsky has shown how such effects can account for not only ground state equilibrium deformations but also the existence of a second minimum in the potential energy curve which is responsible for spontaneous fission shape isomerism. The nature of these shell corrections is that the ground state deformation appropriate to neutron emission is associated with lower than average single particle level densities, and that shell effects will tend to produce a fission barrier at a deformation associated with a higher than average single particle level density. This is illustrated in Fig. VII-11, where the nuclear deformation energy is sketched as a function of deformation leading toward fission. The local single particle level densities at various maxima and minima are also

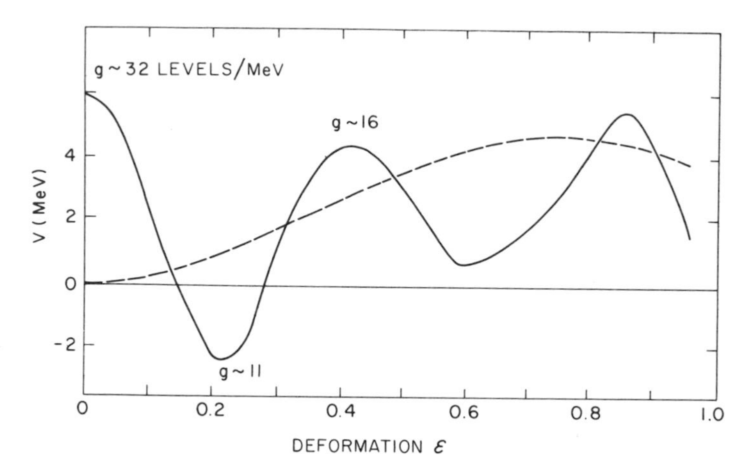

Fig. VII-11. Illustration of the dependence of the local single particle level density on deformation. (—) Shell-corrected potential energy curve of Tsang and Nilsson for ^{242}Pu; (---) liquid drop model potential energy curve. The outer barrier of the shell-corrected curve is lowered when reflection-asymmetric degrees of freedom are included in the description of the nuclear shape. [From Bishop *et al.* (1972).]

indicated. These average level densities were obtained by simply counting the levels in the 2-MeV region closest to the Fermi level. At very low excitation energies we should properly use a smaller averaging interval. This would increase the discrepancy that we see between the g values at the different deformations. Since the level density parameter is proportional to g, it is seen that the expected ratio of a_f to a_n is 1.5 (assuming that a is determined at the first hump in the fission barrier). As the excitation energy is increased, the single particle level density farther away from the Fermi surface becomes increasingly important, and the ratio a_f/a_n will decrease.

The situation at the fission barrier may be more complex than that suggested by the simple one-dimensional representation in Fig. VII-11. The local maximum in the single particle level density at the barrier may be reduced when other degrees of freedom are considered, such as the mass-asymmetry degree of freedom which lowers the outer barrier in heavy elements.

These ideas concerning the deformation and excitation energy dependence of the nuclear level density have been explored more quantitatively in a number of studies (Ramamurthy *et al.*, 1970; Moretto and Stella, 1970a; Vandenbosch and Mosel, 1972; Williams *et al.*, 1972). The starting point for such a calculation is a set of realistic single particle energy levels for each of the relevant deformations. These energy levels may be obtained from either deformable harmonic oscillator potentials or Woods–Saxon-like potentials. [In the case of finite potentials it is necessary to take into account the unbound

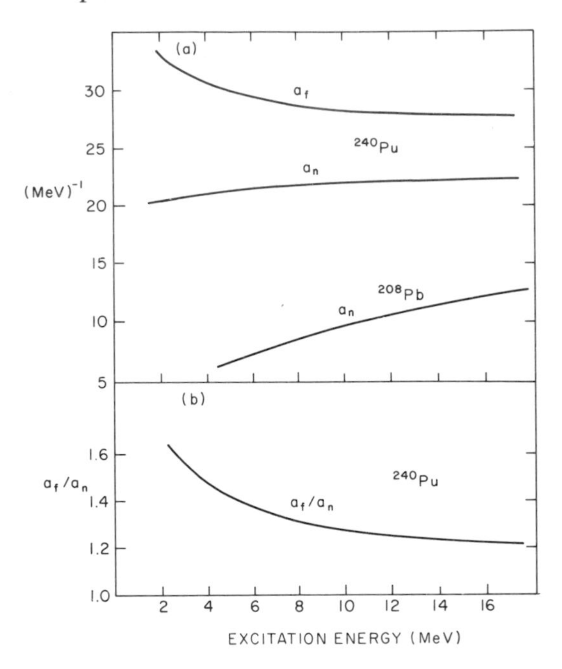

Fig. VII-12. (a) The dependence of the level density parameters a_f and a_n on excitation energy in excess of the fission barrier and neutron binding energy, respectively. The ground state deformation appropriate to neutron emission was $\varepsilon = 0.22$ and the saddle point deformation was $\varepsilon = 0.44$ for ^{240}Pu. Also shown is the strong shell effect on a_n for the double-magic nucleus ^{208}Pb ($\varepsilon = 0$). (b) The ratio a_f/a_n for ^{240}Pu. [From Bishop *et al.* (1972).]

(continuum) states if the total nuclear level density at high energies is to be obtained.] The total nuclear level density can be calculated from the single particle energies by either combinational (Hillman and Grover, 1969; Williams *et al.*, 1972) or statistical (Decowski *et al.*, 1968; Moretto *et al.*, 1970b) techniques. Although the dependence of the nuclear level density on excitation energy no longer has the simple functional form of Eq. (VII-6) appropriate for equally spaced single particle levels, it is convenient to parameterize the results for illustrative purposes by defining effective a values which are now excitation energy dependent. Some examples of the excitation and deformation dependence of these effective a values is given in Fig. VII-12. Also shown is the excitation energy dependence of a_f/a_n for ^{240}Pu. The fact that the values of a_f/a_n do not quickly approach their asymptotic value of unity as the energy U is raised is not to be interpreted as a very slow healing out of shell corrections with increasing excitation energy. It is rather a consequence of our choice of the zero energy reference from which the excitation energy is measured. We have chosen the true, shell-affected ground state and barrier energies. If we were to measure the excitation energy from the liquid drop energy surface, a_f/a_n would approach unity much faster with excitation energy.

The single particle effects illustrated in Fig. VII-12 are actually reduced somewhat when the pairing interaction is incorporated into the level density calculation, The reason for this is that the pairing gap is also correlated with the single particle level density, with a higher single particle level density resulting in a larger pairing gap and a greater lowering of the ground state energy relative to that of the excited states. Thus although a high single particle level density results in an increased number of nuclear levels, the energy at which these appear, relative to the ground state energy, is somewhat higher. Pairing therefore reduces the shell effects on the total nuclear level density in a manner very similar to its damping of the shell modulation on the potential energy surface discussed in Chapter II.

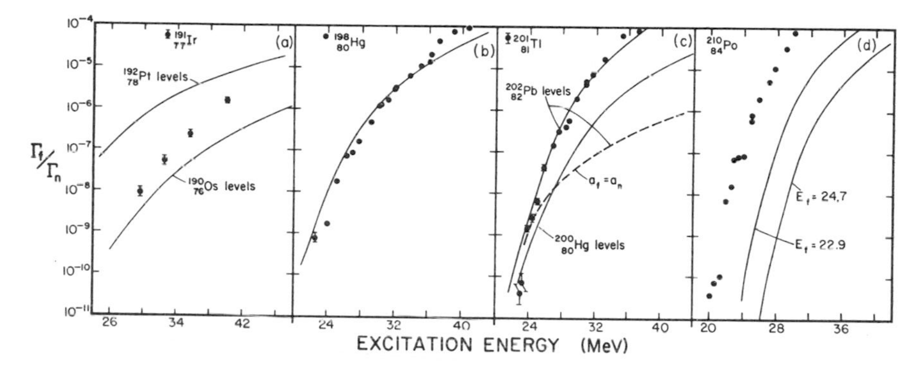

Fig. VII-13. Comparison of theoretical and experimental values of Γ_f/Γ_n. The calculated curves are based on theoretical results and no attempt has been made to vary the parameters to fit the data. The importance of the deformation dependence of the single particle level density is illustrated in (c), where the result expected for $a_f = a_n$ is indicated by the dashed line. The displacement between the calculated and experimental curves in the case of ^{210}Po is due to the fact that the theoretical barrier height is 4 MeV higher than the empirical value. 2 MeV of this is due to the failure of the theory to reproduce the ground state shell correction. The remaining discrepancy has been shown (since the figure was prepared) to be due to the restriction to reflection-symmetric saddle point shapes in the calculation. [From Vandenbosch and Mosel (1972).]

The consequences of the deformation dependence of the nuclear level density on the fission and neutron widths may be obtained by numerically integrating the expressions for the widths given in Eqs. (VII-1) and (VII-2). Some results for the less fissionable elements are illustrated in Fig. VII-13. Similar calculations have been performed for the uranium region and are in good agreement with experiment (Bishop *et al.*, 1972).

C. Effect of Angular Momentum on Γ_n/Γ_f

1. Theoretical Considerations

The angular momentum possessed by the excited nucleus can in principle affect the competition between neutron emission and fission in two ways. The first effect is simply a consequence of the fact that the moment of inertia (perpendicular to the axis of symmetry) of the highly distorted saddle point nucleus is larger than the moment of inertia of the spherical or near spherical shape of stable equilibrium. This means that a smaller amount of energy is tied up in rotation at the saddle point than at the stable equilibrium shape, and the effective barrier for fission *relative* to that for neutron emission is decreased. As a consequence, fission generally competes more favorably with neutron emission as the angular momentum of the system is increased. It should be noted, however, that the effective thresholds of *both* fission and neutron emission are increased by increasing the value of the angular momentum. The effective thresholds are defined relative to the ground state energies. If $E_f \gg B_n$, the effect of angular momentum for excitation energies only slightly above the fission barrier may result in a decrease of Γ_f/Γ_n when, for example, the sum of the rotational energy and the nonrotating fission barrier exceeds the available excitation energy. In general, when the fission threshold is measured relative to the energy of the rotating equilibrium shape, it decreases with increasing values of the angular momentum.

The second effect is that the presence of angular momentum changes the *shape* of the nucleus for both the stable equilibrium (appropriate to neutron emission) and the unstable equilibrium (saddle point). The energy and shape changes associated with the rotation of a uniformly charged liquid drop model have been explored by Pik-Pichak (1958) and Hiskes (1960). The effect of modest amounts of angular momentum on the shape of stable equilibrium is to flatten the sphere into an oblate spheroid, as illustrated in Fig. VII-14. (For very large amounts of angular momentum the axial symmetry of the oblate spheroid is destroyed.) For the shapes of unstable equilibrium, the effect of angular momentum is to decrease the saddle point deformation energy. In the case of necked-in saddle point shapes, the neck tends to fill in as rotation is increased. (Again for very large amounts of angular momentum axial symmetry is destroyed.) The effects of shape changes on the angular momentum dependence of Γ_n/Γ_f are subtle, as both the deformation and the rotational energies change simultaneously. The calculations reported in the literature do not enable us to readily disentangle the two effects.

In order to explore the effects of angular momentum more quantitatively we define several pertinent quantities. The fission threshold in the presence of

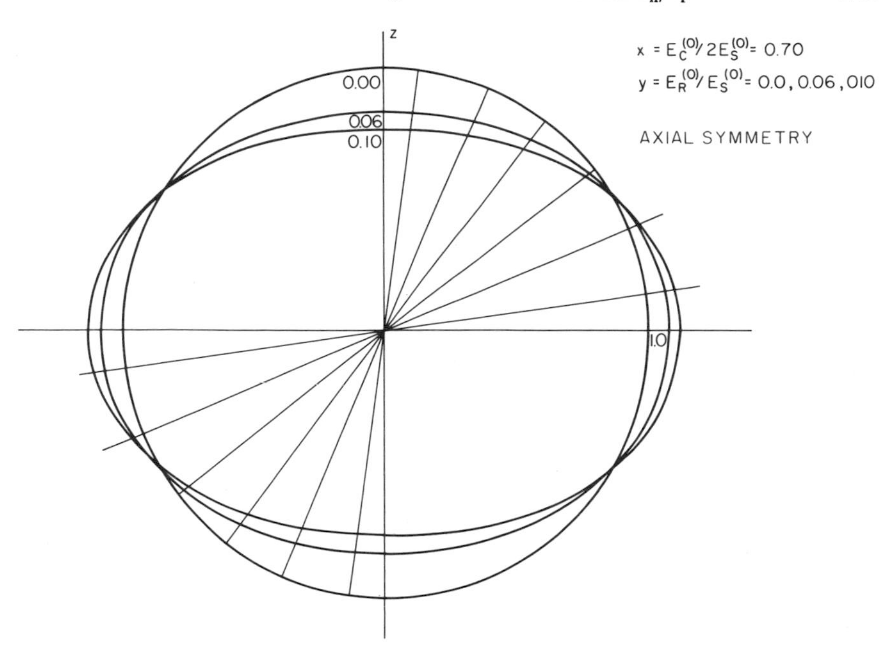

Fig. VII-14. Axially symmetric stable equilibrium configurations for $x = 0.70$ and for various values of the rotation parameter y. [From Hiskes (1960).]

rotation is designated E_f^r in contrast to the lowest threshold E_f for $J = 0$.

$$E_f^r = (M_{sad}^r + R_{sad}^r) - (M_{eq}^r + R_{eq}^r) \qquad \text{(VII-8)}$$

where M_{sad}^r and M_{eq}^r are the masses of the rotating saddle and equilibrium shapes, respectively, in energy units and R_{sad}^r and R_{eq}^r are the corresponding rotational energies. The energy of the rotating equilibrium ground state relative to the energy of the rotating sphere E_H is given by

$$E_H = (M_{eq}^r + R_{eq}^r) - (M_{sph} + R_{sph}) \qquad \text{(VII-9)}$$

where M_{sph} is the mass of the spherical nucleus in energy units and R_{sph} is the corresponding rotational energy.†

The effective fission threshold E_f^{eff} and neutron binding energy B_n^{eff} are defined relative to M_{sph} as follows.

$$E_f^{eff} = M_{sad}^r + R_{sad}^r - M_{sph} = E_f^r + E_H + R_{sph} \qquad \text{(VII-10)}$$

$$B_n^{eff} = M_{eq}^r + R_{eq}^r - M_{sph} + B_n = E_H + R_{sph} + B_n \qquad \text{(VII-11)}$$

where B_n is the normal neutron binding energy.

† A quantity E_{pp}, defined as $(M_{sad}^r + R_{sad}^r) - (M_{sph} + R_{sph})$ or the energy of the rotating saddle shape relative to the energy of a rotating sphere, is found in the literature and is equal to $E_f^r + E_H$.

TABLE VII-9

The Fission Barrier E_f^r of a Rotating Nucleus as a Function of the Fissility Parameter x and Rotational Energy Parameter y[a]

y \ x	0	0.05	0.1	0.15	0.2	0.25
0.0	0.280	0.243	0.221	0.2080	0.1950	0.1800
0.1	0.211	0.186	0.170	0.1506	0.1360	0.1200
0.2	0.152	0.131	0.1155	0.0960	0.0795	0.0625
0.3	0.0972	0.0795	0.0650	0.0506	0.0375	0.0253
0.4	0.0577	0.0455	0.0341	0.0235	0.014	0.0065
0.5	0.0332	0.0235	0.0153	0.0081	0.001	0
0.6	0.0162	0.009	0.0032	0	0	0
0.7	0.0071	0	0	0	0	0

y \ x	0.25	0.3	0.35	0.4	0.45	0.5
0.0	0.1800	0.1695	0.1515	0.1330	0.1155	0.0949
0.05	0.1495	0.1363	0.1165	0.0990	0.0815	0.0594
0.1	0.1200	0.1032	0.0864	0.0678	0.0469	0.0280
0.15	0.0900	0.0725	0.0556	0.0370	0.0190	0.0057
0.2	0.0625	0.0450	0.0304	0.0160	0.0050	0
0.25	0.0406	0.0264	0.0151	0.0052	0	0
0.3	0.0253	0.0144	0.0027	0	0	0
0.35	0.0141	0.0060	0	0	0	0
0.4	0.0065	0.0008	0	0	0	0
0.45	0.0020	0	0	0	0	0
0.5	0	0	0	0	0	0

y \ x	0.5	0.55	0.6	0.65	0.7	0.75	0.8	0.85	0.9	0.95
0.0	0.0949	0.0755	0.0564	0.0382	0.0223	0.0121	0.00588	0.00242	0.00069	0.0001
0.01	0.0873	0.0684	0.0490	0.0306	0.0162	0.00740	0.00267	0.00055	0	0
0.02	0.0801	0.0610	0.0418	0.0235	0.0108	0.00373	0.00071	0		
0.03	0.0730	0.0540	0.0350	0.0178	0.0062	0.00125	0			
0.04	0.0661	0.0470	0.0284	0.0120	0.0025	0				
0.05	0.0594	0.0404	0.0220	0.0065	0					
0.06	0.0528	0.0340	0.0159	0.0020	0					
0.07	0.0465	0.0277	0.0100	0						
0.08	0.0401	0.0217	0.0044	0						
0.09	0.0339	0.0158	0.00024	0						
0.10	0.0280	0.0106	0							
0.11	0.0219	0.0064	0							
0.12	0.0164	0.0025	0							
0.13	0.0122	0								
0.14	0.0085	0								
0.15	0.0057	0								
0.16	0.0035	0								
0.17	0.0016	0								

[a] The values of E_f^r are given in units of the surface energy of a sphere E_{sph}^S; y is in units of the ratio of the rotational energy of a sphere R_{sph} to the surface energy of the sphere, E_{sph}^S. [Plasil and Swiatecki (1972).]

If the neutron to fission competition is calculated with the so-called Fermi gas level densities, then we can account for angular momentum by substituting into Eq. (VII-7) the effective fission threshold and the effective neutron binding energy given by Eqs. (VII-10) and (VII-11), respectively. The rotational energy of a sphere is given by $R_{sph} = (\hbar^2/2\mathscr{I})J(J+1)$. The values of E_f^r and E_H are tabulated in units of E_{sph}^S, the surface energy of a sphere, in Tables VII-9 and VII-10. On inspection of Table VII-9 we see that for each value of x, the fission threshold E_f^r goes to zero as the rotational energy y is increased. For example, for $x = 0.65$ we observe that E_f^r is equal to zero for $y = 0.07$ which corresponds approximately to $J = 90$.

For a constant temperature model, Γ_n/Γ_f is given by Eq. (VII-5). The quantity of interest in this model when rotational energy is present is $E_f^{eff} - B_n^{eff}$. From Eqs. (VII-10) and (VII-11) it can be seen that this difference is simply $E_f^r - B_n$. For $x = 0.65$ (a nucleus in the Bi region where the liquid drop model is reasonably successful), the fission barrier is reduced by more than a factor of 2 for $y = 0.03$. This corresponds approximately to a reduction in the fission barrier from 24.0 MeV for $J = 0$ to 11.2 MeV for $J = 60$. Qualitatively, when the value of y is such that E_f^r becomes approximately equal to B_n, fission is an important decay channel. Angular momentum states with larger values of y will decay predominately by fission and will make little, if any, contribution to the spallation cross section.

If we assume that $M_{sad}^r + R_{sad}^r = M_{sad} + R_{sad}$, where M_{sad} and R_{sad} are the mass (in energy units) and rotational energy of the saddle shape for zero angular momentum, then Eq. (VII-10) may be rewritten as

$$E_f^{eff} = M_{sad} + R_{sad} - M_{sph} = E_f + R_{sad} \qquad \text{(VII-12)}$$

where E_f is the fission threshold for zero angular momentum. If, in addition, we assume that $M_{eq}^r + R_{eq}^r = M_{sph} + R_{sph}$, Eq. (VII-11) reduces to

$$B_n^{eff} = B_n + R_{sph} \qquad \text{(VII-13)}$$

where $R_{sad} \approx \hbar^2 J^2/2\mathscr{I}_\perp$ and $R_{sph} \approx \hbar^2 J^2/2\mathscr{I}_{sph}$. This simple treatment of the angular momentum effects in fission was postulated by Huizenga and Vandenbosch (1962) in order to approximate to first order the effective fission threshold and the effective neutron binding energy. Sikkeland (1964) has used these approximations in the constant temperature form of Γ_n/Γ_f where

$$E_f^{eff} - B_n^{eff} = E_f - B_n + R_{sad} - R_{sph} \qquad \text{(VII-14)}$$

The effect of rotation in this approximation appears simply as the difference in the rotational energy for the saddle and spherical shapes in the absence of rotation. This approximation can be compared numerically with the more detailed theory and for some regions of x this approximation is reasonably accurate (see caption of Fig. VII-15). The effect of angular momentum on

TABLE VII-10

E_H, the Energy of the Rotating Shape of Stable Equilibrium Relative to That of a Rotating Sphere, as a Function of x and y[a]

y \ x	0	0.05	0.1	0.15	0.2	0.25
0	0	0	0	0	0	0
0.1	−0.0057	−0.0058	−0.0060	−0.0061	−0.0062	−0.0063
0.2	−0.0193	−0.0203	−0.0211	−0.0220	−0.0230	−0.0245
0.3	−0.0402	−0.0427	−0.0456	−0.0497	−0.0540	−0.0616
0.4	−0.0755	−0.0812	−0.0899	−0.0988	−0.1090	−0.1200
0.5	−0.1273	−0.1356	−0.1470	−0.1592	−0.1745	−0.1897
0.6	−0.1755	−0.1986	−0.2128	−0.2296	−0.2510	
0.7	−0.255	−0.271	−0.291			
0.8	−0.354					

y \ x	0.25	0.3	0.35	0.4	0.45	0.5
0	0	0	0	0	0	0
0.05	−0.0018	−0.0019	−0.0021	−0.0024	−0.0025	−0.0030
0.1	−0.0063	−0.0070	−0.0076	−0.0083	−0.0091	−0.0095
0.15	−0.0150	−0.0158	−0.0166	−0.0192	−0.0217	−0.0250
0.2	−0.0245	−0.0254	−0.0290	−0.0351	−0.0478	−0.0613
0.25	−0.0387	−0.0438	−0.0532	−0.0622	−0.0845	
0.3	−0.0616	−0.0717	−0.0821	−0.0972		
0.35	−0.0793	−0.1014	−0.1138			
0.4	−0.1200	−0.1340	−0.1503			
0.45	−0.1528	−0.1710	−0.1907			
0.5	−0.1897	−0.2133				

y \ x	0.5	0.55	0.6	0.65	0.7	0.75	0.8	0.85	0.9	0.95
0.0	0	0	0	0	0	0	0	0	0	0
0.01	−0.00012	−0.00014	−0.00016	−0.00018	−0.00020	−0.00024	−0.00029	−0.00036	−0.00065	−0.00089
0.02	−0.00047	−0.00050	−0.00058	−0.00065	−0.00074	−0.00085	−0.00101	−0.00124		
0.03	−0.00100	−0.00105	−0.00124	−0.00138	−0.00156	−0.00179	−0.00275			
0.04	−0.00176	−0.00190	−0.00211	−0.00235	−0.00263	−0.00298				
0.05	−0.00300	−0.00308	−0.00318	−0.00352	−0.00392					
0.06	−0.00374	−0.00410	−0.00444	−0.00488						
0.07	−0.00530	−0.0055	−0.00585	−0.00640						
0.08	−0.00632	−0.0070	−0.00742							
0.09	−0.00790	−0.0085	−0.01022							
0.10	−0.00944	−0.0102	−0.01420							
0.11	−0.0112	−0.0133								
0.12	−0.0130	−0.0178								
0.13	−0.0165	−0.0254								
0.14	−0.0203									
0.15	−0.0250									
0.16	−0.0304									
0.17	−0.0363									
0.18	−0.0423									

[a] The values of E_H are given in units of the surface energy of a sphere E^S_{sph}; y is in units of the ratio of the rotational energy of a sphere, R_{sph}, to the surface energy of the sphere, E^S_{sph}. [Plasil and Swiatecki (1972).]

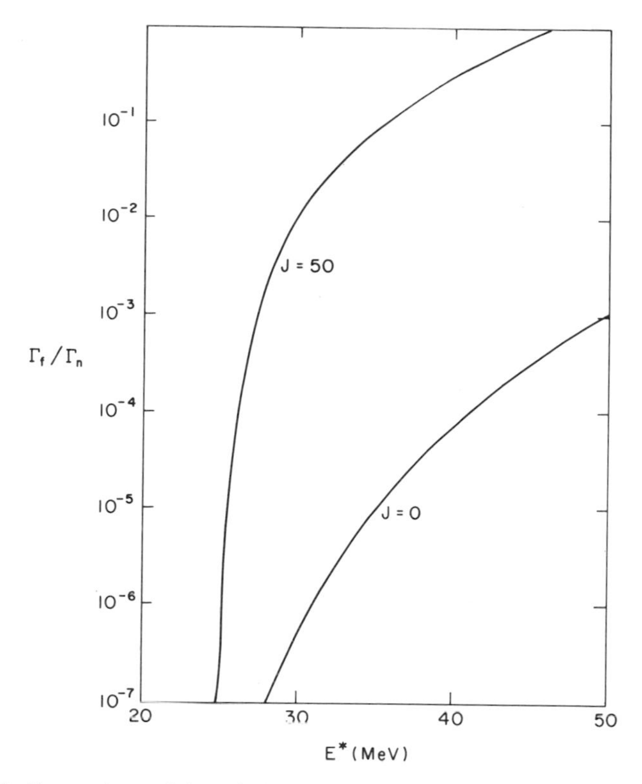

Fig. VII-15. Comparison of the calculated dependence of Γ_f/Γ_n on excitation energy for $J = 0$ and $J = 50$. The fissility parameter was taken as $x = 0.66$, the corresponding liquid drop model fission barrier is $E_f = 21$ MeV, and the ratio of the moment of inertia at the saddle to that of the sphere $\mathscr{I}/\mathscr{I}_{sph}$ is 3.75. The other parameters in this example are $B_n = 8$ MeV, $a_f = 25$ MeV^{-1}, and $a_n = 20$ MeV^{-1}. The value of $E_f^{eff} - B_n^{eff}$ calculated with the approximate Eq. (VII-14) is about 1 MeV smaller than that of the more exact theory.

Γ_f/Γ_n is illustrated in Fig. VII-15, where a comparison is made of the calculated dependence of Γ_f/Γ_n on excitation energy for $J = 0$ and $J = 50$.

In our discussion of the effect of angular momentum on the fission barrier we have assumed that the saddle point configuration is rotating about the axis perpendicular to the nuclear symmetry axis. This is the most energetically favored situation, but we know from fission-fragment angular distributions that the fragments sometimes (although less frequently) appear in the direction parallel to the nuclear symmetry axis. In general, the rotational energy at the saddle is given by

$$R_{sad} = \hbar^2 J^2/2\mathscr{I}_\perp + \hbar^2 K^2/2\mathscr{I}_{eff}$$

where $1/\mathscr{I}_{eff} = (1/\mathscr{I}_\parallel) - (1/\mathscr{I}_\perp)$, and $\mathscr{I}_\perp$ and $\mathscr{I}_\parallel$ are the moments of inertia about an axis perpendicular and parallel to the nuclear symmetry axis, respectively. The quantity K is the projection of J on the same axis. From statistical considerations it can be shown that the mean square value of K is given by

$$K_0{}^2 = \mathscr{I}_{\text{eff}}\, T/\hbar^2$$

where T is the nuclear temperature.

The average rotational energy becomes $R_{\text{sad}} = \hbar^2 J^2/2\mathscr{I}_\perp + T/2 \approx \hbar^2 J^2/2\mathscr{I}_\perp$ for large values of J.

Thus the fraction of the available energy which goes into rotation about the axis parallel to the symmetry axis is only $T/2$, which is generally small compared to the available excitation energy.

2. Comparison with Experiment

The most direct way in which the effect of angular momentum on Γ_f/Γ_n can be observed is to study fission competition in the same nucleus at the same excitation energy but with differing amounts of angular momentum. This can be achieved by suitable choice of projectile and target pairs. Sikkeland (1964) has formed the compound nucleus ^{181}Re in three different ways, and deduced Γ_f/Γ_n as a function of excitation energy. The angular momentum deposition increases with the mass of the projectile, and fission is also observed to compete more effectively with increasing projectile mass, as illustrated in Fig. VII-16. Sikkeland's theoretical fits were made with $\mathscr{I}_{\text{sph}}/\mathscr{I}_{\text{sad}} = 0.5$, somewhat larger than expected on the basis of the liquid drop model. The angular momentum deposition in the different reactions is somewhat uncertain also.

3. Direct Fission

From compound nucleus theory we expect fission-fragment angular distributions for particle-induced fission of heavy nuclei to be symmetric about a plane 90° to the beam direction. Although the anisotropy (which is established by the projection of the angular momentum on the symmetry axis of the fissioning nucleus, K) may be large (Chapter VI), the compound nucleus lifetime is long enough so that the nucleus has rotated several times about the angular momentum axis prior to fission. Such an argument leads to the prediction that the fission-fragment yields should be symmetric in the center-of-mass system about a plane 90° to the beam direction.

If sufficient energy could be transferred to the nucleus by the projectile such that the critical energy for fission is exceeded, fission may occur without the intermediate stage of a compound nucleus. We might regard such a

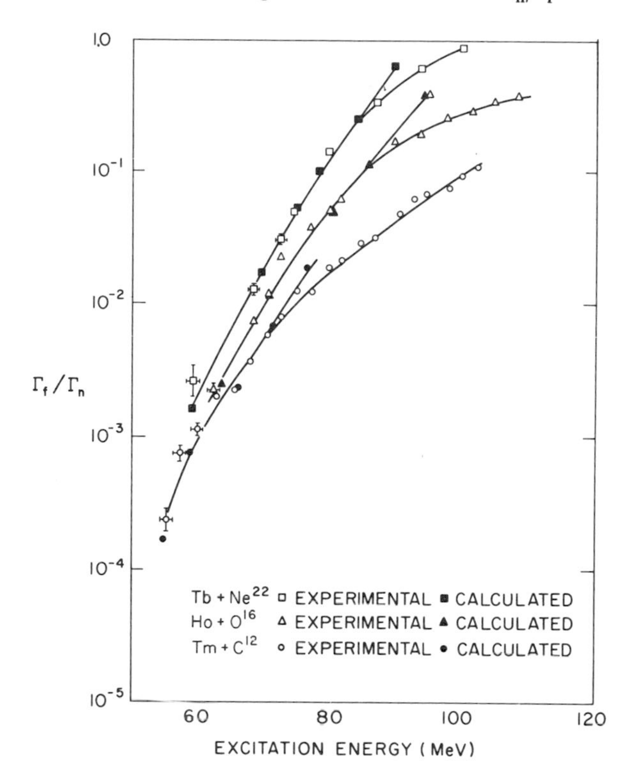

Fig. VII-16. Dependence of Γ_f/Γ_n on excitation energy for the compound nucleus ^{181}Re prepared in three different reactions. The calculated points assumed only first-chance fission; hence, they are expected to deviate when Γ_f/Γ_n is no longer small compared to unity. [From Sikkeland (1964).]

mechanism as direct fission,† since it would take place in approximately 10^{-21} sec, the characteristic time for a nucleon to move across the nucleus. The direct fission of a nucleus may occur if a sufficiently large amount of angular momentum is transferred to it.

Since the time necessary for such a direct fission process may be short compared to the rotational period, the heavy (or light) fragment may preferentially come off in the forward direction. In several experiments, limits have been placed on the amount of fore-aft asymmetry observed. After correcting the experimental fission-fragment angular distributions from the laboratory to the center-of-mass frame of reference, the calculated intensities of the

† Direct fission is to be differentiated from the process, described in Section V,D, in which a direct interaction is sequentially followed by fission. In this process fission proceeds by the normal compound nucleus mechanism.

symmetric and asymmetric groups of fragments in the parallel and antiparallel directions to the beam are equal for fission induced in heavy targets ($Z \geqslant 90$) by 22-MeV protons ($\pm 2\%$), 43-MeV helium ions ($\pm 5\%$), and 14-MeV neutrons ($\pm 5\%$). From an experimental point of view we should look for direct fission in a heavy ion bombardment of a target with a low fission threshold, since we expect direct fission in such a system to be much more probable than in the foregoing systems. The rotational energy required to reduce the fission threshold to zero for nuclei with various values of the fissility parameter x are listed in Table VII-9.

References

Auchampaugh, G. F., Farrell, J. A., and Bergen, D. W. (1971). *Nucl. Phys. A* **171**, 31.

Babenko, Yu. A., Ippolitov, V. T., Nemilov, Yu. A., Selitskii, Yu. A., and Funshtein, V. B. (1969). *Sov. J. Nucl. Phys.* **10**, 133.

Bishop, C. J., Halpern, I., Shaw, R. W., Jr., and Vandenbosch, R. (1972). *Nucl. Phys. A* **198**, 161.

Blatt, J. M., and Weisskopf, V. F. (1952). "Theoretical Nuclear Physics," pp. 386–390. Wiley, New York.

BNL 325 (1965) 2nd ed. Suppl. 2.

Bohr, N., and Wheeler, J. A. (1939). *Phys. Rev.* **56**, 426.

Bowman, C. D., Auchampaugh, G. F., and Fultz, S. C. (1964). *Phys. Rev.* **133**, B 676.

Burnett, D. S., Gatti, R. C., Plasil, F., Price, P. B., Swiatecki, W. J., and Thompson, S. G. (1964). *Phys. Rev.* **134**, B 952.

Cheifetz, E., Fraenkel, Z., Galin, J., Lefort, M., Peter, J., and Tarrago, X. (1970). *Phys. Rev. C* **2**, 256.

Cramer, J. D., and Britt, H. C. (1970). *Nucl. Sci. Eng.* **41**, 177.

Davey, W. G. (1968). *Nucl. Sci. Eng.* **32**, 35.

Decowski, P., Grochulski, W., Marcinkowski, A., Siwek, K., and Wilhelmi, Z. (1968). *Nucl. Phys. A* **110**, 129.

Fomushkin, E. F., and Gutnikova, E. K. (1970). *Sov. J. Nucl. Phys.* **10**, 529.

Fomushkin, E. F., Gutnikova, E. K., Maslov, A. N., Novoselov, G. V., and Panin, V. I. (1972). *Sov. J. Nucl. Phys.* **14**, 41.

Halpern, I. (1959). *Annu. Rev. Nucl. Sci.* **9**, 245.

Halpern, I., and Nicholson, W. J. (1960). Annu. Prog. Rep. Univ. of Washington Cyclotron Res.

Hillman, M., and Grover, J. R. (1969). *Phys. Rev.* **185**, 1303.

Hiskes, J. R. (1960). Rep. UCRL–9275. Univ. of California, Berkeley.

Howerton, R. J. (1958). Semi empirical neutron cross sections. UCRL–5351. Univ. of California, Livermore.

Huizenga, J. R. (1958). *Phys. Rev.* **109**, 484.

Huizenga, J. R., and Vandenbosch, R. (1962). *In* "Nuclear Reactions", Vol. II, p. 42. North-Holland Publ., Amsterdam.

Huizenga, J. R., Chaudhry, R., and Vandenbosch, R. (1962). *Phys. Rev.* **126**, 210 (1962).

Hyde, E. K. (1964). "The Nuclear Properties of the Heavy Elements." Vol. III "Fission Phenomena." Prentice-Hall, Englewood Cliffs, New Jersey.

Jackson, J. D. (1956). *Can. J. Phys.* **34**, 767.
Katz, L., Baerg, A. P., Brown, F. (1958). *Proc. U.N. Int. Conf. Peaceful Uses At. Energy, 2nd, September 1958*, **15**, p. 188. United Nations, Geneva.
Khodai-Joopari, A. (1966). Thesis Rep. UCRL–16489. Univ. of California, Berkeley.
Lessler, R. M. (1958). Thesis, Rep. UCRL–8439. Univ. of California, Berkeley.
Lindner, M., and Turkevich, A. (1960). *Phys. Rev.* **119**, 1632.
Moore, M. S., and Keyworth, G. A. (1971). *Phys. Rev.* **C 3**, 1656.
Moore, M. S., McNally, J. H., and Baybarz, R. D. (1971). *Phys. Rev. C* **4**, 273.
Moretto, L. G., and Stella, R. (1970a). *Phys. Lett. B* **32**, 558.
Moretto, L. G., Stella, R., and Crespi, V. C. (1970b). *Energ. Nucl.* **17**, 436.
Myers, W. D., and Swiatecki, W. J. (1967). *Ark. Fys.* **36**, 343.
Nat. Bur. Stand. (*U.S.*) (1968). *Spec. Publ.* No. 299, **1**, 567.
Pik–Pichak, G. A. (1958). *Sov. Phys. JETP* **7**, 238.
Plasil, F. (1963). Rep. UCRL–11193. Univ. of California, Berkeley.
Plasil, F., and Swiatecki, W. J. (1972). Private communication of unpublished results.
Raisbeck, G. M., and Cobble, J. W. (1967). *Phys. Rev.* **153**, 1270.
Ramamurthy, V. S., Kapoor, S. S., and Kataria, S. K. (1970). *Phys. Rev. Lett.* **25**, 386.
Ramler, W. J., Wing, J., Henderson, D. J., and Huizenga, J. R. (1959). *Phys. Rev.* **114**, 154.
Sikkeland, T. (1964). *Phys. Rev.* **135**, B 669.
Sikkeland, T., Clarkson, J. E., Steiger-Shafrir, N. H., and Viola, V. E. (1971). *Phys. Rev. C* **3**, 329.
Sov. J. At. Energy (1961). **8**, 125.
Vandenbosch, R., and Huizenga, J. R. (1958). *Proc. Int. Conf. Peaceful Uses At. Energy, 2nd*, **15**, p. 284. United Nations, Geneva.
Vandenbosch, R., and Mosel, U. (1972). *Phys. Rev. Lett.* **28**, 1726.
Vandenbosch, R., *et al.* (1958). *Phys. Rev.* **111**, 1358.
Vorotnikov, P. E., Dubrovina, S. M., Otroschenko, G. A., Chistyakov, L. V., Shigin, V. A., and Shubko, V. M. (1970). *Sov. J. Nucl. Phys.* **10**, 419.
Williams, F. C., Jr., Chan, G., and Huizenga, J. R. (1972). *Nucl. Phys. A* **187**, 225.
Wing, J., Ramler, W. J., Harkness, A. L., and Huizenga, J. R. (1959). *Phys. Rev.* **114**, 163

CHAPTER

Fission Barrier Heights

A. Experimental Values of Fission Barrier Heights

In this chapter we collect fission barrier heights as deduced from a somewhat diverse body of experimental data. We discuss briefly the kinds of data from which barrier heights can be deduced and the uncertainties involved. We also indicate the appropriate sections where a fuller account of the processes involved are given. Where possible, the fission barrier height is designated by that energy at which the penetration is equal to 0.5. For a parabolic barrier this is equal to the height of the barrier. A summary of experimental barrier heights is given in Table VIII-1.

B. Methods of Determining Fission Barrier Heights from Reactions

1. Barrier Heights from Excitation Energy Dependence of Γ_n/Γ_f

For the lighter elements where the fission barrier is typically several times larger than the neutron binding energy, the only way to estimate the fission barrier is through a theoretical fit to the absolute value and energy dependence of Γ_n/Γ_f. Because of the low fission cross sections, experimental data are often not available for the last several MeV as one goes down in excitation energy

TABLE VIII-1

Summary of Fission Barrier Heights[a]

Nuclide	Barrier height (MeV)	Ref.
^{173}Lu	28.7 ± 3	Raisbeck and Cobble (1967)
^{179}Ta	27 ± 3	Raisbeck and Cobble (1967)
^{186}Os	23.4 ± 0.5	Moretto *et al.* (1972)
^{187}Os	22.7 ± 0.5	Moretto *et al.* (1972)
^{188}Os	24.2 ± 0.5	Moretto *et al.* (1972)
^{189}Ir	22 ± 3	Raisbeck and Cobble (1967)
^{191}Ir	23 ± 3	Raisbeck and Cobble (1967)
^{198}Hg	21.8 ± 0.7	Khodai-Joopari (1966)
^{201}Tl	22.3 ± 0.5	Khodai-Joopari (1966), Burnett *et al.* (1964)
^{207}Bi	21.2 ± 0.5	Khodai-Joopari (1966)
^{209}Bi	22.6 ± 0.5	Khodai-Joopari (1966)
^{210}Po	20.4 ± 0.5	Khodai-Joopari (1966)
^{211}Po	19.7 ± 0.5	Moretto *et al.* (1972)
^{212}Po	18.6 ± 0.5	Khodai-Joopari (1966)
^{213}At	16.8 ± 0.5	Khodai-Joopari (1966)
^{226}Ra	8.5 ± 0.5	Zhagrov *et al.* (1968)
^{227}Ra	8.5 ± 0.5	
^{231}Th	6.2	
^{232}Th	6.0	
^{233}Th	6.4	
^{234}Th	6.1	
^{232}Pa	6.3	
^{233}U	5.7 ± 0.3	
^{234}U	6.0	
^{235}U	5.8	
^{236}U	5.9	
^{237}U	6.1	
^{238}U	5.8	
^{239}U	6.2	
^{240}U	5.7	
^{237}Np	5.6 ± 0.3	
^{238}Np	6.0	
^{239}Pu	5.8	
^{240}Pu	5.9	
^{241}Pu	5.9	
^{242}Pu	5.8 ± 0.3	
^{243}Pu	5.8	
^{244}Pu	5.4 ± 0.3	
^{245}Pu	5.4	
^{241}Am	5.9 ± 0.3	
^{242}Am	6.4	
^{244}Am	6.3	
^{245}Cm	6.2	
^{247}Cm	5.9	
^{249}Cm	5.5	
^{249}Bk	5.9	
^{253}Cf	5.3 ± 0.3	

[a] In principle these are the energies where the penetration is equal to $\frac{1}{2}$ for the lowest transition state. In practice, for the heavy elements they are often the energy at which the fission cross section has reached half of its plateau value. For nuclei that exhibit a double barrier, the higher barrier is listed. Barrier heights for which no reference is given are based on our evaluation of the available data. Where no errors are indicated, the values are believed to be uncertain by ± 0.2 MeV.

towards the barrier. Thus, we are dealing with a theoretically guided extrapolation over several MeV from the last data point to the presumed barrier height. Another complication, particularly for those nuclides closest to the $Z = 82$, $N = 126$ closed shell, are the large shell effects on the level densities governing the neutron competition. The uncertainties are typically 1 MeV. The relevant theory and some examples are given in Section VII, B, 2.

2. Barrier Heights from (n, f), (t, pf), (t, df), and (d, pf) Excitation Functions

For heavy nuclei where the barrier height and neutron binding energies are fairly comparable, the fission excitation function exhibits a fairly characteristic threshold rise followed by a fairly flat plateau (see Fig. VII-1). The most precise values of the barrier height can be obtained from a channel analysis which quantitatively takes into account the γ-ray and neutron competition with fission (see Chapter V). For odd-A nuclei it is sometimes possible to determine the energies of individual transition states at the barrier if they have different spins. In the absence of a channel analysis the energy at which the cross section has risen to $\frac{1}{2}$ of its plateau value can be taken as a reasonable estimate of the barrier height. It must be remembered that for an odd-A or odd-odd nucleus the particular reaction employed may not exploit the lowest possible barrier because the spin state happening to have the lowest barrier height is not populated in the reaction. That one spin state can have a lower barrier than another spin state is a consequence of the "specialization energy"

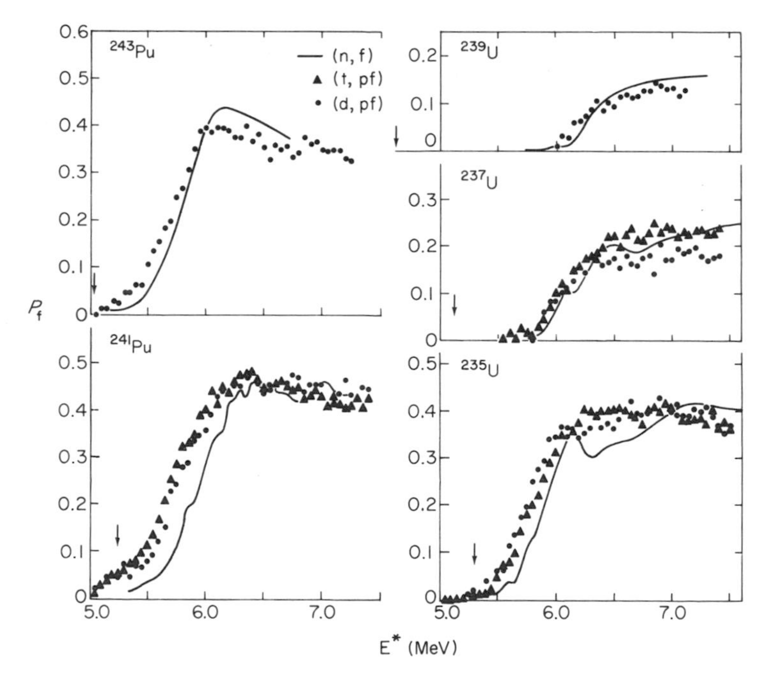

Fig. VIII-1. Comparison of fission probabilities for three different reactions. Arrows indicate the binding energy of the last neutron. The fission probabilities P_f were obtained from the ratio of the fission cross section to the reaction cross section for neutrons, and from the ratios $\sigma(t, pf)/\sigma(t, p)$ or $\sigma(d, pf)/\sigma(d, p)$. In the latter case a correction for deuteron breakup has been made. [From Britt and Cramer (1970).]

effect discussed in connection with spontaneous fission (see Section III, C). Some examples of this effect are shown in Fig. VIII-1, where the fission probability excitation functions from (n, f) reactions are compared with those from (d, pf) and (t, pf) reactions. For some targets the thresholds for the (d, pf) and (t, pf) reactions are several tenths of an MeV lower than for the (n, f) reaction. This can be attributed to the fact that in the threshold region the average angular momentum transfer is about $1.5\hbar$ (for neutron energies of about $\frac{3}{4}$ MeV) in the (n, f) reaction, whereas for the (d, pf) and (t, pf) reactions they are 3 to $4\hbar$. Thus, a larger variety of spin states can contribute for the latter reactions. When the threshold occurs at a higher neutron energy, as in the case of ^{239}U, the average angular momentum transfer for the (n, f) reaction becomes more comparable to that in the other reactions and the likelihood of not populating the spin state with the lowest barrier in the (n, f) reaction becomes less.

3. Barrier Heights from Photofission Excitation Functions

Extraction of barrier heights from photofission data is complicated by a number of factors. In the first place, the dependence of the photon absorption cross section on energy is poorly known. In the second place, only states with restricted angular momentum and parity values are excited due to the nature of the photoabsorption process. Finally, most of the experimental excitation function data were obtained by unfolding yield curves obtained with bremsstrahlung radiation (see Section V, B).

4. Barrier Heights for Nuclei Exhibiting a Double Barrier

It is of considerable interest to characterize both the inner and outer barriers of nuclei in the actinide region that exhibit a double barrier. This is often difficult to do in an unambiguous and model-independent way. Apart from structure due to resonant penetration, fission probability excitation functions are most sensitive to the height of the higher of the two barriers. The presence of resonance structure at a given excitation energy requires that both barriers be comparable to or higher than that excitation energy. Care must be exercised, however, in that resonances for one particular spin state may be observed at an excitation energy higher than the lowest barrier of another spin state. Anisotropy measurements are useful in characterizing the K values associated with a particular resonance.

Further information can be obtained by considering the yields and half lives of spontaneous fission isomers, as discussed in Section III, D. Such analyses are less ambiguous for those nuclides where the inner barrier is considerably higher than the outer barrier and where γ decay of the isomer can be neglected. Britt *et al.* (1973) have performed an analysis of isomer yield data and determined the outer barrier heights given in the last column of Table VIII-2. The inner barriers have been taken from other studies, primarily direct reaction fission probability excitation functions. The values assumed for the inner barrier are also given in the table.

TABLE VIII-2

Fission Barrier Heights of Nuclei Exhibiting a Double Barrier[a]

Nuclide	Inner barrier height (MeV)	Outer barrier height (MeV)
^{239}Pu	6.27	5.2
^{240}Pu	6.0	5.4
^{238}Am	6.4	5.4
^{239}Am	5.8	4.6
^{240}Am	6.45	5.5
^{241}Am	5.7	4.5
^{242}Am	6.4	5.0
^{243}Am	5.8	4.9
^{244}Am	6.2	4.8
^{243}Cm	5.8	4.0
^{245}Cm	6.3	4.4

[a] Only those nuclei for which both barriers have been characterized are listed. These barriers are based on the analysis of Britt *et al.* (1973) of fission isomer results. We estimate the uncertainties in the barrier heights to be approximately 0.3 MeV. The barrier height of the highest barrier differs in some cases from those listed in Table VIII-1 because of differing evaluations of the available data.

References

Britt, H. C., Bolsterli, M., Nix, J. R., and Norton, J. L. (1973). *Phys. Rev. C* **7**, 801.

Britt, H. C., and Cramer, J. D. (1970). *Phys. Rev. C* **2**, 1758 (1970).

Burnett, D. S., Gatti, R. C., Plasil, F., Price, P. B., Swiatecki, W. J., and Thompson, S. G. (1964). *Phys. Rev.* **134**, B 952.

Khodai-Joopari, A. (1966). Rep. UCRL–16489. Univ. of California Radiation Lab. Berkeley (unpublished).

Moretto, L. G., Thompson, S. G., Routi, J., and Gatti, R. C. (1972). *Phys. Lett. B* **38**, 471.

Raisbeck, G. M., and Cobble, J. W. (1967). *Phys. Rev.* **153**, 1270.

Zhagrov, E. A., Nemilov, Yu. A., and Selitskii, Yu. A. (1968). *Sov. J. Nucl. Phys.* **7**, 183.

CHAPTER

IX

Motion from Saddle to Scission: Theories of Mass and Energy Distributions

A. Introduction

One of the earliest observations regarding the fission process was the strong preference for heavy elements at low excitation energy to fission into fragments of unequal mass. Asymmetric mass distributions have proved to be one of the most persistent puzzles in the fission process. Although many suggestions as to the origin of this effect have been offered, no theoretical model has been proposed which has been explored in a complete enough manner or has been sufficiently free of parameter fitting to be generally accepted.

Most models of fragment mass and energy distributions can be divided into two groups, depending on whether or not the motion from saddle to scission is assumed to be adiabatic with respect to the particle degrees of freedom. If the collective motion toward scission is sufficiently slow or the coupling to internal degrees of freedom sufficiently weak so that the single particle degrees of freedom can easily readjust to each new deformation as the distortion proceeds, an adiabatic approximation may be valid. In this situation the decrease in potential energy from saddle to scission appears in the collective degrees of freedom at scission, primarily as kinetic energy associated with the relative motion of the nascent fragments. If, however, the motion is somewhat faster or the coupling somewhat stronger, there will be transfer of collective energy into nucleonic excitation in a manner analogous to viscous heating. If there is sufficient nonadiabatic mixing of the energy among the single particle degrees of freedom by the time the scission point is approached, a statistical model may be a reasonable approximation. The possibility of transfer of collective energy to the internal degrees of freedom can be thought of as leading to a nuclear viscosity.

The distinction between slow and rapid collective motion can be illustrated in terms of a level crossing diagram. Consider two single particle energy levels which would cross as a function of deformation if it were not for some residual interaction, as illustrated in Fig. IX-1. Initially for small deformations at the left of the diagram the lower level is filled and the upper level is empty. If the deformation proceeds sufficiently slowly and the coupling is sufficiently weak, the lower level will remain occupied throughout the deformation process. If, however, the deformation proceeds very rapidly, a jump will occur and the upper level will be occupied and the lower level unoccupied when the deformation has reached values at the right of the diagram.

It is well to call attention to an apparent contradiction in the relevant time scales for the two approximations and to the likelihood of feedback effects. The adiabatic approximation relies on the motion being *slow* enough or the coupling weak enough so that readjustment can occur without nucleonic excitation at level crossings. The consequence of motion proceeding according to this approximation is that most of the decrease in potential energy between saddle and scission will appear as kinetic energy in the fission direction; hence, the motion will become *faster* as fission proceeds. Similarly, if the

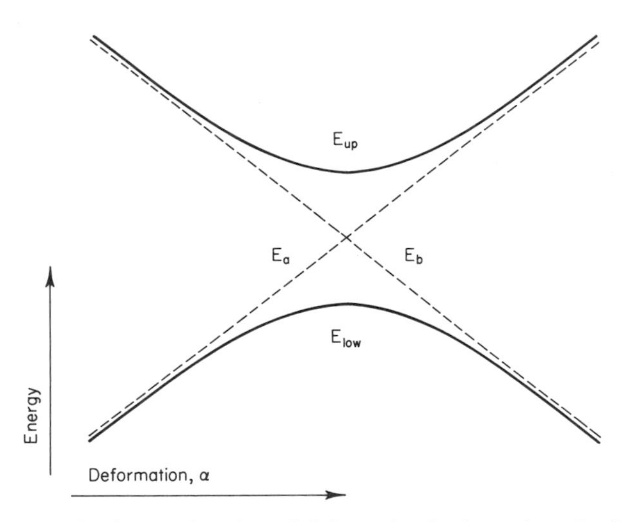

Fig. IX-1. Energy levels as a function of deformation in the region of a level crossing where the levels are split by a nondiagonal residual interaction. [From Hill and Wheeler (1953).]

motion is *fast* enough to induce nucleonic excitations, less energy will be available for kinetic energy of motion in the fission mode as fission proceeds. Thus, there is a feedback mechanism tending to prevent either approximation from becoming increasingly valid as the motion from saddle to scission proceeds.

It is difficult to make theoretical estimates which are sufficiently reliable to decisively indicate whether an adiabatic approximation is valid. Wilets (1964) has examined this question from the point of view of a level crossing model and concluded that although the process is unlikely to be adiabatic, it is also unclear whether the nonadiabatic mixing is sufficiently strong to assure the validity of the statistical model. Wilets has emphasized that the large energy gap associated with pairing may cause the fission of even-even nuclei at low excitation energy to be most nearly adiabatic. The insensitivity of the fragment mass distributions to nuclear type and the smoothly varying changes in these distributions with excitation energy suggest that the role of the gap is not dominant. Fong (1964) has attempted to show by use of time-dependent perturbation theory that the condition for the validity of the adiabatic approximation cannot be satisfied. The condition for adiabaticity is that the change in potential energy over the time interval during which nucleonic excitation can occur is small compared with the energy change involved in the particle excitation. This can be expressed as $(\tau/\Delta E)(dV/dt) \ll 1$, where ΔE is the energy spacing at the ground state. Letting $\tau = h/\Delta E$ gives $[h/(\Delta E)^2](dV/dt) \ll 1$. Fong's estimate of the numerical value of dV/dt assumes that the perturbation potential is simply the Coulomb potential at the scission point. We have already established (see Chapter II) that prior to scission there is a large cancellation between the surface and Coulomb components of the potential energy. Therefore his estimate of $\Delta E \gg 8.4$ MeV for the energy spacing required to ensure adiabaticity is only valid after the neck has broken, a time which is too late in the fission process for justification of a statistical theory appropriate for determination of mass and energy distributions. On the other hand, insertion of a more reasonable estimate of dV/dt does not result in a clear satisfaction of the condition for adiabaticity. In reality the descent from saddle to scission is undoubtedly very complicated, with near adiabaticity occurring for some degrees of freedom and significant nonadiabatic mixing for other degrees of freedom.

Before leaving this discussion involving the time scale between saddle and scission, it is worthwhile to assure ourselves that the opposite limit, the sudden approximation, is inappropriate. That is, is the descent so rapid that the nucleonic motion cannot readjust at all to the change in potential? A condition for the sudden approximation to be valid is that the time for descent from saddle to scission is short relative to the time for a nucleon to travel a nuclear dimension. The time for a nucleon to cross a nucleus is of the order of

2×10^{-22} sec, a time period which is considerably shorter than estimates of the saddle to scission time. For example, an estimated value of about 3×10^{-21} sec has been derived on the basis of a nonviscous liquid drop model calculation described later (Section B,1). We conclude that the descent from saddle to scission is slow enough for the impulse approximation to be invalid. However, the actual tearing of the neck and the collapse of the nascent fragments may be sufficiently rapid so that an impulse approximation at this stage of fission is appropriate.

As we have shown in the foregoing, it is not possible to make a conclusive theoretical judgment about the validity of the adiabatic or statistical models of fission. It is also not possible to make a choice between these two extreme models on the basis of experimental data on fragment mass and energy distributions. This is true because even within the framework of a particular model there are sufficient uncertainties in the application of the model to real nuclei to preclude definitive predictions. There is, however, one experimental result which bears on this problem in a more direct way. In Chapter XII we compare the neutron yield, a measure of the fragment excitation energy, and the total kinetic energy for spontaneous fission with those for thermal neutron fission. From the comparison of these two types of fission, it appears that approximately $\frac{3}{4}$ of the additional energy in "over-the-barrier" neutron-induced fission appears as excitation energy rather than kinetic energy. This implies that most of the additional potential energy made available in the descent from the top of the barrier to the point of emergence from the barrier in spontaneous fission has gone into excitation energy of the fragments. If the adiabatic approximation were valid, we would expect the larger fraction of the additional potential energy to have appeared as kinetic energy of the fragments. The excitation energy dependence of the variation of the fragment yield and kinetic energy release with mass ratio complicates the interpretation of such experiments. There is evidence at still higher excitation energy for a small decrease in average kinetic energy. This decrease in the average kinetic energy results from a larger decrease in kinetic energy with excitation energy for mass divisions in which one fragment is close in mass number to the $A = 132$ double closed shell as compared to the increase in kinetic energy release for more symmetric and very asymmetric mass divisions (Bishop *et al.*, 1970). On the other hand, there is some less direct evidence in support of the adiabatic model. From the characteristics of α-accompanied fission, several workers have concluded that the fragments possess considerably more kinetic energy at scission than would be expected if the statistical model were valid (see Chapter XIV). In view of the uncertain state of our knowledge as to which approach might be most valid, we discuss both adiabatic and nonadiabatic models.

B. Adiabatic Models

1. The Liquid Drop Model

If the nucleus is idealized as a liquid drop assumed to be comprised of a nonviscous fluid, then as the nucleus distorts from saddle to scission, all of the potential energy appears in the collective degrees of freedom and the motion is adiabatic with respect to the single particle degrees of freedom. Fairly extensive studies of both the statics and dynamics of a charged liquid drop have been made. We discuss them in considerable detail, as a large number of quantitative predictions can be made with a minimum of assumptions. The model also provides a useful orientation for the consideration of more realistic models.

Let us first recall what is known about the statics of the saddle point shapes as given by the liquid drop model (LDM). For all x values greater than about 0.4 the symmetric saddle point state is lower in energy than the asymmetric saddle point state and is stable with respect to mass asymmetry. [Strutinsky (1964a, b) has claimed that a loss of stability against asymmetry occurs at $x \simeq 0.8$. The investigations of Cohen and Swiatecki (1963) and Nix (1967) do not substantiate this claim.] For x values less than about 0.6 or 0.65 the neck is well developed at the saddle and the scission configuration is fairly similar to the saddle configuration. The original preference for a symmetric configuration is not likely to be destroyed during the short passage from saddle to scission.

For heavy nuclei the saddle point is not close to the scission point and it is necessary to consider the dynamics of the descent from saddle to scission. Early dynamic calculations of Hill (1958) and Hill and Wheeler (1953) gave a hint that there might be a dynamic origin for the observed mass asymmetry in the fission of heavy nuclei. More complete calculations have shown, however, that this hint was illusory and that the most probable mass division is a split into two equal parts.

The most extensive calculations of the dynamic descent from saddle to scission have been performed by Nix (1969). In this study, the shape of the nuclear surface was specified in terms of two spheroids connected by a smoothly joined quadratic surface of revolution (e.g., a hyperboloidal neck). The success of such a parameterization in reproducing a wide range of saddle point shapes is illustrated in Fig. IX-2. A total of six degrees of freedom, associated with the nascent fragments' separation, eccentricities, and relative masses, were considered. The procedure consists of the following steps.

(1) The potential energy is mapped as a function of the deformation coordinates. The saddle point is located and its properties studied.

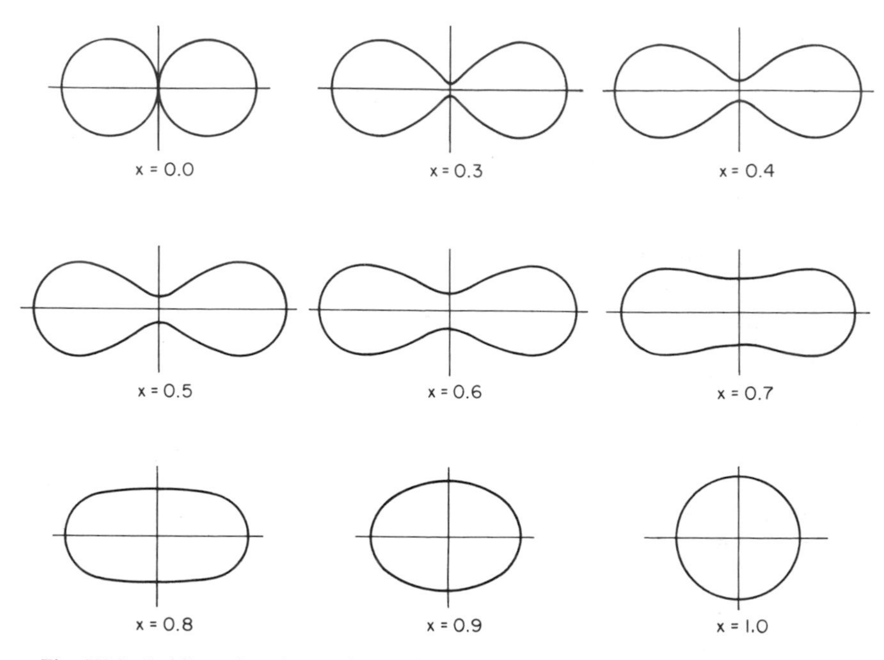

Fig. IX-2. Saddle point shapes for various values of the fissility parameter x. Those shapes approximated by two spheroids connected by a quadratic surface of revolution are nearly indistinguishable from the more exact shapes calculated by Cohen and Swiatecki (1963) by the dashed lines [From Nix and Swiatecki (1965).]

(2) The kinetic energy is calculated as a function of the deformation coordinates and their time derivatives. Irrotational nonviscous hydrodynamic flow is assumed.

(3) The frequencies and eigendisplacements of the normal modes of oscillation of the system about its saddle point shape are determined.

(4) The transition state method is used to calculate the probability for finding the system in a given state of motion as it passes through the saddle point. This assumes that statistical equilibrium is established by the time the system reaches the vicinity of the saddle point. The probability distributions of the initial states are calculated by quantum statistical mechanics.

(5) Hamilton's equations of motion are solved classically for a given set of initial conditions near the saddle point. This final step converts the probability distributions for the states of motion near the saddle point into the observable mass and energy probability distributions of the fission fragments at infinity. It should be emphasized that in this procedure there are no adjustable parameters, the constants of the theory being taken from analyses of nuclear masses.

The potential energies of symmetric and asymmetric saddle shapes as a

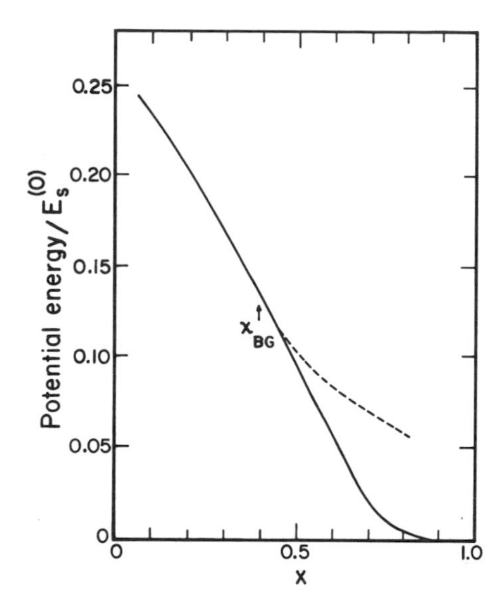

Fig. IX-3. The potential energies (in units of the surface energy of a sphere) of saddle point shapes as functions of the fissility parameter x. (—) Energy of the symmetric saddle point shape, (---) energy of the asymmetric saddle point shape. The critical Businaro–Gallone point is indicated by the arrow. [After Nix (1969).]

function of the fissility parameter are illustrated in Fig. IX-3. As indicated in the introduction to this section, for x greater than the Businaro–Gallone critical point $x = 0.4$ the asymmetric saddle point configuration is higher in energy than the symmetric saddle point configuration. For x values typical of heavy elements the asymmetric saddle point is at more than twice the energy of the symmetric configuration. An example of symmetric and asymmetric saddle shapes from an earlier study of Strutinsky (1964b) is shown in Fig. IX-4.

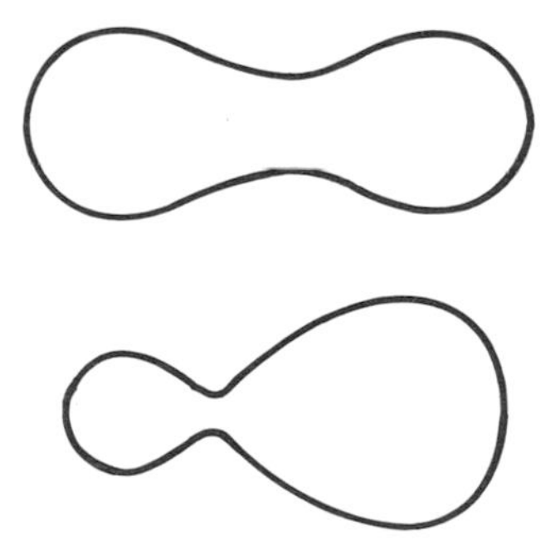

Fig. IX-4. Symmetric and asymmetric saddle shapes for $x = 0.65$. The energy of the asymmetric saddle is twice the energy of the symmetric saddle. [From Strutinsky (1964b).]

TABLE IX-1

Characteristics of Normal Modes of Oscillation about Symmetric Saddle Point Shapes

n	Mode	Symmetry	Stability
1	Center-of-mass shift	Asymmetric	Neutral
2	Fission	Symmetric	Unstable
3	Mass asymmetry	Asymmetric	Stable for $x > 0.4$; unstable for $x < 0.4$
4	Stretching	Symmetric	Stable
5	Distortion asymmetry	Asymmetric	Stable
6		Symmetric	Stable

In order to discuss the *motion* of the system in the vicinity of the saddle point it is convenient to identify the independent normal modes of oscillation about the symmetric saddle point shape. The characteristics of the six normal modes considered in Nix's (1969) study, including the descriptive names which have been given to five of them, are presented in Table IX-1.

The dynamics can be followed for any particular choice of initial conditions. Several examples of the descent from saddle to scission for 1 MeV of kinetic energy in the fission mode and 1 MeV in the mass asymmetry mode are

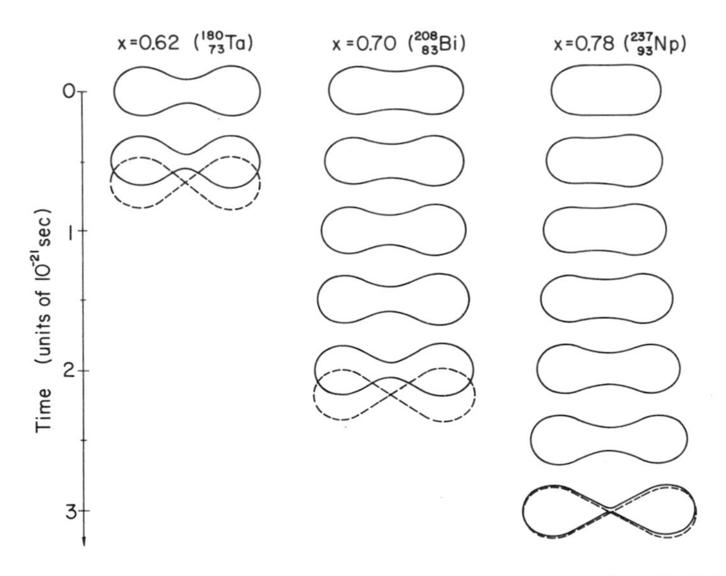

Fig. IX-5. Shape of the drop during the descent from saddle to scission. The initial conditions correspond to starting from the saddle point with 1 MeV of kinetic energy in the fission mode and 1 MeV in the mass asymmetry mode. The scission shapes are drawn with dashed lines. [From Nix (1969).]

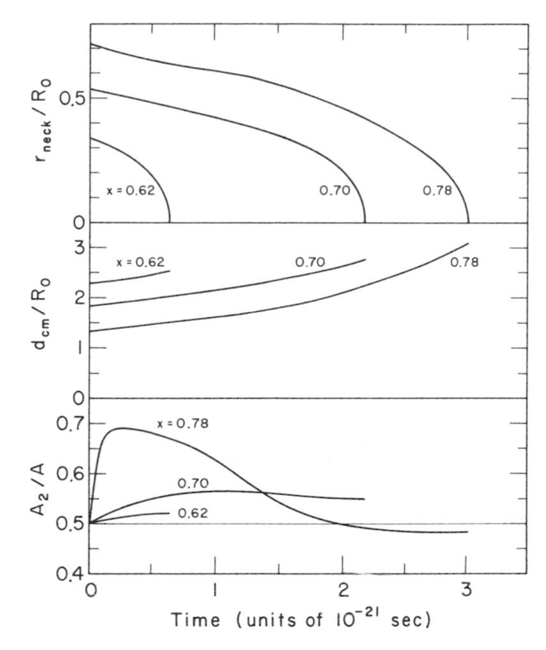

Fig. IX-6. Examples of the time dependence of the minimum neck radius (top), the distance between the centers of mass of the nascent fragments (middle), and the mass of the right-hand nascent fragment (bottom). The initial conditions are the same as those in Fig. IX-5. R_0 is the radius of the sphere of equal volume. The dividing point between the two nascent fragments is taken to be the position of the minimum neck radius. [From Nix (1969).]

illustrated in Figs. IX-5 and IX-6. It can be seen from Fig. IX-5 that irrespective of the x value the initial motion in the mass asymmetry mode does not result in a significant final mass asymmetry. If the dividing point between the two nascent fragments is taken to be the minimum neck radius, significant mass-asymmetric oscillations can occur for shapes with large necks, but (at least for the example in Fig. IX-6) by the time of scission the motion has stabilized close to the symmetric configuration. The calculations of Hill (1958) mentioned previously were terminated, due to computational difficulties in integrating the equations of motion, at the time 0.3×10^{-21} sec. (The calculation of Hill was for $x = 0.74$.) This corresponds to the time when the asymmetry oscillation has reached its maximum value, a value much larger than at the time of scission. Nix (1969) concludes from these solutions and solutions for many other initial conditions that the most probable mass division is a split into two equal parts.

The most probable final translational kinetic energies for initial conditions corresponding to starting from rest at the saddle point are shown in Fig. IX-7. For those x values for which the saddle point is fairly close to the scission

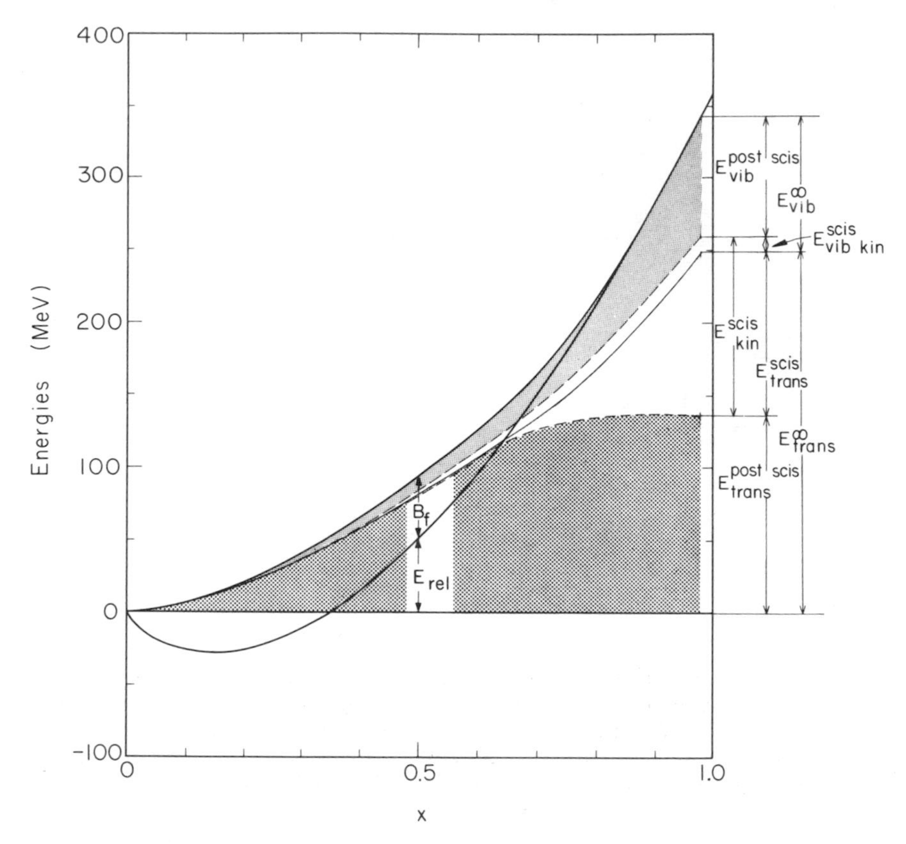

Fig. IX-7. The most probable fission-fragment energies in MeV for nuclei along the line of β stability, as functions of the fissility parameter x. The final translational kinetic energies E_{trans}^{∞} comprise the translational kinetic energy acquired prior to scission E_{trans}^{scis} and that acquired subsequent to scission $E_{trans}^{postscis}$. The final internal vibrational energies of the fragments E_{vib}^{∞} also comprise that acquired prior to scission $E_{vib\,kin}^{scis}$ and that acquired after scission $E_{vib}^{postscis}$. [After Nix (1969).]

point almost all of the kinetic energy is acquired by postscission acceleration of the fragments in their mutual Coulomb fields. For heavier nuclei this model (assuming nonviscous irrotational flow) predicts that a significant fraction of the final total kinetic energy has been acquired prior to scission. There is some rather indirect evidence in support of this prediction from analysis of the energy and angular distributions of long-range α particles emitted during fission (see Fig. IX-9 and also Chapter XIV).

An interesting effect occurs in this model after scission has taken place. The fragments oscillate fairly rapidly between prolate and oblate shapes as they separate in their mutual Coulomb field, as illustrated in Fig. IX-8. This

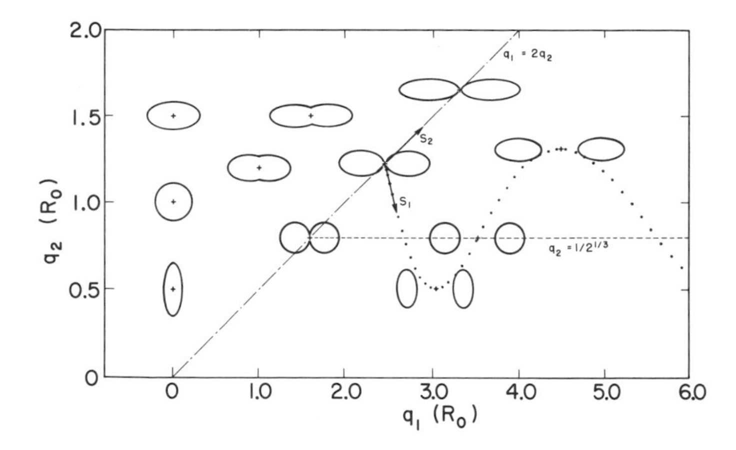

Fig. IX-8. Separation and shape oscillation in the two-spheroid approximation for $x=0.7$. In this approximation the saddle occurs for tangent spheroids for this x value. q_1 is the separation between the centers of the two spheroids and q_2 is the semimajor axis of either of the spheroids. Each point along the trajectory, corresponding to starting from rest, is equally spaced in time at intervals of 5×10^{-23} sec. [From Nix (1963).]

results in the appearance of part of the Coulomb interaction energy of the scission configuration as vibrational energy of the final fragments. This result is a consequence of the nonviscous flow assumed, and would be lessened (resulting in *higher* translational energies) if the flow were somewhat viscous.

Experimental values of the most probable translational kinetic energies for cases where the mass distributions are symmetric are compared in Fig. IX-9 with the calculated total translational kinetic energies. The trend with the fissility parameter x is reproduced quite well, with, however, a systematic tendency to underestimate the absolute value of the energies by about 5%.

In order to find the distribution in final energies and mass divisions it is necessary to determine the relative probabilities for different initial starting conditions. The transition state method is used for this purpose. This method consists of dividing a system of N degrees of freedom into two systems at the saddle point: a system having a single degree of freedom that represents unstable motion (e.g., $n=2$, the fission mode) and a second system associated with the remaining $N-1$ degrees of freedom. The $n=1$ mode can be neglected, as it corresponds to a center-of-mass shift. The probability distributions for the $n=2$ fission mode have not been considered, as to lowest order the final mass and energy distributions are independent of the *initial* values of the fission coordinate and fission momentum in this model. The quantal probabilities for the remaining normal modes at the saddle point ($n=3$–6) are Gaussian distributions, since in the determination of potential and kinetic energies of the normal modes by expansion about the saddle point, terms higher than

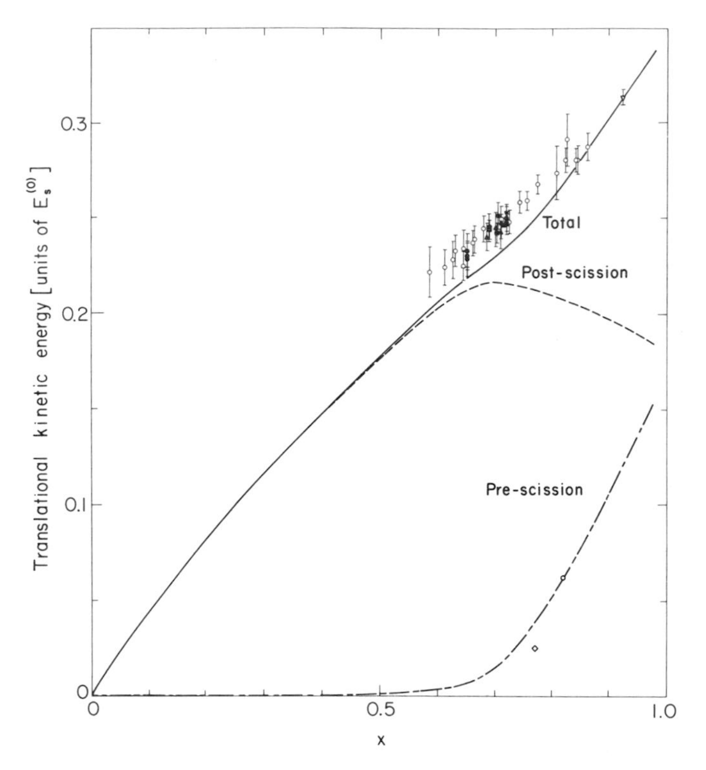

Fig. IX-9. Comparisons of calculated and experimental most probable fission-fragment translational kinetic energies, expressed in units of the surface energy of a sphere. The dot-dashed curve gives the calculated most probable translational kinetic energy acquired by the fragments prior to scission, the broken-line curve that acquired after scission, and the solid curve the total at infinity. The two points in the vicinity of the prescission curve are estimates from the energy and angular spectra of α particles accompanying fission. [From Nix (1969).]

quadratic were neglected. The magnitudes of the frequencies for the normal modes are illustrated in Fig. IX-10. The coordinate (Q) and momenta (P) distributions are given by

$$N(Q_n) = (2\pi\sigma_n{}^2)^{-1/2} \exp[-Q_n{}^2(2\sigma_n{}^2)]$$

$$N(P_n) = (2\pi\sigma_n'^2)^{-1/2} \exp[-P_n{}^2/(2\sigma_n'^2)]$$

where the temperature-dependent variances $\sigma_n{}^2$ and $\sigma_n'^2$ are given explicitly by

$$\sigma_n{}^2 = \tfrac{1}{2}(\hbar\omega_n/K_n)\coth(\hbar\omega_n/2\theta) \qquad \text{and} \qquad \sigma_n'^2 = \tfrac{1}{2}M_n\hbar\omega_n\coth(\hbar\omega/2\theta_n),$$

respectively, and K_n and M_n are the normal mode stiffness and inertia parameters, respectively. The nuclear temperature θ is dependent on the internal

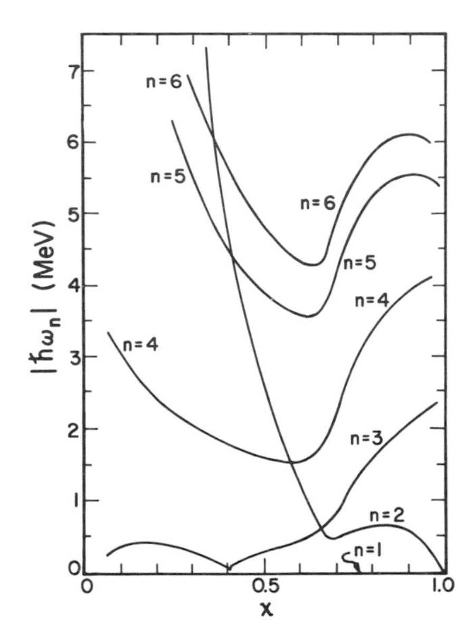

Fig. IX-10. Magnitudes of the normal mode frequencies in MeV for nuclei along the line of β stability calculated by Nix using the three-quadratic-surface parameterization. [After Nix (1969).]

excitation energy at the saddle point. The temperature-dependent variances reduce to the classical Boltzmann factor at high temperature and to quantum mechanical zero point motions at low temperature.

The comparisons of the calculated *widths* of the kinetic energy and mass distributions with experiment are difficult to display because of the dependence on both excitation energy and fissility. It is clear, however, that the theory fails completely for the mass distributions of heavy elements at low energies. The experimental data show asymmetric fission is strongly favored, whereas the theory predicts symmetric fission to be most probable. As one goes to lighter elements the situation improves considerably, with symmetric distributions observed. For all cases, however, the calculated width of the mass distribution remains narrower than the observed width.

A comparison of the experimental and calculated temperature dependences of the mass and kinetic energy distributions can best be done for medium weight nuclei where the complication of fission following neutron emission is minimized and where the most probable mass division is symmetric at all excitation energies investigated. Such comparisons are shown in Figs. IX-11 and IX-12. The general trend is qualitatively reproduced, with the theory consistently underestimating the absolute values of the widths. (An earlier LDM estimate of the widths of the mass and charge distribution using a more

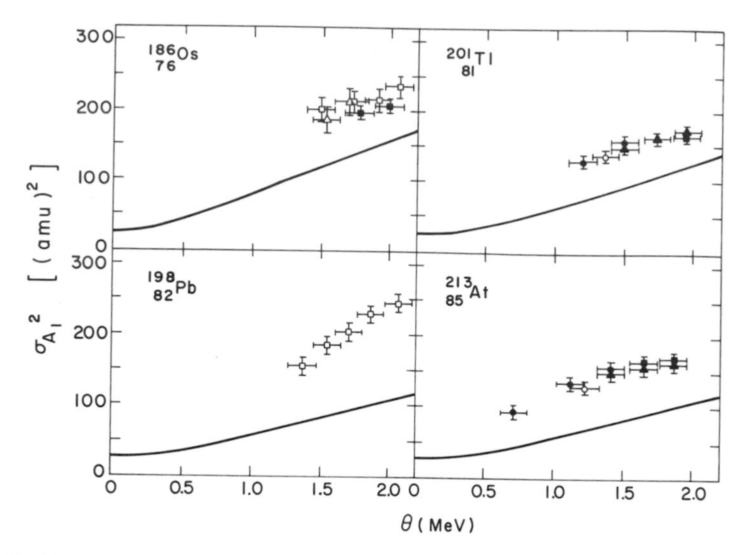

Fig. IX-11. Comparisons of calculated and experimental variances of fission-fragment mass distributions for four compound nuclei as functions of the nuclear temperature θ at the saddle point. [From Nix (1969).]

restricted parameterization of touching spheroids for the saddle configuration was in better agreement with experiment, but this success must be considered fortuitous.)

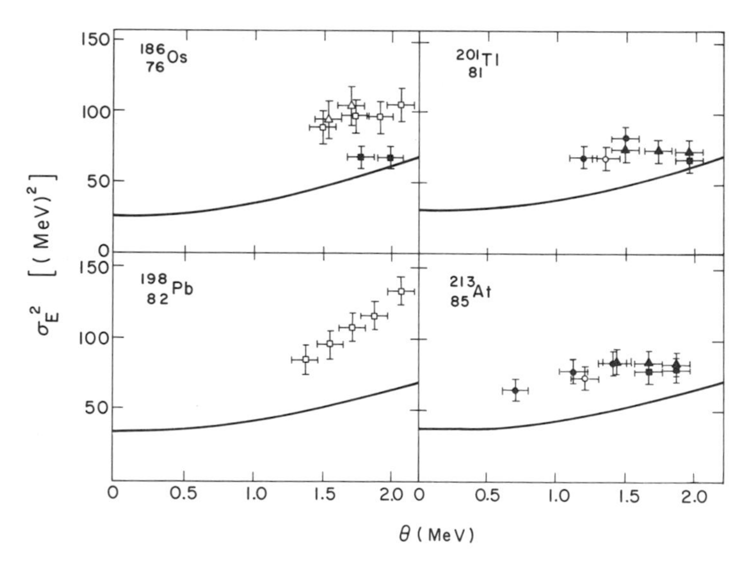

Fig. IX-12. Comparisons of Nix's calculated and Plasil *et al.*'s experimental variances of fission-fragment total translational kinetic energy distributions for four compound nuclei as functions of the nuclear temperature θ at the saddle point. [From Nix (1969).]

What conclusions can be drawn from these comparisons of experiment with the calculations for division of an idealized nonviscous liquid by irrotational flow? The underestimation of both the widths and the most probable kinetic energies may indicate that the flow is either viscous or rotational (or both). Neither of these possibilities is particularly surprising for nuclear matter. The most significant failure of the theory is the failure to account for asymmetric mass distributions. It is not possible to rule out this failure as being a consequence of the constraints of the shape expansion employed, but the likelihood of this being the difficulty seems extremely small. Instead, it seems more reasonable to believe that the mass asymmetry is in some way associated with single particle motion. In principle one way to investigate this possibility would be to treat single particle effects as a perturbation of the liquid drop model Hamiltonian.

2. Single Particle Effects on the Potential Energy Surface and on the Dynamics of the Descent from Saddle to Scission

We have seen that a fairly complete theory for the descent from the saddle for a nonviscious liquid drop qualitatively accounts for many of the observed characteristics of mass and energy distributions for sufficiently light or sufficiently excited nuclei. The importance of single particle effects on the potential energy surface at deformations up to and including the saddle deformation is, however, clear from such properties as nonspherical equilibrium deformations and the existence of the second minimum associated with spontaneous fission shape isomerism. It has also been known for some time from quasi-equilibrium scission models that rather modest shell effects at scission can be amplified into quite large effects on final fragment kinetic energies and excitation energies. What we would like to do, in the spirit of the approach described in the previous sections, is first to calculate the single particle modifications to the potential energy surface, and then to solve the equations of motion including any shell effects on the inertial paramcters. Unfortunately neither of these objectives has been achieved yet. Only very recently have the first realistic estimates been made of single particle effects on the potential energy surface for shapes sufficiently general to include reflection-asymmetric shapes. No dynamic calculations for those potential surfaces have been performed. There has been one dynamic calculation for a potential energy surface with much cruder shell effects, and this calculation will be described briefly in subsection c.

a. Potential Energy Surface Mapping with Inclusion of Odd-Multipole Shape Distortions

Möller and Nilsson (1970) determined shell structure corrections to the liquid drop potential energy surface by calculation of single particle energies in a deformed harmonic oscillator potential and application of the Strutinsky normalization method. In order to make the calculations computationally feasible it is necessary to limit the number of degrees of freedom which are to be considered. It has been noted (Cohen and Swiatecki, 1963; Nix, 1967) that the liquid drop energy surface is soft to certain asymmetric shape distortions when the symmetric distortions are large. In order to obtain a large softening with respect to asymmetric distortions, it is necessary to include P_5 as well as P_3 deformations, with the coefficients of P_5, namely ε_5, generally being of the order of $-0.5\varepsilon_3$. It has also been found in investigations of the shell correction for large symmetric distortions (in the P_2–P_4 space) that the path to scission is approximately given if the coefficient ε_4 has the values $\varepsilon_4 = 0.1767\varepsilon_2 - 0.0308$. Möller and Nilsson have plotted the shell corrected potential energy surfaces as a function of these limited symmetric and asymmetric degrees of freedom. Their potential energy surface for ^{236}U is illustrated in Fig. IX-13. For large distortion in the vicinity of the second barrier the energy is reduced by several MeV for asymmetric distortions. The shape of the nucleus at the asymmetric distortion which minimizes the energy for fixed symmetric distortion associated with the maximum of the second barrier is shown in Fig. IX-14. The increased relative importance of allowing the asymmetric degree of freedom at the second barrier is to be expected, since the second barrier is a result of a large energetic-

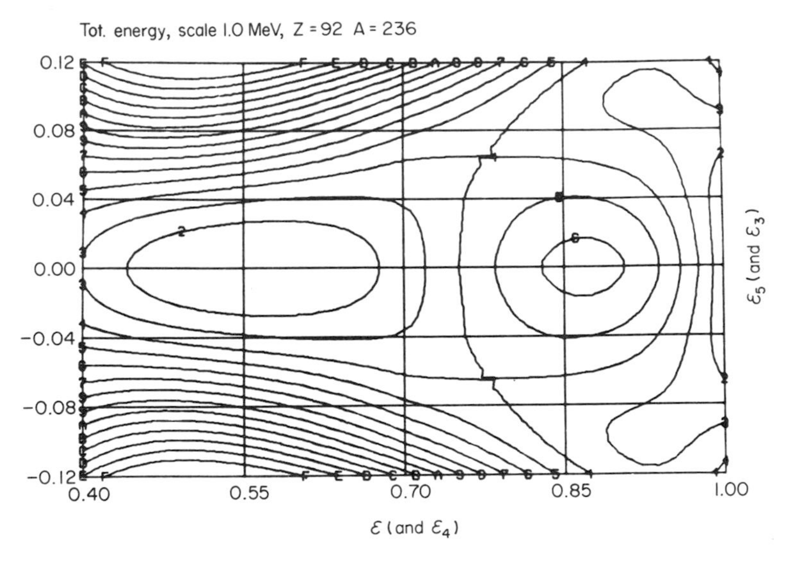

Fig. IX-13. The shell structure-corrected potential energy surface for ^{236}U as a function of the symmetric distortion parameter ε_2 ($\varepsilon_4 = 0.1767\varepsilon_2 - 0.0308$) and asymmetric distortion parameter ε_5 ($\varepsilon_3 = -2\varepsilon_5$). [From Möller and Nilsson (1970).]

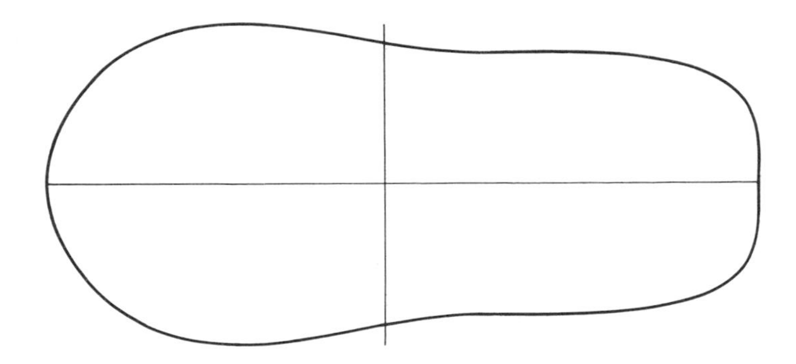

Fig. IX-14. The nuclear shape (at $\varepsilon = 0.85$, $\varepsilon_3 = -0.16$, $\varepsilon_4 = 0.12$, and $\varepsilon_5 = 0.08$) acquired by ^{236}U near its second saddle point. [From Möller and Nilsson (1970).]

ally unfavorable shell correction at this distortion. One point must be emphasized about the surface illustrated in Fig. IX-13. For the still larger distortions leading to scission, the symmetric shape again becomes stabilized. This feature does not appear in other calculations that more accurately describe near-scission configurations, as discussed in a later paragraph in this section.

Möller and Nilsson have performed similar calculations for the lighter element ^{210}Po. In this case, they do not find a lowering of the barrier for asymmetric disortions, which is somewhat encouraging in view of the symmetric mass yield distribution observed. This result, however, must be considered provisional because the saddle point shape for such a light nucleus corresponds to a nuclear neck which may be thin enough so that the harmonic oscillator representation is questionable. The real nuclear potential may actually be appreciably shallower in the neck region than in other regions, a feature which is not reproduced by a one-center oscillator. The number of degrees of freedom employed in the shape parameterization is also inadequate, as is evidenced by the squarish end of the shape illustrated in Fig. IX-14.

The potential energy surface has also been calculated for ^{252}Fm and exhibits very little reduction in height due to the asymmetric degree of freedom. This is partly a consequence of the stiffening of the liquid drop potential energy surface with respect to asymmetric distortions as Z^2/A is increased.

One-center oscillator potentials become an increasingly unsatisfactory representation for deformations larger than that of the second barrier. Recently a two-center oscillator model has been developed (Scharnweber *et al.*, 1970, 1971). Such a potential has the correct asymptotic behavior for separated fragments. In its simplest form the equipotential surfaces correspond to separated or overlapping spheres, as illustrated in Fig. IX-15. This model

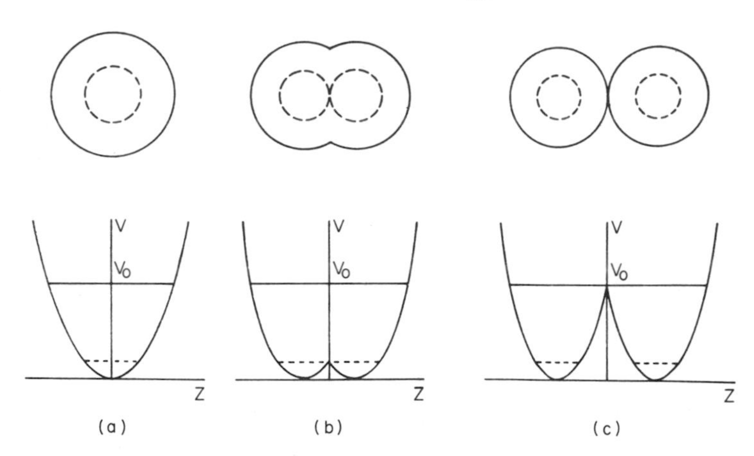

Fig. IX-15. Equipotentials in a two-center oscillator model of overlapping equal-mass spheres. In the upper part of the figure the equipotentials for the nuclear surface ($V = V_0$) are indicated by the solid lines and the equipotentials for a lower V value (indicated by dashed lines in the lower part) are shown by dashed lines. The values of Z_0 for (a), (b), and (c) are 0, 3, and 5.9 F, respectively. [From Scharnweber *et al.* (1971).]

can be extended to overlapping spheroids and to shapes which are smoothed in the joining region. Perhaps the most interesting result obtained thus far with this oscillator model is the evidence for preformation of the final shell structure while there is still appreciable overlap of the nascent fragments. This is illustrated in Fig. IX-16 by the energy level diagram for shapes corresponding to overlapping or separated equal-mass spheres. For small deformations (almost complete overlap) the level scheme resembles the usual one-center Nilsson diagram. For complete separation of the fragments we see that the smaller final fragments have a larger oscillator spacing. In addition, each level is doubly degenerate due to the identical noninteracting final nuclei. At intermediate values of the deformation where the neck develops, the lower levels first become degenerate, reflecting the fact that a barrier is developing between the two fragments. States near the Fermi surface still show an appreciable splitting.

Other calculations have been performed using finite Woods–Saxon-like potentials (Bolsterli *et al.*, 1969, 1972; Pashkevich, 1971; Pauli *et al.*, 1971). In the calculations of Bolsterli *et al.* (1969, 1972) the nuclear shape is specified in terms of smoothly joined portions of three quadratic surfaces of revolution (e.g., two spheroids connected by a hyperboloidal neck). Examples of shapes for various symmetric distortions and some equipotential curves for these shapes are illustrated in Fig. IX-17. Results reported thus far have been primarily for symmetric deformations. The stability of the potential energy with respect to mass asymmetry has been investigated at selected symmetric

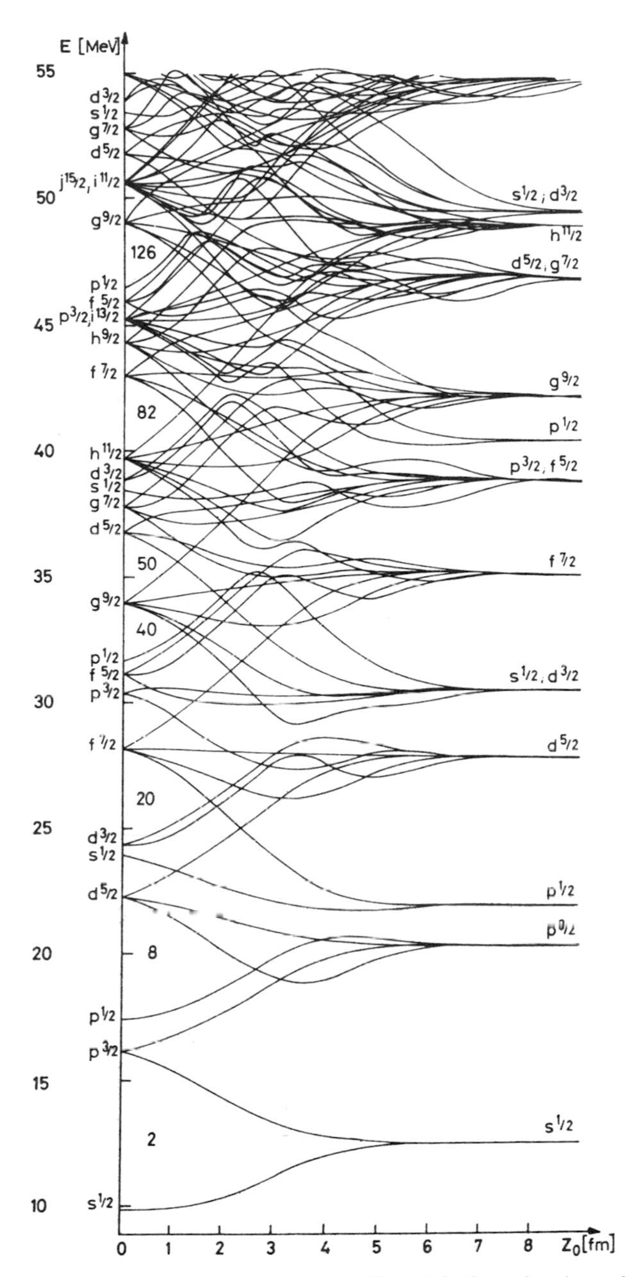

Fig. IX-16. Level structure for a two-center shell model of overlapping spheres as illustrated in Fig. IX-15. This diagram, which includes spin-orbit and l^2 terms, is appropriate for neutrons for ^{236}U. [From Scharnweber *et al.* (1971).]

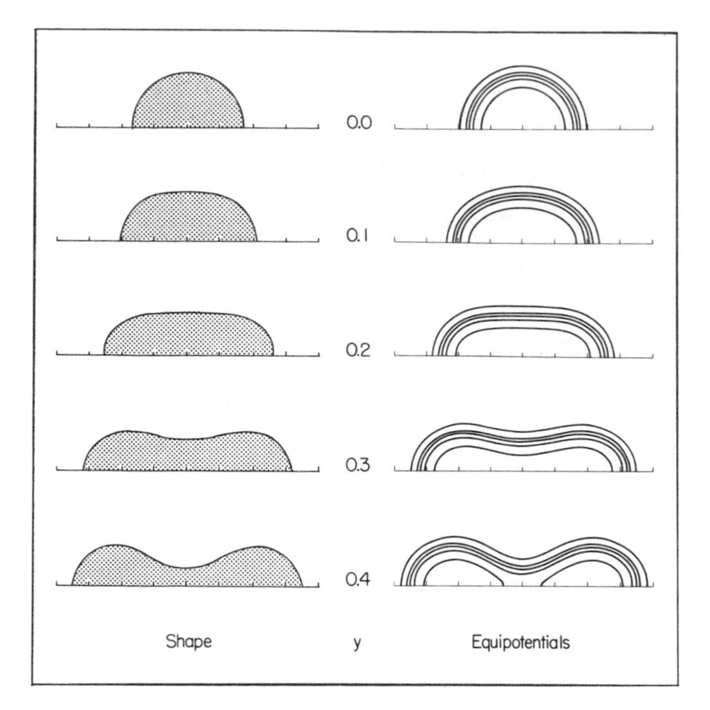

Fix. IX-17. Shapes described by the symmetric deformation coordinate *y*, and resulting equipotentials for the diffuse-surface potential. The equipotential curves shown are for 10, 30, 50, 70, and 90% of the maximum well depth. [From Bolsterli *et al.* (1972).]

distortions. For ^{240}Pu the first barrier and the second minimum are stable with respect to asymmetric distortions, while the second barrier is lowered considerably when asymmetric distortions are allowed. These results are in agreement with those obtained by Möller and Nilsson (1970) and Pauli *et al.* (1971). The lighter nucleus ^{210}Po, whose saddle corresponds to a shape with a more pronounced neck, is also found to be slightly unstable with respect to mass asymmetry, contrary to the findings of Möller and Nilsson and Pauli *et al.*

One of the more extensive calculations of the potential energy surface for deformations up to scission and for reflection-asymmetric shapes has been performed by Pashkevich (1971). The nuclear shape is generated by an expansion about Cassinian ovaloids. A Woods–Saxon type of potential is defined which results in a reduced nuclear density in the neck region compared to the nascent fragment centers. There is no discontinuity in the joining region as exhibited by the simple two-center harmonic oscillator potential. Reflection-asymmetric shapes are generated by taking the first term in a Legendre poly-

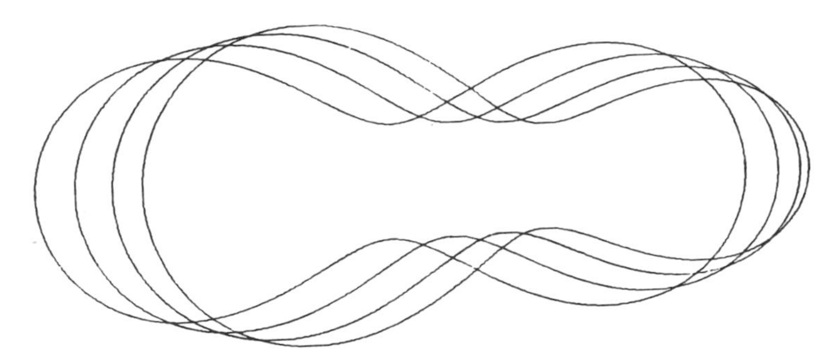

Fig. IX-18. Cassinian ovals perturbed by "dipole" deformation α_1. The four figures with increasing asymmetry correspond to $\alpha_1 = 0.0$, 0.1, 0.2, and 0.3, respectively. The symmetric distortion parameter α is 0.95 for all ovals. [From Pashkevich (1971).]

nomial expansion. Examples of some of the shapes considered are shown in Fig. IX-18. The single particle energies are obtained for various combinations of the symmetric distortion parameter α and the asymmetric distortion parameter α_1. The potential energy surface is then generated using the Strutinsky shell correction method described in Chapter II.

Contour diagrams of the potential energy surfaces obtained for ^{208}Pb and ^{230}Th are shown in Fig. IX-19. The potential energy surface is very flat with respect to α_1 for small values of α. This is a consequence of the fact that for small values of α, an increase in α_1 results primarily in a center-of-mass shift with little change in shape. The first barriers and the secondary minima associated with spontaneous fission isomers are seen to be stable with respect to asymmetric distortions. The saddle associated with the second barrier occurs for reflection-asymmetric shapes. As one progresses toward scission configurations the potential valley is always located at finite asymmetry except in the case of ^{208}Pb, where the valley returns to symmetry for large distortions. For ^{208}Pb, a secondary, higher valley occurs for large asymmetry. Examination of the individual shell corrections for neutrons and for protons shows that the final asymmetry exhibited by the heavier elements is due to stabilization of nascent fragments with nucleon numbers of the order of 6 nucleons more than the 50, 82, and 126 magic numbers. Thus, after allowing for the "neck" nucleons, the shell effects of the final fragments come to play a determining role at large deformations.

The distortion parameters of the bottom of the valley leading from the second minimum over the second saddle point and on to scission are plotted for several different nuclei in Fig. IX-20. A scale relating the mass ratio to the distortion parameter α_1 is shown at $\alpha = 0.9$. Scission is expected for values

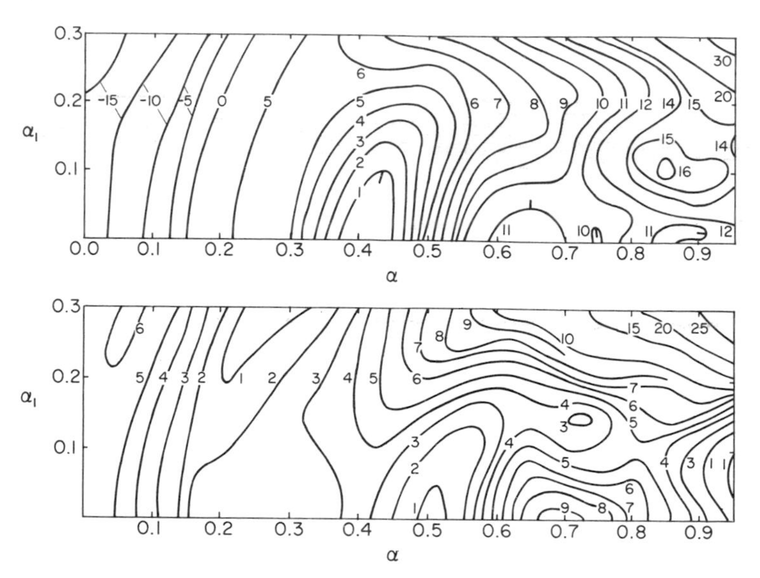

Fig. IX-19. Contour map (contour levels in MeV) as a function of symmetric distortion α and asymmetric distortion α_1 for ^{208}Pb (top) and ^{230}Th (bottom). The apparent instability of the ground state with respect to α_1 is removed when hexadecapole distortions are included. [From Pashkevich (1971).]

slightly larger than this. If the descent from the second saddle to scission were to follow the valley up to a distortion close to scission, the expected mass asymmetry would be in fairly good agreement with experimental observations.

b. Potential Energy Surface and Mass Asymmetry

Potential energy surfaces including the odd-multipole shape distortions (ε_3 and ε_5) as well as the even-multipole shape distortions were described in the previous section. Recent calculations of such surfaces by Möller and Nilsson (1970), Bolsterli *et al.* (1972), Pauli *et al.* (1971), and Brack *et al.* (1972) show that the saddle point corresponding to the second fission barrier has an asymmetric shape for heavy nuclei. This is in contrast to the first barrier and the secondary minimum, which have reflection symmetric (but not necessarily axially symmetric) shapes at their lowest energies.

Tsang and Wilhelmy (1972) have used the asymmetric saddle point shapes at the second (outer) barrier to calculate the mass asymmetry in fission as characterized by the peak-to-valley mass yield ratio. In the calculation, the peak-to-valley ratio (asymmetric to symmetric fission) is determined from the

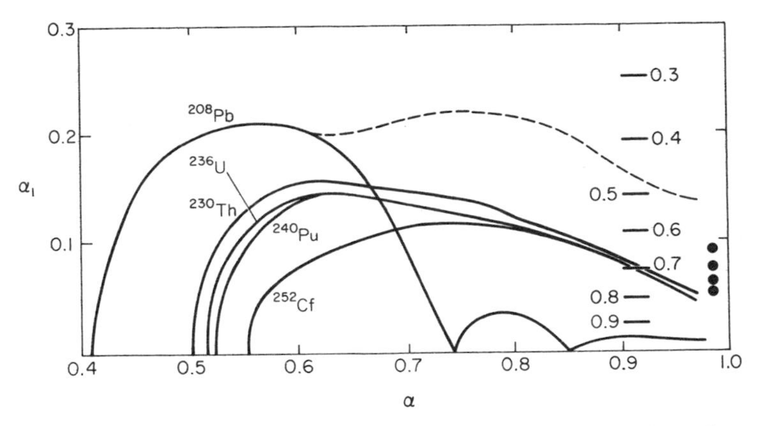

Fig. IX-20. Distortion coordinates of the bottom of the valley leading from the second saddle point to scission and from the second saddle point to the second minimum. For ^{208}Pb there are two troughs; the higher one is shown with a broken line. For $\alpha = 0.9$ a scale relating mass ratio to α_1 is given. ● Experimental most probable mass ratio for ^{230}Th, ^{236}U, ^{240}Pu, and ^{252}Cf (from top to bottom). The experimental value for ^{208}Pb would be at the abscissa, $\alpha_1 = 0$. [From Pashkevich (1971).]

difference in magnitude of the potential energies for symmetric and asymmetric shapes at the second barrier. Specifically, the probabilities of fissioning into symmetric and asymmetric products are assumed to be proportional to the number of levels or the barrier penetrability for both the symmetric and asymmetric barriers. The dynamic descent from the saddle point to scission and the shell effects of the final fragments are completely ignored in this model calculation. Three cases are considered:

(a) the excitation energy exceeds both the symmetric and asymmetric barriers of the second barrier;
(b) the excitation energy exceeds the asymmetric barrier and is below the symmetric barrier; and
(c) the excitation energy is below both the symmetric and asymmetric barriers.

For case (a), the peak-to-valley ratio is directly proportional to the number of levels at the symmetric and asymmetric barriers (the simple Fermi gas model without the preexponential energy dependence is used to calculate the level density). For case (c), barrier penetration formulas are employed for both barriers, while in case (b) the level density is used for the asymmetric barrier, and the barrier penetrability formula is used for the symmetric barrier.

Results of calculations with this model have been compared to experimental data for several nuclei, some at more than one energy. A marked correlation between calculated and experimental peak-to-valley ratios is reported (Tsang and Wilhelmy, 1972), although unrealistic values of the level density parameter a are required in a number of cases. The model is also in agreement with the recent peak-to-valley data measured for the spontaneous fission of ^{256}Fm and the neutron-induced fission of ^{257}Fm. The calculated results are very sensitive to the choice of potential energy surface, with the best agreement obtained with the energy surface of Möller and Nilsson (1970). If the potential energy surfaces of Bolsterli *et al.* or Pashkevich were used, then fission of the lighter elements ^{208}Pb and ^{210}Po would be predicted to fission asymmetrically, contrary to observation.

Whether or not such a simple model, based solely on the magnitudes of the internal excitation energies for the asymmetric and symmetric saddle shapes at the second barrier (and neglecting both the dynamics from the second barrier to scission and the influence of the shells in the fragments), provides an explanation of mass asymmetry must await more accurate evaluations of potential energy surfaces and a more detailed understanding of the level density of nuclei with such deformed shapes.

c. *Dynamics of Descent on a Particular Shell Structure-Corrected Potential Surface*

The only calculation attempting to explore the consequence of descent on a potential energy surface which is lower at some deformations for asymmetric shapes than for symmetric shapes has been performed by Hasse (1969).

The shell correction used in the region between just past the saddle and the scission configurations was based on the Myers–Swiatecki shell correction for separated spherical fragments. The shell correction at the saddle was assumed to vanish and hence lead to a symmetric saddle configuration, while the shell effects appropriate to the final spherical fragments were assumed to become operative just past the saddle, reaching their full value at scission. It is unlikely that the particular shell degeneracies of the final spherical fragments are relevant at distortions close to the saddle, and it is fairly certain that the shell effects of the quite highly deformed nascent fragments at scission are not the same as for the spherical fragments. This surface also is in striking contrast to that of Möller and Nilsson, the former having a symmetric saddle and asymmetric scission minimum potential energy, whereas the latter has an asymmetric saddle with the symmetric configuration having the lowest energy as one moves toward scission. Thus, the shell structure correction used in Hasse's calculation must be considered not only crude but probably incorrect in even its qualitative features. The calculation is instructive, however, as a guide to the sensitivity of the dynamics to structure in the potential energy surface.

Hasse investigated the trajectories of 100 events with randomly distributed initial coordinates. The resulting mass distribution peaked at a mass ratio of approximately 1.2 rather than 1.44 as observed experimentally. The calculated final mass ratio simply reflects the minimum in the potential energy surface at this mass ratio near scission due to the influence of the $A = 132$ closed shell. Perhaps the lesson to be learned from this calculation is that even modest (2–3 MeV) shell effects on the potential energy surface between saddle and scission can be felt during the descent. This is not particularly encouraging for an explanation of asymmetric fission based on the Möller–Nilsson potential, but enhances the significance of the more realistic calculations of Pashkevich.

A final warning concerning potential energy surfaces and their effects on the dynamic descent is perhaps in order. As has been emphasized by Wilets (1964), the appearance of a potential energy surface is dependent on the choice of the parameters against which it is plotted. Certain features, particularly local maxima, minima, and saddle points, remain invariant under coordinate transformation. Other features are not preserved: for example, it is possible to change a sloping trough into a sloping ridge by an appropriate coordinate transformation. The appearance of a potential energy surface has more meaning if the inertial parameters have comparable values for all the coordinates employed.

C. Nonadiabatic Models of Fission

1. Statistical Theory of Fission

The statistical model assumes the existence of a well-defined scission configuration, up to which strong nonadiabatic effects maintain statistical equilibrium, and beyond which no nuclear interactions occur between the fragments. In this theory, the relative probability of different scission configurations (characterized by mass, charge, deformation, and kinetic energy of the nascent fragments) is proportional to the density of translational and excitational states for said configuration. The usual statistical expression given (Fong, 1956) is that obtained after an implicit integration over fragment deformations, so that the relative probability of a particular configuration characterized by fragment charges Z_1 and Z_2 and fragment masses A_1 and A_2 is given by

$$Y(Z_1, A_1, Z_2, A_2) \propto \int_0^{E_m} \omega(K)\, dK \int_0^{E_m - K} \omega_1(E_1)\, \omega_2(E_m - K - E_1)\, dE_1$$

where $\omega(K)$ is the density of translational states for an amount K of kinetic energy at scission; $\omega_1(E)$ and $\omega_2(E)$ are the level densities of the fragments at excitation energy E; and E_m is the maximum total available energy, kinetic plus excitation energy, for the most energetically favorable fragment deformations at the scission point consistent with the values of Z_1, A_1, Z_2, and A_2 under consideration.

There are two problems with this model, one fundamental and one practical. The fundamental problem is that the potential energy does not exhibit a minimum at scission but continues to decrease as one passes through scission. A scission configuration can only be defined by the introduction of an arbitrary restraint such as touching fragments. The second problem is that evaluation of the foregoing expression requires fragment deformation energies and level densities at deformations where direct experimental information is not available. Many applications of this formula have involved empirical parameter adjusting. Perhaps the most noteworthy feature of the statistical formula is the exponential dependence of the level density ω_1 and ω_2 on excitation energy and hence on the maximum energy E_m. Thus, the probability for differing mass splits is very sensitive to the mass defects of the deformed fragments, which are in turn likely to be strongly dependent on the shell structure of the final fragments. In addition, the shell structure of the fragments also affects the absolute value of the level density for a particular excitation energy. The statistical model is therefore capable in principle of amplifying small shell effects into large factors in mass and charge distributions. It is also capable of accounting for the tendency of improbable mass divisions to become more probable as the excitation energy of the fissioning nucleus is increased, since a given shell effect becomes relatively less important as the total available excitation energy increases. Indeed, increasing excitation energy probably destroys shell effects much more rapidly than is implied by conventional applications of the statistical formalism by virtue of its effect on the relative contributions of different deformations at scission and the dependence of the level density on deformation.

Since there are many more degrees of freedom associated with particle motion than with translation, the kinetic energy of relative motion of the fragments at scission is only of the order of kT, or approximately 1 MeV. The final kinetic energy of the fragment energies is therefore essentially determined by the Coulomb interaction potential between the nascent fragments, this energy being converted into fragment kinetic energy as the fragments separate.

In addition to mass and kinetic energy distributions, the statistical model is capable in principle of predicting many other properties of fission, such as the most probable charge for a particular mass split, the dispersion of the charge distribution curve, and neutron yields and energy spectra from individual light and heavy fragments. Numerous attempts have been made over the years to use a statistical approach to account for one or another feature of fission properties, usually employing adjustable parameters at some point. The reasonableness of the parameters obtained may then be some measure of the validity of the statistical approach. Many features of fission do seem to be qualitatively consistent with a statistical approach when shell effects are incorporated.

As an example of a fairly sophisticated statistical model calculation, we discuss briefly the work of Ignatyuk (1969). The scission configuration is represented by two touching deformed fragments. Only quadrupole deformations are considered. The dependence of the potential energy on deformation for each fragment is obtained by using the Strutinsky shell correction method discussed in Chapter II. The potential energy of the scission configuration is given by the sum of deformation energies of each of the fragments and the Coulomb interaction energy between the fragments. If the potential energy of the scission configuration is minimized with respect to the deformation of the fragments, the variation of potential energy with fragment mass illustrated in Fig. IX-21 is obtained. The dependence of the nuclear level densities on both deformation and excitation energy was obtained from the appropriate Nilsson single particle energy levels. The mass yield curve is then obtained by integration over fragment deformation and fragment charge. The result is compared with experiment in Fig. IX-22. The inhibition against symmetric fission is qualitatively reproduced. The most problable calculated value of the heavy

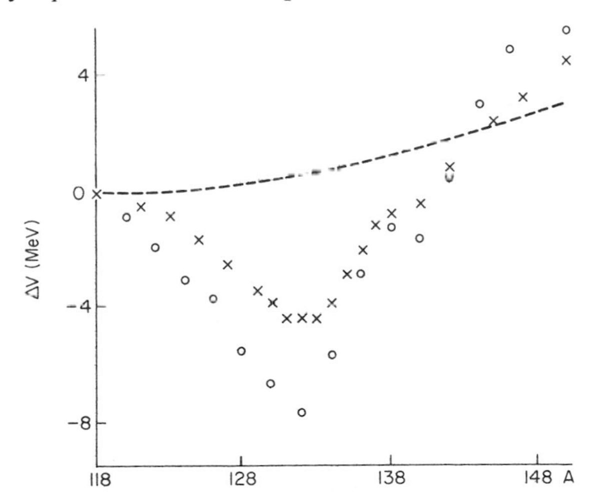

Fig. IX-21. The difference in potential energy of the scission configuration as a function of heavy fragment mass A for the fissioning nucleus ^{236}U. (---) Dependence expected in the absence of shell effects, o result obtained by inclusion of the shell correction, × final result after inclusion of both shell and pairing effects. [After Ignatyuk (1969).]

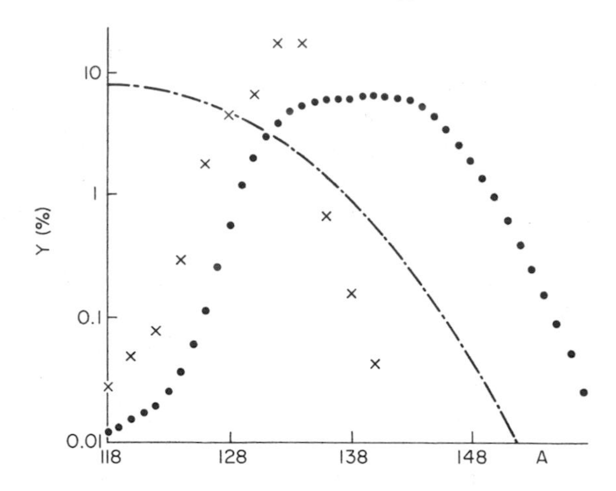

Fig. IX-22. Comparison of calculated (×) and experimental (●) mass distributions. The dot–dash curve is the prediction in the absence of shell effects. [From Ignatyuk (1969).]

fragment mass is at $A \sim 132$ rather than $A \sim 140$, and the calculated width of the mass distribution is too small. It is suggested that these discrepancies may be a consequence of the neglect of the existence of a neck between the fragments. The dependence of the average kinetic energy of the fragments on mass division is also qualitatively reproduced, as are the fragment deformation energies implied by neutron yields. The calculated dependence of the ratio of symmetric to asymmetric fission on the excitation energy of the fissioning nucleus is also in general agreement with observations.

One final comment should be made concerning the relationship of the statistical model to the nonviscous LDM model discussed previously. It is that we should not be surprised if in the final outcome many of the predictions of the two models turn out to be similar. In both models the potential energy surface as one approaches scission will be important. In the LDM it will influence the outcome through the dynamics, and in the statistical model by virtue of minima in the potential energy surface giving maxima in the available excitation energy and hence in the density of states.

2. A Model of Kinetic Dominance

Griffin (1969) has suggested that the collective motion may be sufficiently fast that the path followed in configuration space is that corresponding to a minimal collective inertia. Paths with many level crossings are shown to be associated with large inertial parameters. Thus, the adiabatic path between saddle and scission, which generally involves many level crossings, may not be the optimal path for rapid collective motions. In the limit of rapid collective motions every crossing will result in nucleonic excitation. The sums of these excitations can be taken to define a "rapid" potential energy surface which may have a quite different structure in configuration space than the adiabatic potential energy surface. In the limit of rapid collective motion the valleys in the "rapid" potential energy surface will determine the path to scission. A rather schematic estimate of this path has been made, and the implied mass ratios as a function of Z^2/A are illustrated in Fig. IX-23. The dependence of mass ratio on $x(\propto Z^2/A)$ is in rather poor accord with the experimental observations of a most probable mass ratio of unity for low x, a discontinuity at $x \approx 0.7$ to a value of about 1.4, and a decrease in mass ratio with x for further increases in x.

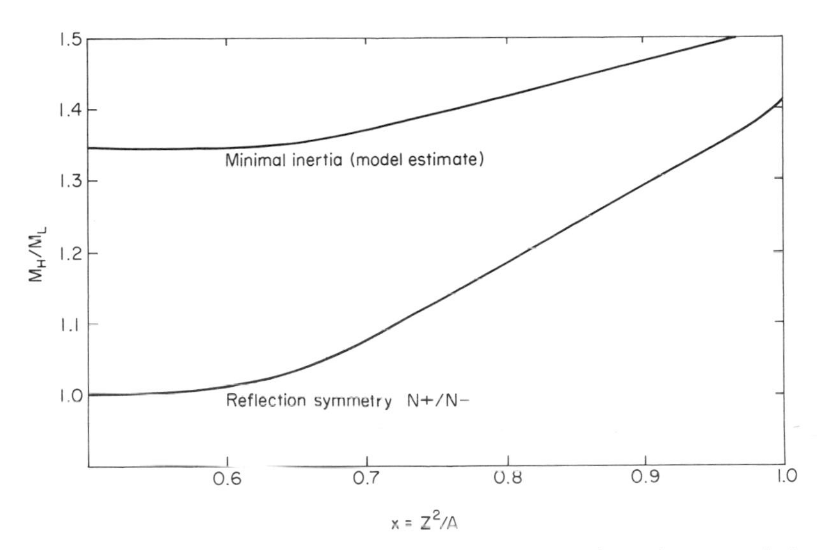

Fig. IX-23. Mass ratios for rapid descent. The upper curve shows the mass ratio implied by a criterion of following a path of minimum inertia as estimated from a schematic model. This corresponds to following a path on a potential surface obtained assuming "jumps" occur at every level crossing. The lower curve is an earlier estimate assuming "jumps" to occur whenever levels of different symmetry cross. [From Griffin (1969).]

References

Bishop, C. J., Vandenbosch, R., Aley, R., Shaw, R. W., Jr., and Halpern, I. (1970). *Nucl. Phys. A* **150**, 129.

Bolsterli, M., Fiset, E. O., and Nix, J. R. (1969). *Proc. IAEA Symp. Phys. Chem. Fission, 2nd, Vienna, 1969*, p. 183. IAEA, Vienna.

Bolsterli, M., Fiset, E. O., Nix, J. R., and Norton, J. L. (1972). *Phys. Rev. C* **5**, 1050.
Brack, M., Damgaard, Jens, Jensen, A. S., Pauli, H. C., Strutinsky, V. M., and Wong, C. Y. (1972). *Rev. Mod. Phys.* **44**, 320.
Cohen, S., and Swiatecki, W. J. (1963). *Ann. Phys.* (*New York*) **22**, 406.
Fong, P. (1956). *Phys. Rev.* **102**, 434.
Fong, P. (1964). *Phys. Rev.* **135**, B 1338.
Griffin, J. J. (1969). *Proc. IAEA Symp. Phys. Chem. Fission, 2nd, Vienna, 1969*, p. 3. IAEA, Vienna.
Hasse, R. W. (1969). *Nucl. Phys. A* **128**, 609.
Hill, D. L. (1958). *Proc. U. N. Int. Conf. Peaceful Uses At. Energy, 2nd*, **15**, p. 244. United Nations, Geneva.
Hill, D. L., and Wheeler, J. A. (1953). *Phys. Rev.* **89**, 1102.
Ignatyuk, A. V. (1969). *Sov. J. Nucl. Phys.* **9**, 208.
Möller, P., and Nilsson, S. G. (1970). *Phys. Lett. B* **31**, 283.
Nix, J. R. (1963). *Proc. Conf. Reactions between Complex Nuclei, 3rd, Asilomar, California, 1963* (A. Ghiorso, R. M. Diamond, and H. E. Conzett, eds.), p. 366. Univ. of California Press, Berkeley.
Nix, J. R. (1967). *Ann. Phys.* (New York) **41**, 1.
Nix, J. R. (1969). *Nucl. Phys. A* **130**, 241.
Nix, J. R., and Swiatecki, W. J. (1965). *Nucl. Phys.* **71**, 1.
Pashkevich, V. V. (1971). *Nucl. Phys. A* **169**, 275.
Pauli, H. C., Ledergerber, T., and Brack, M. (1971). *Phys. Lett. B* **34**, 264.
Scharnweber, D., Mosel, U., and Greiner, W. (1970). *Phys. Rev. Lett.* **24**, 601.
Scharnweber, D., Greiner, W., and Mosel, U. (1971). *Nucl. Phys. A* **164**, 257.
Strutinsky, V. M. (1964a). *Sov. Phys. JETP* **18**, 1305.
Strutinsky, V. M. (1964b). *Sov. Phys. JETP* **18**, 1298.
Swiatecki, W. J. (1970). *Proc. Int. Conf. Nucl. Reactions Induced by Heavy Ions, Heidelberg, July 1969* (R. Bock and W. R. Hering, eds.), p. 729. North-Holland Publ., Amsterdam.
Tsang, C. F., and Wilhelmy, J. B. (1972). *Nucl. Phys. A* **184**, 417.
Wilets, L. (1964). "Theories of Nuclear Fission." Oxford Univ. Press, London and New York.

CHAPTER

Kinetic Energy Release in Fission

A. Dependence of the Total Kinetic Energy on the Charge and Mass of the Fissioning Nucleus

Of the available energy release in fission, by far the largest fraction goes into the kinetic energy of the fragments. A simple empirical correlation in which the total kinetic energy varies linearly with $Z^2/A^{1/3}$ was first noted for low energy fission by Terrell (1959). A dependence of this form can be qualitatively understood by assuming that the kinetic energy of the fragments is the result of the Coulomb repulsion of two spheres in contact

$$V_c = \frac{Z_1 Z_2 e^2}{R_1 + R_2} \tag{X-1}$$

Considering symmetric fission, $Z_1 = Z_2 = Z/2$ and $A_1 = A_2 = A/2$, and expressing the radius in terms of the mass number, $R = r_0 A^{1/3}$, we have

$$V_c = \frac{Z^2 e^2}{8(\frac{1}{2})^{1/3} r_0 A^{1/3}} \tag{X-2}$$

This equation accounts for the functional dependence of the kinetic energy on Z and A, although the magnitude of r_0 required to fit the absolute values of the kinetic energies is much too high. This is to be expected, since the fragments are assumed to be spherical rather than deformed. We should be cautious, however, and not read too much into this correlation, in that the observed dependence of kinetic energy release on mass division deviates greatly from that expected on the basis of Eq. (X-1). Also, theoretical calculations indicate that for high Z^2/A values a significant fraction of the final total kinetic energy has been developed prior to scission.

A summary of most probable total kinetic energy data, averaged over all possible mass divisions, is given for a large variety of fissioning nuclei in Table X-1. These results are also illustrated in Fig. X-1. The values given are

TABLE X-1

Total Preneutron Emission Kinetic Energy Release in Fission[a]

Compound nucleus	Target	Projectile	$Z^2/A^{1/3}$	E^*	Reported $[E_k]$		Corrected $[E_k]$
^{157}Ho	^{141}Pr	166-MeV ^{16}O	832.1	126	111.2	±6.6	114.7
^{171}Lu	^{159}Tb	125-MeV ^{12}C	908.2	106	117.5	±4.8	120.9
^{175}Ta	^{159}Tb	166-MeV ^{16}O	952.7	131	124.3	±4.4	127.7
^{177}Ta	^{165}Ho	125-MeV ^{12}C	949.1	104	122.0	±5.2	125.4
^{181}Re	^{169}Tm	125-MeV ^{12}C	994.4	102	122.3	±4.2	125.7
		166-MeV ^{16}O		129	127.3	±5.4	130.7
^{186}Os	^{174}Yb	109-MeV ^{12}C	1012	88	127	±6	—
		125-MeV ^{12}C		103	129	±6	—
	^{170}Er	120-MeV ^{16}O		88	124	±6	—
		136-MeV ^{16}O		103	124	±6	—
		151-MeV ^{16}O		116	128	±6	—
		166-MeV ^{16}O		129	127	±6	—
^{185}Ir	^{169}Tm	166-MeV ^{16}O	1041	127	132.2	±4.0	135.5
^{187}Ir	^{175}Lu	125-MeV ^{12}C	1037	101	131.4	±3.6	134.7
^{191}Au	^{175}Lu	166-MeV ^{16}O	1084	126	137.8	±3.9	141.1
^{200}Tl	^{197}Au	25.5-MeV ^{3}He	1122	36	140.3	±2.0	143.7
^{201}Tl	^{197}Au	43-MeV ^{4}He	1120	41	138	±4	140.0
		70-MeV ^{4}He		67	142.1	±4	—
^{198}Pb	^{182}W	102-MeV ^{16}O	1154	68	144	±5	—
		115-MeV ^{16}O		80	144	±5	—
		127-MeV ^{16}O		92	146	±5	—
		144-MeV ^{16}O		106	146	±5	—
		165-MeV ^{16}O		126	147	±5	—
^{208}Bi	nat. Tl	25.5-MeV ^{3}He	1163	36	141.7	±3.0	145.1
^{209}Bi	nat. Tl	43-MeV ^{4}He	1161	39	143	±5	145.0
^{209}Po	^{206}Pb	25.5-MeV ^{3}He	1189	35	145.5	±2.0	148.8
^{211}Po	nat. Pb	43-MeV ^{4}He	1185	34	146	±5	148.0
^{211}Po	^{209}Bi	21.5-MeV ^{2}H	1185	29	143	±5	145.0

TABLE X-1 (continued)

Compound nucleus	Target	Projectile	$Z^2/A^{1/3}$	E^*	Reported $[E_k]$		Corrected $[E_k]$
^{209}At	^{197}Au	125-MeV ^{12}C	1218	85	147.2	±3.6	150.5
^{212}At	^{209}Bi	22.1-MeV ^{3}He	1212	27	146.5	±2.0	149.8
		25.5-MeV ^{3}He		31	147.3	±2.0	150.9
^{213}At	^{209}Bi	42-MeV ^{4}He	1210	31	150	±3	152.0
		43-MeV ^{4}He		32	148	±4	150.0
		65-MeV ^{4}He		54	150.1	±4	—
^{213}Fr	^{197}Au	166-MeV ^{12}C	1267	124	155.6	±3.4	158.6
^{221}Ac	^{209}Bi	125-MeV ^{12}C	1310	92	158.7	±3.0	161.6
^{228}Ac	^{226}Ra	9.8-MeV ^{2}H	1297	18	155.8	±2.0	159.0
		11.7-MeV ^{2}H		20	154.9	±2.0	158.1
		14.0-MeV ^{2}H		22	154.9	±2.0	158.1
^{229}Th	^{226}Ra	20.9-MeV ^{3}He	1324	30	157.7	±2.0	160.9
		23.9-MeV ^{3}He		32	158.9	±2.0	162.1
^{230}Th	^{226}Ra	22.1-MeV ^{4}He	1322	17	159.7	±2.0	162.9
		27.1-MeV ^{4}He		22	160.3	±2.0	163.5
		30.8-MeV ^{4}He		26	165	±3	166.5
		33.7-MeV ^{4}He		29	165	±4	166.5
^{225}Pa	^{209}Bi	166-MeV ^{16}O	1362	115	166.3	±3.0	169.1
^{231}Pa	^{230}Th	6.8-MeV ^{1}H	1350	12	162.4	±2.0	165.6
		8.0-MeV ^{1}H		13	162.6	±2.0	165.8
^{232}Pa	^{230}Th	12.0-MeV ^{2}H	1348	20	166.3	±2.0	—
		14.0-MeV ^{2}H		22	167.3	±2.0	—
^{234}U	^{233}U	Thermal n	1374	5.9	169.4	±1.7	—
	^{230}Th	25.7-MeV ^{4}He		20	167.5	±2.0	—
				20	169.6	±2.0	—
		29.5-MeV ^{4}He		24	166.0	±2.0	—
				24	169.0	±2.0	—
^{236}U	^{235}U	Thermal n	1370	5.3	168.3	±1.7	—
	^{232}Th	21.4-MeV ^{4}He		16	172.5	±3.2	—
		21.8-MeV ^{4}He		17	169.1	±2.0	—
		22.1-MeV ^{4}He		17	171.4	±2.0	—
				18.9	170.5	±3.5	—
		25.7-MeV ^{4}He		21	168.2	±2.0	—
					171.1	±2.0	—
					171.3	±3.2	—
		29.5-MeV ^{4}He		24	167.0	±2.0	—
				24	170.7	±2.0	—
		33.0-MeV ^{4}He		28	170.5	±3.5	—
		65.0-MeV ^{4}He		59	168.0	±4.5	—
^{237}Pu	^{233}U	21.8-MeV ^{4}He	1428	16	176.3	±2.0	—
		22.1-MeV ^{4}He		16	175.5	±2.0	—
		25.5-MeV ^{4}He		20	170.3	±2.0	173.4

TABLE X-1 (continued)

Compound nucleus	Target	Projectile	$Z^2/A^{1/3}$	E^*	Reported $[E_k]$		Corrected $[E_k]$
		25.7-MeV ^{4}He		20	174.9	±2.0	—
				20	174.6	±2.0	—
		29.5-MeV ^{4}He		24	174.2	±2.0	—
		29.7-MeV ^{4}He		24	173.9	±2.0	—
^{240}Pu	^{239}Pu	Thermal n	1422	6.5	175.0	±1.7	—
^{242}Pu	^{238}U	25.7-MeV ^{4}He	1418	20	176.0	±3.2	—
		27.8-MeV ^{4}He		22	175	±5	—
		29.4-MeV ^{4}He		24	173	±4	—
		30.4-MeV ^{4}He		25	175	±5	—
		32.6-MeV ^{4}He		27	175	±5	—
		33.0-MeV ^{4}He		28	174.7	±3.5	—
		34.2-MeV ^{4}He		29	177	±5	—
		35.6-MeV ^{4}He		30	176.5	±5	—
		42.0-MeV ^{4}He		36	171	±4	173.3
		65.0-MeV ^{4}He		59	173.0	±4.0	—
^{248}Cm	SF		1467	0	179	±2	181.7
^{248}Cf	SF		1529	0	188.7	±1.3	189.9
^{250}Cf	SF		1525	0	185	±3	187.7
	^{238}U	125-MeV ^{12}C		95	183.2	±4.0	185.5
^{252}Cf	SF		1521	0	185.7	±0.1	—
	SF			0	185.5	±1.0	—
	SF			0	186.5	±1.2	—
^{254}Cf	SF		1517	0	185	±2	187.7
	SF			0	186.1	±2.8	187.3
^{253}Es	SF		1550	0	188	±3	190.7
^{252}Fm	^{240}Pu	125-MeV ^{12}C	1583	98	185.6	±4.0	187.9
^{254}Fm	SF		1579	0	189	±2	191.7
	^{238}U	166-MeV ^{16}O		118	185.6	±2.0	187.9
^{256}102	^{240}Pu	166-MeV ^{16}O	1639	114	192.7	±4.6	194.9

[a] When possible the data have been corrected to a common energy standard based on Whetstone's $[E_k] = 185.7$ MeV value for ^{252}Cf. Corrections have also been made for neutron emission and center-of-mass motion. The symbol $[E_k]$ designates the most probable total kinetic energy. [From Viola (1966), where the references to the original work may be found.]

for the sum of the fragment energies prior to neutron emission, which for low energy fission are typically 1–2% higher than postneutron emission kinetic energies. The inclusion of data for lighter elements in this figure has shown that an improved empirical correlation is obtained if a constant term is added

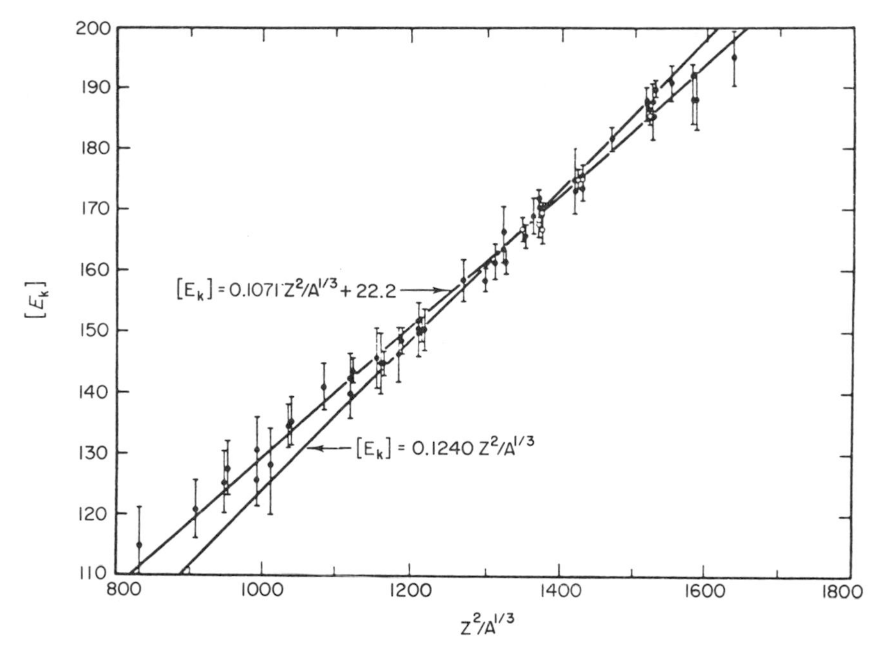

Fig. X-1. Dependence of $[E_k]$, the most probable value of the total fission-fragment kinetic energy, on the Coulomb energy parameter $Z^2/A^{1/3}$. Z and A are the charge and mass number of the fissioning (compound) nucleus. ● Semiconductor data, ○ time-of-flight data. Plotted points are corrected values (or reported values, if no correction was necessary) listed in Table X-1. Where measurements have been made on the same compound nucleus at more than one excitation energy by the same experimenters, only the average of these values is plotted. [From Viola (1966).]

to the term proportional to $Z^2/A^{1/3}$. A more recent summary of kinetic energy data for low energy fission of heavy elements has been compiled by Unik and Gindler (1971).

There are sizeable variations in the kinetic energy release from event to event. These variations simply reflect the distribution of nuclear elongations at the scission point. A typical value of the rms deviation from the most probable or average total kinetic energy for a fixed mass ratio is 8 MeV. This corresponds to a rms deviation in the fragment separation distance of only about 5% or about 1 F. The width of the total kinetic energy distribution curve is found to increase with both the excitation energy of the fissioning nucleus and the Z^2/A value of the fissioning nucleus. The latter increase is largely a consequence of the strong variations in the average energy release with mass ratio observed for the heavier elements. A comparison of experimental values of the widths of the total kinetic energy distributions with predictions of a liquid drop model have been presented in Section IX, A.

B. Dependence of the Total Kinetic Energy on the Mass Division

Recently very extensive measurements of the dependence of the total kinetic energy on mass division have been performed. For low energy fission of uranium and plutonium these measurements confirm a large dip in kinetic energy for symmetric mass divisions and a prominent peak in the kinetic energy release for asymmetric fission leading to heavy fragments with $A \simeq 132$. An example of such a result is shown in Fig. X-2. The dip is approximately 20 MeV relative to the peak at $A \simeq 132$. This is much larger than the 2 MeV differences in the maximum available energy release between symmetric and asymmetric fission as estimated from mass equations (see Fig. X-2a). The dependence on mass split predicted by the simple touching-spheres model is also illustrated in this figure. The falloff in kinetic energy release for very asymmetric mass splits has a slope similar to that of the available energy release curve, suggesting that simple energetic limitations have become relevant for these divisions. The single fragment kinetic energies and the rms width of the total kinetic energy distributions are illustrated, along with the mass yield curve, in Fig. X-3. The results of a similar experiment for spontaneous fission of ^{252}Cf are illustrated in Fig. X-4. The energy deficit

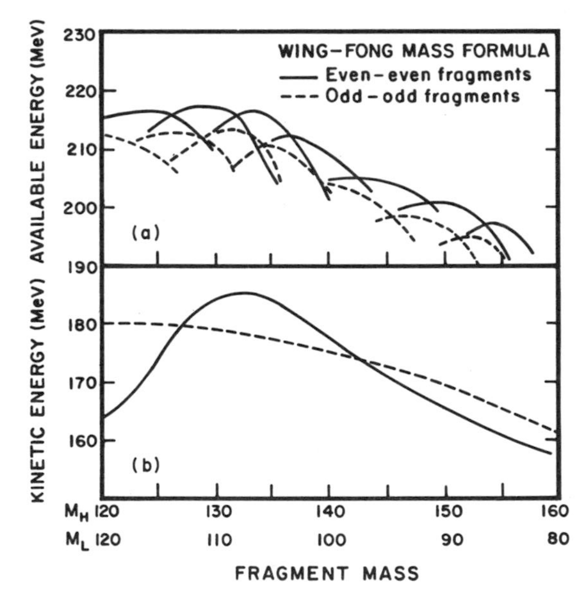

Fig. X-2. Total kinetic energy (b) and available energy release (a) as a function of fragment mass for the thermal neutron-induced fission of ^{239}Pu. [After Neiler *et al.* (1966).] The dependence on mass predicted by $E_k \propto Z^2/(A_1^{1/3}+A_2^{1/3})$ is indicated by the broken line (with arbitrary normalization).

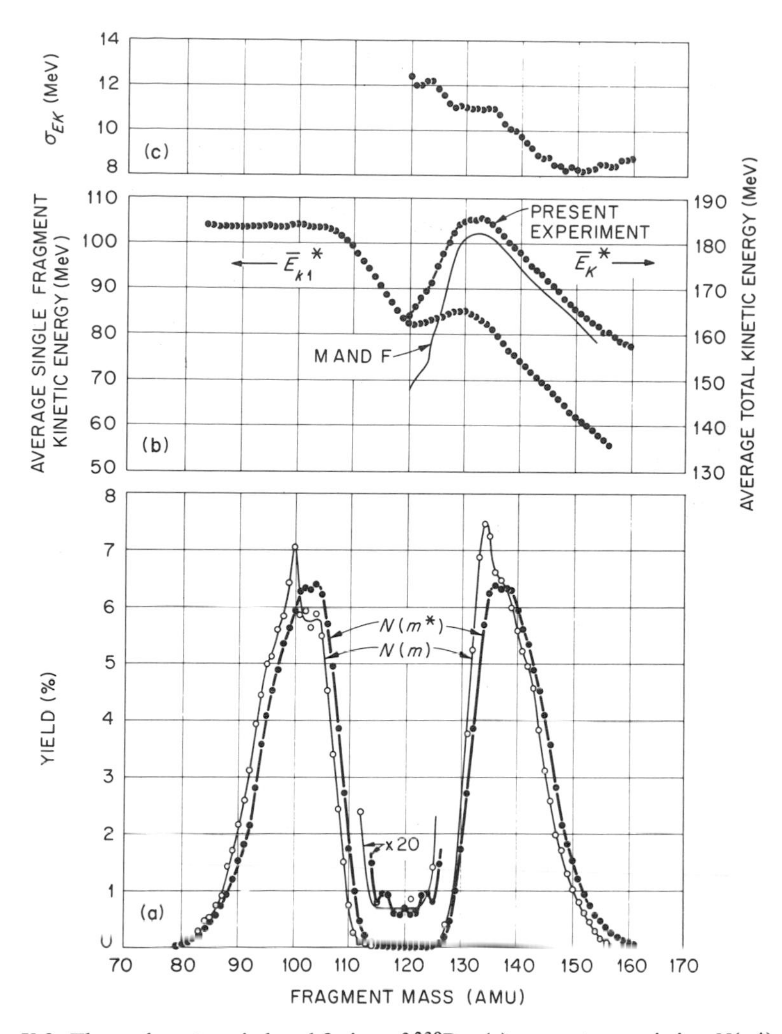

Fig. X-3. Thermal neutron-induced fission of ^{239}Pu: (a) preneutron emission $N(m^*)$ and postneutron emission $N(m)$ mass distributions; (b) average single fragment and total kinetic energies (the thin solid line refers to earlier time-of-flight measurements of Milton and Fraser); and (c) rms width of total kinetic energy distribution. [From Neiler *et al.* (1966).]

at symmetry is much less for ^{252}Cf, with, however, the maximum in the kinetic energy release again occurring for $A \sim 132$. Recent results for spontaneous and neutron-induced fission of ^{257}Fm (Balagna *et al.*, 1971; John *et al.*, 1971) indicate that as the fissioning mass approaches twice 132, the total kinetic energy release for symmetric mass divisions will be very large. The remarkable stability of the heavy fragment mass number for which the kinetic energy has

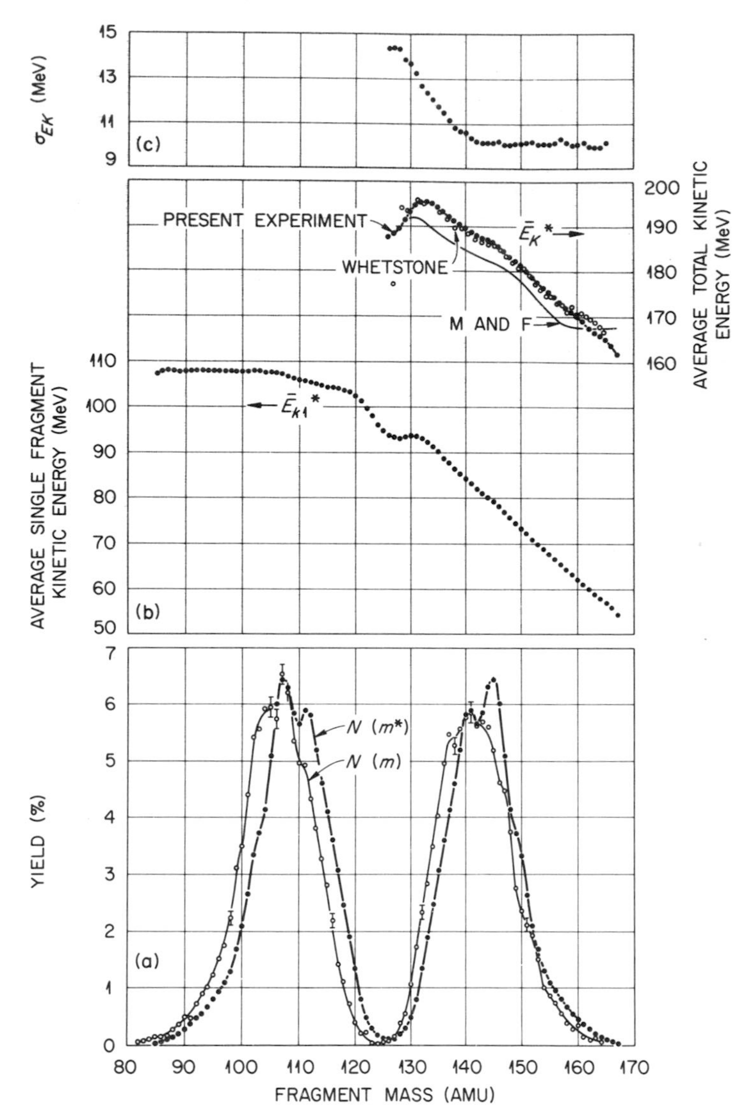

Fig. X-4. Spontaneous fission of ^{252}Cf. See caption to Fig. X-3. [From Schmitt *et al.* (1966).]

a maximum is illustrated in Fig. X-5. This strongly suggests that the shell structure of this fragment is perturbing the scission shape so as to result in a higher Coulomb interaction energy.

The effect of increasing excitation energy is to decrease the structure in the variation of the kinetic energy release with excitation energy, as illustrated in

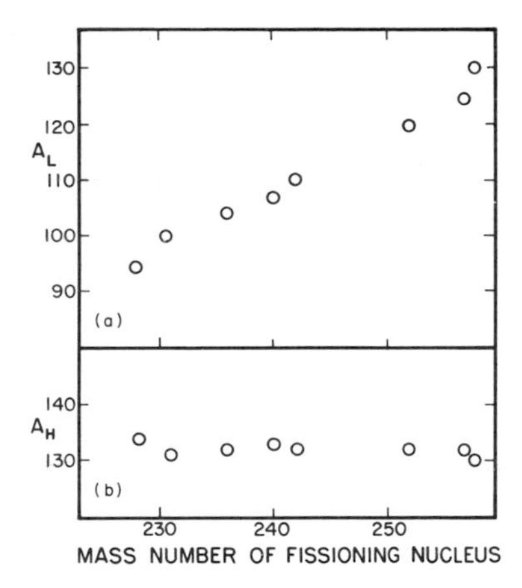

Fig. X-5. Mass numbers of the complementary light (a) and heavy (b) fragments for which the total kinetic energy is a maximum as functions of the mass of the fissioning nucleus.

Fig. X-6. Since the effect of excitation energy is to destroy special properties having to do with certain numbers of neutrons and protons, the fact that the kinetic energy release both decreases for $A \simeq 132$ and increases for symmetric divisions suggests that shell effects are operating which can make nascent fragments both more and less stable with respect to distortion than "average" nuclei.

For targets lighter than radium which exhibit symmetric mass yield curves, the maximum kinetic energy release occurs for symmetric mass divisions and falls off with increasing mass asymmetry at a rate dependent on the excitation

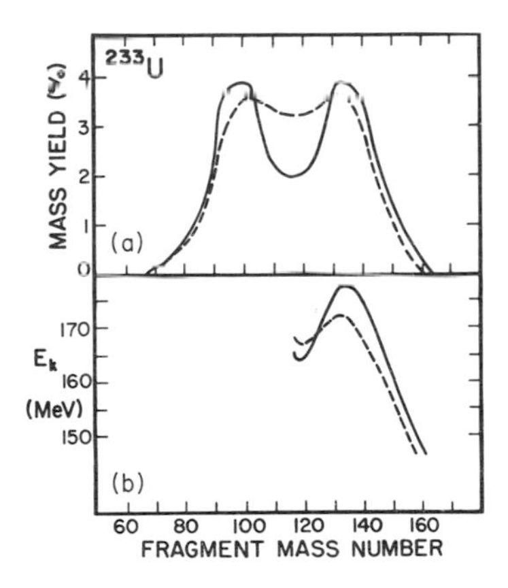

Fig. X-6. Excitation energy dependence of the mass distribution and the variation of kinetic energy release with mass division. Curves represent 9.5-MeV (—) and 20-MeV (---) proton-induced fission of ^{233}U. [From Bishop *et al.* (1970).]

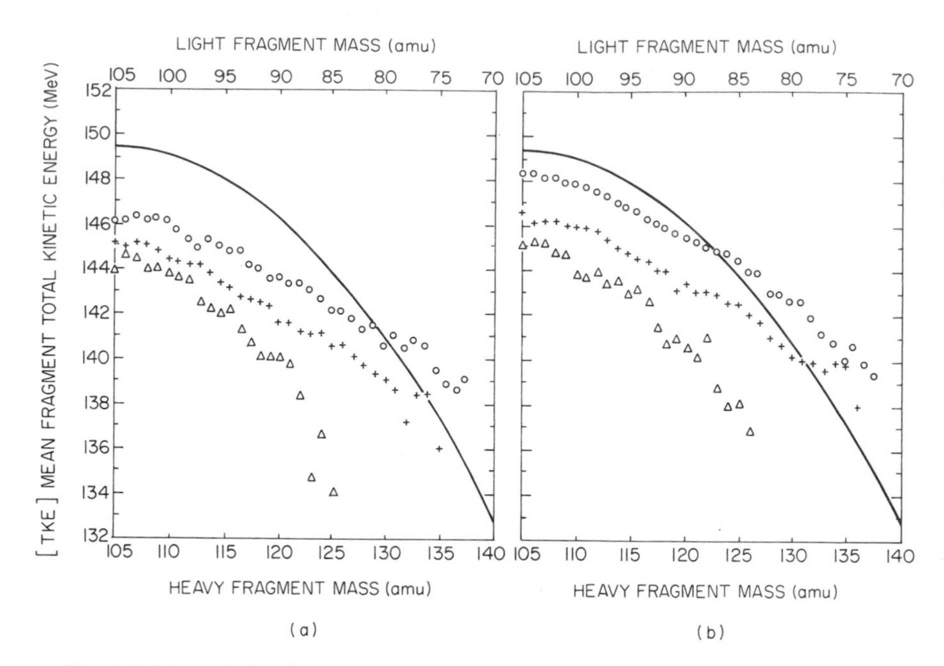

Fig. X-7. Mean fission-fragment total kinetic energy as a function of fragment mass for fission of ^{210}Po at three excitation energies (E^*). (a) $^{209}\text{Bi}+\text{p}\rightarrow{}^{210}\text{Po}^* \xrightarrow{f}$ ($\triangle$: $E^* = 31$ MeV); (b) $^{206}\text{Pb}+\alpha\rightarrow{}^{210}\text{Po}^* \xrightarrow{f}$ ($+$: $E^* = 44$ MeV, ○: $E^* = 57$ MeV). Solid curves are from the theoretical work of Nix and Swiatecki. [From Unik *et al.* (1969).]

energy. This is illustrated in Fig. X-7 for the compound nucleus ^{210}Po. No significant increase in kinetic energy for fragments with $A \simeq 132$ is observed, although this may simply reflect the fact that the yield of fragments in this mass region does not become sufficiently large to determine the kinetic energy release until the initial compound nuclear excitation energy is very large.

The variation of the width of the kinetic energy distribution with mass division is poorly understood. A peak in the width is usually observed in the region between symmetric fission and fission leading to fragments with $A \simeq 132$. This is illustrated by the solid points in the upper part of Fig. X-3. This is just the transition region where shell effects favoring unusually large distortions (mid-shell fragments) are reversing to shell effects favoring small distortions ($A \sim 132$; closed shell).

Britt *et al.* (1963) have had some success in relating the variation in the width with mass ratio to the proportions of fission arising from each of two "modes." This approach has been applied primarily to fission of nuclei, such as ^{226}Ra, which exhibit mass distributions with both symmetric and asymmetric peaks. It is assumed that each of the two modes is characterized by a breaking distance D, which is constant for all mass ratios for which that mode contributes. The average kinetic energy is assumed to be given by $Z_1 Z_2 e^2/D$. The D values for the symmetric and asymmetric modes are deduced by fitting the observed kinetic energies for symmetric and very asymmetric splits, respectively. The breaking distance is found to be larger for the symmetric mode, resulting in a lower total kinetic energy release for the symmetric mode than for the asymmetric mode. This is the only characteristic distinction implied by the use of the term "mode" in the present context. The rise in the measured values of the variance of the kinetic energy distribution for mass ratios in the transition region is attributed to the superposition of two distributions with the same variance but different average kinetic energies. The mass distributions are decomposed into components associated with the symmetric and asymmetric modes, as illustrated in Fig. X-8a. If it is assumed that the kinetic

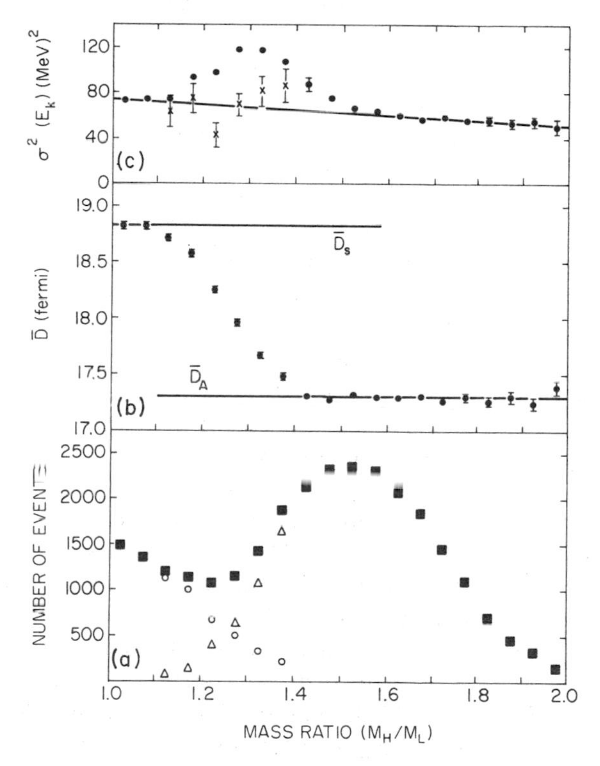

Fig. X-8. The yield (a), separation distance $\bar{D}$ (b), and variance of the total kinetic energy distributions (c) as functions of the mass ratio of the fragments from 27.1-MeV ^{4}He-induced fission of ^{226}Ra. (a) ■: Measured yield values, ○: yield values attributed to the symmetric mode, △: yield values attributed to the asymmetric mode. (c) Measured values (●) and corrected values (×) for $\sigma_i^2(E_K)$. The success of the decomposition into the two modes is the extent to which the crosses follow the smooth line in (c). [From Britt *et al.* (1963).]

energy distributions for a given mass ratio for each mode have a Gaussian shape and that they have the same variances $\sigma_i^2(E_k)$, then the contribution to the measured values $\sigma^2(E_k)$ because of the two distributions is given by $\sigma^2(E_k) - \sigma_i^2(E_k) = F(1-F)(E_k^a - E_k^s)$ where F is the fraction of fissions due to the symmetric mode and E_k^s and E_k^a are the average total kinetic energy releases for the symmetric and asymmetric modes, respectively. When the experimental variances are corrected for this contribution, the points indicated by the crosses ($\times$) in Fig. X-8c are obtained. If we attempt to extend this approach to fission of thorium and uranium targets, it is found that the separation distance D is no longer independent of mass ratio for the asymmetric (and dominant) mode.

C. Dependence of the Total Kinetic Energy on Excitation Energy

The relative insensitivity of the kinetic energy release to excitation energy was observed very early in the history of fission studies by comparison of the ranges of fission products from 350-MeV proton-induced fission with those from thermal neutron-induced fission. The near equality of the ranges, and hence of the kinetic energies, shows that most of the additional available excitation energy does not go into collective motion in the fission degree of freedom. This simply reflects the fact that there are many more particle degrees of freedom which can also be excited. The near constancy of the kinetic energy release also implies that the mean distance between fragments at scission is nearly independent of excitation energy.

More precise studies of the dependence of kinetic energy release on excitation energy for heavy element fission have indicated small variations in both directions in the vicinity of the fission barrier. Unik and Loveland (1971) have measured the rate of change of the total kinetic energy (TKE) with excitation energy (E^*) for the fissioning nucleus ^{240}Pu over the excitation energy interval between 4.8 and 9.0 MeV. The values of $\{d(\text{TKE})/dE^*\}$ depend on the reaction employed, with a value of -0.31 ± 0.04 MeV for the $^{239}Pu(d, pf)$ reaction and a value of -1.1 ± 0.1 MeV for the $(\alpha, \alpha'f)$ reaction. The difference between the two reactions may be related to the fact that different spin states are populated in the two reactions. There is also evidence (Okolovich and Smirenkin, 1963) that the kinetic energy release in spontaneous fission of ^{240}Pu is 1.5 ± 0.5 MeV lower than for thermal neutron fission of ^{239}Pu. This last result might suggest that the fissioning nucleus in the case of over-the-barrier fission has picked up only 1 or 2 MeV of kinetic energy by the time it reaches the point between saddle and scission corresponding to the emergence from the barrier in spontaneous fission. The interpretation of these results is complicated by the facts that the dependence of the total kinetic energy on excitation energy is different for different mass splits, and that the mass distribution itself changes with excitation energy.

An investigation of the dependence of the kinetic energy on excitation energy for lighter elements where the mass distribution exhibits a symmetric peak has revealed an increase of the kinetic energy release with excitation energy. Unik *et al.* (1969) found an increase in the average total kinetic energy, after correction for angular momentum effects, of 0.07 MeV per MeV increase in excitation energy. Even this variation is larger than expected on the basis of a statistical distribution of the available energy, which would give a kinetic energy of $0.5T$ in the fission mode at the saddle point. This may be due in part to complications arising from the dependence of the kinetic energy release on mass ratio. The increase in kinetic energy with increasing excitation energy is larger for asymmetric mass divisions than for symmetric divisions. This may reflect the limitations on the available energy release as the mass asymmetry becomes large. The Q values are much smaller for very asymmetric mass divisions than for symmetric divisions. The more rapid increase in the average total kinetic energy with excitation energy than expected on purely statistical grounds may also be partly due to changes in the nuclear tensile strength associated with increased excitation energy. If the tensile strength decreases with increasing excitation energy, the neck will rupture at smaller distortions where the Coulomb forces are stronger. This would result in an increase of kinetic energy with excitation energy (Unik and Huizenga, 1964).

The dependence of the widths of the total kinetic energy and mass distributions on angular momentum have also been examined in the aforementioned study. No significant variation of the widths with angular momentum were observed.

D. Dependence of the Total Kinetic Energy on Angular Momentum

Very little is known experimentally about the dependence of the kinetic energy release on angular momentum. There has been one study (Unik *et al.*, 1969) in which the compound nucleus ^{210}Po was formed at the same excitation energy by three different reactions. The average squared angular momentum was quite different for the various reactions, being smallest for proton bombardment of ^{209}Bi and largest for ^{12}C bombardment of ^{198}Pt. The results of this study are illustrated in Fig. X-9. Only in the case where the energies of the three projectiles were chosen to produce ^{210}Po at 58 MeV of excitation energy were the reactions studied with three different angular momentum distributions. These results suggest that the total kinetic energy is proportional to the average angular momentum. Such a dependence is reasonable if we assume that the final total kinetic energy is the resultant of both the Coulomb and centrifugal forces at scission. The magnitude of the effect is appreciably larger, however, than might be expected on the basis of the simplest considerations. As shown in Section VII,C,1, the average rotational energy at the saddle point is given by

$$R_s^0 = (\hbar^2 \langle J^2 \rangle / 2\mathscr{I}_\perp) + T/2$$

For this less fissionable nucleus the nuclear shape at scission may not be much different from that of the saddle point. The slopes of the lines in Fig. X-9 give $\hbar^2/2\mathscr{I} = 0.006$ MeV, which is several times larger than would be expected for the elongated nucleus at the saddle point.

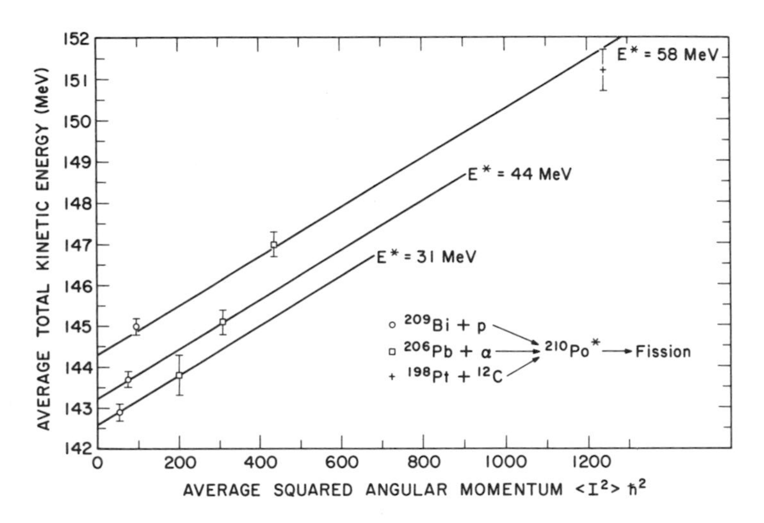

Fig. X-9. Dependence of the average total kinetic energy on the average squared angular momentum. The angular momentum associated with the $^{12}C + ^{198}Pt$ reaction is somewhat uncertain. [From Unik *et al.* (1969, 1971).]

There are several possible factors which may account for this discrepancy. As mentioned in Section VII,C,1, the presence of angular momentum results in a saddle configuration which is less distorted than for no angular momentum. If this is also true at scission, or if the descent from saddle to scission is fairly adiabatic, the Coulomb forces will be larger, resulting in higher total kinetic energies. Another contributing factor may be the weakening of the pairing forces by angular momentum due to the Coriolis forces (Mottelson and Valatin, 1960). Such an effect might cause the neck to snap earlier, in which case a higher kinetic energy release is possible. Other effects of the angular momentum on the dynamics are also likely.

E. Dependence of the Total Kinetic Energy Release on Charge Division

It is possible by fragment range measurements to determine the kinetic energy release for charge splits which differ from the most probable split. Since improbable charge splits are associated with smaller energy release, a decrease in kinetic energy is expected. Measurements (Niday, 1961; Nakahara *et al.*, 1969) of the kinetic energy decrease have been made for the shielded nuclides ^{86}Rb and ^{136}Cs in thermal neutron-induced fission of ^{233}U and ^{235}U, with energy deficits of 5–10 MeV. Part of this deficit, particularly in the case of ^{136}Cs, is attributed to the fact that these species are formed in events in which a larger than average number of neutrons are emitted.

References

Balagna, J. P., Ford, G. P., Hoffman, D. C., and Knight, J. D. (1971). *Phys. Rev. Lett.* **26,** 145.

Bishop, C. J., Vandenbosch, R., Aley, R., Shaw, R. W., Jr., and Halpern, I. (1970). *Nucl. Phys. A* **150,** 129.

Britt, H. C., Wegner, H. E., and Gursky, J. C. (1963). *Phys. Rev.* **129,** 2239.

John, W., Hulet, E. K., Lougheed, R. W., and Wesolowski, J. J. (1971). *Phys. Rev. Lett.* **27,** 45.

Mottelson, B. R., and Valatin, J. G. (1960). *Phys. Rev. Lett.* **5,** 511.

Nakahara, H., Harvey, J. W., and Gordon, G. E. (1969). *Can. J. Phys.* **47,** 2371.

Neiler, J. H., Walter, F. J., and Schmitt, H. W. (1966). *Phys. Rev.* **149,** 894.

Niday, J. B. (1961). *Phys. Rev.* **121,** 1471.

Okolovich, V. N., and Smirenkin, G. N. (1963). *Sov. Phys. JETP* **16,** 1313.

Schmitt, H. W., Neiler, J. H., and Walter, F. J. (1966). *Phys. Rev.* **141,** 1146.

Terrell, J. (1959). *Phys. Rev.* **113,** 527.

Unik, J. P., and Gindler, J. E. (1971). Rep. ANL–7748. Argonne Nat. Lab., Argonne, Illinois.

Unik, J. P., and Huizenga, J. R. (1964). *Phys. Rev. B* **134,** 90.

Unik, J. P., and Loveland, W. D. (1971). Private communication.

Unik, J. P., Cuninghame, J. G., and Croall, I. F. (1969), *Proc. IAEA Symp. Phys. Chem. Fission, 2nd, Vienna, 1969*, p. 717. IAEA, Vienna.

Unik, J. P., Cunninghame, J. G., and Croall, I. F. (1971). Private communication.

Viola, V. E., Jr. (1966). *Nucl. Data Sect. A* **1,** 391.

CHAPTER

XI

Distribution of Mass and Charge in Fission

A. Mass Distributions

1. Introduction

The mass distribution is probably the most important characteristic of the fission process. Although a large amount of experimental data on the mass distribution in fission for various nuclei under a variety of conditions has been available for a number of years, no suitable theory yet exists which explains all the observations. It is well known experimentally, for example, that the mass distribution for a particular system changes with excitation energy and that the mass distribution depends on the mass number A of the system undergoing fission. A number of theoretical models have been suggested to explain these data (without complete success, as we discussed in Chapter IX).

In this chapter we present some of the experimental information on the fission mass distribution. The terminology to be used in this section is summarized in the following sentences. The term primary fission fragment refers to the nuclear species before the emission of prompt neutrons. The term primary fission product, on the other hand, refers to the nucleus formed after neutron emission but before any β decay has occurred, while a secondary fission product is one which is formed from a primary fission product by at least one β decay. The general term fission product is used when referring to either primary or secondary fission products or a combination of such products.

2. Spontaneous Fission

Radiochemical data on the mass distributions associated with the spontaneous fission of ^{238}U, ^{240}Pu, ^{242}Cm, and ^{252}Cf have been summarized by von Gunten (1969). In the case of ^{252}Cf, a symmetric yield was established with a resulting peak-to-valley ratio of about 750. In all the other cases only an upper limit on the symmetric yield was established. More recently, radiochemical measurements on the mass yields from the spontaneous fission of ^{244}Cm and ^{256}Fm have been reported by Flynn *et al.* (1972a, b). The fission product yields for ^{244}Cm and ^{256}Fm are shown in Figs. XI-1 and XI-2, respectively. In each case, the fission product yields are compared with those of ^{252}Cf. In the case of ^{244}Cm the upper limit for the yield of ^{121}Sn is 10^{-3}%. This is the lowest symmetric fission product yield known (and in this case, is

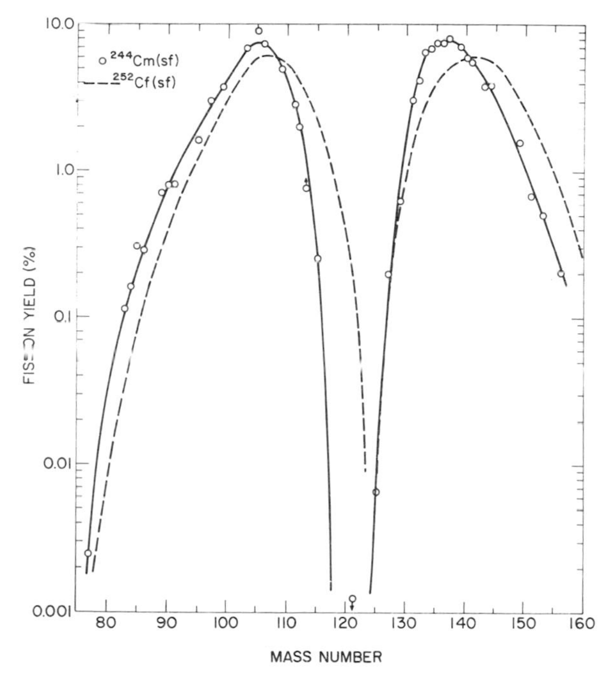

Fig. XI-1. Fission product yield (%) as a function of mass number for the spontaneous fission of ^{244}Cm. The data are from radiochemical determinations except for the yields of isotopes of Kr and Xe, which were determined mass spectrometrically. Fission product yields as a function of mass number for the spontaneous fission of ^{252}Cf is included for comparison. [After Flynn *et al.* (1972a).]

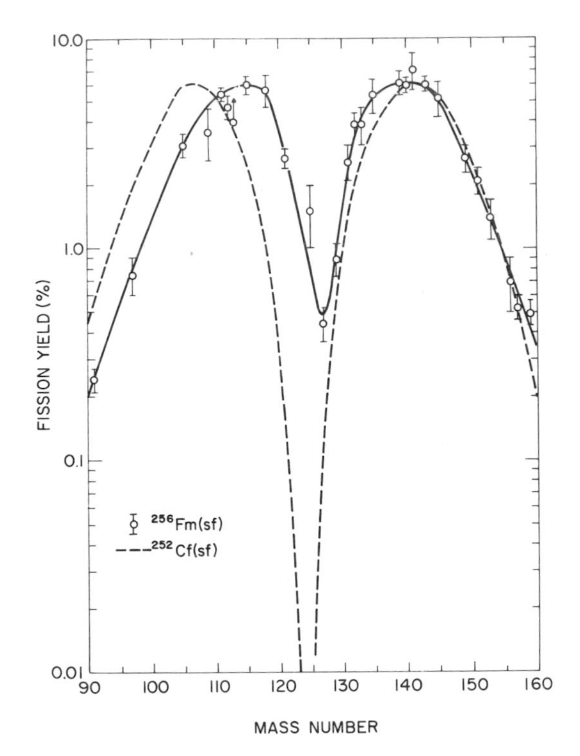

Fig. XI-2. Fission product yield (%) as a function of mass number for the spontaneous fission of ^{256}Fm. The data were obtained by radiochemical techniques. Fission product yields as a function of mass number for the spontaneous fission of ^{252}Cf is included for comparison. [After Flynn *et al.* (1972b).]

only an upper limit). The fission product yields near symmetry for spontaneous fission of ^{256}Fm are orders of magnitude larger than similar yields for the other nuclides examined by radiochemical techniques. The peak-to-valley yield as shown in Fig. XI-2 is approximately 12.

Primary fission product yields have been reported by Balagna *et al.* (1971) for the spontaneous fission of ^{257}Fm. These results were obtained by coincident kinetic energy measurements of the primary fission products (after neutron emission). The postneutron emission masses can be converted to the primary fission-fragment masses (preneutron emission masses) by making some assumptions about the neutron yield as a function of fragment mass. The resulting yield curves are shown in Fig. XI-3 for two assumptions about the neutron yield. As can be seen from the results plotted in Fig. XI-3, the mass yield near symmetry deduced from the double kinetic energy technique is quite sensitive to the assumption made about the neutron yield.

The remarkable change in the peak-to-valley ratio with increasing A for three fermium isotopes is illustrated in Fig. XI-4. Although the most symmetric distribution is associated with thermal neutron-induced fission (John *et al.*, 1971) of ^{257}Fm, it is included in this figure to show the trend toward symmetric fission with increasing A and/or excitation energy.

3. Particle-Induced Fission

Recently, a compilation of yields of fission products for several fissionable nuclides at different incident neutron energies was published (Flynn and Glendenin, 1970). In Table XI-1 we reproduce the total chain yields for neutron fission of ^{235}U at three energies. These yields are based on information obtained from both radiochemical and mass spectrometric techniques. The estimated errors in the fission yield values range from 5% to 20%. As we might expect, the accuracy is highest in cases such as mass spectrometric measure-

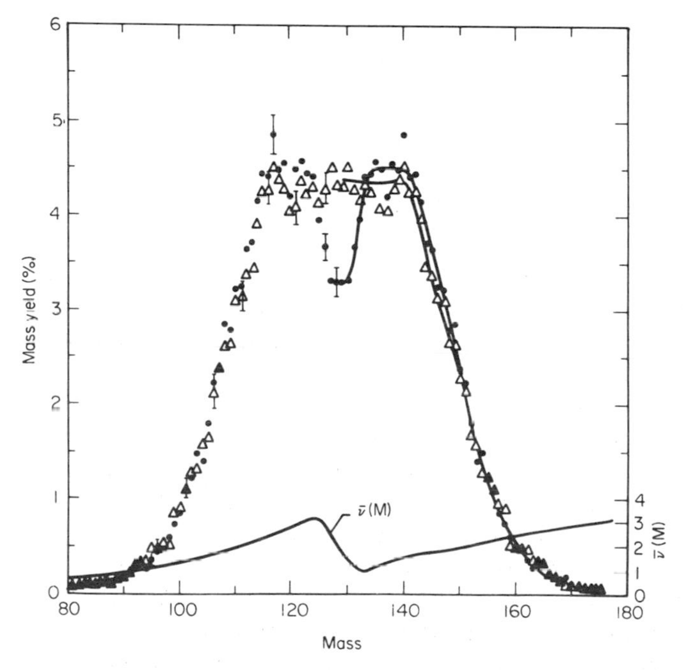

Fig. XI-3. Primary fission-fragment yields (preneutron emission mass) as a function of mass number for the spontaneous fission of ^{257}Fm. Results were deduced from coincident kinetic energy measurements. The primary fission-fragment yields are displayed for two assumptions about the neutron yields. The open triangles are based on the assumption of a constant average neutron yield as a function of mass, $\bar{\nu} = 2$. The solid circles are based on the neutron yield function plotted at the bottom of the graph. [After Balagna *et al.* (1971).]

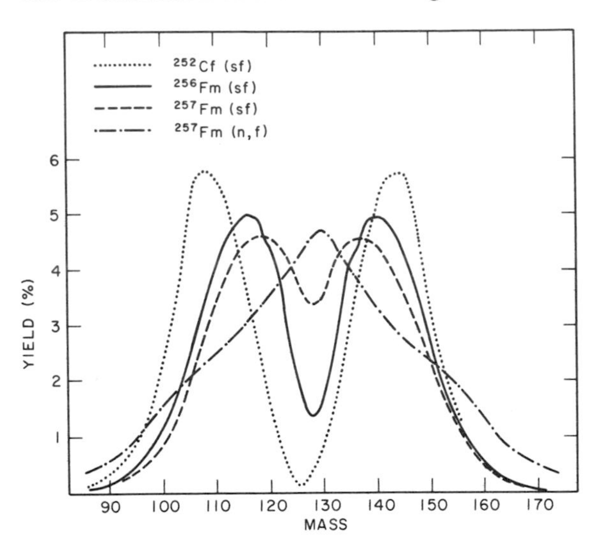

Fig. XI-4. Primary fission-fragment yields (preneutron emission masses) as a function of mass number for the spontaneous fission of ^{252}Cf, ^{256}Fm, and ^{257}Fm and for the thermal neutron-induced fission of ^{257}Fm. The results were deduced from coincident kinetic energy measurements. This figure displays the trend toward symmetric fission. [After Glendenin (1971).]

ments of rare gas (krypton and xenon) yields and in select cases of a few radiochemical measurements where determinations based on absolute fission counting have been made by several independent workers. The fission product in column 2 of Table XI-1 indicates the nucleus which was measured for a particular mass chain. This fission product nuclide (or nuclides for some mass numbers) is assumed to give the integrated yield for this mass number. The numbers in parentheses represent interpolated chain yields for those masses where no measurements exist. For thermal neutron-induced fission of ^{235}U the sums over the chain yields for masses in the light and heavy groups are 99.8% and 100.9%, respectively, very close to the desired value of 100%.

The mass yield curves for the fission of ^{235}U with thermal neutrons, fission-spectrum neutrons, and 14-MeV netrons are shown in Figs. XI-5, XI-6, and XI-7, respectively. One of the most striking features of low energy fission (defined here as spontaneous and thermal neutron-induced fission) is the asymmetric double-peaked mass distribution, as illustrated in Fig. XI-5. As the energy of the neutrons which are inducing fission is increased, the symmetric mass yield increases. In going from thermal neutrons to fission-spectrum neutrons the symmetric yield has increased by a factor of approximately 3, as can be seen by comparing Figs. XI-5 and XI-6. With 14-MeV neutrons, the symmetric yield has increased by about a factor of 100 (cf. Figs. XI-5 and

TABLE XI-1

Total Chain Yields for Neutron Fission of ^{235}U[a]

Mass number	Fission product	Yield (%) Thermal neutrons	Fission-spectrum neutrons	14-MeV neutrons
72	^{72}Zn	0.000016		0.0063
73	^{73}Ga	0.00010		(0.010)
74	^{74}Ga	0.00035		(0.016)
75	^{75}Ge	0.0008		(0.028)
76		(0.0025)		(0.046)
77	^{77}As	0.0083	(0.012)	(0.078)
78	^{78}As	0.020	(0.025)	(0.13)
79	^{79}As	0.056	(0.050)	(0.24)
80		(0.104)	(0.097)	(0.39)
81	^{81}Se	0.148	(0.195)	(0.59)
82		(0.32)	(0.315)	(0.84)
83	^{83}Br, ^{83}Kr	0.544	0.615	1.23
84	^{84}Kr	0.97	1.07	(1.5)
85	^{85}Rb	1.30	1.49	(1.90)
86	^{86}Kr	1.89	1.93	(2.35)
87	^{87}Rb	2.53	2.66	(2.85)
88	^{88}Sr	3.58	3.63	(3.4)
89	^{89}Sr	4.76	5.0	4.16
90	^{90}Sr	5.83	5.24	4.5
91	^{91}Zr	5.90	6.1	4.71
92	^{92}Zr	5.98	(6.2)	(5.05)
93	^{93}Zr	6.39	(6.4)	5.4
94	^{94}Zr	6.44	(6.5)	(5.45)
95	^{95}Zr, ^{95}Mo	6.41	6.47	4.80
96	^{96}Zr	6.29	(6.6)	(5.6)
97	^{97}Zr, ^{97}Mo	6.21	6.55	5.29
98	^{98}Mo	5.86	6.04	(5.45)
99	^{99}Mo	6.16	5.9	5.25
100	^{100}Mo	6.44	6.35	(4.85)
101	^{101}Ru	5.02	5.46	(4.5)
102	^{102}Ru	4.17	4.65	(4.0)
103	^{103}Ru	3.0	3.5	3.35
104	^{104}Ru	1.81	2.35	(2.6)
105	^{105}Ru, ^{105}Rh	0.90	1.50	2.09
106	^{106}Ru	0.399	0.97	1.76

TABLE XI-1 (continued)

Mass number	Fission product	Yield (%) Thermal neutrons	Fission-spectrum neutrons	14-MeV neutrons
107	^{107}Rh	0.19	(0.50)	(1.55)
108		(0.075)	(0.17)	(1.4)
109	^{109}Pd	0.030	0.146	1.17
110		(0.020)	(0.078)	(1.21)
111	^{111}Ag	0.019	0.0456	1.11
112	^{112}Pd	0.010	0.039	0.93
113	^{113}Ag	(0.012)	0.0342	1.15
114	^{114}Cd	(0.011)	0.0342	(1.03)
115	^{115}Cd	0.0104	0.032	0.97
116	^{116}Cd	(0.0104)	0.0359	(1.00)
117	^{117}Cd	0.011	(0.032)	(1.01)
118		(0.0105)	(0.032)	(1.02)
119		(0.0107)	(0.033)	(1.05)
120		(0.0113)	(0.034)	(1.08)
121	^{121}Sn	0.014	(0.036)	1.09
122		(0.013)	(0.040)	(1.19)
123	^{123}Sn	0.015	(0.045)	(1.27)
124		(0.017)	(0.056)	(1.38)
125	^{125}Sb	0.0291	0.073	1.45
126		(0.064)	(0.20)	(1.7)
127	^{127}Sb	0.137	(0.38)	2.09
128	^{128}Sn	0.46	(0.71)	(2.5)
129	^{129}I, ^{129}Sb	1.0	(1.40)	(2.40)
130	^{130}Sb	2.0	(2.4)	(3.95)
131	^{131}I, ^{131}Xe	2.93	3.17	4.23
132	^{132}Te, ^{132}Xe	4.33	4.45	4.70
133	^{133}Xe, ^{133}Cs	6.69	6.69	5.6
134	^{134}Xe	7.92	7.09	5.9
135	^{135}Xe, ^{135}Cs	6.43	6.54	5.7
136	^{136}Xe	6.46	5.93	(5.7)
137	^{137}Cs	6.18	6.25	5.9
138	^{138}Ba	6.71	6.60	(5.5)
139	^{139}Ba	6.48	(6.45)	5.0
140	^{140}Ba, ^{140}Ce	6.34	6.21	4.61
141	^{141}Ce	6.1	6.3	3.8
142	^{142}Ce	5.90	5.82	(4.25)

TABLE XI-1 (continued)

Mass number	Fission product	Yield (%) Thermal neutrons	Fission-spectrum neutrons	14-MeV neutrons
143	^{143}Ce, ^{143}Nd	5.91	5.80	3.81
144	^{144}Ce, ^{144}Nd	5.40	5.15	3.20
145	^{145}Nd	3.88	3.85	(2.50)
146	^{146}Nd	2.95	3.00	(1.98)
147	^{147}Nd, ^{147}Sm	2.19	2.3	1.64
148	^{148}Nd	1.67	1.75	(1.16)
149	^{149}Pm, ^{149}Sm	1.04	1.16	(0.85)
150	^{150}Nd	0.65	0.832	(0.62)
151	^{151}Sm	0.42	0.438	(0.45)
152	^{152}Sm	0.24	0.309	(0.31)
153	^{153}Sm, ^{153}Eu	0.158	0.21	0.22
154	^{154}Sm	0.064	0.098	(0.153)
155	^{155}Eu	0.031	(0.050)	(0.105)
156	^{156}Eu	0.0134	0.025	0.08
157	^{157}Eu	0.0066		(0.038)
158	^{158}Eu	0.0037		(0.023)
159	^{159}Gd	0.00105	0.0034	0.0127
160				(0.0076)
161	^{161}Tb	0.000082	0.00046	0.0056

[a] Flynn and Glendenin (1970).

XI-7). Several characteristics, including the symmetric yields and peak-to-valley ratios, of the mass distributions for neutron-induced fission of several nuclei are listed in Table XI-2. Many interesting features are contained in this summary. The median mass number at half-maximum height of the heavy group is approximately a constant, whereas the same quantity for the light group varies to account for the total mass of the fissioning system. The nearly constant position of the light side of the heavy peak has been observed for a large number of fissioning nuclei. The pronounced stability of this position is undoubtedly influenced by the closed-proton ($Z = 50$) and closed-neutron ($N = 82$) shell structures found in this mass region. The constancy in position of the light side of the heavy peak is illustrated graphically for the fission of four nuclei in Fig. XI-8A (Wahl, 1965). The symmetric yield for thermal neutron-induced fission has a minimum around ^{235}U. For lighter or heavier nuclei, the symmetric yield goes up (see Table XI-2). In Fig. XI-8B are plotted

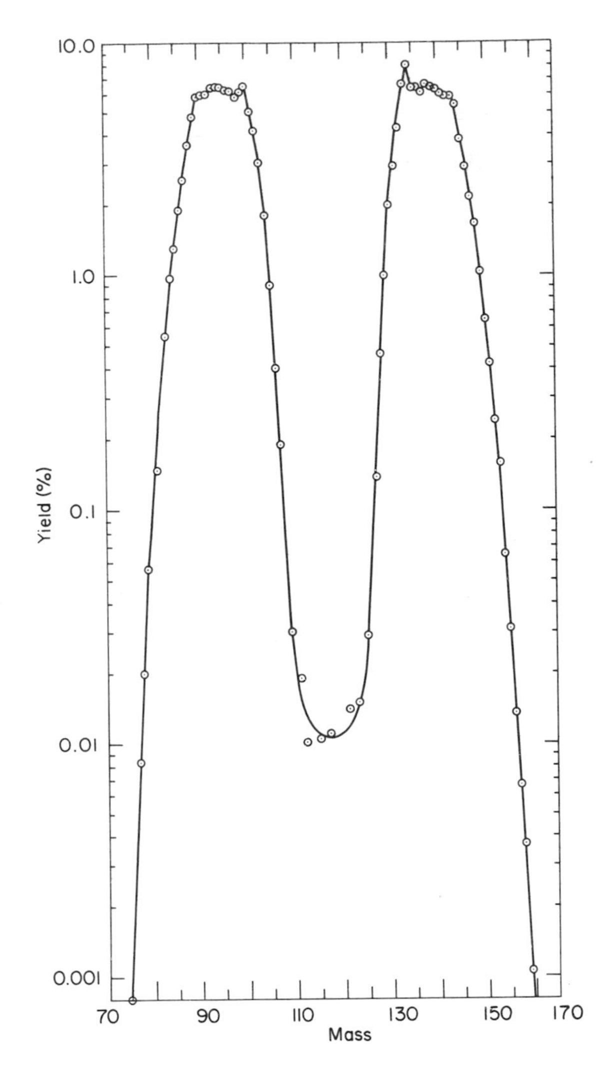

Fig. XI-5. Fission product mass yield curve for thermal neutron-induced fission of ^{235}U. [After Flynn and Glendenin (1970).]

the average masses of the light and heavy fission product groups as a function of the masses of the fissioning nucleus. This is a rather striking illustration of the result that the heavy group fission product mass remains approximately fixed at mass 139 ± 1, while the light group fission product mass increases linearly with the mass of the fissioning nucleus. It is of interest that the very heaviest fissioning nuclei show some deviations from these systematic patterns.

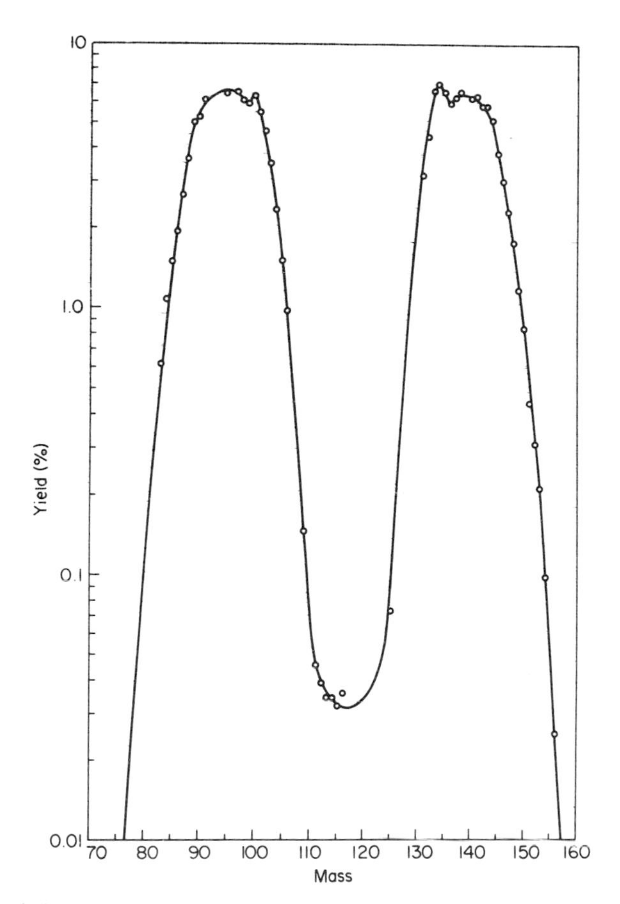

Fig. XI-6. Fission product mass yield curve for the fission of ^{235}U induced with fission-spectrum neutrons. [After Flynn and Glendenin (1970).]

The twin-peaked asymmetric mass distributions of heavy nuclei are characteristic not only of neutron-induced fission but of photo- and charged particle-induced fission as well. In Fig. XI-9 are shown the fission product mass distribution curves obtained from the fission of ^{238}U with ^{4}He ions of various energies. These data were determined by radiochemical techniques (Colby *et al.*, 1961). Just as was seen earlier for neutron-induced fission of ^{235}U, the symmetric yield for ^{4}He-ion-induced fission of ^{238}U also increases with energy. The average excitation energy at which fission occurs does not increase as fast as the projectile energy due to contributions from second- and higher-chance fission.

Target nuclei with $Z \leqslant 83$ exhibit, in general, mass distributions which are symmetric. The primary fragment mass distribution obtained with 42-MeV

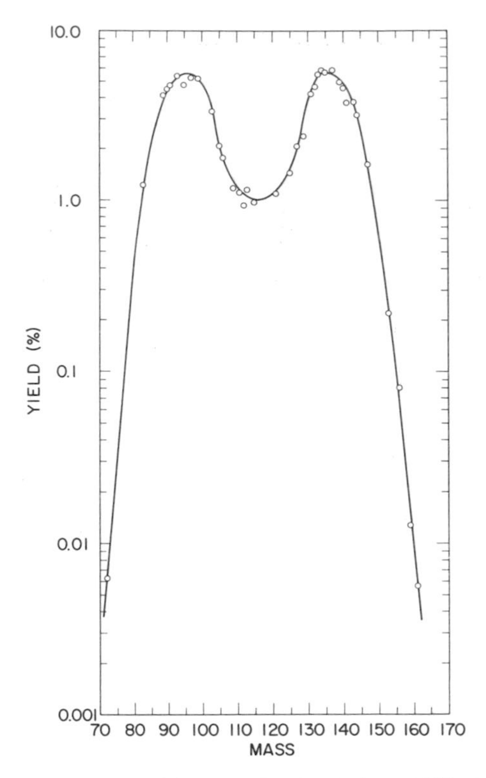

Fig. XI-7. Fission product mass yield curve for the fission of ^{235}U induced with 14-MeV neutrons. [After Flynn and Glendenin (1970).]

^{4}He ions incident on ^{209}Bi is shown in Fig. XI-10. The postneutron and preneutron emission mass distributions for helium-ion-induced fission of ^{238}U can be compared by examining Fig. XI-9 and Fig. XI-10. Radium occupies a position intermediate between those nuclei which fission either symmetrically or predominately asymmetrically. The primary fragment mass distributions obtained with radium as a target and 31- and 39-MeV ^{4}He ions are shown in Fig. XI-10. As can be seen, significant contributions of both symmetric and asymmetric fission occur.

A triple-peaked mass distribution was obtained for radium initially by Jensen and Fairhall (1958) with 11-MeV proton projectiles. Similar mass distributions have since been observed for radium fission with photons and a number of different light particle projectiles. Konecny and Schmitt (1968) have shown that for proton-induced fission of ^{226}Ra the ratio of symmetric to asymmetric fission increases with energy. With 8-MeV protons asymmetric fission is highly favored and with 11-MeV protons symmetric and

TABLE XI-2

Characteristics of Mass Distributions for Neutron-Induced Fission[a]

Fissile target	Median mass number at half-maximum height		Peak width (mass units)		Symmetric yield (%)	Peak-to-valley ratio
	Light group	Heavy group	Half maximum	Tenth maximum		
			Thermal-neutron fission			
^{227}Th	89	138.5	12	20	0.035	230
^{229}Th	88	140	12	18	0.017	500
^{233}U	93.5	138	15	23	0.015	440
^{235}U	95	138.5	15	22	0.0105	620
^{239}Pu	99.5	137.5	15	25	0.04	150
^{241}Pu	101	138	15	24	0.029	230
^{245}Cm	104.5	137.5	13	26.5	0.045	155
^{249}Cf	106	139	16	28	<0.2	>30
			Fission-spectrum-neutron fission			
^{232}Th	91	139.5	14	20	0.045	170
^{231}Pa	91	139	13	22	0.07	100
^{233}U	93	138	14	21	0.06	110
^{235}U	95	138.5	15	23	0.032	205
^{238}U	97	138.5	16	24	0.04	160
^{237}Np	97	137	14	22	0.04	175
^{239}Pu	100	137.5	15	24.5	0.06	115
			14-MeV-neutron fission			
^{232}Th	91.5	138	13	—	1.3	5
^{235}U	95.5	137	16	—	1.0	6
^{238}U	97	137.5	16	—	0.8	7

[a] Flynn and Glendenin (1970).

asymmetric fission are about equally likely. More recently, Perry and Fairhall (1971) have studied the primary fission-fragment mass distributions obtained for proton- and deuteron-induced fission of radium as a function of bombarding energy. They conclude that the first-chance fission of ^{228}Ac at 24-MeV excitation energy is symmetric.

The enhancement of symmetric fission with energy for the heavy nuclei can be understood in terms of several models, including a model with an aysmmetric second fission barrier (see Chapter IX). The observed mass distribution at the higher energies is, of course, a result of a compilation of the distributions of several fissioning nuclei arising from multichance fission. An explanation of the triple-peaked mass distribution of ^{226}Ra in terms of saddle shapes

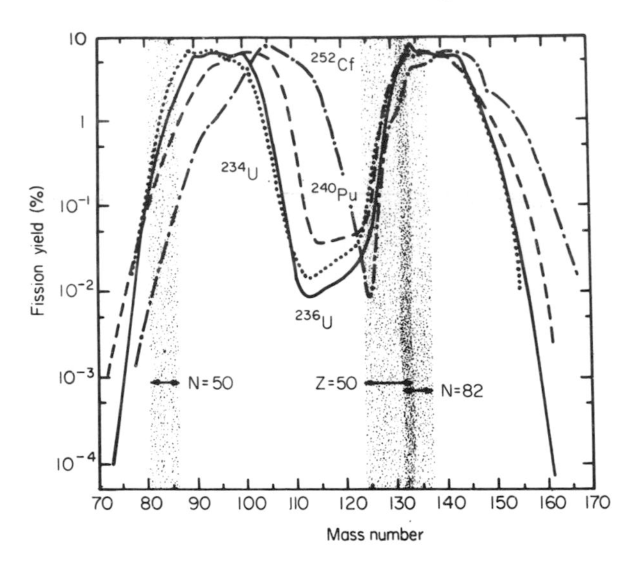

Fig. XI-8A. Fission product mass yield curves for thermal neutron-induced fission of ^{233}U, ^{235}U, and ^{239}Pu and spontaneous fission of ^{252}Cf. Shaded areas indicate approximate positions of nuclear shell edges. Curves are denoted by the symbol of the fissioning nucleus. [After Wahl (1965).]

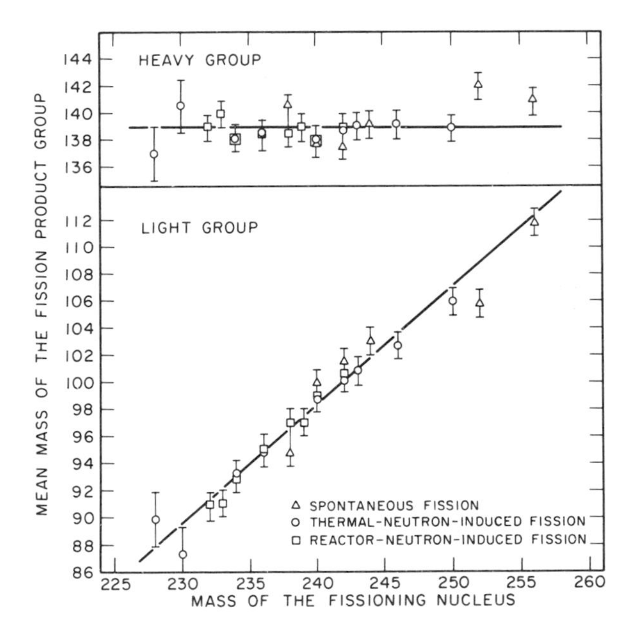

Fig. XI-8B. Average masses of the light and heavy fission product groups as functions of the masses of the fissioning nucleus. [After Flynn *et al.* (1972b).]

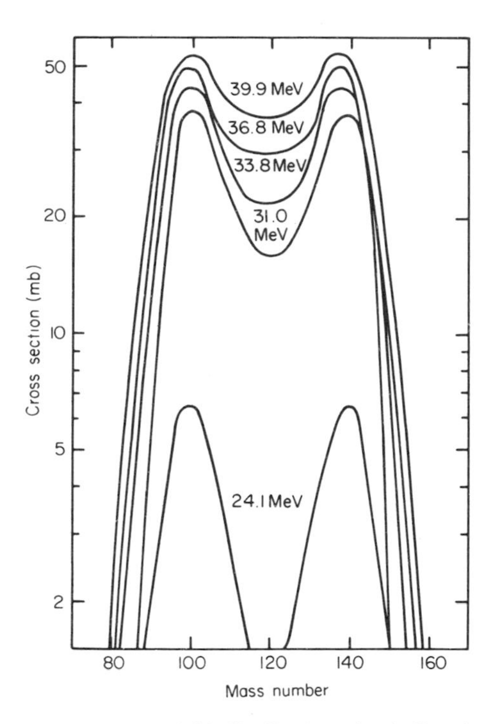

Fig. XI-9. Fission product mass yield distributions for helium-ion-induced fission of ^{238}U. The ^{4}He projectile energies are given in the figure. The mass distributions were measured radiochemically. [After Colby *et al.* (1961).]

requires rather special conditions. One possibility is that the outer barrier has two maxima at some asymmetric deformation. Symmetric fission then occurs between the two maxima and asymmetric fission occurs around the outside of the maxima (Möller and Nilsson, 1971). However, with such a barrier scheme, it is not possible to account for the energy dependence of the mass distribution (Tsang and Wilhelmy, 1972). The reported result of symmetric fission for ^{228}Ac raises the question whether the barrier shape goes from symmetric to asymmetric for a 1-unit change in A. If so, then triple-peaked mass distributions as observed for the fission of radium might be ascribed to mixtures of two components associated with two or more chance fission.

4. Fine Structure in Primary Fission-Fragment and Fission Product Mass Distributions

Fine structure in the fission product mass distributions from spontaneous and thermal neutron-induced fission has been known to exist for some time. These results were determined by high resolution radiochemical and mass

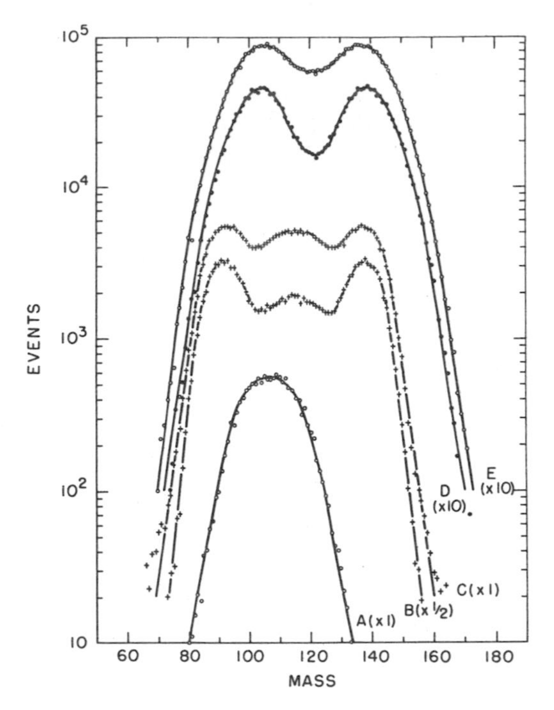

Fig. XI-10. Primary fission-fragment mass yield distributions for helium-ion-induced fission of ^{209}Bi, ^{226}Ra, and ^{238}U. For convenience of display, the number of events for each system has been multiplied by the scale factor given in parentheses. (A) ^{209}Bi (42.0-MeV ^{4}He, f); (B) ^{226}Ra (30.8-MeV ^{4}He, f); (C) ^{226}Ra (38.7-MeV ^{4}He, f); (D) ^{238}U (29.4-MeV ^{4}He, f); (E) ^{238}U (42.0-MeV ^{4}He, f). [After Unik and Huizenga (1964).]

spectrometric techniques. The measurements of Farrar and Tomlinson (1962) for thermal neutron-induced fission of ^{235}U are plotted in Fig. XI-11 and serve to illustrate the irregularities in the fission product mass distributions.

Historically, fission product mass distributions (postneutron emission masses) have been determined radiochemically and mass spectrometrically. With the great improvement in techniques to measure energies and velocities of fission products, measurements of fission product mass distributions are now determined by correlated energy and velocity measurements of single fragments as well (Schmitt *et al.*, 1965). Primary fission-fragment distributions (preneutron emission masses) are determined from double velocity and coincident energy measurements. The use of these new techniques has established that fine structure exists in both the postneutron and preneutron emission mass distributions for a number of fissile nuclei (Pappas *et al.*, 1969). The primary fission-fragment yields for thermal neutron-induced fission of ^{233}U, ^{235}U, and ^{239}Pu are plotted for the heavy fragments in Fig. XI-12. Thomas and

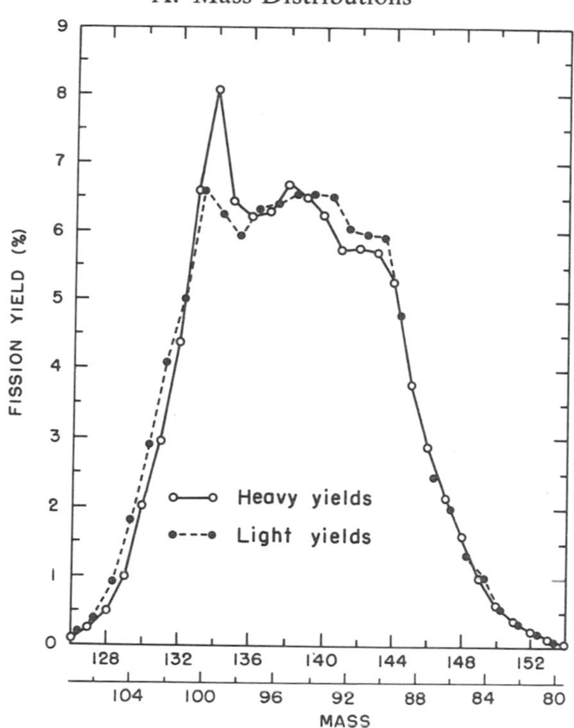

Fig. XI-11. Comparison of light and heavy cumulative fission product yield distributions corrected for the effects of delayed neutrons. [After H. Farrar and R. H. Tomlinson (1962). *Canadian Journal of Physics*, **40**, pp. 943–953.]

Vandenbosch (1964) pointed out a correlation between the fine structure in the primary fragment yields and structure in the energy release in fission as calculated from semiempirical mass equations. The origin of the structure rests in the fact that the energy release is greater for even-even fragments (due to nuclear pairing) than for odd-mass fragments, and that there is a preference for forming even-even primary fragments in the fission of an even-even compound nucleus. As can be seen in Fig. XI-12, the structure has a periodicity of about 5 mass units. This is true because the mass number of the most stable nuclide for a given Z changes about 2.5 units for a unit change in Z.

The structure in the primary fission-fragment yields (preneutron emission) is directly related to the structure in the fission product yields (postneutron emission). This is illustrated in Fig. XI-13, where the resolution-corrected preneutron data for the thermal neutron-induced fission of ^{235}U have been transformed to postneutron data (Reisdorf *et al.*, 1971) to allow a direct comparison with radiochemical data. Excellent agreement is obtained between the coincident energy semiconductor data and the radiochemical data. The

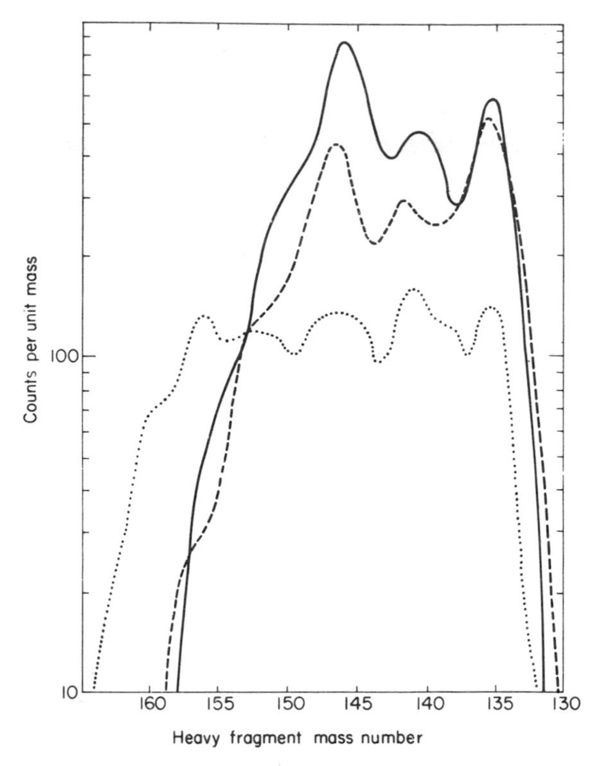

Fig. XI-12. Experimentally observed fine structure in the primary fission fragments for thermal neutron-induced fission of ^{233}U (—), ^{235}U (---), and ^{239}Pu (···). [After Thomas and Vandenbosch (1964).]

$\nu(M)$ curve of Fig. XI-13 used in the transformation of the preneutron data to the postneutron data is a rather smooth function. Hence, the main effect of the neutron emission on the structure is to shift it to a lower mass number. It is, of course, possible that the number of neutrons emitted as a function of mass is somewhat irregular, causing additional structure in the fission product (postneutron emission) mass distribution.

5. Mass Distribution for Spontaneous Fission Isomers

The primary fission product mass distributions have been determined recently for several spontaneous fission isomers by measurement of the kinetic energies of primary fission product pairs (Ferguson *et al.*, 1971). Resulting mass distributions for several spontaneous fission isomers are shown in Fig. XI-14 and mean values and widths of such distributions are listed in Table XI-3 (Ferguson *et al.*, 1971). The fragment distributions for

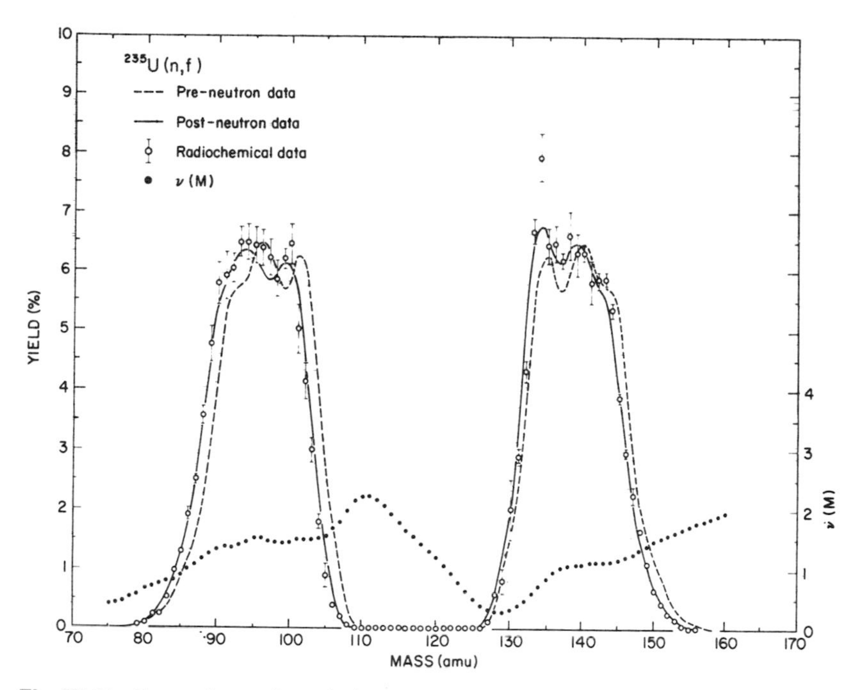

Fig. XI-13. Comparison of resolution-corrected primary fission-fragment (preneutron emission) and fission product (postneutron emission) mass distributions with radiochemical fission product mass distributions for thermal neutron-induced fission of ^{235}U. The $\nu(M)$ curve (●) represents the neutron emission data used in the calculation of the postneutron emission mass distribution. [After Reisdorf *et al.* (1971).]

spontaneous fission isomers are consistent with those for low excitation prompt fission (with possible exceptions deep in the valley at symmetry where no data exist).

TABLE XI-3

Mean Values and Widths of Fission-Fragment Distributions for Spontaneous Fission Isomers[a]

	^{239m}Am[b]	^{239m}Am[c]	^{237m}Pu	^{238m}U	^{236m}U	^{236}U[d]
$\langle m_L \rangle$	101.6	101.8	100.4	100.5	96.9	96.8
$\langle m_H \rangle$	137.8	138.3	136.2	138.5	138.8	139.2
σ_m	6.1	5.9	5.3	6.1	5.4	5.36
$\langle E_k \rangle$	183.2	181.0	182.3	172.5	172.1	170.0
σ_{E_k}	13.0	12.3	11.9	11.7	10.0	10.9

[a] Ferguson *et al.* (1971).
[b] $^{242}Pu(p, 2n)^{239m}Am$.
[c] $^{239}Pu(d, 2n)^{239m}Am$.
[d] Values for $^{235}U(n_{th}, f)$, from Schmitt *et al.* (1966) (without neutron emission corrections).

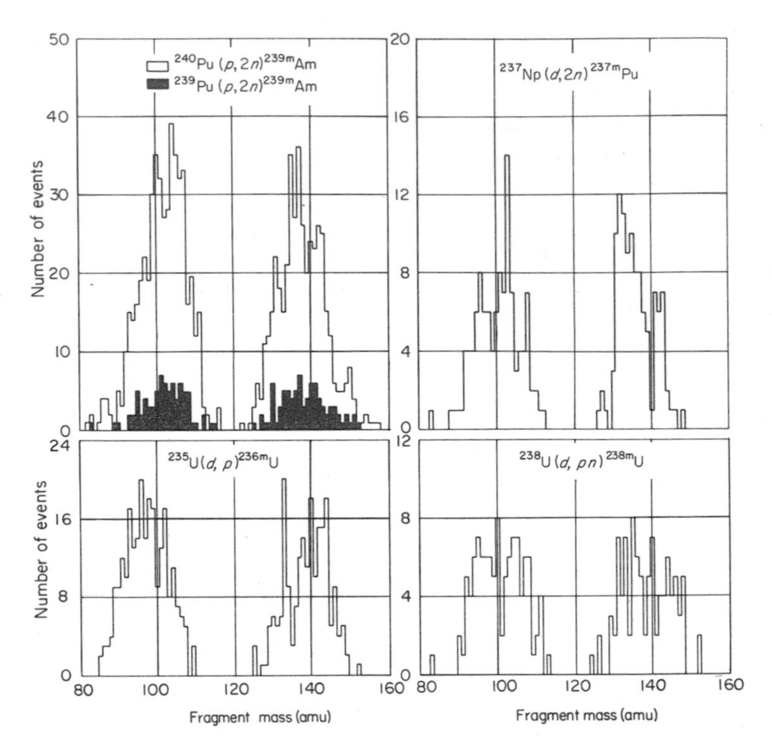

Fig. XI-14. Primary fission product mass distributions for spontaneous fission isomers formed in the reactions indicated. The results were obtained by coincident kinetic energy measurements. No corrections for neutron emission are included. [After Ferguson *et al.* (1971).]

B. Charge Distributions

1. Charge Distribution in Low Energy Fission

An important parameter describing the fission process is the division of nuclear charge between the two fragments. From the point of view of increasing our understanding of the fission process the quantity of most interest is the charge-to-mass ratio of the fragment just produced at scission prior to the emission of prompt neutrons.

A complete characterization of the charge division in a mass split corresponding to a particular final mass requires a knowledge of the independent yield of each isobar of that particular mass chain. In practice such detailed information is difficult to obtain and we compromise by assuming the charge distribution to have a specific functional form, i.e., a Gaussian distribution, and then characterize the charge division by determining the most probable charge and the width of the distribution. We then explore the dependence of the most probable charge and possibly the distribution width on the mass of the fragment.

Until relatively recently most of the experimental data on nuclear charge division were obtained by radiochemical determinations of independent fission yields. By independent yield is meant the yield for a particular isobar not resulting from the β decay of the adjacent isobar. Since the fissioning nuclei generally have a lower charge-to-mass ratio than the stable nuclei in the fission product region, the primary fission products are neutron rich, decaying by $\beta-$ emission toward the stable species of a given isobaric chain. In fact, most of the primary yield is to primary products so neutron rich and β unstable that their β-decay half lives are too short for radiochemical yield determinations. Generally most of the yield of a particular radioactive or stable product is from β decay of the more neutron-rich isobars formed in greater yield. Such a yield is said to be the cumulative yield if all of the more neutron-rich precursors have had time to β decay to the species in question. The independent yield can be determined, however, either if the adjacent isobar has a sufficiently long half life that its yields and, hence, its decay contribution to the species of interest can be measured, or if the radioactive species is completely shielded from β decay by a stable isobar. In general, shielded radioactive species lie close to the line of β stability and have quite small independent yields. Some information about charge division can be obtained from cumulative yields for a species that lies far enough away from the line of β stability so that its cumulative yield is significantly less than the total yield for that mass.

There are a number of mass chains where it has been possible to determine the independent yields of more than one isobar. An example of such a chain is illustrated in Fig. XI-15. The independent yields of ^{93}Rb, ^{93}Sr, and ^{93}Y, as well as the cumulative yield of ^{93}Kr, have been determined (Wahl, 1965). The yields are consistent with a Gaussian distribution about a most probable charge $Z_p = 37.30$. The functional dependence of the independent yields is usually expressed as $P(Z) = (c\pi)^{-1/2} \exp[-(Z-Z_p)^2/c]$, with $c = 0.82$ in this example. Results for other mass chains exhibit a similar distribution, with the width of the distribution generally varying little with fission product mass number. The average value of c is 0.80 ± 0.14, with the largest deviations occurring in the vicinity of closed shells. This width is that of the primary fission products after neutron emission. The width of the charge distribution for primary fragments is expected to be less, due to the removal of fluctuations in the number of neutrons emitted.

It is sometimes useful to parameterize the charge distribution in terms of the

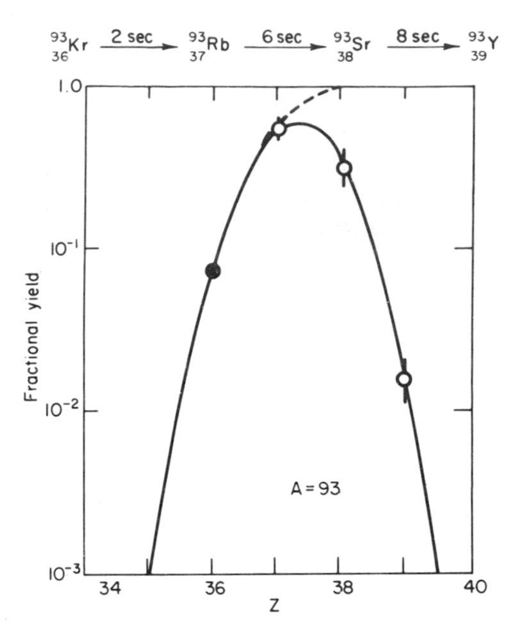

Fig. XI-15. Charge dispersion for products with $A = 93$ from thermal neutron fission of ^{235}U: ○ independent yield, ● cumulative yield, (—) Gaussian for the independent yields, (---) for cumulative yields. [After Wahl (1965).]

fractional cumulative yield as defined by

$$\sum_{0}^{Z} P_n = (1/\sigma(2\pi)^{1/2}) \int_{-\infty}^{Z+\frac{1}{2}} \exp[-(n-Z_p)^2/2\sigma^2]\, dn$$

The distribution characterized by the width parameter σ is considered to be continuous and when integrated over all charges gives unity. The earlier-mentioned width parameter c, corresponding to a normalization where summation at unit intervals of charge gives a value of unity, is related to the parameter σ by Sheppard's correction, $c = 2(\sigma^2 + \frac{1}{12})$.

The near constancy of the width of the charge distribution curve has led to the assumption of a universal charge distribution curve, enabling a deduction of the most probable charge for any mass chain for which a single independent yield is known. In this manner Z_p has been determined for most high yield mass splits in the thermal neutron-induced fission of ^{235}U. The resulting dependence of Z_p on mass number A is usually presented in such a way as to exhibit the displacement of the most probable charge from the charge expected if the charge-to-mass ratio of the fissioning nucleus were preserved in the fragments, $A_{frag}(Z/A)_F$. Such a plot is illustrated in Fig. XI-16. It can be seen that the actual charge division is such that the lighter fragment receives

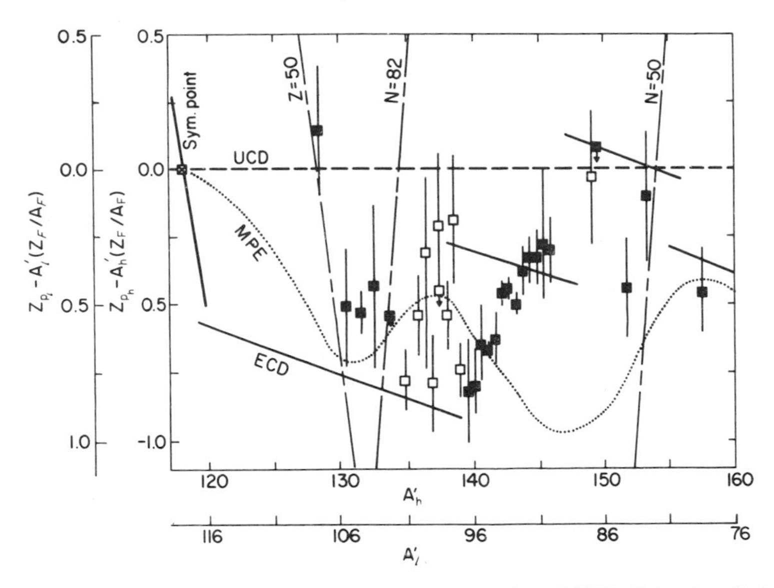

Fig. XI-16. Empirical Z_p values for thermal neutron fission of ^{235}U. Subscripts F, l, and h refer to the fissioning nucleus, light fragments, and heavy fragments, respectively; A' denotes mass number before neutron emission; UCD, unchanged charge distribution; MPE, minimum potential energy; ECD, equal charge displacement. [From Wahl *et al.* (1962).]

a larger fraction of the available charge than would be expected on the basis of an unchanged charge distribution (UCD). It was suggested very early in the investigation of charge distributions that the observed charge division follows much more nearly that expected on the basis of an equal charge displacement (ECD) postulate (Glendenin *et al.*, 1951). This postulate assumes that the difference between the most probable charge Z_p and the charge of the most stable isobar of that mass chain Z_A will be equal for the light and heavy complementary fragments, $(Z_p - Z_A)_l = (Z_p - Z_A)_h$. The predictions of this postulate are indicated by the heavy line segments in Fig. XI-16, the discontinuities resulting from shell discontinuities in the Z_A values. This postulate qualitatively accounts for the observed charge division if Z_p is assumed to cross gradually from one curve to the other in the vicinity of shell discontinuities in the Z_A function. Much less complete radiochemical data are available for other fissioning systems, but the general trend is consistent with the equal charge displacement postulate. The extent to which the available charge distribution data can be correlated in terms of a continuous Z_p curve drawn through the data in Fig. XI-16, and the assumption of a common width parameter for all mass chains are illustrated in Fig. XI-17.

In more recent years several physical techniques have been applied to the

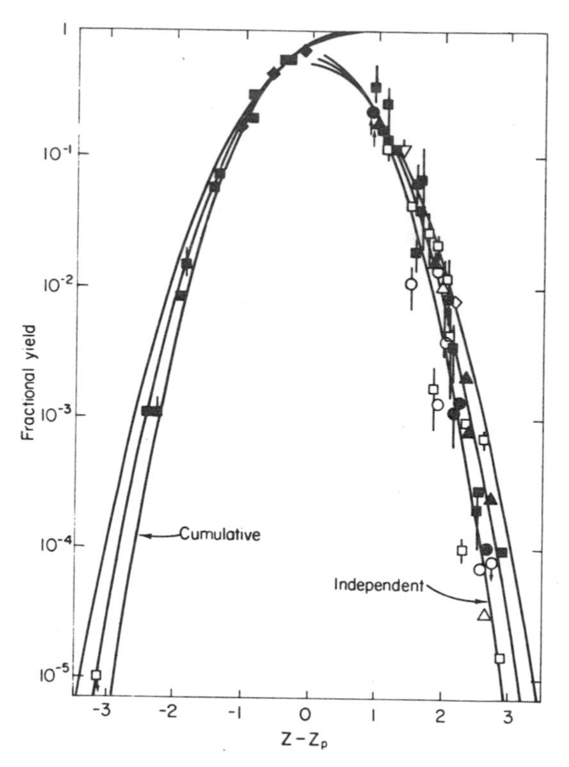

Fig. XI-17. Conventional charge distribution plot: (○) $^{233}U(n, f)$; (□) $^{235}U(n, f)$; (△) $^{239}Pu(n, f)$; (▽) $^{242}Cm(SF)$; (◇) $^{252}Cf(SF)$. Shaded symbols represent products with a composition neither close to nor complementary to a major nuclear shell edge. The curves are calculated with $\sigma = 0.62 \pm 0.06$. [From Wahl *et al.* (1962).]

measurement of charge distribution in fission. One such technique is to determine the mean chain length by observing the mean number of β particles emitted from mass-separated fission products (Sistemich *et al.*, 1969). Another method is to determine the Z of the fragments by observation of the characteristic K x rays in coincidence with fission fragments (Reisdorf *et al.*, 1971). The K x rays arise from internal conversion associated with the deexcitation of the primary fission products. Only a negligible fraction of the x rays are associated with vacancies created by the shakeoff of electrons during the fission process, as evidenced by the relatively long time delay ($\approx 10^{-10}$ sec) between fission and x-ray emission (see Section XIII, D). This method is limited by the poor mass resolution (3–4 amu) due to the energy resolution of the solid state detectors used to measure the fragment kinetic energies and the fluctuations in neutron emission. It is also necessary to understand the dependence of the K x-ray yield on the N and Z of the fragments. The strong dependence of the

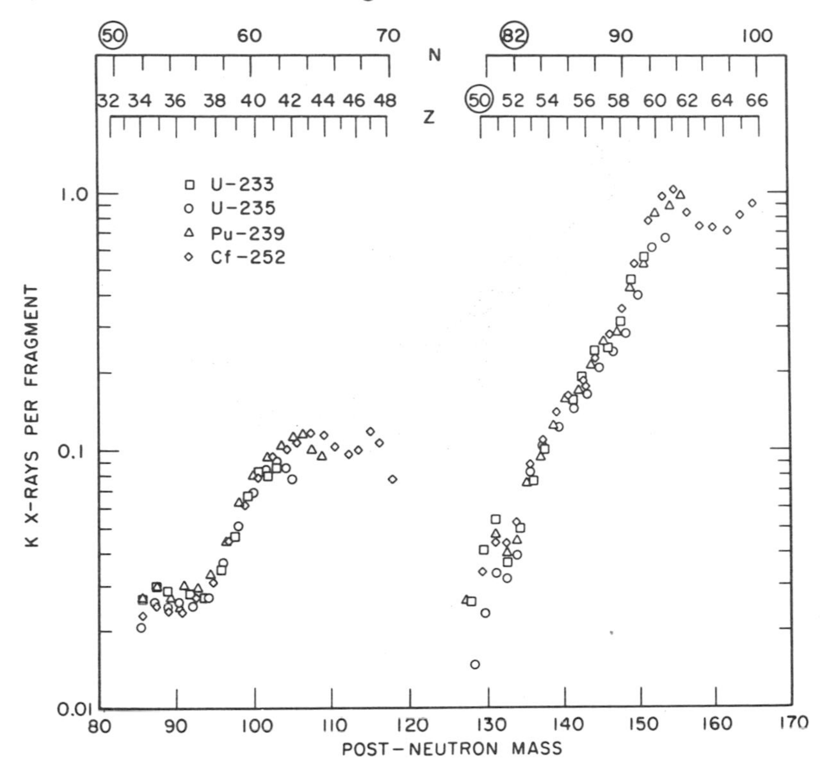

Fig. XI-18. K x rays per fragment emitted within 1 nsec after fission as a function of post-neutron mass for ^{233}U(n, f), ^{235}U(n, f), ^{239}Pu(n, f), and ^{252}Cf(SF). Approximate average proton and neutron numbers (for ^{252}Cf) are given on the upper abscissa for orientation. [From Reisdorf *et al.* (1971).]

K x-ray yield on fragment mass is shown in Fig. XI-18. The general trend of the results of the physical methods is in agreement with the radiochemical data, as illustrated in Fig. XI-19, although the structure observed in the radiochemical experiments is not seen. It should be remarked that some of the structure seen in the radiochemical results may be introduced by anomalies in the neutron emission in the vicinity of shells rather than by the primary charge division.

It is of interest to try to relate the two charge distribution variables $Z_p(A)$ and c to a model of the fission process. The fact that $Z_p(A)$ is different from that expected if the charge-to-mass ratio of the original nucleus were preserved implies that there is charge redistribution prior to separation. As an extreme we could assume that there is equilibrium of the internal charge distribution at the time of scission. Then in the spirit of Fong's statistical model Z_p would be determined by the charge division for which the total fragment excitation

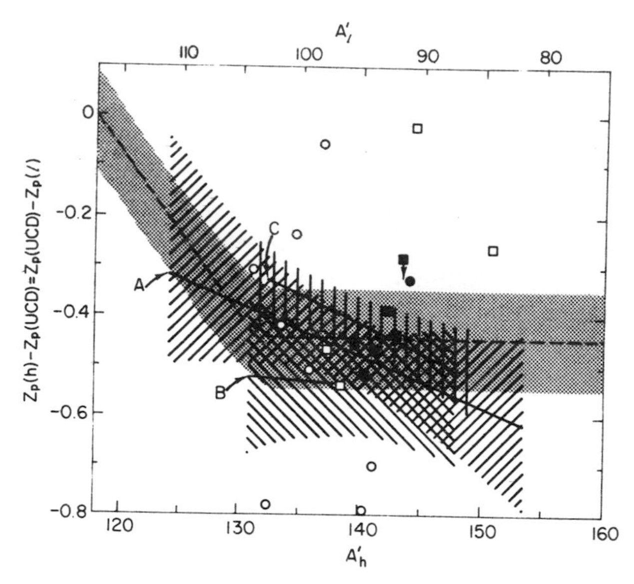

Fig. XI-19. Z_p values from radiochemical measurements for mass chains in which more than one yield was determined (○, ●, □, ■) are compared with Z_p values determined by the *K* x-ray method (curves A and B) and by the counting of β particles from mass-separated fission products (curve C). [From Wahl *et al.* (1969).]

energy at scission is a maximum. This criterion is nearly but not exactly equivalent to a criterion that the potential energy is a minimum (MPE). Halpern (1959) has pointed out that if we overlook for the moment the dependence of the Coulomb interaction energy on the division of Z, we would expect the most probable division to be one where the β-decay chain to the stable valley is shorter for the light fragment. (The isobaric mass parabolas are steeper for lighter fragments than for heavier fragments.) To take the dependence of the Coulomb interaction energy on the Z division into account, it is assumed that the separation d at which scission takes place does not depend on the Z division. Then the Coulomb energy, proportional to Z_1Z_2/d, is less and the excitation energy more for smaller charges of the light fragment. The combination of the two effects turns out to predict charge divisions similar to that of the empirical equal charge displacement postulate. One particular estimate of Z_p values corresponding to MPE is illustrated by the dotted line in Fig. XI-16.

The second feature of the charge distribution of interest is the rather narrow width, which results in very small independent yields for Z values only a few units away from the most probable charge. This can be qualitatively understood in terms of the sizeable changes in the available energy as a function of charge division. The available energy release in fission as calculated from the difference in the mass of the fissioning nucleus and that of the two fragments changes by approximately 6 MeV as one goes from the most energetically favored charge split to a split with 2 units more charge on one fragment and 2 units less charge on the complementary fragment. The dependence of the available energy release on charge division is illustrated in Fig. XI-20. Thus on the basis of a statistical model the yield for $(Z - Z_p) = 2$ compared to $Z = Z_p$ would be expected to be of the order of a Boltzmann factor $\exp(-\Delta E/T)$ or, for $T = 1$, $\exp(-6) = 2 \times 10^{-3}$. This is somewhat smaller than observed but illustrates the sensitivity of the yield to the energy cost of unfavorable charge divisions. Wing and Fong (1967) have made a more detailed estimate based on the statistical model, obtaining a value of $\sigma = 0.51$ for the standard deviation of the Gaussian in the peak yield mass regions. This value is for the charge distribution prior to neutron emission (primary fragments). The value implied by the radiochemical results after neutron emission (primary products) is 0.56 ± 0.06 after correcting the observed width for discrete charges to that expected for a continuous distribution using Sheppard's correction. The K x-ray method allows an estimate of the width which takes into account to lowest order the effects of neutron emission; hence, it provides a value more directly comparable to the calculated value. Results of Reisdorf *et al.* (1971) give a value $\sigma = 0.40 \pm 0.05$, which is even narrower than predicted.

What lesson is to be learned from the simplest features of the charge division? First, we might observe that there are no surprises. The qualitative aspects of the charge division, unlike the mass division, are easily understood. Second, there is fairly clear evidence that there is a small but significant redistribution of charge prior to scission. The redistribution is essentially as

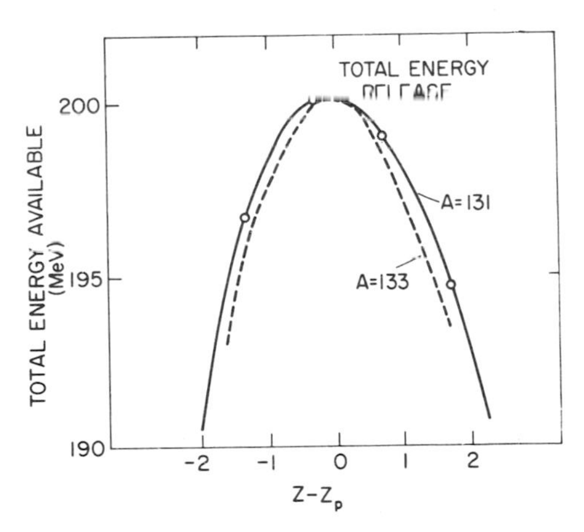

Fig. XI-20. Total energy release as a function of charge division for the $A = 131$ and $A = 133$ chains as given by Cameron's mass equation.

large as expected on the basis of statistical equilibrium at the scission point. The result may, however, also be understood if fission proceeds adiabatically down a valley toward a configuration of minimum potential energy at scission. The degree of freedom associated with charge redistribution is essentially that associated with the giant dipole resonance. The characteristic time associated with the empirically observed $\hbar\omega$ values of about 15 MeV is 10^{-22} sec. This very short characteristic time suggests that this may be the degree of freedom most easily exploited during the motion from saddle to scission.

In addition to the general features of charge distribution discussed previously, more detailed features have also been studied. These include variations in the width of the charge distribution curve in the region of closed shells and the dependence of the charge division on the kinetic energy of the fragments. Charge distribution curves have been measured as a function of fragment kinetic energy using a mass spectrometer (Konecny *et al.*, 1967). The mass resolution was sufficient to separate adjacent isobars. For odd mass number isobars there was little dependence of the charge distribution curve on the kinetic energy of the fragments, as illustrated by Fig. XI-21. For even mass numbers the yield of odd-Z fragments decreased significantly with increasing kinetic energy, as can be seen in Fig. XI-22. The differences between the two

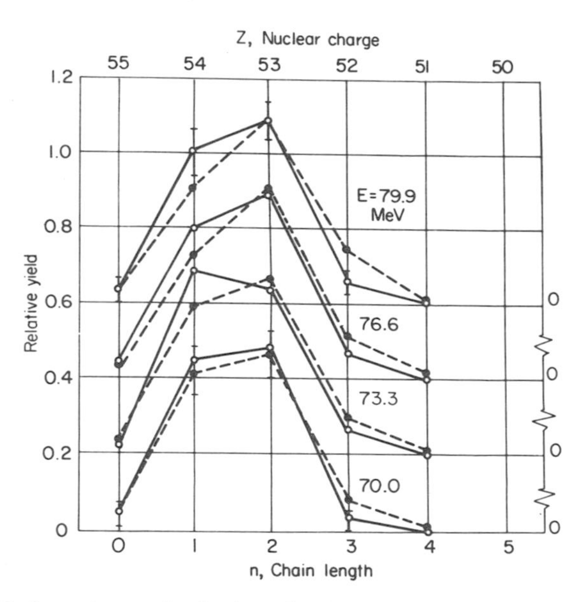

Fig. XI-21. Nuclear charge distributions for fission fragments of mass $A = 137$ and kinetic energies 70.0, 73.3, 76.6, and 79.9 MeV. Each distribution is normalized to 1.0. ○,(—) Data corrected for conversion elections; ●,(---) uncorrected values. [From Konecny *et al.* (1967).]

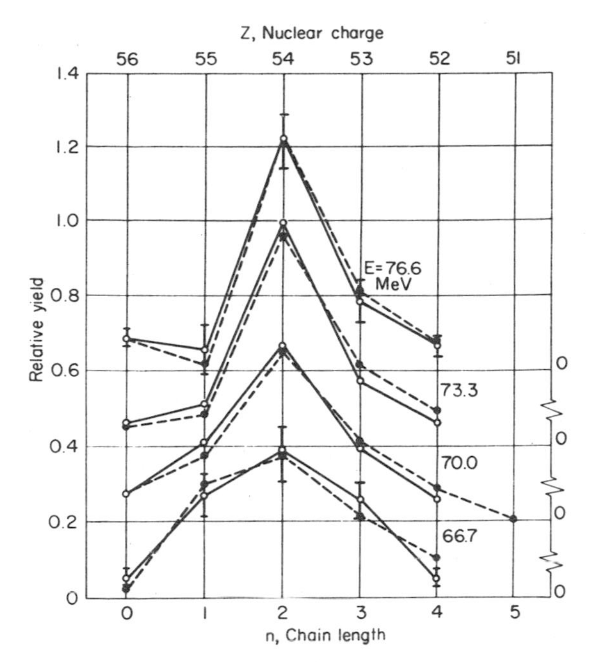

Fig. XI-22. Nuclear charge distributions for fission fragments of mass $A = 138$ and kinetic energies 66.7, 70.0, 73.3, and 76.6 MeV. See caption of Fig. XI-21. [From Konecny *et al.* (1967).]

kinds of isobars is exhibited more clearly in Fig. XI-23. The favoring of even-even splits over odd-odd splits for high kinetic energies (and therefore low scission excitation energies) can be attributed to the significantly larger Q values for division leading to even-even nuclei due to pairing effects (Thomas and Vandenbosch, 1964). The apparent disappearance of the enhancement of even-even final fragments with decreasing kinetic energy (increasing excitation energy) may reflect the consequences of increased neutron emission rather than a disappearance of the preference for fragments whose preneutron emission character is even-even. The observation that the width of the charge distribution for odd-A nuclei does not narrow appreciably with increasing kinetic energy is consistent with the dominance of quantum mechanical zero-point effects discussed later in connection with the dependence of the width of the charge distribution on total excitation energy.

2. Dependence of Charge Division on Excitation Energy

The available information on charge division at higher excitation energies is not very extensive and sometimes conflicting. The qualitative indications are

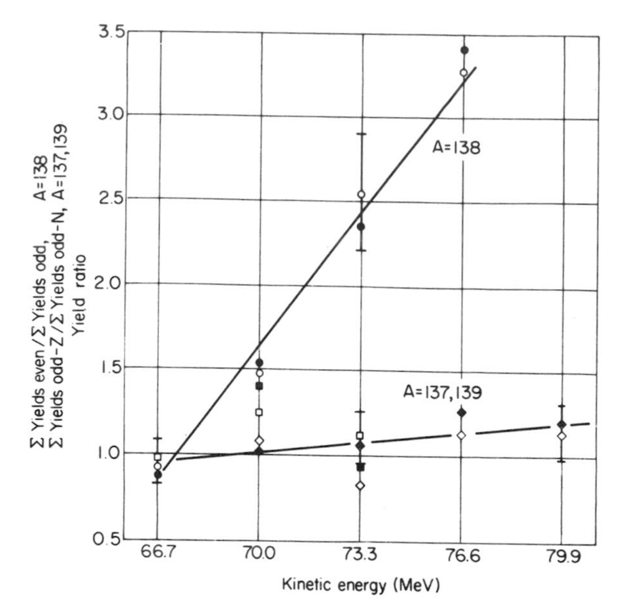

Fig. XI-23. Ratio of the sums of yields of doubly even nuclei to doubly odd nuclei for $A = 138$ (○, ●) and of odd-proton to odd-neutron nuclei of $A = 137$ (◇, ◆) and $A = 139$ (□, ■) as functions of the kinetic energies of the fragments. Open points and solid lines refer to data corrected for conversion elections, broken lines and closed points to uncorrected values. [From Konecny *et al.* (1967).]

that the charge division at high excitation energies more nearly resembles that expected on the basis of an unchanged charge distribution (UCD) than that of equal charge displacement (ECD) observed at lower energies. An alternative description, however, which is quite successful, is that of ECD if we abandon the Z_A functions which trace out all the shell-influenced discontinuities in the ground state masses of stable nuclei and use instead a smooth Z_A function which ignores pronounced shell effects. Alternatively, the charge division can be correlated fairly successfully at high excitation energies using an MPE treatment.

The width of the charge distribution curve is surprisingly insensitive to the excitation energy for excitation energies less than about 40 MeV. This is illustrated in Fig. XI-24 for the $A = 135$ chain. We might expect some broadening due to increased neutron emission, which can occur both before and after fission. Swiatecki and Blann (1961) have pointed out that the charge distribution width may be rather constant at not too high excitation energies if the quantum mechanical zero-point contribution dominates over the temperature effects in a thermodynamic equilibrium model. (The functional form of the dependence is identical to that discussed previously for LDM mass and kinetic energy distributions, although the temperature at which the quantum effects disappear depends on the characteristic frequencies and stiffness parameters.)

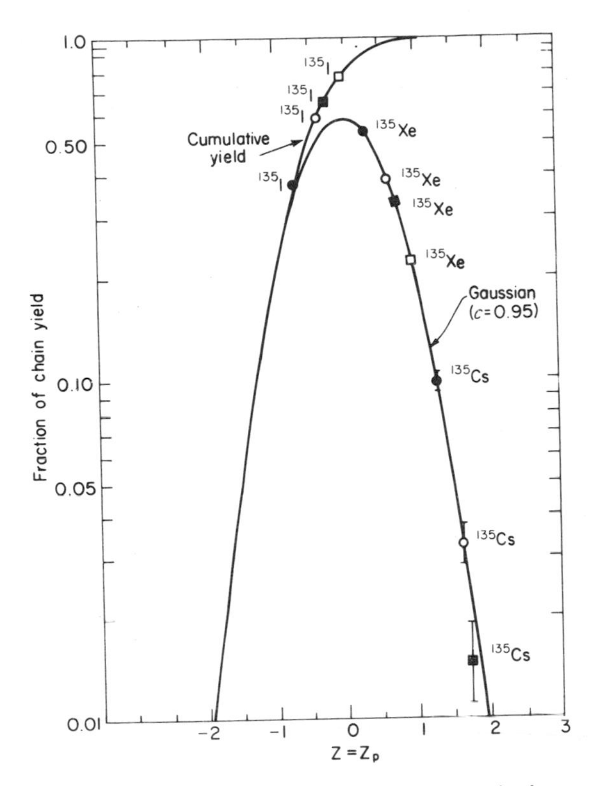

Fig. XI-24. Charge distribution curve for ^{236}U at different excitation energies: (●) 39.3 MeV, (○) 26.8 MeV, (■) 23 MeV, (□) 15–10 MeV. [After McHugh and Michel (1968).]

References

Balagna, J. P., Ford, G. P., Hoffman, D. C., and Knight, J. D. (1971). *Phys. Rev. Lett.* **26**, 145.

Colby, L. J., Jr., Shoaf, M. L., and Cobble, J. W. (1961). *Phys. Rev.* **121**, 1415.

Farrar, H., and Tomlinson, R. H. (1962). *Can. J. Phys.* **40**, 943.

Ferguson, R. L., Plasil, F., Alam, G. D., and Schmitt, H. W. (1971). *Nucl. Phys. A* **172**, 33.

Flynn, K. F., and Glendenin, L. E. (1970). Rep. ANL–7749. Argonne Nat. Lab., Argonne, Illinois.

Flynn, K. F., Srinivasan, B., Manuel, O. K., and Glendenin, L. E. (1972a). *Phys. Rev. C* **6**, 2211.

Flynn, K. F., Horwitz, E. P., Bloomquist, C. A., Barnes, R. F., Sjoblom, R. K., Fields, P. R., and Glendenin, L. E. (1972b). *Phys. Rev. C* **5**, 1725.
Glendenin, L. E. (1971). Private communication.
Glendenin, L. E., Coryell, C. D., and Edwards, R. (1951). *In* "Radiochemical Studies: The Fission Products" (C. D. Coryell and N. Sugarman, eds.), Nat. Nucl. Energy Ser., Plutonium Project Rec., Paper 52. McGraw-Hill, New York.
Halpern, I. (1959). *Annu. Rev. Nucl. Sci.* **9**, 245.
Jensen, R. C., and Fairhall, A. W. (1958). *Phys. Rev.* **109**, 942.
John, W., Hulet, E. K., Lougheed, R. W., and Wesolowski, J. J. (1971). *Phys. Rev. Lett.* **27**, 45.
Konecny, E., and Schmitt, H. W. (1968). *Phys. Rev.* **172**, 1226.
Konecny, E. Gunther, H., Siegert, G., and Winter, L. (1967). *Nucl. Phys. A* **100**, 465.
McHugh, J. A., and Michel, M. C. (1968). *Phys. Rev.* **172**, 1160.
Möller, P., and Nilsson, S. G. (1971). Private communication.
Pappas, A. C., Alstad, J., and Hagebø, E. (1969). *Proc. IAEA Symp. Phys. Chem. Fission, 2nd, Vienna, 1969*, p. 669. IAEA, Vienna.
Perry, D. G., and Fairhall, A. W. (1971). *Phys. Rev. C* **4**, 977.
Reisdorf, W., Unik, J. P., Griffin, H. C., and Glendenin, L. E. (1971). *Nucl. Phys. A* **177**, 337.
Schmitt, H. W., Kiker, W. E., and Williams, C. W. (1965). *Phys. Rev.* **137**, B 837.
Schmitt, H. W., Neiler, J. H., and Walter, F. J. (1966). *Phys. Rev.* **141**, 1146.
Sistemich, K., Armbruster, P., Eidens, J., and Roeckl, E. (1969). *Nucl. Phys. A* **139**, 289.
Swiatecki, W. J., and Blann, H. M. (1961). Private communication.
Thomas, T. D., and Vandenbosch, R. (1964). *Phys. Rev.* **133**, B 976.
Tsang, C. F., and Wilhelmy, J. B. (1972). *Nucl. Phys. A* **184**, 417.
Unik, J. P., and Huizenga, J. R. (1964). *Phys. Rev.* **134**, B 90.
von Gunten, H. R. (1969). *Actinides Rev.* **1**, 275.
Wahl, A. C. (1965). *Proc. IAEA Symp. Phys. Chem. Fission, Salzburg, 1965*, Vol. **I**, p. 317. IAEA, Vienna.
Wahl, A. C., Ferguson, R. L., Nethaway, D. R., Troutner, D. E., and Wolfsberg, K. (1962). *Phys. Rev.* **126**, 1112.
Wahl, A. C., Norris, A. E., Rouse, R. A., and Williams, J. C. (1969). *Proc. IAEA Symp. Phys. Chem. Fission, 2nd, Vienna*, 1969, p. 813. IAEA, Vienna.
Wing, J., and Fong, P. (1967). *Phys. Rev.* **157**, 1038.

CHAPTER

Prompt Neutrons from Fission

A. Introduction

The major fraction of the energy release not appearing as kinetic energy of the fragments is dissipated by emission of neutrons. To the extent that neutrons are emitted from the fragments subsequent to scission (as appears to be the case for the vast majority of the neutrons), the number of neutrons emitted is a direct consequence of the amount of energy stored in the nascent fragments at scission. This energy may either be in the form of internal (heat) energy, or be stored in the form of deformation energy and converted to excitation energy as the fragments collapse to their equilibrium shape. Thus the variation of neutron yield with such parameters of the fission process as mass division, compound nuclear excitation energy, and kinetic energy release provides important insights into the partition of energy at scission.

B. Total Neutron Yield in Low Energy Fission

1. Time Scale of Neutron Emission

A lower limit to the time scale on which neutron emission occurs may be deduced from the angular correlation of the neutrons with respect to the fragment direction. If the neutrons are emitted at times longer than the

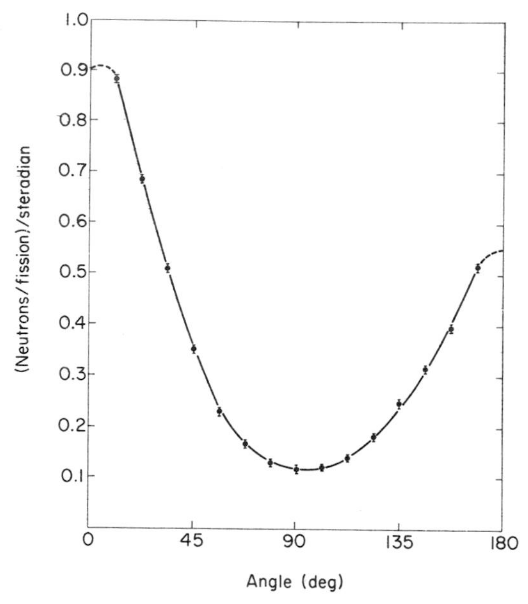

Fig. XII-1. Number of prompt neutrons versus the angle with respect to the light fragment direction in spontaneous fission of ^{252}Cf. [From Bowman *et al.* (1962).]

acceleration times of the fission fragments, the neutrons will appear in directions strongly correlated with the fragment direction. This is indeed what is observed experimentally in low energy fission as illustrated in Fig. XII-1. Since the fragments have become nearly fully accelerated in approximately 10^{-20} sec, most of the neutrons are emitted at times longer than this. More detailed analyses indicate that of the order of 10% of the neutrons exhibit an angular distribution indicative of emission from a stationary nucleus. One such analysis (Milton and Fraser, 1965) indicates that as many as 30% of the neutrons are isotropic in the laboratory system. It is believed that these neutrons are emitted at scission by a mechanism similar to that responsible for the emission of α and other light charged particles. They are usually designated "scission" neutrons or the "central" component. As the excitation energy of the fragments is increased due to an increase in the excitation energy of the fissioning nucleus, the lifetime for neutron emission is expected to decrease and eventually become shorter than the acceleration time of the fragments. Eismont (1965) has used a statistical model to explore this problem, and estimates that when the fragment excitation energy is about 12 MeV, which is approximately the mean fragment excitation energy for thermal neutron fission of ^{235}U, the lifetime is about 10^{-18} sec. This is sufficient time for the fragments to have achieved 99% of their final velocity. When the fragment energy reaches 50 MeV, emission of the first neutron will occur in about 2×10^{-21} sec, a time interval in which the fragments have attained only one half their final velocity. This is difficult to verify experimentally, as when such excitation energies are produced it is likely that significant evaporation of neutrons prior to fission has also occurred. This effect must be considered, however, in the interpretation of neutron yields and angular distributions when the excitation energy is large.

There is experimental information at the upper end of the time scale which shows (Fraser, 1952) that most of the neutrons are emitted at times shorter than 4×10^{-14} sec, as expected. There are a very few neutrons emitted at much later times, i.e., about 1 sec after fission. These delayed neutrons follow β decay of the primary fission fragments in rare situations where the β-decay energy of the parent nucleus is larger than the neutron binding energy of the daughter nucleus. Although the delayed neutrons are very important in the dynamics and control of nuclear reactors, their characteristics do not bear directly on an understanding of the fission process. Further information is contained in the review article by Amiel (1969).

2. Average Neutron Yield $\bar{\nu}$ for Various Fissioning Species

The average number of neutrons per fission is known for a fairly large number of fissioning systems, primarily for spontaneous fission and thermal neutron-induced fission. A summary of the available data is given in Table XII-1. The trend with A and Z can be seen more clearly if the values for thermal neutron fission are corrected to the value expected for zero excitation energy on the basis of an average value of $d\bar{\nu}/dE$ (see Section B, 4). These "corrected" values are illustrated in Fig. XII-2. The general increase in $\bar{\nu}$ with mass number primarily reflects the increase in the available energy release in fission as the mass of the fissioning nucleus increases. Part of the increase in available energy appears as an increase in the kinetic energy of the fragments, and part as an increased number of emitted neutrons. A small part of the increase in the number of evaporated neutrons can be attributed to the fact that the heavier nuclei are more neutron rich, hence, their fragments have lower neutron binding energies, as a result of which they emit slightly more neutrons for a given excitation energy.

3. Distribution of Neutron Emission Numbers

The fact that the average neutron yields are nonintegral implies at least some variation in neutron yield from one fission event to another fission event. The relative probabilities of emitting various numbers of neutrons,

TABLE XII-1

Values of $\bar{\nu}$ (Normalized to $\bar{\nu} = 3.764$ for ^{252}Cf)

Fissioning nucleus	Thermal neutron value	Spontaneous fission value	Ref.
^{230}Th	2.08 ±0.02		*a, p*
^{233}U	3.13 ±0.06		*b*
^{234}U	2.478±0.007		*b, c*
^{236}U	2.405±0.005		*b, c*
^{238}U		2.00±0.08	*f, h*
^{236}Pu		2.22±0.2	*i*
^{238}Pu		2.28±0.08	*i*
^{239}Pu	2.892±0.027		*b, p*
^{240}Pu	2.884±0.007		*b, c*
^{240}Pu		2.16±0.02	*j, l, m*
^{242}Pu	2.91 ±0.02		*b*
^{242}Pu		2.15±0.02	*i, m*
^{244}Pu		2.30±0.19	*s*
^{242}Am	3.22 ±0.04		*b, p*
^{243}Am	3.26 ±0.02		*b*
^{242}Cm		2.59±0.09	*i*
^{244}Cm	3.43 ±0.05		*b*
^{244}Cm		2.76± .07	*i, j*
^{246}Cm	3.83 ±0.03		*b, p*
^{246}Cm		3.00±0.20	*o, r*
^{248}Cm		3.15±0.06	*r, s*
^{250}Cm		3.31±0.08	*s*
^{249}Bk		3.64±0.16	*k*
^{246}Cf		2.85±0.19	*k*
^{250}Cf		3.53±0.09	*s*
^{252}Cf		3.764	*d, e*
^{254}Cf		3.88±0.14	*k, s*
^{254}Fm		3.99±0.20	*g*
^{256}Fm		3.83±0.18	*q*
^{257}Fm		4.02±0.13	*n*

[a] Lebedev and Kalashnikova (1958)
[b] Jaffey and Lerner (1970)
[c] Boldeman and Dalton (1967)
[d] Fillmore (1968)
[e] Westcott *et al.* (1965)
[f] Asplund-Nilsson *et al.* (1963)
[g] Choppin *et al.* (1956)
[h] Shev and Geroy (1960)
[i] Hicks *et al.* (1956)
[j] Diven *et al.* (1956)
[k] Pyle (1964)
[l] Colvin and Sowerby (1965)
[m] Boldeman (1968)
[n] Cheifetz *et al.* (1971)
[o] Thompson (1970)
[p] Kroshkin and Zamgatnia (1970)
[q] Dakowski *et al.* (1972).
[r] Stoughten *et al.* (1973).
[s] Orth (1971)

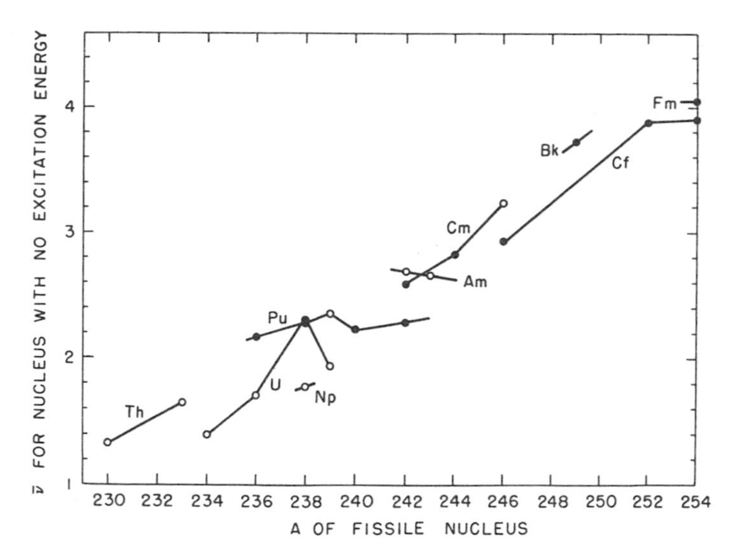

Fig. XII-2. Dependence of the average number of prompt neutrons for a compound nucleus with no excitation energy on the nuclear parameters A and Z. (●) Spontaneous fission; thermal neutron values (○) have been corrected to zero excitation energy assuming $d\bar{\nu}/dE = 0.12\ \text{MeV}^{-1}$. [From Huizenga and Vandenbosch (1962).]

ranging from zero to six or eight, are known for a number of neutron-induced reactions and for several spontaneously fissioning nuclides. The distributions for the different types are very similar, as can be seen by plotting the neutron emission probability P_ν as a function of $\nu - \bar{\nu}$. Such a plot is shown in Fig. XII-3. The discrete neutron emission probabilities exhibit a variance $\sigma_d^2 = \sum_{\nu=0}^{\infty} P_\nu(\nu - \bar{\nu})^2$ which increases only slightly from thermal neutron fission of ^{233}U to spontaneous fission of ^{252}Cf. The smooth curve in Fig. XII-3 results (Terrell, 1965) from the integrals of a continuous Gaussian distribution

$$\sum_{n=0}^{\nu} P_n = (2\pi\sigma^2)^{-1/2} \int_{\infty}^{(\nu - \bar{\nu} + 1/2 + b)/\sigma} \exp[-(\nu - \bar{\nu})^2/2\sigma^2]\, d(\nu - \bar{\nu})$$

with $\sigma = 1.08$. The quantity b is a small correction ($b < 0.01$) to compensate for the fact that negative values of ν are not allowed here. This width corresponds, by Sheppard's correction, to a variance of the discrete distribution of $\sigma_d^2 = 1.25$. A summary of more recent neutron emission parameters is given in Table XII-2.

The width of the continuous Gaussian distribution is expected to be closely related to the width of the total excitation energy distribution of the fragments. Using an average value of $d\bar{\nu}/dE = 0.13\ \text{MeV}^{-1}$, the width of the excitation

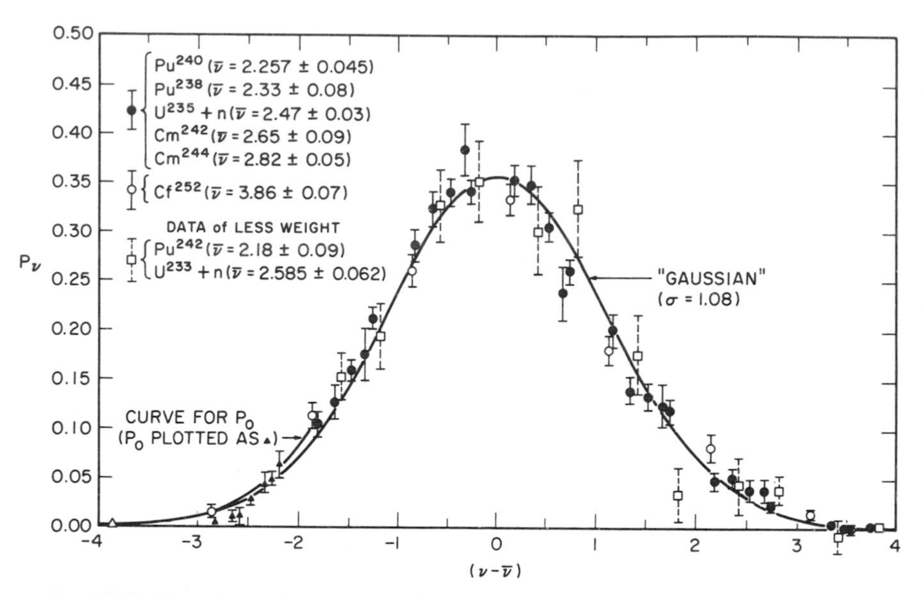

Fig. XII-3. Experimental neutron emission probabilities P_ν for emitting ν neutrons. [From Terrell (1965).]

energy distribution becomes $\sigma/0.13 \simeq 8$ MeV. Any effects due to a variation in the amount of excitation energy appearing as γ rays has been neglected in this estimate. The excitation energy distribution width is of about the same magnitude as that of the fragment kinetic energy distribution. This would be qualitatively expected since for a fixed energy release the available energy is partitioned between kinetic and excitation energies. This can be demonstrated by observing the correlation between the average neutron yield and total kinetic energy, as illustrated in Fig. XII-4. This correlation does not require that the widths of the two distributions be equal, as there are variations in the energy release due to differing charge and mass splits as well.

4. Dependence of Neutron Yield on Excitation Energy

If the excitation energy of the fissioning nucleus is increased, the neutron yield also increases. The magnitude of the increase is similar for all fissioning species, and is qualitatively that expected for the energy cost of neutron emission in terms of the sum of the neutron binding and kinetic energies. That most of the excitation energy increase of the fissioning nucleus appears in the form of enhanced neutron emission is to be expected from the very weak dependence of kinetic energy release on excitation energy. The variation with excitation energy of the division of the available energy between kinetic energy

TABLE XII-2

Neutron Emission Parameters for Thermal Neutron Fission of Four Nuclides and Spontaneous Fission of ^{252}Cf [a,b]

$$\bar{\nu} = \sum_{\nu=0}^{\nu=10} \nu P \quad \langle \nu^2 \rangle = \sum_{\nu=0}^{\nu=10} \nu^2 P_\nu, \quad \text{var} = \sum_{\nu=0}^{\nu=10} P_\nu(\nu-\bar{\nu})^2, \quad R = \frac{\langle \nu^2 \rangle - \bar{\nu}^2}{(\bar{\nu})^2}$$

Nuclide	Neutron Emission Parameters				
	^{233}U	^{235}U	^{239}Pu	^{241}Pu	^{252}Cf
$\bar{\nu}$	2.492 ±0.008	2.416 ±0.008	2.904 ±0.008	2.947 ±0.007	3.784 ±0.000
$\langle \nu^2 \rangle$	7.416 ±0.034	7.073 ±0.032	9.838 ±0.040	10.063 ±0.037	15.925 ±0.007
var	1.099 ±0.004	1.112 ±0.004	1.185 ±0.005	1.173 ±0.004	1.268 ±0.002
R	0.7932±0.0013	0.7979±0.0013	0.8221±0.0016	0.8190±0.0010	0.8479 ±0.0005
P_0	0.0259±0.0010	0.0313±0.0060	0.0094±0.0010	0.0097±0.0010	0.00197±0.00008
P_1	0.1526±0.0020	0.1729±0.0016	0.0990±0.0027	0.0877±0.0025	0.02447±0.00025
P_2	0.3289±0.0034	0.3336±0.0029	0.2696±0.0034	0.2636±0.0030	0.1229 ±0.0005
P_3	0.3282±0.0035	0.3078±0.0029	0.3297±0.0035	0.3343±0.0032	0.2707 ±0.0008
P_4	0.1320±0.0017	0.1232±0.0016	0.1982±0.0030	0.2099±0.0035	0.3058 ±0.0010
P_5	0.0252±0.0020	0.0275±0.0020	0.0924±0.0040	0.0811±0.0040	0.1884 ±0.0007
P_6	0.0045±0.0020	0.0038±0.0015	0.0119±0.0020	0.0112±0.0020	0.0677 ±0.0006
P_7					0.0160 ±0.0003
P_8					0.0021 ±0.0002

[a] From Boldeman and Dalton (1967).

[b] The errors quoted for the various P_ν have been estimated from experimental reproducibility only.

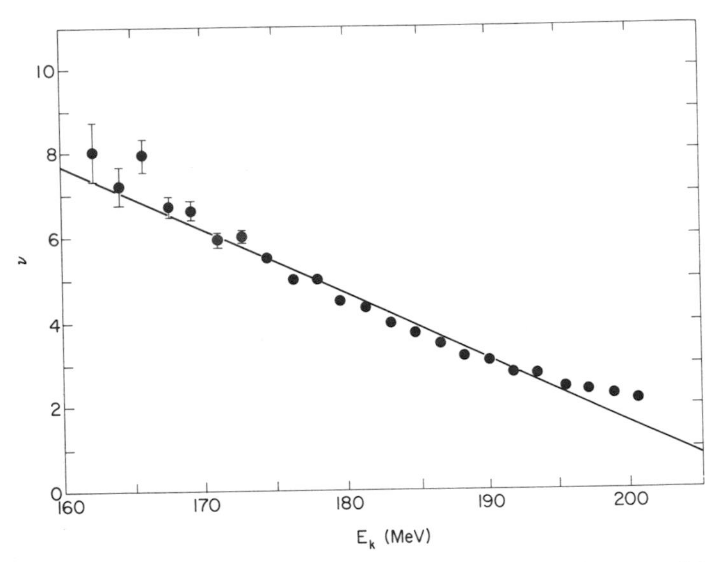

Fig. XII-4. Variation of the number of neutrons $\bar{\nu}$ with total kinetic energy of fragments. The straight line corresponds to a slope of 6.6 MeV per neutron, or $d\bar{\nu}/dE = 0.15$. [From Bowman *et al.* (1963).]

and fragment excitation energy can in fact usually be determined with greater precision by deducing the change in the average fragment excitation energy from the neutron yield than by measuring the difference in the much larger kinetic energy.

The most complete information on the dependence of neutron yield on excitation energy is that for the compound nucleus ^{240}Pu. The neutron yield, illustrated in Fig. XII-5, has been measured for spontaneous fission and for neutron-induced fission of ^{239}Pu over a large range of neutron energies. For neutron energies greater than about 6 MeV some of the neutrons observed are emitted from the compound nucleus prior to fission. The overall slope corresponds to about 7.5 MeV per neutron, somewhat larger than the estimated energy cost of 6.5 MeV for a neutron from a fragment. There are also small but definite deviations from the overall trend, both in this case and for neutron fission of ^{235}U. The variations in the slope with neutron energy are given for selected energy regions in Table XII-3. The slope for the neutron energy range between 0 and 15 MeV is in good agreement with that expected on the basis of the energy cost of a neutron from a fragment. The fragment excitation yield implied by the neutron yield for spontaneous fission of ^{240}Pu is, however, 1.5 MeV greater than expected. This is consistent with the observation that the kinetic energy release in spontaneous fission of ^{240}Pu is 1.5 ± 0.5 MeV lower than for thermal neutron fission of ^{239}Pu (Okolovich and Smirenken,

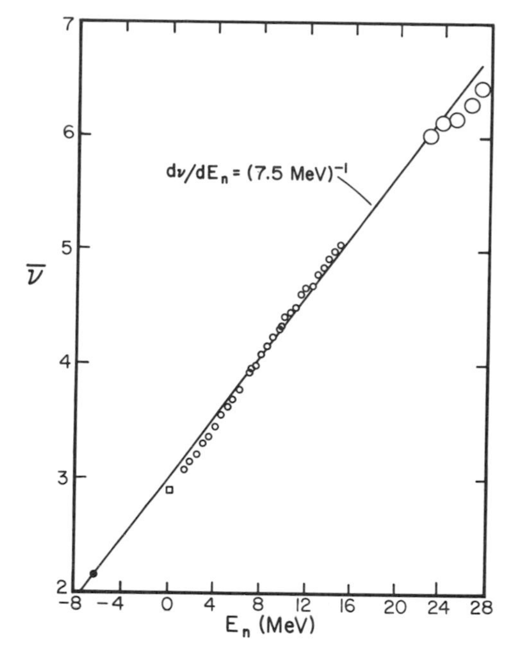

Fig. XII-5. Prompt neutron yield for neutron-induced fission of ^{239}Pu (○, □) and for spontaneous fission of ^{240}Pu (●). The relative errors are approximately the size of the plotted symbols. [Data from Soleilhac *et al.* (1969, 1970); Baron *et al.* (1966).]

1963). The neutron yield for spontaneous fission of ^{246}Cm is also larger than might be expected from an energy cost of 6.5 MeV per neutron and the observed thermal neutron-induced fission $\bar{\nu}$ value.

There is also evidence that $\bar{\nu}$ is lower than might be expected for the highest energies shown in Fig. XII-5. This may reflect the fact that the cost of prefission neutrons is higher by an MeV or so than that of postfission neutrons. Such an effect, however, is expected to first occur at about 6-MeV incident neutron energy. There are indications in the case of neutron-induced fission of uranium that $\bar{\nu}$ experiences perturbations when the thresholds for the (n, n′f) and (n, 2nf) reactions are surpassed.

There have been suggestions of structure in the dependence of $\bar{\nu}$ versus neutron energy in the first MeV or so of bombarding energy which is related to the properties of specific saddle point channels. The more recent data, however, indicate that there is less structure than originally indicated. It has also been suggested that variations in the slope are related to a component of "scission" neutrons whose yield decreases with excitation energy. This also does not seem to be indicated by the more recent data, particularly in the case of ^{240}Pu. In any event the excitation energy dependence of the yield of scission

TABLE XII-3

Dependence of Neutron Yield on Incident Neutron Energy for Selected Energy Intervals for the Compound Nuclei ^{240}Pu and ^{236}U

^{240}Pu			^{236}U		
E_n range (MeV)	$(d\bar{\nu}/dE_n)^{-1}$	Ref.	E_n range (MeV)	$(d\bar{\nu}/dE_n)^{-1}$	Ref.
−6.45(S.F.)–0	8.7 ±0.3	*a, b*	0–2	8.8±0.7	*d*
1.36–5.06	6.6 ±0.1	*c*	1.36–5.06	7.7±0.3	*c*
6.08–11.9	6.6 ±0.1	*c*	5.06–7.48	5.0±0.3	*c*
1.36–14.8	6.67±0.04	*c*	7.48–14.8	7.4±0.1	*c*

[a] Boldeman (1968).
[b] Boldeman and Dalton (1967).
[c] Soleilhac *et al.* (1969).
[d] Boldeman and Walsh (1970).

alphas is very weak, and presumably this is also the case with scission neutrons. The most likely cause of deviations in the neutron yield from that predicted by the simple considerations developed earlier is the influence of initial excitation energy on the partition of energy between kinetic and particle excitation due to the disappearance of shell effects. The excitation energy dependence of the mass distribution will also perturb the neutron yield. An energy balance calculation taking into account the dependences of the neutron yield and kinetic energy on mass division as well as the variation in kinetic energy release and mass division on neutron energy has been performed (Vorobeva *et al.*, 1970). These effects have been demonstrated to account fairly well for the observed irregularities in $\bar{\nu}$ as a function of neutron energy.

5. The Neutron Energy Spectrum

The average neutron kinetic energy in the laboratory system is approximately 2 MeV. Since most of the neutrons are evaporated from the moving fragments, this energy is a composite of the emission energy in the center of mass and the energy given to the neutron by virtue of the fragment's kinetic energy. The latter corresponds to approximately $\frac{2}{3}$ MeV, hence, is about $\frac{1}{3}$ of the total neutron kinetic energy observed.

The form of the emission spectrum in the center of mass is expected to be approximately Maxwellian, $\varepsilon_n \exp(-\varepsilon_n/T)$, where T is the nuclear temperature. If we assume an emission spectrum of this form, and assume that emission occurs from fragments all having the same kinetic energy per nucleon E_f, we

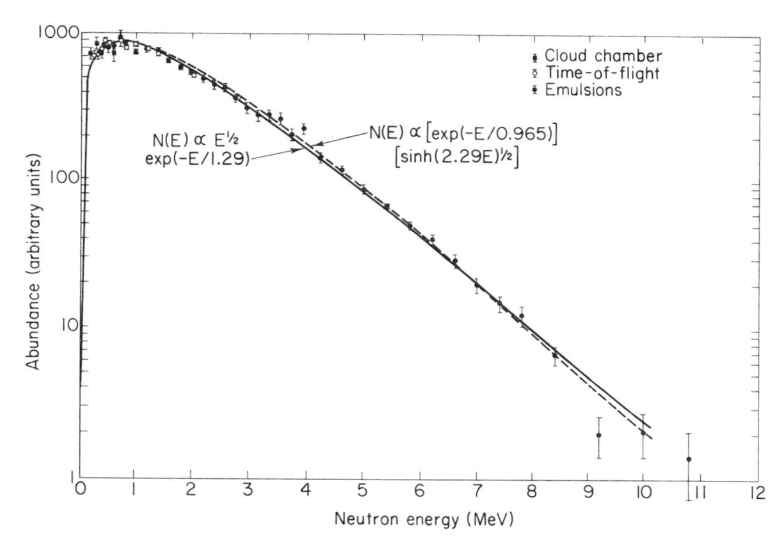

Fig. XII-6. Laboratory neutron kinetic energy spectrum for thermal neutron fission of ^{235}U. [From Leachman (1956).]

expect a laboratory spectrum of the Watt form

$$e^{-E/T} \sinh(4EE_f/T^2)^{1/2}$$

where E is the neutron energy in the laboratory system. This expression is compared with some experimental data in Fig. XII-6. The form of the spectrum turns out to be very little different from that of a Maxwellian. The parameters required to obtain the fit are $T = 0.965$ and $E_f = 0.54$. The latter value is smaller than expected and may reflect the simple assumptions of a single emission spectrum of fixed T and a single fragment energy. In reality we expect the total spectrum to be a superposition of spectra of different temperatures transformed with varying fragment energies. We will defer further consideration of the neutron spectrum until later, where we discuss the neutron yield as a function of fragment mass and energy, in which case the center-of-mass spectrum can be obtained by direct transformation of the experimental spectrum.

C. Dependence of Neutron Yield on Fragment Mass

One of the most interesting aspects of neutron emission is the striking variation in neutron yields as a function of fragment mass. Since the neutrons are primarily emitted from the fully accelerated fragments, there is a strong angular correlation between the neutrons and the fragments from which the

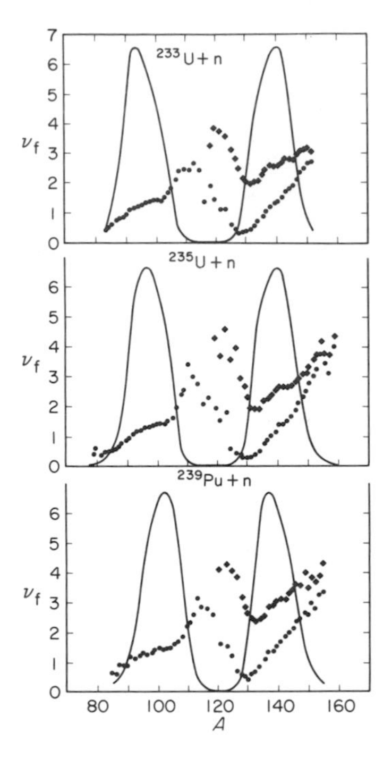

Fig. XII-7. The average number of neutrons emitted per fragment as a function of fragment mass (●) and the average total number of neutrons per fission as a function of heavy fragment mass (◆) for thermal neutron fission of ^{233}U, ^{235}U, and ^{239}Pu. [After Apalin *et al.* (1965).]

neutrons came. This makes possible an experimental determination of the neutron yield as a function of fragment mass. The first evidence for structure in the dependence of neutron yield on fragment mass was obtained by Fraser and Milton (1954), and considerably more detailed information is presently available. Some results for thermal neutron fission are illustrated in Fig. XII-7. Not only are there very unequal numbers of neutrons from the two complementary fragments for particular mass splits, but even the total number of neutrons emitted varies as a function of mass split.

It is also possible to determine the dependence of neutron yield on fragment mass by a comparison of initial and final mass yields. The results are in agreement with those obtained by direct neutron counting. The superimposed dependences of neutron yield on fragment mass for a number of fissioning systems are shown in Fig. XII-8. This comparison emphasizes the similarity of the neutron yields from the different systems for a given fragment mass, particularly the low neutron yields for fragments in the vicinity of closed shells. This led Terrell (1962, 1965) to the conclusion that the neutron yields

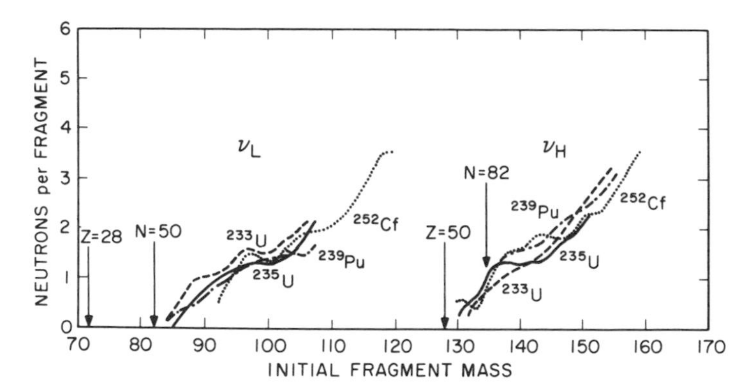

Fig. XII-8. Neutron yields as a function of fragment mass for four types of fission, as determined from mass–yield data. [From Terrell (1965).]

are closely related to the deformabilities of the nascent fragments. Closed-shell and near-closed-shell nuclei prefer spherical shapes and are unusually resistant to deformation. If the complementary fragment of such nuclei is not close to a closed shell, most of the deformation energy at scission will be in the complementary fragment, with the closed-shell fragment having little deformation. After scission and separation of the fragments, the neutron yields will be dependent on the amount of energy stored in deformation at scission. If both fragments are mid-shell nuclei which are unusually soft with respect to deformation, the scission configuration will have a large amount of deformation energy, as well as a large center-to-center separation which will result in low kinetic energies. This appears to be the case for symmetric fission of the uranium and plutonium isotopes, which exhibit a large total neutron yield and a large deficit in kinetic energy. The effect is much less pronounced in the case of ^{252}Cf because of the proximity of the $Z = 50$ shell for symmetric fragments. These ideas have been explored more quantitatively and will be discussed in further detail later.

The apparent near-universality of the neutron yield dependence on fragment mass implied in Fig. XII-8 must be somewhat qualified. If the Apalin *et al.* (1964) results for the symmetric region of ^{235}U had been included in the figure, the neutron yield for fragments with $A \sim 118$ from ^{235}U would be seen to not be as large as for ^{252}Cf. This is evidence that part of the large yield at $A \sim 120$ for ^{252}Cf is due to the influence of the complementary closed-shell fragment, whereas in the case of ^{235}U for $A \sim 120$ the complementary fragment is also mid-shell and does not exploit the deformability of its partner to the same extent that a closed-shell nucleus does. It must be remembered, however, that the total neutron yield is higher for ^{252}Cf than for ^{235}U, accounting in part for the larger yield at $A \sim 120$ for ^{252}Cf.

There is much less experimental data on the variation of the neutron yield with fragment mass for the lighter elements. Results obtained by an indirect method for proton-induced fission of ^{226}Ra (Konecny and Schmitt, 1968) show $\bar{\nu}(A)$ rising with increasing A until $A \sim 122$, with a broad minimum at $A = 135$–145. The same technique has been applied to α particle-induced fission of ^{209}Bi (Plasil *et al.*, 1969), and $\bar{\nu}(A)$ has been found to increase monotonically over the mass region studied, $A = 80$ to $A = 125$. In both of these studies the kinetic energy release was also studied, and the total energy release deduced exhibits significant discrepancies with that expected from mass equations. These indirect results on $\bar{\nu}(A)$ should therefore be considered somewhat preliminary.

The strong "saw-toothed" variation of neutron yield with fragment mass observed in low energy fission is fairly rapidly washed out with increasing excitation energy, as illustrated in Fig. XII-9. There is more structure for ^{238}U

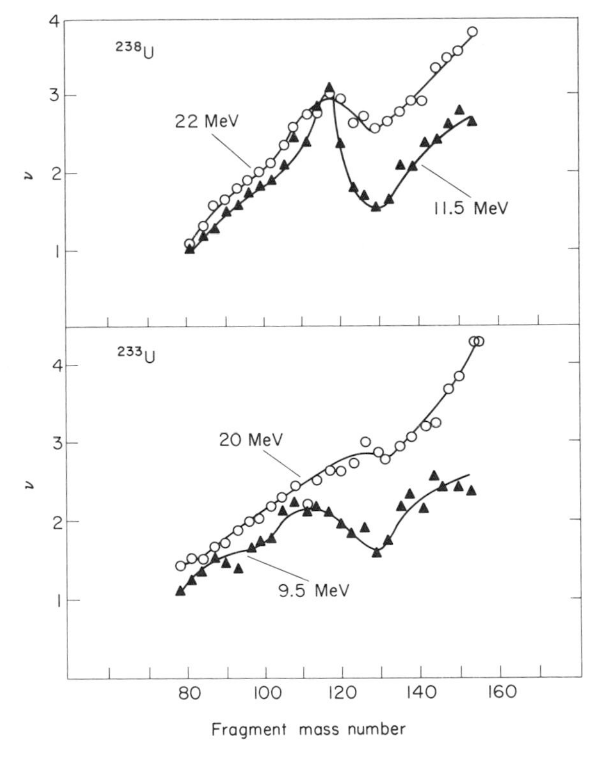

Fig. XII-9. Dependence of neutron yield on fragment mass for proton-induced fission of ^{233}U and ^{238}U. The incident proton energy is indicated for each curve. These results were obtained for neutrons emitted at 0° with respect to a fragment, hence include very few prefission neutrons. [From Bishop *et al.* (1970).]

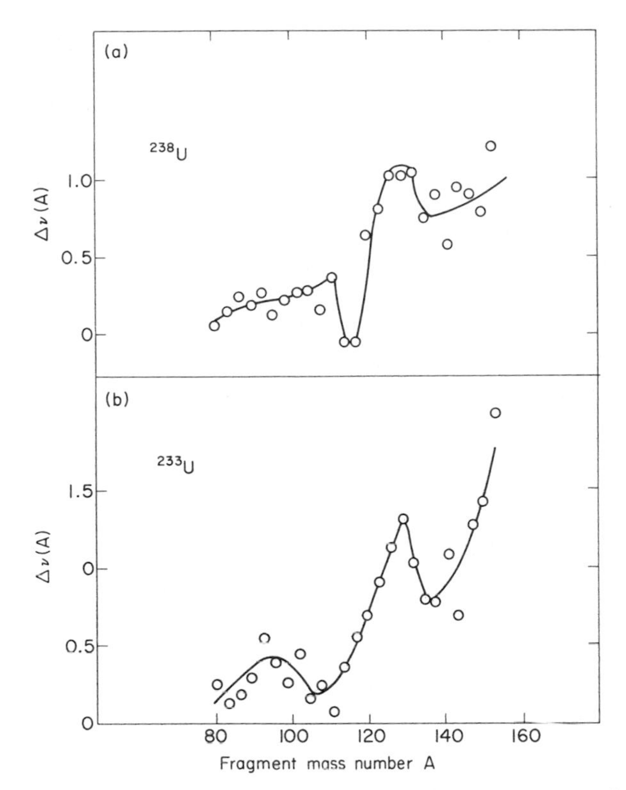

Fig. XII-10. (a) The difference between the neutron yields at proton energies of 11.5 and 22 MeV for ^{238}U, and (b) the difference between the yields at proton energies of 9.5 and 20 MeV for ^{233}U. [From Bishop *et al.* (1970).]

than for ^{233}U due to the larger fraction of second-chance, lower energy fission for the former. The effects of excitation energy on $\bar{\nu}(A)$ can be seen more clearly in Fig. XII-10, where we plot $\Delta\bar{\nu}(A)$, the difference in $\bar{\nu}(A)$ at two bombarding energies.

The disappearance of the saw-toothed structure with increasing excitation energy is associated with a large increase in the neutron yield at $A \approx 130$, and with an abnormally small increase for the complementary fragment. While the maximum in $\Delta\bar{\nu}$ occurs at about the same mass in both cases, the minimum in $\Delta\bar{\nu}$ occurs at a lower fragment mass for the lighter target nucleus. The anomalously small neutron yield for fragments with $A \approx 132$ has been attributed to the special stability of the spherical configuration for this mass region and the accompanying resistance to deformation. As the excitation energy is increased, shell effects disappear, resulting in a more stretched scission configuration, a lower total kinetic energy release, and a higher total excitation energy of the fragments. It is interesting to note, however, that essentially all of the increased excitation energy appears in the heavy fragment, largely because it can now accept a more equal share of the deformation energy of the scission configuration.

In addition to the dissolution of shell structure for fragments with $A \approx 130$ and for their partners, the $\Delta\bar{\nu}$ curves in Fig. XII-10 indicate that away from the regions of closed shells the heavy fragment still receives a higher fraction of the increase in the total excitation energy. This can be at least partly attributed to a tendency toward thermal equilibrium in the nascent fragments, which results in approximately the same excitation energy per nucleon in each fragment, or a higher excitation energy in the heavier fragment. A more quantitative examination of this hypothesis can be obtained by examining the $\Delta\bar{\nu}(A)$ values for a particular mass split away from closed shells. The $\Delta\bar{\nu}$ value for the light and heavy fragments is a measure of the division of the increment in excitation energy between the two fragments. (This use of $\Delta\bar{\nu}$ rather than $\bar{\nu}$ avoids the problem of how much energy was tied up in deformation at scission, as long as it remains constant with excitation energy. The assumption of the constancy of the deformation energy is consistent with the constancy of the kinetic energy with excitation energy for this mass split.) If we assume the fragments are in thermal equilibrium as scission is approached, then $T_{\text{H}} = T_{\text{L}}$, with the nuclear temperatures given by $(E/a)^{1/2}$. Since the level density parameter a is expected to vary linearly with A for fragments away from closed shells,

$$(\Delta E_{\text{H}}/A_{\text{H}})^{1/2} = (\Delta E_{\text{L}}/A_{\text{L}})^{1/2} \qquad \text{or} \qquad \Delta E_{\text{L}}/\Delta E_{\text{H}} = A_{\text{L}}/A_{\text{H}}$$

where ΔE_{L} and ΔE_{H} are the difference in fragment excitation energies for two different bombarding energies. The quantity $\Delta E_{\text{L}}/\Delta E_{\text{H}}$ can be determined from the experimentally observed $\Delta\bar{\nu}$ values and estimates of the mean energy required to emit a neutron, $\bar{B}_{\text{n}} + \varepsilon$, which is somewhat different for light and heavy fragments. For the $A_{\text{L}} \approx 92$ and $A_{\text{H}} \approx 142$ mass split for ^{233}U

$$\frac{\Delta E_{\text{L}}}{\Delta E_{\text{H}}} \approx \frac{[\Delta\bar{\nu}(A_{\text{L}})(\bar{B}_{\text{n}} + \varepsilon)]_{A_{\text{L}} \approx 92}}{[\Delta\bar{\nu}(A_{\text{H}})(\bar{B}_{\text{n}} + \varepsilon)]_{A_{\text{H}} \approx 142}} \approx 0.5 \pm 0.2$$

as compared to the expected value of

$$A_{\text{L}}/A_{\text{H}} = 0.65$$

For proton-induced fission of ^{238}U, the observed ratio of excitation energy increments is 0.35 ± 0.2 for the same mass ratio. Thus in the case of ^{233}U most of the smaller yield of neutrons from the light fragment can be attributed to simply sharing the energy so as to give equal excitation energy per nucleon. In the case of ^{238}U, however, an additional enhancement of the neutron yield from the heavy fragment is indicated, for reasons that are not readily apparent.

Finally, we might wonder why the very low neutron yields observed in the vicinity of $A \approx 80$, which is where the $N = 50$ closed shell appears, do not increase more rapidly with increasing excitation energy. A possible explanation for the smallness of the increase here may be related to the observation earlier that the increment in excitation energy is not distributed equally between light and heavy fragments. The $A \approx 80$ nascent fragment, if in near thermal equilibrium with its partner at scission, will get a small fraction of the increased excitation energy because of its very heavy complementary fragment. This is to be contrasted with the situation for the $A \approx 130$ closed-shell fragment, which has a light complementary partner and hence competes more successfully for the increment in excitation energy. Thus the $A \approx 80$ nascent fragment may not acquire sufficient excitation energy to destroy the shell effects.

D. Correlation between Neutron Yields of Complementary Fragments

A potentially interesting feature of the neutron yield distribution is that of a possible correlation between the number of neutrons emitted by one fragment and the number of neutrons emitted by the complementary fragment. That is to say, if one fragment happens to emit a larger than average number of neutrons, is the complementary fragment most likely to emit more than, less than, or its average number of neutrons? The answer often assumed is that the complementary fragment is most likely to emit an average number of neutrons; i.e., that the individual fragment yields are uncorrelated. (If the kinetic energy and mass ratio were fixed, there would certainly be an anticorrelation. We are concerned here, however, with the general case of no constraint on the total kinetic energy.)

There are two recent experiments which provide some tentative evidence that the neutron yields are anticorrelated, i.e., that if one fragment emits a more than average number of neutrons, the complementary fragment will most likely emit a less than average number of neutrons. Blinov *et al.* (1970) have measured the mean product $\langle \nu_L \nu_H \rangle$ to be 1.53 ± 0.08 for thermal neutron fission of ^{235}U. This is smaller than a value of 1.82 ± 0.10 calculated on the basis of no correlation in yields. This comparison would imply a fairly large anticorrelation, but unfortunately this technique of measuring $\langle \nu_L \nu_H \rangle$ is not very sensitive to the presence of correlations.

In a more direct experiment Nifenecker *et al.* (1969) used a high-efficiency neutron detector to directly measure the dispersion in the number of neutrons emitted from light or heavy fragments traveling in the direction of the neutron detector. They found for ^{252}Cf that

$$\sigma_L^2 = 1.108 \qquad \text{and} \qquad \sigma_H^2 = 1.446$$

The sum of these two variances, 2.554, is much larger than the variance in the total number of neutrons per fission, 1.55 ± 0.04. This gives a value of the correlation coefficient μ, defined by

$$\sigma_T^2 = \sigma_L^2 + \sigma_H^2 + 2\mu$$

of -0.55. Again a fairly strong anticorrelation is indicated. There is one difficulty which should be mentioned concerning the manner in which the light and heavy fragments were identified in this experiment. The method was based on the charge as determined by the measurement of K x rays in coincidence with fragment neutrons. There is therefore some possibility for bias in the mass divisions studied due to the strong variation of x-ray yield with fragment mass.

The evidence that the fragment yields are anticorrelated implies that if one fragment has a higher than average excitation energy, its partner will have a lower than average excitation energy. The final fragment excitation energy originates from both the fragment excitation energy at scission and any distortion energy at scission. If the distortion energy at scission were small compared with the internal excitation energy, the statistical model of fission would predict a positive correlation. This is because the two fragments are assumed to have equal temperatures for the most probable fission events. Since, however, the internal excitation energy at scission is probably comparable to or smaller than the distortion energy, it is necessary to consider how the distortion energy might be divided between the fragments.

This question has been explored in the framework of the liquid drop model by Nix and Swiatecki (1965). They found for ^{213}At (the two-spheroid model employed is only appropriate for low values of Z^2/A) that the excitation energies are expected to be anticorrelated. This result is dependent on the relative amplitudes of the pure stretching mode and the distortion asymmetry mode. The former corresponds to correlated fragment excitation energies, while the latter gives anticorrelated excitation energies. They found that the potential energy in the vicinity of the saddle point (which is fairly similar to the scission configuration for low Z^2/A) is stiffer with respect to the stretching mode than with respect to the distortion-asymmetry mode, giving rise to larger distortion-asymmetry amplitudes and hence to anticorrelation of fragment excitation energies.

E. Dependence of Neutron Kinetic Energy on Fragment Mass

The dependence of the average center-of-mass neutron kinetic energy on fragment mass has been determined for a number of systems. Since most of

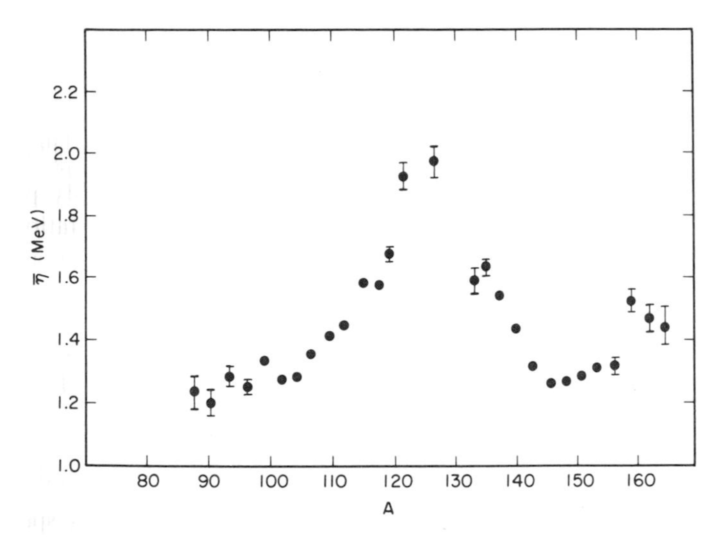

Fig. XII-11. The average center-of-mass neutron kinetic energy as a function of fragment mass for spontaneous fission of ^{252}Cf. [From Bowman *et al.* (1963).]

the neutrons are emitted after separation of the fragments and conversion of their deformation energy into excitation energy, the neutron kinetic energies are a reflection of the temperatures of the final separated fragments. The results for spontaneous fission of ^{252}Cf are illustrated in Fig. XII-11. Although there is considerable structure in this curve, the neutron kinetic energies are not as simply correlated with the fragment structure as the neutron yield. In particular, the average energy is nearly symmetric with respect to mass 126 (symmetric mass division), whereas the neutron yield is very asymmetric about $A = 126$. Fragments around mass 120, which emit an average of four neutrons, and fragments around mass 132, which emit less than one neutron, both evaporate neutrons with very nearly equal energies. This apparently is another manifestation of the shell effects. The shell effects which are responsible for near-closed-shell nuclei acquiring only small amounts of deformation energy at scission (hence, little final excitation energy) also result in these same fragments having an unusually high temperature for a given excitation energy. On the other hand, mid-shell nuclei produced in fission will exhibit lower than average temperatures for a given excitation energy. Thus there may be a fortuitous near cancellation of the asymmetry in excitation energy division by the mass dependence of the temperature on excitation energy. The peak in the average neutron kinetic energy at symmetry in Fig. XII-11 can be qualitatively attributed to the increase in total neutron yield (hence, in fragment excitation energy) near symmetry. The total neutron yield exhibits a broad minimum at mass splits in the vicinity of the $A = 148$, 104 mass split.

It is possible to more quantitatively relate the neutron kinetic energies with the neutron yields in order to isolate the nuclear level density parameter defining the relationship between temperature and fragment excitation energy. Lang (1963) has derived an expression for an evaporation cascade which relates the average neutron energy $\bar{\varepsilon}$, the average binding energy $\bar{B}_n$, the neutron yield v, and the level density parameter a. For a Fermi gas, Lang's formulation gives $E = aT^2 - \frac{5}{4}T$. (The second term is small and has been neglected in our previous discussion of the division of excitation energy between the fragments.) The initial fragment excitation energy E_i is $v(\bar{B}_n + \bar{\varepsilon})$, and the average neutron energy averaged over the cascades as well as the energy spectrum is $\frac{4}{3}T_i$, where T_i is the temperature for the first emission. This leads to the expression

$$\bar{v}(\bar{B}_n + \bar{\varepsilon}) = a(\tfrac{3}{4}\bar{\varepsilon})^2 - \tfrac{15}{16}\bar{\varepsilon}$$

For systems in which $\bar{v}$ and $\bar{\varepsilon}$ have been determined experimentally as a function of fragment mass, values of a as a function of initial fragment mass can be determined by using $\bar{B}_n$ values from available mass formulas. The results of such an analysis for several fissioning nuclei are illustrated in Fig. XII-12. Although the general trend is given by $a = \frac{1}{8}A$, there are variations with mass number corresponding to low values of a close to closed shells and high values at mid-shell mass values. (It must be remembered that the a values are plotted here as a function of initial fragment mass, so that the average fragment masses determining the neutron emission spectrum are several mass units smaller than indicated on the abscissa.) It is clear that as the average

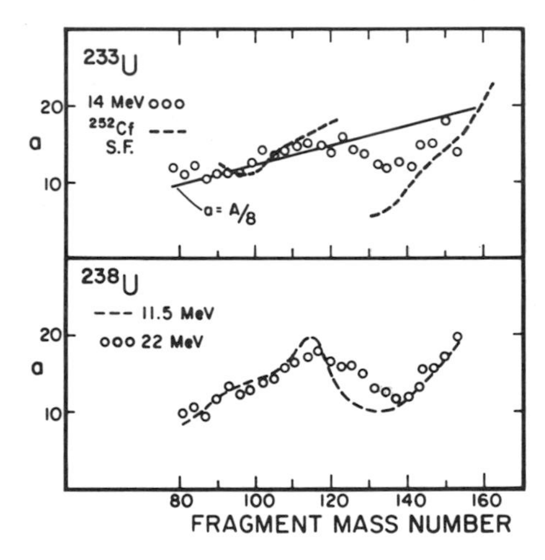

Fig. XII-12. Level density parameter a as a function of initial fragment mass. The broken line for ^{252}Cf is from Lang's (1963) analysis of the Bowman *et al.* (1962) data. The values for uranium are from Bishop *et al.* (1970).

fragment excitation energy is increased, the shell effects become less prominent. This is shown not only by the increase of the *a* values near closed shells but also by a decrease in the *a* values for mid-shell nuclei. Both of these effects are expected, as the single particle excitations are spread out farther from the Fermi surface and the local variations in single particle level density due to shell degeneracies are averaged out.

The smaller variation of *a* with mass number in the case of ^{233}U as compared to ^{238}U is attributed to the higher average initial excitation energy of the fragments in the former case. The fragments from the fission of ^{233}U have a higher initial excitation energy because a larger fraction of the fission events are associated with first chance fission.

References

Amiel, S. (1969). *Proc. IAEA Symp. Phys. Chem. Fission, 2nd, Vienna, 1969*, p. 569. IAEA, Vienna.

Apalin, V. F., Gritsyuk, Yu. N., Kutikov, I. E., Lebedev, V. I., and Mikaelyan, L. A. (1964). *Nucl. Phys.* **55**, 249.

Apalin, V. F., Gritsyuk, Yu. N., Kutikov, I. E., Lebedev, V. I., and Mikaelyan, L. A. (1965). *Proc. IAEA Symp. Phys. Chem. Fission, Salzburg, 1965*, **1**, 586. IAEA, Vienna [as arranged by J. E. Gindler, and J. R. Huizenga (1968). *In* "Nuclear Chemistry" (L. Yaffe, ed.), Vol. 2. Academic Press, New York].

Asplund-Nilsson, I., Conde, H., and Starfelt, N. (1963). *Nucl. Sci. Eng.* **15**, 213.

Baron, E., Frehaut, J., Ouvry, F., and Soleilhac, M. (1966). *Proc. IAEA Conf. Nucl. Data for Reactors*, **2**, p. 57. IAEA, Vienna.

Bishop, C. J., Vandenbosch, R., Aley, R., Shaw, R. W., Jr, and Halpern, I. (1970). *Nucl. Phys. A* **150**, 129.

Blinov, M. V., Kazarimov, N. M., Krisyuk, I. T., and Kovalenko, S. S. (1970). *Sov. J. Nucl. Phys.* **10**, 533.

Boldeman, J. W. (1968). *J. Nucl. Energy* **22**, 63.

Boldeman, J. W., and Dalton, A. W. (1967). *Aust. At. Energy Comm. Rep. AAEC/E 172.*

Boldeman, J. W., and Walsh, R. L. (1970). *J. Nucl. Energy* **24**, 191.

Bowman, H. R., Thompson, S. G., Milton, J. C. D., and Swiatecki, W. J. (1962). *Phys. Rev.* **126**, 2120.

Bowman, H. R., Milton, J. C. D., Thompson, S. G., and Swiatecki, W. J. (1963). *Phys. Rev.* **129**, 2133.

Cheifetz, E., Bowman, H. R., Hunter, J. B., and Thompson, S. G. (1971). *Phys. Rev. C* **3**, 2017.

Choppin, G. R., Harvey, B. G., Hicks, D. A., Ise, J., Jr., and Pyle, R. V. (1956). *Phys. Rev.* **102**, 766.

Colvin, D. W., and Sowerby, M. G. (1965). *Proc. IAEA Symp. Phys. Chem. Fission, Salzburg, 1965*, **2**, p. 25. IAEA, Vienna.

Dakowski, M., Lazarev, Yu. A., and Oganesyan, Yu. Ts. (1972). Dubna Preprint, pp. 15–6518.

Diven, B. C., Martin, H. C., Taschek, R. F., and Terrell, J. (1956). *Phys. Rev.* **101**, 1012.

Eismont, V. F. (1965). *Sov. At. Energy* **19**, 1000.

Fillmore, F. L. (1968). *J. Nucl. Energy* **22**, 79.

Fraser, J. S. (1952). *Phys. Rev.* **88**, 536.
Fraser, J. S., and Milton, J. C. D. (1954). *Phys. Rev.* **93**, 818.
Hicks, D. A., Ise, J., Jr., and Pyle, R. V. (1956). *Phys. Rev.* **101**, 1016.
Huizenga, J. R., and Vandenbosch, R. (1962). *In* "Nuclear Reactions" (P. M. Endt and P. B. Smith, eds.), Vol. II. North-Holland Publ., Amsterdam.
Jaffey, A. H., and Lerner, J. L. (1970). *Nucl. Phys. A* **145**, 1.
Konecny, E., and Schmitt, H. W. (1968). *Phys. Rev.* **172**, 1213.
Kroshkin, N. I., and Zamgatnia, Y. S. (1970). *Sov. At. Energy* **29**, 790.
Lang, D. W. (1963). *Nucl. Phys.* **53**, 113.
Leachman, R. B. (1956). *Proc. Int. Conf. Peaceful Uses At. Energy, 1956*, **2**, p. 193. United Nations, New York.
Lebedev, V. I., and Kalashnikova, V. I. (1958). As quoted by Bondarenko, I. I., Kuzminov, B. D., Kutsayeva, L. S., Prokhorova, L. I., and Smirenkin, G. N. *Proc. U.N. Int. Conf. Peaceful Uses At. Energy, 2nd*, **15**, p. 352. United Nations, Geneva.
Milton, J. C. D., and Fraser, J. S. (1965). *Proc. IAEA Symp. Phys. Chem. Fission, Salzburg, 1965*, **2**, p. 39. IAEA, Vienna.
Nifenecker, H., Frehaut, J., and Soleilhac, M. (1969). *Proc. IAEA Symp. Phys. Chem. Fission, 2nd, Vienna, 1969*, p. 491. IAEA, Vienna.
Nix, J. R., and Swiatecki, W. J. (1965). *Nucl. Phys.* **71**, 1.
Okolovich, V. N., and Smirenken, G. N. (1963). *Sov. Phys. JETP* **16**, 1313.
Orth, C. J. (1971). *Nucl. Sci. Eng.* **43**, 54.
Plasil, F., Ferguson, R. L., and Schmitt, H. W. (1969). *Proc. IAEA Symp. Phys. Chem. Fission, 2nd, Vienna, 1969*, p. 505. IAEA, Vienna.
Pyle, R. V. (1964). Unpublished results as given in E. K. Hyde, "The Nuclear Properties of the Heavy Elements, III: Fission Phenomena." Prentice–Hall, Englewood Cliffs, New Jersey.
Shev, R., and Geroy, J. (1960). *J. Nucl. Energy Part A* **12**, 101.
Soleilhac, M., Frehaut, J., and Gauriau, J. (1969). *J. Nucl. Energy* **23**, 257.
Soleilhac, M., Frehaut, J., and Gauriau, J. (1970). Private communication.
Stoughten, R. W., Halperin, J., Bemis, C. E., and Schmitt, H. W. (1973). *Nucl. Sci. Eng.* **50**, 169.
Terrell, J. (1962). *Phys. Rev.* **127**, 880.
Terrell, J. (1965). *Proc. IAEA Symp. Phys. Chem. Fission, Salzburg, 1965*, **2**, p. 3. IAEA, Vienna.
Thompson, M. C. (1970). *Phys. Rev. C* **2**, 763.
Vorobeva, V. G., Dyachenko, P. P., Kuzminov, B. D., Sergachev, A. I., and Smirenko, L. D. (1970). *Sov. At. Energy* **29**, 835.
Westcott, C. H., Ekberg, K., Hanna, G. C., Pattenden, N. J., Sanatini, S., and Attree, P. M. (1965). *At. Energy Rev.* **3**, 1.

CHAPTER

XIII

Gamma Rays from Primary Fission Products

A. Introduction

We will be concerned here with γ rays originating from the excited fragments prior to β decay. Since the shortest β-decay half lives are of the order of milliseconds, we can assume that all γ rays which are emitted within 0.1 msec of fission originate from the primary fission products. The qualitative characteristics of these γ rays can be quickly summarized. More than one half of these γ rays have half lives shorter than 2×10^{-11} sec. There are typically about 8 photons per fission with an average of 1 MeV per photon, resulting in a total γ-ray decay energy of approximately 8 MeV. Both the average number and average energy per photon are functions of fragment mass.

As will be discussed more fully later, even the qualitative features, such as the γ-ray multiplicity and average energy, are rather different from simple statistical model expectations or from comparison with γ rays from neutron capture. It will be shown that this difference can be attributed to the presence of rather large amounts of angular momentum in the fragments. Thus the characteristics of the fission γ rays, which are emitted in a rather late stage of the fission process, can provide interesting information about a property of the fragments determined at a considerably earlier stage of the fission process.

Recently, techniques have improved sufficiently to make it possible to perform detailed studies of γ rays associated with individual fission fragments. These spectroscopic studies lead to very interesting results regarding the properties of nuclides far from the line of β stability. It is, however, beyond the scope of this work to review this aspect of γ decay, and we will concentrate our attention on those properties which are more characteristic of the fission process than the fission products.

We first review those properties of the γ rays corresponding to integration over all mass divisions, and then the dependence of these properties on mass division where such information exists. We conclude with a discussion of theoretical estimates of the amount of angular momentum given to the fragments during fission.

B. Gamma Rays from Fragments of All Masses

1. The Time of Emission

It is possible to measure electronically, by the delayed coincidence method, time differences between fission and γ-ray emission of 10^{-9} sec or greater. Somewhat shorter times may be reached by exploiting the velocity of the fission fragments, which is approximately $\sim 10^9$ cm/sec. This technique, pioneered by Skliarevski *et al.* (1957), replaces a time measurement by a measurement of

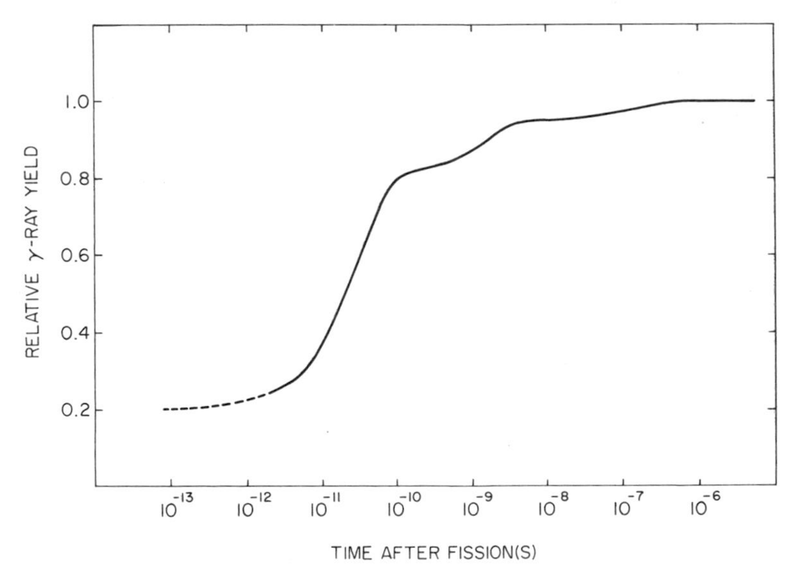

Fig. XIII-1. Fraction of γ-ray yield emitted up to a certain time after fission as a function of time after fission. [From Maier-Leibnitz *et al.* (1965a).]

the distance traveled by the fragment before emission of the γ ray. This distance measurement is achieved by suitable collimation of the detector with respect to the flight path of the fragment.

The various time measurements are summarized in a curve constructed by Maier-Leibnitz *et al.* (1965a) and shown in Fig. XIII-1. It has been assumed that there are no significant differences between thermal neutron fission of ^{235}U and spontaneous fission of ^{252}Cf in the construction of this curve. Comparison of the average lifetime with the Weisskopf estimate for 1-MeV radiation indicates that only dipole and quadrupole radiation are important for the majority of the γ rays.

2. The Gamma-Ray Energy Spectrum

The γ-ray energy spectrum has been measured for thermal neutron fission of ^{235}U and spontaneous fission of ^{252}Cf and found to be rather similar. The energy spectrum of ^{235}U is shown in Fig. XIII-2. The most recent investigation of ^{235}U (Peelle and Maienschein, 1971) gives 7.45 ± 0.3 photons per fission and 7.2 ± 0.3 MeV per fission for photons between 0.14 MeV and 10.0 MeV. Less precise measurements for spontaneous fission of ^{252}Cf give 10 photons per fission and a total photon energy release of 8.6 MeV (Bowman and Thompson, 1958; Smith *et al.*, 1958). These numbers give an average γ-ray energy of slightly less than 1 MeV. The average total γ-ray energy for 2.8- and 14-MeV neutron-induced fission of ^{235}U is the same as for thermal neutron-induced fission to within 15% (Protopopov and Shiryaev, 1958).

The total γ-ray yield is considerably larger than would be expected on the basis of simple statistical arguments if it is assumed that γ-ray emission does not occur when it is energetically possible to emit neutrons. In view of the large dispersion in the number of neutrons emitted from one fission event to another, residual excitation energies between zero and the neutron binding energy are equally probable after neutron emission is no longer energetically possible. Thus each fragment will have an average excitation energy equal to one half of the neutron binding energy after neutron emission is no longer possible, and the total γ-ray energy expected for both fragments is approximately equal to the average neutron binding energy, or 5–6 MeV. This is several MeV less than the 7–9 MeV of total γ-ray energy observed, and the discrepancy is attributed to the fact that the fragments are formed with sizeable amounts of angular momentum. The emitted neutrons do not carry away much of this angular momentum. Hence, the neutron emission probability of the last possible neutron may be inhibited due to a paucity of high spin states near the ground state of the residual nucleus. Thus, neutron emission is hindered by an angular momentum barrier, and γ-ray decay is

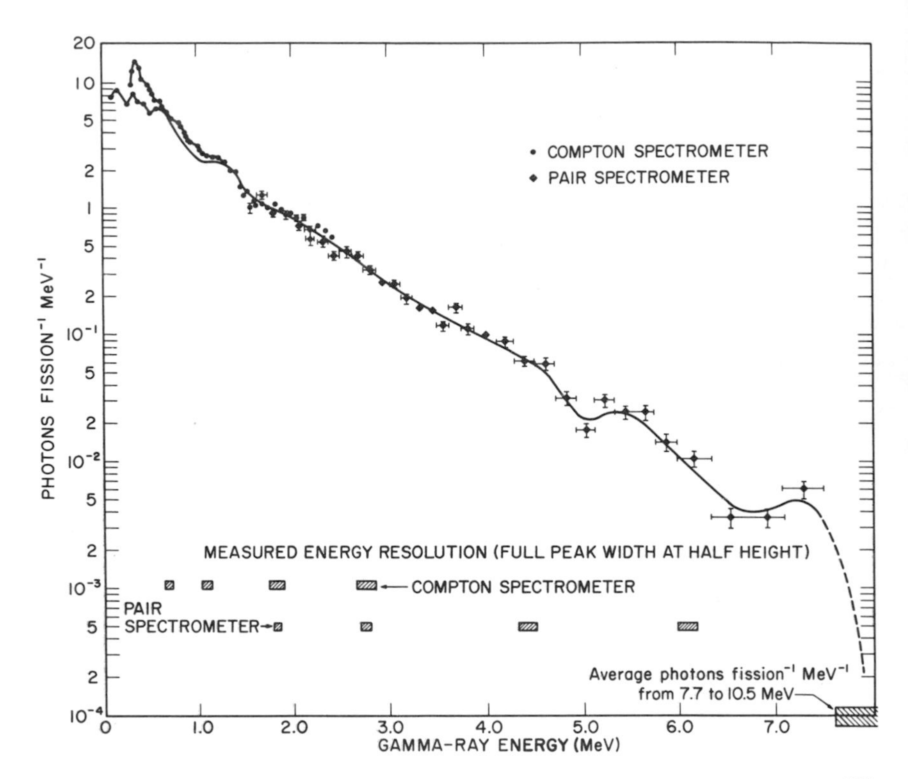

Fig. XIII-2. The energy spectrum of prompt γ rays from thermal neutron fission of ^{235}U as measured by Maienschein *et al.* (1958) and Rau (1963) (low energy portion of the curve). [After Maier-Leibnitz *et al.* (1965a).]

able to compete successfully because the γ ray can decay to higher lying states possessing higher angular momentum. These ideas have been supported by semiquantitative calculations of Thomas and Grover (1967). They made estimates of the spin values of the low lying states on the basis of the shell model and used a spin-dependent level density expression at higher excitation energies. Values of the radiation width were obtained from slow neutron resonance spectroscopy. Assuming each fragment to have an initial root-mean-square (rms) angular momentum of $8.4\hbar$, they found the total γ-ray energy for a typical pair of fragments to be 7.1 MeV, or approximately 50% more than if they neglected inhibition of neutron emission due to angular momentum. Inclusion of angular momentum effects was also shown to decrease the average photon energy. Thus the angular momentum of the fragment is also responsible for the average energy of the photons from fission being less than that for neutron capture γ rays.

3. Angular Distribution of Fission Gamma Rays

An angular correlation between the fission fragments and the prompt γ rays has been observed, with emission in the direction of the fragments more probable by about 15% than for emission perpendicular to the fragment direction. The results of Hoffman (1964) for thermal neutron fission of three targets is shown in Fig. XIII-3. The sign of the anisotropy is that expected for quadrupole radiation if the axis of rotation of each fragment lies in a plane perpendicular to the fragment direction. This is expected if the fragment angular momentum is a result of a bending mode at scission. Any angular momentum present in the original fissioning system appears later in the final fragments (but with the axis of rotation tending to lie in the direction of fragment motion) and in the relative orbital angular momentum between the fragments. For the low values of K (K is the projection of the angular momentum along the nuclear symmetry axis) expected for thermal neutron fission most of the angular momentum in the initial nucleus is expected to appear

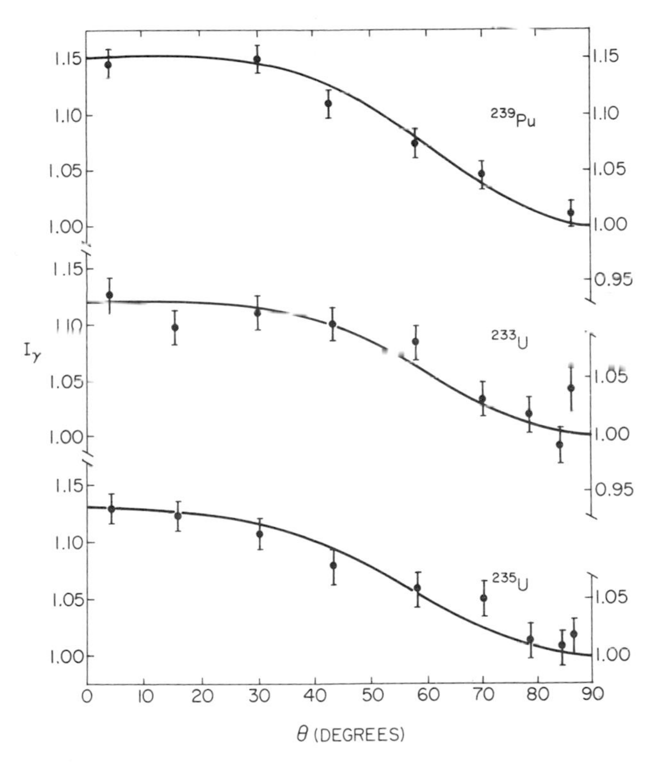

Fig. XIII-3. Relative number of γ rays as a function of angle θ with respect to the fragment direction for those γ rays with an energy greater than 0.25 MeV. [From Hoffman (1964).]

as orbital angular momentum between the fragments. This argument, together with the relative insensitivity of the anisotropy to the target spin ($I = \frac{7}{2}$ for ^{235}U and $I = \frac{1}{2}$ for ^{239}Pu), implies that the angular momentum responsible for the anisotropy is induced in the fragments during the fission process.

Strutinsky (1960) has pointed out that if the nucleus is not symmetric about the fission axis at scission, the presence of a transverse component of the Coulomb repulsion between the fragment stubs will cause the fragments to rotate after fission in opposite directions about axes perpendicular to the fission direction. Strutinsky has derived an expression for the anisotropy of the radiation assuming a statistical model for the deexcitation cascade. The angular distribution with respect to the fragment direction is found to be

$$W(\theta) = 1 + K_{\mathrm{L}}(\hbar^2 j/\mathscr{I}T)^2 \sin^2\theta$$

where $K_{\mathrm{L}} = +\frac{1}{8}$ for dipole radiation, $K_{\mathrm{L}} = -\frac{3}{8}$ for quadrupole radiation, and $K_{\mathrm{L}} = -\frac{81}{64}$ for octupole radiation, j is the angular momentum of the fragment, $\mathscr{I}$ is the moment of inertia of the fragment, and T is the nuclear temperature. If the γ rays are emitted while the fragments are still in flight, corrections must be made for the effect of fragment motion on solid angle and γ-ray energy (Skarsvåg, 1967). The product of the moment of inertia and the temperature in the equation above is related to the spin dependence of the level density. It is this spin-dependent level density which determines the relative fractions of the radiative decays which populate states with spin higher and lower than the initial fragment spin. The values of the fragment spin deduced from this relation and the experimental anisotropies are rather sensitive to the values chosen for the nuclear temperature and moment of inertia and range from 8 to 15 units of angular momentum.

It is not clear that the assumptions often made in the use of this expression are valid. In the first place, the usual assumption is to look at the sign of the anisotropy seen experimentally to deduce the multipolarity of the radiation. When the sign is negative, it is assumed that *all* of the radiation is quadrupole, whereas undoubtedly there is a mixture of dipole and quadrupole radiation. The assumption of pure quadrupole emission leads to an *underestimate* of the initial fragment spin. Second, the assumption that the relative probabilities of decay to states of higher or lower spins than the emitting state are dependent on a statistical level density weighting factor is surely wrong for the last steps of the cascade where there are no final states of higher spin available. The expected value of the anisotropy $W(\theta = 0°)/W(\theta = 90°)$ for a gamma ray which is a member of a "stretched" cascade (i.e., where a state of initial spin J_{i} decays to a state of final spin $J_{\mathrm{f}} = J_{\mathrm{i}} - L$) is in the classical limit equal to 1.6 for quadrupole radiation (Halpern *et al.*, 1968). This means that the assumption of a statistical cascade leads to an *overestimate* of the initial angular momentum. That there must be a considerable number of "stretched" transitions is clear from a comparison of the known total number of γ rays per fragment, approximately four, and an estimate of the difference between the initial spin and the final spin, the latter of which must be close to zero. This argument can be used to exclude values of the initial fragment spin much higher than 10 in low energy fission. It is not possible to circumvent this conclusion by invoking octupole radiation, as the expected lifetimes are too long for octupole radiation to be found in prompt coincidence with fission fragments.

A weak positive correlation of the γ-ray anisotropy with the total kinetic energy of the fragments has been observed (Val'skii *et al.*, 1969; Ivanov *et al.*, 1967). Events with high total kinetic energy correspond to situations where the neck breaks while the fragments are closer together than on the average. In such cases, not only is the Coulomb repulsion greater, resulting in higher kinetic energies, but the Coulomb torque is also greater, resulting in larger induced angular momenta. Another contribution to the increased anisotropy may be a decrease in the smearing of the initial alignment of the angular momentum relative to the fragment direction by neutron emission. Fewer neutrons are emitted in fission events of high kinetic energy.

A dependence of the angular anisotropy of fission γ rays on the energy of the neutron inducing fission has also been observed. Jéki *et al.* (1969) find an anisotropy relative to the direction of fragment motion twice as large for 14-MeV neutrons as for thermal neutron fission. If this is taken to imply that the fragment angular momentum is twice as large as for thermal neutron fission, it would be a somewhat surprising result. Although 14-MeV neutrons bring in an average angular momentum almost as large as that estimated to appear in fragments from thermal neutron fission, the two angular momenta are not expected to be simply additive, as the two components are not expected to have the same projection with respect to the fragment direction. The large increase probably reflects the increasing importance of nonstatistical "stretched" transitions.

C. Dependence of Gamma-Ray Yield and Energy on Fragment Mass

The average energy of the individual γ rays and the number of γ rays per fission have been found to be a smooth, slowly varying function of fragment mass ratio (Maier-Leibnitz *et al.*, 1965b; Milton and Fraser, 1958; Val'skii *et al.*, 1969; Pleasonton *et al.*, 1972). There is some evidence of a decrease in the number of γ rays and an increase in the average energy of the γ rays for mass splits corresponding to a heavy fragment mass of about $A \approx 130$. It is a more difficult experimental problem to determine the dependence of γ-ray yield and

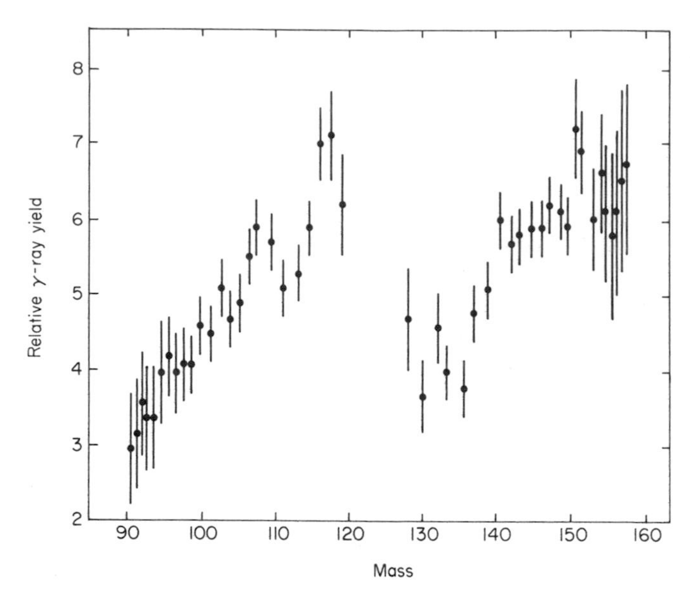

Fig. XIII-4. Relative number of γ rays emitted as a function of fragment mass for the spontaneous fission of ^{252}Cf. Note the displaced zero of the ordinate. [From Johansson (1964).]

energy on the individual fragment masses. Two methods for measuring γ rays from individual fragments have been devised. Both methods exploit the fact that under suitable experimental conditions the γ rays are emitted from moving fragments. The first method makes use of the fact that a significant fraction of the gammas are delayed sufficiently for the fragment to have moved far enough away from the source so that a γ detector can be collimated to see γ rays from that fragment alone. The results of such a measurement for ^{252}Cf are shown in Fig. XIII-4. The second method uses the relativistic change in solid angle with fragment direction for fragments emitting in flight. A forward-backward anisotropy of γ rays in the direction of flight of the fragments results. A sawtooth dependence of the γ-ray yield on fragment mass, similar to that observed for ^{252}Cf, has been observed for thermal neutron fission of ^{235}U (Maier-Leibnitz *et al.*, 1965b; Val'skii *et al.*, 1969; Pleasonton *et al.*, 1972).

The dependence of the γ-ray yield as well as the average total γ-ray energy per fragment on fragment mass is illustrated in Fig. XIII-5. The average energy of each photon can be obtained by dividing the total γ-ray energy by the number of γ rays. This energy is found to increase significantly for fragments close to the $A = 132$ double closed shell and the $N = 50$ closed neutron shell at $A \sim 82$, as illustrated in Fig. XIII-6.

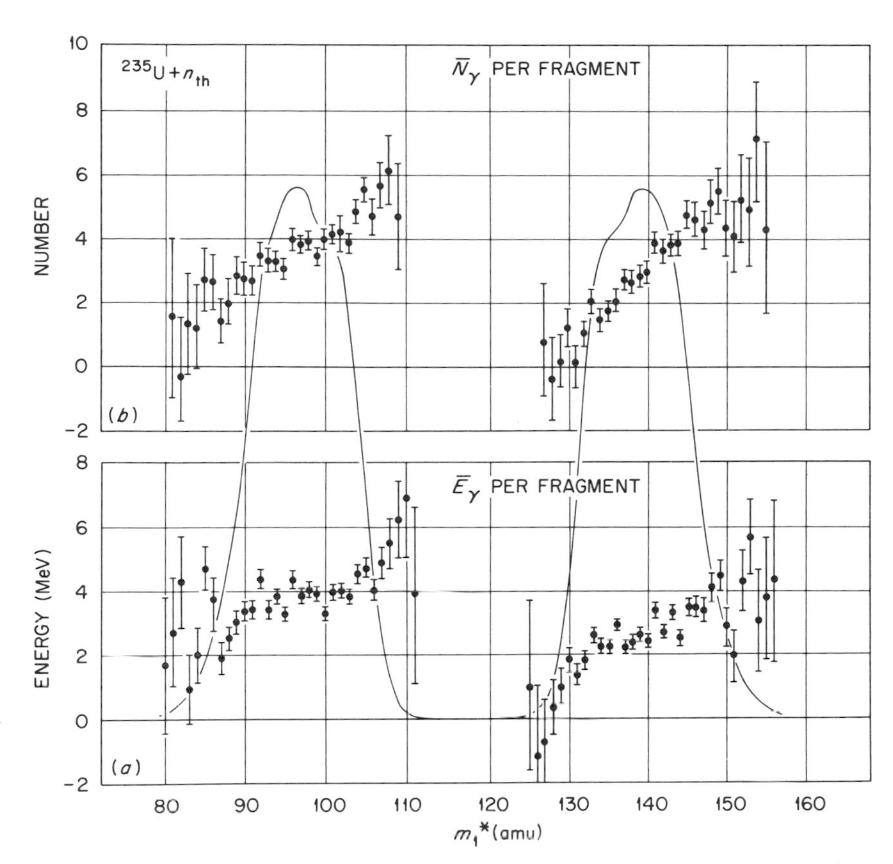

Fig. XIII-5. The average number $\bar{N}_\gamma$ and the average total energy $\bar{E}_\gamma$ of γ rays emitted as functions of fragment mass. The binary yield is sketched in for reference. [From Pleasonton *et al.* (1972)]

The dependence of γ-ray yield on fragment mass is sometimes interpreted as evidence for the primary fragment angular momentum increasing with increasing number of γ rays. Thus, fragments between closed shells would have larger angular momentum than fragments at or near closed shells. Although there may be some dependence of this kind, it is well to remember that, for a given angular momentum, the character of the γ-ray cascade is expected to depend on the nuclear structure of the fragments. Nuclei near closed shells have larger level spacings, and we might expect fewer γ rays of higher average energy. A trend of this sort can be seen in the γ-ray spectra obtained in neutron capture, where the initial spin is always rather small. This is illustrated in Fig. XIII-7, where γ-ray yields and average energies are plotted as functions of fragment mass. The region of the double closed shell at $A = 132$, so important

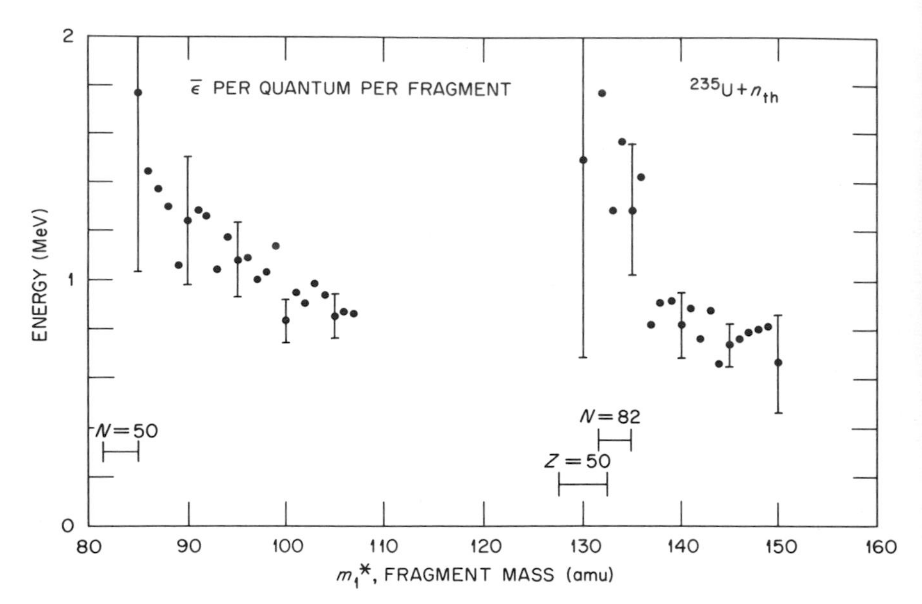

Fig. XIII-6. Average γ-ray energy as a function of fragment mass. [From Pleasonton *et al.* (1972).]

in fission, is unfortunately inaccessible by neutron capture. In view of the dependence of the character of the γ-ray cascade on fragment mass for small initial angular momentum, it is difficult to infer much about fragment angular momenta from γ-ray yield measurements alone.

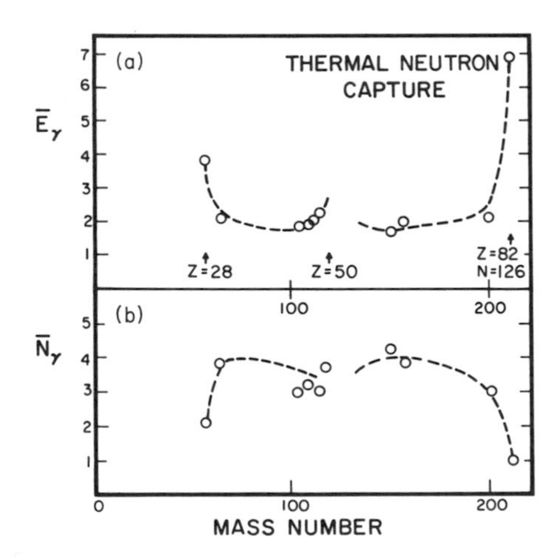

Fig. XIII-7. Dependence of the average γ-ray energy (a) and number of γ rays (b) on mass number. [Constructed from data of Groshev *et al.* (1959).]

D. *K* x Rays and Conversion Electrons

For low energy transitions internal conversion becomes competitive with photon emission, and we expect to observe not only conversion electrons but also *K* x rays resulting from filling of the electron vacancies caused by the conversion process. In the spontaneous fission of ^{252}Cf an average *K* x-ray yield of 0.55 per fragment has been determined for x rays emitted in $\leqslant 10^{-9}$ sec after fission (Glendenin *et al.*, 1965). The average time of emission of the x rays varies between 10^{-10} and 10^{-9} sec, depending on the mass of the fragment, as illustrated in Fig. XIII-8. These lifetimes confirm the foregoing mechanism for the production of the x rays, as their lifetimes are much too long to be associated with vacancies produced in the disruption of the electron cloud of the fissioning nucleus. One does not expect vacancies in the *K* shell associated with either the disruption of the electron cloud during scission or the stopping of the fragments, since the fragment velocities are much less than the velocities of the *K* electrons.

The dependence of the *K* x-ray yield on fragment mass was illustrated in Fig. XI-18. The small yields near closed shells are a consequence of the higher transition energies between the low lying states in these nuclei. As one moves into the mass region of deformed nuclei, rotational deexcitations with low

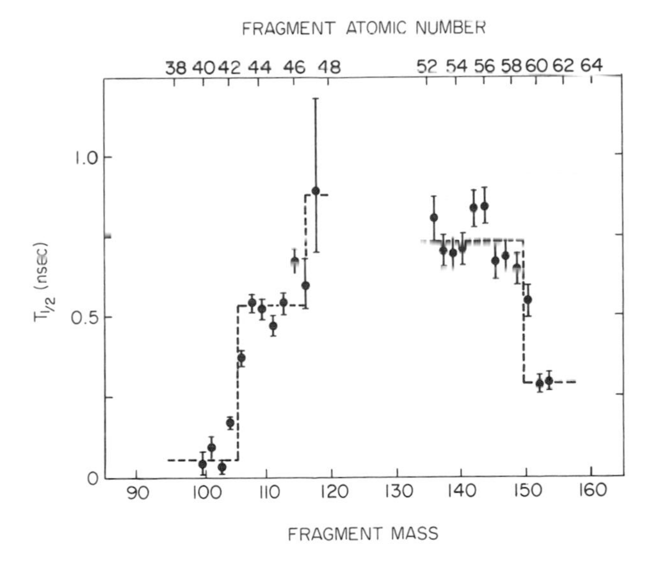

Fig. XIII-8. The average half life for *K* x-ray emission versus the final masses and atomic numbers of the fragments, as estimated from the analysis of the intensity ratios of the x rays on the assumption that the x-ray emission has an exponential decay with a single decay constant. [From Bowman *et al.* (1965).]

energy and high conversion coefficients result in large increases in the K x-ray yield.

The electron yield behaves in a manner qualitatively similar to the K x-ray yield. However, the available electron data are not as representative a sample of the mass dependence of the total internal conversion yield as is the K x-ray yield, due to difficulties in measuring both the shorter lifetime components and the lower energy electrons.

E. Isomeric Yield Ratios and Rotational State Populations as a Measure of Fragment Angular Momentum

1. Isomeric Yield Ratios

Most of the angular momentum present in the initial fission fragments is dissipated at the later stages of the γ-ray cascade. If the final nucleus has isomeric states of different spin, the relative population of the high and low spin isomers can be related to the initial fragment angular momentum. It is essential that the independent yields of the isomers be measured, i.e., that no contribution from β decay be present. This means that the isomeric pair must be shielded by a stable nucleus or by a nucleus sufficiently long-lived for the contribution from β decay to be evaluated and corrected for. The deduction of the initial fragment angular momentum requires a knowledge of the angular momentum carried away by neutrons and γ rays prior to population of the isomeric states. A statistical model is used for this analysis. If the parameters of such a calculation are taken from similar calculations which reproduce isomeric yield ratios of nonfission reactions where the initial angular momentum population is known, the results become less sensitive to the detailed validity of the assumptions made in the calculation.

A summary of the available data is given in Table XIII-1. We have included in the table only data where there is considerable confidence that the independent yields have been measured. We have also included only data for reactions where the angular momentum brought in by the inducing particle is smaller than that expected to finally appear in the fragments. We estimate an uncertainty of at least 2 units in the angular momenta derived due to uncertainties in the statistical model analyses. The experiments have been performed for rather diverse combinations of targets and projectiles, making it impossible to discern any trends with such variables as fragment mass or excitation energy. The values of the fragment angular momenta implied by the isomer ratios are in qualitative agreement with those deduced from other information and with the theoretical expectations discussed in Section F.

TABLE XIII-1

Summary of the rms Initial Angular Momenta of Selected Fragments as Deduced from Isomer Yield Ratio Studies

Reaction	Isomeric nucleus	Initial fragment angular momentum	Ref.
^{233}U + (0–16)-MeV γ rays	^{134}Cs	7.5	*a*
^{239}Pu + thermal neutrons	^{81}Se, ^{83}Se	8	*c*
^{235}U + thermal neutrons	^{131}Te, ^{133}Te	6	*d*
^{235}U + 10-MeV protons	^{95}Nb	12	*b*

[a] Warhanek and Vandenbosch (1964).
[b] Rudy *et al.* (1968).
[c] Croall and Willis (1963).
[d] Sarantites *et al.* (1965).

Isomer ratio studies on reactions induced by energetic charged particles (Loveland and Shum, 1971) indicate that there is considerably more angular momentum appearing in the fragments than in low energy fission. This is probably a consequence of both the increased angular momentum brought in by the inducing particle and the larger amount of excitation energy available at scission.

2. Ground State Band Populations

The relative population of different members of the ground state band of an even-even nucleus contains information about the initial fragment angular momentum in a way similar to that contained in isomeric yield ratios. The method may be "calibrated" by comparison with ground state band populations measured in (particle, xn) reactions where the initial angular momentum distribution is known from optical model calculations. Wilhelmy *et al.* (1972) have applied this technique to the spontaneous fission of ^{252}Cf. Depending on whether the product nucleus is rotational or vibrational, the populations of up to five spin levels (the 2+, 4+, 6+, and 8+ from the γ-ray intensities and the 0+ ground state from the independent yields of the products) have been obtained. The average fragment angular momentum is about $7\hbar$, which corresponds to an rms angular momentum of $9\hbar$ for the angular distribution given in Section F. The detailed statistical model analyses of the populations of the ground state bands in 33 even-even nuclides (^{94}Sr to ^{158}Sm) show that the heavy fission fragments have a higher primary angular momentum than the light fragments and that the more symmetric the mass division, the lower the initial angular momentum. These trends are illustrated in Fig. XIII-9.

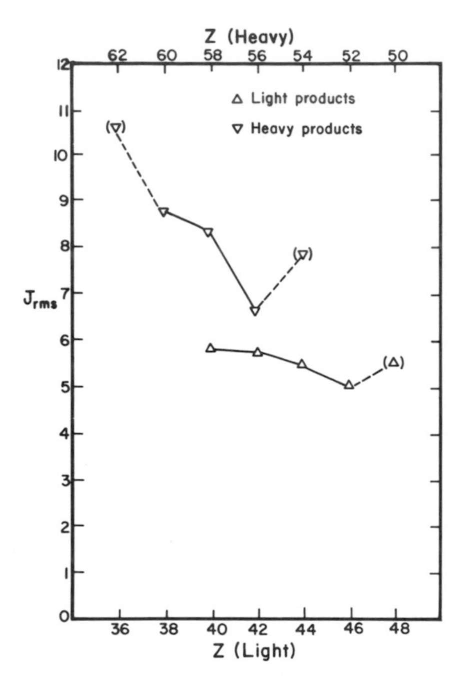

Fig. XIII-9. A plot of the rms primary fragment angular momentum for spontaneous fission of ^{252}Cf as deduced from statistical model fits to population of ground state bands. Each data point represents an average of J_{rms} as determined from various measured isotopes of that element. The data in parentheses joined by dashed lines represent determinations from limited experimental data and are less certain. The plot is presented so that complementary elements lie on the same abscissa. [After Wilhelmy *et al.* (1972).]

3. Anisotropies of Specific Deexcitation Gamma Rays

A mechanism for producing angular momentum in the fission fragments by Coulomb torques just after scission was mentioned earlier. In Section F the contribution to the fragment angular momenta due to the excitation of bending modes at scission is discussed. Both of these closely related mechanisms result in the initial fragment angular momenta being aligned in a plane perpendicular to the fragment direction. The neutrons and γ rays emitted will lead to a partial destruction of this alignment, but a correlation of specific γ rays, such as those for the $2+ \rightarrow 0+$ ground state transition, with the fragment direction is expected. The magnitude of the anisotropy is a function of both the initial fragment angular momenta and the characteristics of the deexcitation process prior to entering the ground state band. Once the ground state band is entered, the anisotropies of succeeding deexcitations are identical, i.e., the anisotropy of the radiation corresponding to a $4+ \rightarrow 2+$ transition following a 6+ → 4+ transition is the same as for the radiation associated with the 6+ → 4+ transition. Wilhelmy *et al.* (1972) have reported the anisotropy for the 2+ → 0+ transition in ^{144}Ba, and after allowance for the modification of the initial alignment by neutrons and γ rays, deduce an initial fragment rms angular momentum of approximately 6. This value is somewhat lower than, but in reasonable agreement with, the estimate based on ground state band populations.

F. Theoretical Estimates of Fission-Fragment Angular Momentum

The first order-of-magnitude estimate of the primary fragment angular momentum was made by Strutinsky (1960). His estimate of 10 units of angular momentum per fragment was based on the angular momentum induced by Coulomb forces for a nonaxially symmetric scission configuration.

More detailed investigations of this problem have taken into account explicitly the contribution due to angular momentum already present in the fragments before scission due to excitation of bending and wriggling modes. Nix and Swiatecki (1965) have explored a model where the saddle point is approximated by two tangent or overlapping spheroids. The dynamics are treated classically, although the quantal zero-point contributions are included in the consideration of the statistical probabilities of exciting different modes at the saddle point, where thermal equilibrium among the collective modes is assumed. This model, which is most appropriate for lighter elements, gives an rms angular momentum at scission of $6\hbar$ for a fragment from fission of ^{213}At in the zero-temperature limit. The angular momentum increases to a value of 12.5 for a temperature of 1.13 MeV, corresponding to an excitation energy in excess of the barrier of 40–50 MeV. These values must be increased by the contribution from angular momentum generated subsequent to scission by the Coulomb torque. If the fragments were to keep their deformed shape throughout the Coulomb acceleration, the final angular momentum would be increased by approximately 50%. The increase is less than 20% if the fragments are assumed to vibrate nonviscously as they separate.

Rasmussen *et al.* (1969) have made a quantal estimate of the fragment angular momentum at scission, employing a somewhat different model. The existence of a scission barrier sufficiently high to minimize energy in the collective modes orthogonal to the fission mode was assumed. Therefore only the ground state of a rotational mode analogous to the bending and wriggling mode of Nix and Swiatecki was considered. This corresponds to the zero-temperature limit of the calculation of Nix and Swiatecki, although in the calculations of Rasmussen *et al.* equilibrium among the collective degrees of freedom at scission was assumed, whereas Nix and Swiatecki assume equilibrium only at the saddle point. The wave function for the ground state of this special rotational mode was expanded in angular momentum components to give the angular momentum distribution at scission. One numerical example was worked out, and an rms angular momentum at scission of $7\hbar$ was obtained. After adding the contribution from Coulomb excitation, assuming damped vibrations during acceleration, they obtained an rms fragment angular momentum of about 8 units.

The functional form of the fragment angular momentum distribution obtained (Rasmussen *et al.*, 1969) was

$$P(J) \propto (2J+1) \exp[-(J+\tfrac{1}{2})^2/b^2]$$

The classical analog of this expression had been obtained in the study of Nix and Swiatecki (1965). The average angular momentum for this distribution is given in terms of the rms angular momentum by

$$\bar{J} = 0.88 J_{\text{rms}} - \tfrac{1}{2}$$

The foregoing theoretical estimates are in good agreement with the values required to explain such diverse phenomena as the anomalous total γ-ray energy and the isomeric and rotational state populations.

References

Bowman, H. R., and Thompson, S. G. (1958). *Proc. U.N. Int. Conf. Peaceful Uses At. Energy, 2nd,* **15**, p. 212. United Nations, Geneva.

Bowman, H. R., Thompson, S. G., Watson, R. L., Kapoor, S. S., and Rasmussen, J. O. (1965). *Proc. IAEA Symp. Phys. Chem. Fission, Salzburg, 1965,* **2**, p. 125. IAEA, Vienna.

Croall, I. F., and Willis, H. H. (1963). *J. Inorg. Nucl. Chem.* **25**, 1213.

Glendenin, L. E., Unik, J. P., and Griffin, H. C. (1965). *Proc. IAEA Symp. Phys. Chem. Fission, Salzburg, 1965,* **1**, p. 369.

Groshev, L. V., Demidov, A. M., Lutsenko, V. N., and Pelekhov, V. I. (1959). "Atlas of γ-Ray Spectra from Radiative Capture of Thermal Neutrons." Pergamon, Oxford.

Halpern, I., Shepherd, B. J., and Williamson, C. F. (1968). *Phys. Rev.* **169**, 805.

Hoffman, M. M. (1964). *Phys. Rev.* **133**, B 714.

Ivanov, O. I., Kushnir, Yu, A., and Smirenkin, G. N. (1967). *JETP Lett.* **6**, 327.

Jéki, L., Kluge, G., and Lajtai, A. (1969). *Proc. IAEA. Symp. Phys. Chem. Fission, 2nd, Vienna, 1969,* p. 561. IAEA, Vienna.

Johansson, S. A. E. (1964). *Nucl. Phys.* **60**, 378.

Loveland, W. D., and Shum, Y. S. (1971). *Phys. Rev. C* **4**, 2282.

Maienschein, F. C., Peelle, R. W., Zobel, W., and Love, T. A. (1958) *Proc. U.N. Int. Conf. Peaceful Uses At. Energy, 2nd,* **15**, p. 366. United Nations, Geneva.

Maier-Leibnitz, H., Armbruster, P., and Specht, H. J. (1965a). *Proc. IAEA Symp. Phys. Chem. Fission, Salzburg, 1965*, **2**, p. 113. IAEA, Vienna.
Maier-Leibnitz, H., Schmitt, H. W., and Armbruster, P. (1965b). *Proc. IAEA Symp. Chem. Fission, Salzburg, 1965*, **2**, p. 143. IAEA, Vienna.
Milton, J. C. D., and Fraser, J. S. (1958). *Phys. Rev.* **111**, 877.
Nix, J. R., and Swiatecki, W. J. (1965). *Nucl. Phys.* **71**, 1.
Peelle, R. W., and Maienschein, F. C. (1971). *Phys. Rev. C* **3**, 373.
Pleasonton, F., Ferguson, R. L., and Schmitt, H. W. (1972). *Phys. Rev. C* **6**, 1023.
Protopopov, A. N., and Shiryaev, B. M. (1958). *Sov. Phys. JETP* **7**, 231.
Rasmussen, J. O., Nörenberg, W., and Mang, H. J. (1969). *Nucl. Phys. A* **136**, 465.
Rau, F. E. W. (1963) *Ann. Phys.* (*Paris*) **10**, 252.
Rudy, C., Vandenbosch, R., and Ratcliffe, C. T. (1968). *J. Inorg. Nucl. Chem.* **30**, 365.
Sarantites, D. G., Gordon, G. E., and Coryell, C. D. (1965). *Phys. Rev.* **138**, B 353.
Skarsvåg, K. (1967). *Nucl. Phys. A* **96**, 385.
Skliarevskii, V. V., Fomenko, D. E., and Stepanov, E. P. (1957). *Sov. Phys. JETP* **5**, 220.
Smith, A., Fields, P., Friedman, A., Cox, S., and Sjoblom, R. (1958). *Proc. U.N. Int. Conf. Peaceful Uses At. Energy, 2nd*, **15**, p. 392. United Nations, Geneva.
Strutinsky, V. M. (1960). *Sov. Phys. JETP* **10**, 613.
Thomas, T. D., and Grover, J. R. (1967). *Phys. Rev.* **159**, 980.
Val'skii, G. V., Petrov, G. A., and Pleva, Y. S. (1969). *Sov. J. Nucl. Phys.* **8**, 171.
Warhanek, H., and Vandenbosch, R. (1964). *J. Inorg. Nucl. Chem.* **26**, 669.
Wilhelmy, J. B., Cheifetz, E., Jared, R. C., Thompson, S. G., Bowman, H. R., and Rasmussen, J. O. (1972). *Phys. Rev. C* **5**, 2041.

CHAPTER

Ternary Fission

A. Light-Particle-Accompanied Fission

In a certain sense almost all fission events are accompanied by the emission of a light particle if we include the neutrons evaporated from the fragments. Our concern here, however, is with light particles emitted at or near the instant of scission, i.e., those appearing on a time scale considerably shorter than evaporation times. That there are such particles has been known since the discovery of "long-range" α particles accompanying fission. These particles appear predominantly at right angles to the fission fragments, hence, are not evaporated from the accelerated fragments. To account for the sidewise peaking, these particles must be emitted from the region between the fragments while the fragments are still close together. Although these particles are rather rare, occurring in only one out of every 300–500 fissions, there has been considerable interest in studying the characteristics of these particles as probes of the scission process. Once such charged particles are formed, their motion under the influence of the Coulomb forces exerted by the fragments can be followed in trajectory calculations. The final energies and angular distributions of charged particles emitted in fission are quite sensitive to the separation and speed of the nascent fragments at the time of formation of the particles.

1. Mechanism for Charged Particle Emission during Fission

As has been mentioned, the angular distributions for the light charged particles indicate that they are produced in the neck region between the fragments. Estimates have also been given for the energy required to release particles into the region between the fragments, and this energy turns out to be a very large fraction of the available energy. It has also been suggested that this energy is stored in deformation of the fragments. It must be stored in some readily convertible form involving only a few degrees of freedom in order for it to be transferred to the escaping particle on a sufficiently fast time scale. We now wish to explore a possible mechanism for the rapid transfer of deformation energy into release energy.

Halpern (1965) has suggested that the energy transfer to the light particle takes place through the sudden snap of the neck stubs after scission. The particle release cannot be considered as a slow thinning of the neck on either side of the particle followed by a slow retraction of the neck stubs into the fragments. For such a process most of the deformation energy would appear as excitation energy of the fragments due to their acquisition of the deformation energy of the slowly collapsing stubs. There is insufficient energy to both leave the α particle in the region of high potential energy and leave a lot of deformation in the fragments also.

If the process of rupture of the neck and collapse of the stubs is sufficiently sudden, the rapid change in the nuclear potential of the fragments as felt by the light particle can result in a large change of the potential energy of the light

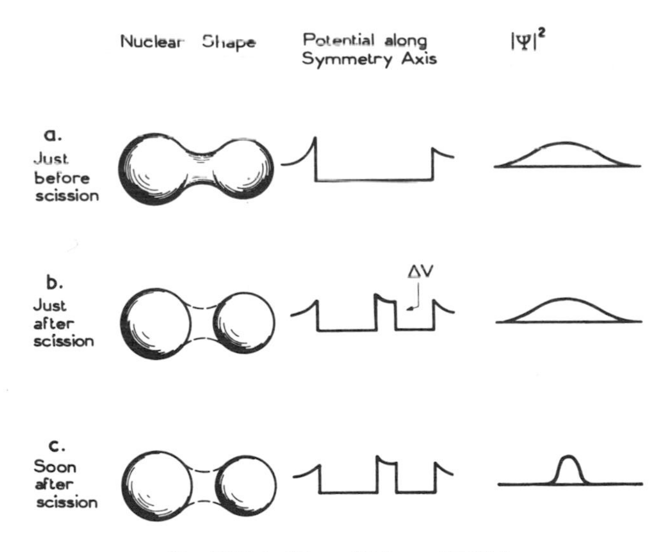

Fig. XIV-1. [From Halpern (1965).]

particle with little corresponding change in the kinetic energy of the light particle. This process, as viewed from the point of view of wave mechanics, is illustrated in Fig. XIV-1. In this limiting example the potential changes associated with scission are assumed to be instantaneous. The validity of the sudden approximation depends on the collapse of the neck in times much shorter than the time required for the light particle being formed to travel a distance of the order of the neck dimension. Estimates of the shortest possible collapse time based on the speed of the fastest nucleons show that the ratio of this time to that of the light particle period is smaller than, but not negligible compared to, unity. Thus the sudden approximation is probably qualitatively but not quantitatively valid.

It is clear that there is nothing in this process which depends on the light particle being charged, and Fuller (1962) has investigated this mechanism for producing scission neutrons. Since the neutron has a low mass, the potential change appears less sudden to it than to a heavier particle. On the other hand, the energy required to release the neutron is much less, and the process may be fairly probable.

2. Dependence of Yield on Nuclear Species and Excitation Energy

By far the most complete data on light-particle-accompanied fission is that for α particle-accompanied fission. The yield of α particles is greater than for any other light charged particle by about an order of magnitude. We will therefore concentrate first on fission accompanied by α particles, and then consider the yield and other characteristic differences for other light particles.

On the basis of available data it is somewhat difficult to separate the dependence of yield on A and Z from the dependence on excitation energy. This is due in part to the fact that for the heaviest elements only spontaneous fission has been studied, whereas for the lightest elements only energetic particle-induced fission has been studied.

The excitation energy dependence appears fairly weak, with the light particle yield decreasing perhaps 25% for thermal neutron fission ($E^* \sim 7$ MeV) compared to spontaneous fission ($E^* = 0$). At excitation energies over 15 MeV there is some evidence for an increase in the yield, as illustrated in Fig. XIV-2.

The relative insensitivity of the probability of light charged particle emission to excitation energy confirms the earlier assertion that these particles are not evaporated. Since the total kinetic energy of the fragments is approximately constant with initial excitation energy, the additional increment of excitation energy of the fissioning nucleus is present as internal excitation energy at scission. If these particles were evaporated, their yield would depend sensitively

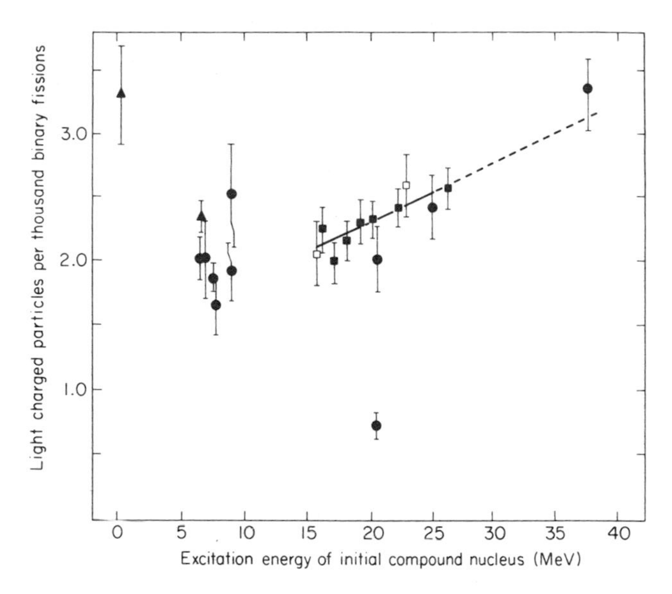

Fig. XIV-2. Dependence of number of light charged particles per thousand fissions on the excitation energy of the initial compound nucleus [(●) ^{236}U; (▲) ^{240}Pu; (□, ■) ^{239}Np]. [From Thomas and Whetstone (1966).]

on the excitation energy. The near constancy of the kinetic energy and the character of the angular distributions also implies that the average deformation at scission is independent of excitation energy. The near constancy of both the distortion at scission and the yield of light charged particles with increasing excitation energy is consistent with a mechanism for the ejection of these particles which depends on converting some of the deformation energy at scission into separation energy, as discussed in the previous section.

Nobles (1962) has observed a correlation of yield with the fissility parameter Z^2/A. The result of such a correlation, with the inclusion of data at higher excitation energy for the lighter elements, is shown in Fig. XIV-3. An estimated correction for the small enhancement of light particle yield in spontaneous fission has been incorporated into this presentation. A significant increase in the yield with the fissility parameter is observed. This is consistent with the idea that the yield is dependent on the amount of deformation at scission. Liquid drop model calculations have shown an increase in the deformation energy at scission with increasing Z^2/A. The calculated increase (see Fig. IX-7) is approximately 50% over the Z^2/A interval spanned by the data in Fig. XIV-3. It should be borne in mind, however, that the data span a rather narrow range of Z^2/A values and exhibit considerable scatter, so that the correlation observed might reflect a correlation simply with Z rather than with Z^2/A.

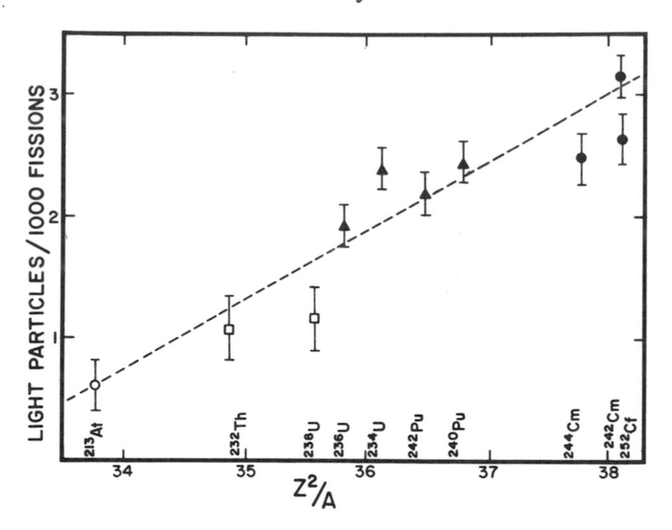

Fig. XIV-3. Light particle yield as a function of Z^2/A. The spontaneous fission values (●) have been "corrected" to the same excitation energy as the thermal neutron values (▲), assuming the decrease observed for ^{240}Pu and ^{242}Pu is common to all nuclei. [Data from Nobles (1962).] (□) 2.5-MeV neutron-induced fission (Nagy *et al.*, 1969) normalized to data of Nobles; (○) 40-MeV α-induced fission of ^{209}Bi normalized to low energy fission of ^{238}U (Loveland *et al.*, 1967).

3. Relative Yields of the Different Light Particles

Alpha particles are by far the most abundant light particles emitted. The yields of other species are down by at least another order of magnitude, with tritons being the next most abundant particle. A summary of the yields of light charged particles for several fissioning systems is given in Table XIV-1. In addition to the species listed in this table, B, C, N, and O particles have been observed (Raisbeck and Thomas, 1968; Natowitz *et al.*, 1968). The absolute yields of some species are not well known due to experimental difficulties in observing all possible kinetic energies. There is some evidence, as can be seen in Fig. XIV-4, of a rough correlation between the observed yields and the energy required to form the particle in the region between the two fragments by separating it from one of the fragments. This energy can be shown to be the sum of the separation (binding) energy of the particle (as calculated for removing a particle to infinity from an isolated fragment) plus the difference in Coulomb energy for a pair of fragments at scission as compared with a configuration of a pair from which a particle has been removed and placed midway between the fragments. This gives the expression

$$E_R = B_i + \frac{z(Z_i - z)e^2}{D/2} + \frac{z(Z - Z_i)e^2}{D/2} + \frac{(Z - Z_i)(Z_i - z)e^2}{D} - \frac{Z_i(Z - Z_i)e^2}{D}$$

TABLE XIV-1

Relative Yields of Different Third Particles (in % of α Particle Yield)[a]

Nucleus:	^{234}U	^{236}U					^{252}Cf			
Method[b]:	PC+M	PC+M	PC	C	PC	PC	PC	PC	PC	PC
Reference:	*c*	*d*	*e*	*f*	*g*	*h*	*i*	*j*	*k*	*l*
p			1.2				1.8	1.6		1.5
d	0.4	0.4	0.5				0.7	0.6		0.5
t	4.6	6.3	6.2				8.5	5.9		6.6
^{3}He		$< 5 \times 10^{-3}$						<1		
^{4}He	100	100	100	100	100	100	100	100	100	100
^{6}He	1.4	1.4	1.1				2.6	2.4		1
^{8}He	0.036	0.033					0.09	0.2		0.06
^{6}Li		$< 5 \times 10^{-4}$			∧	∧	∧	∧	∧	∧
^{7}Li	0.037	0.036			0.12	0.1	0.13	0.12	0.3	0.12
^{8}Li	0.02	0.014								
^{9}Li	0.036	0.011			∨	∨	∨	∨	∨	∨
^{7}Be				$< 10^{-6}$	∧	∧	∧	∧	∧	∧
^{9}Be	0.037	0.02			0.37	0.34	0.21	>0.01	0.65	0.2
^{10}Be	0.43	0.30			∨	∨	∨	∨	∨	∨

[a] From Halpern (1971).
[b] PC is particle counter, M is magnet, and C is radiochemistry.
[c] Vorobiev *et al.* (1969a).
[d] Vorobiev *et al.* (1969b).
[e] Dakowski *et al.* (1967).
[f] Roy (1961).
[g] Blocki *et al.* (1969).
[h] Andreev *et al.* (1969).
[i] Cosper *et al.* (1967).
[j] Whetstone and Thomas (1967).
[k] Gazit *et al.* (1970).
[l] Raisbeck and Thomas (1968).

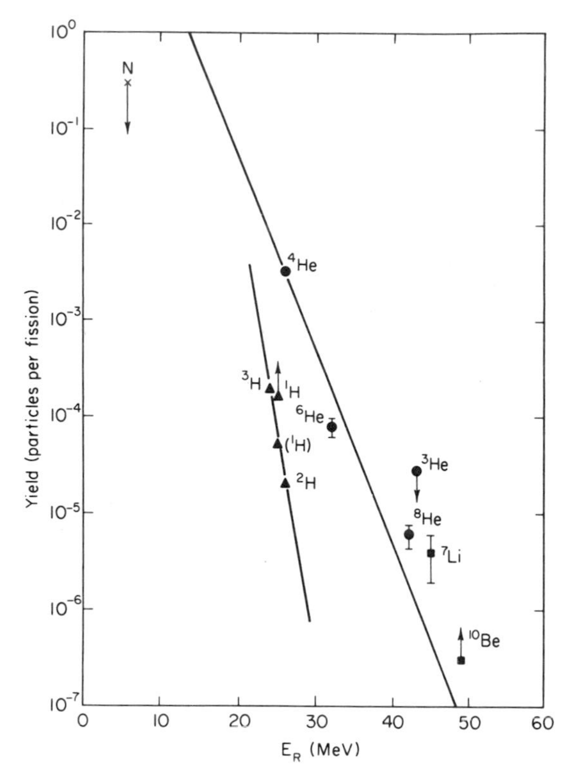

Fig. XIV-4. Estimated total particle yields as a function of the removal energy E_R required for release of the particles into the space between the fission fragments. E_R in this graph includes 2 MeV of kinetic energy, assumed constant for all particles. [From Whetstone and Thomas (1967).]

where z is the light particle atomic number, Z_i is the atomic number of the fragment to which the particle was bound with binding energy B_i, Z is the atomic number of the fissioning nucleus, and D is the separation of the fragments at the time of scission. This expression can be rewritten as

$$E_R = B_i + zV_0[(Z_i+Z-2z)/Z_i(Z-Z_i)]$$

where

$$V_0 = \frac{Z_i(Z-Z_i)e^2}{D}$$

is the Coulomb potential energy of the fragments at the time of scission. This removal energy is approximately 20 MeV or more for all light charged particles, and is comparable to the total distortion energy at scission for an ordinary fission event. Thus the emission of particles with removal energies larger than the average deformation energy must be associated with fission events involving greater than average distortion. In fact, unless the mechanism for converting deformation energy is very efficient, light particle emission can only occur for the relatively rare events associated with unusually large distortion energy. This factor alone may account for the low absolute yield of light charged particles in fission.

The correlation of yields in Fig. XIV-4 is improved somewhat if a mass-dependent rather than constant initial light particle kinetic energy is used. Such a dependence is expected on the basis of the uncertainty principle if the particles are constrained to originate from the neck region. The simplest application of the uncertainty principle would predict an initial energy which is inversely proportional to the mass of the particle ejected. If V is the volume of the neck region, the minimum initial kinetic energy is given by $E_0 = 3\pi^2\hbar^2/2mV^{2/3}$ where m is the mass of the light particle. Considerable care, however, must be exercised in considering the effective emission volume due to the finite times involved. The lightest particles (e.g., the proton) can get quite far from the fragment axis in the time during which the just-formed distorted fragments are collapsing to their spherical shape. This implies a larger effective volume for the lightest particles, which tends to cancel the mass dependence mentioned earlier. There is no evidence from the trajectory calculations to be discussed later for a mass-dependent initial kinetic energy. In considering the spatial localization in the direction perpendicular to the fragment direction (the direction most relevant to the initial kinetic energies involved in the trajectory calculations), it may be necessary to consider also the amplitude and frequency of the bending mode which is responsible for inducing the angular momentum acquired by the fragments.

In addition to the kinds of light charged particles listed in Table XIV-1, scission neutrons are also expected. Their existence is much more difficult to establish experimentally because of the large number of neutrons evaporated from the fragments. There is some evidence for such neutrons from analyses of the energy and angular distributions of neutrons from spontaneous and thermal neutron fission (Bowman *et al.*, 1962; Skarsväg and Bergheim, 1963; Kapoor *et al.*, 1963; Milton and Fraser, 1965). The laboratory distributions are better fit if it is assumed that a small fraction of the neutrons, perhaps 0.5 neutron per fission, are assumed to be emitted isotropically in the laboratory system, rather than from the moving fragments. The average energy of these neutrons appears to be somewhat higher than that of the evaporation neutrons (Bowman *et al.*, 1962). It is not possible to characterize these neutrons more fully without considerably more detailed and precise experimental data.

4. Angular and Energy Distributions of the Light Charged Particles

As has been mentioned previously, the light charged particles are emitted preferentially in the direction perpendicular to the fission fragments. For example, the angular distribution of α particles from ^{252}Cf is quite sharply peaked even when averaged over all mass divisions, having a full width at half maximum of approximately 40°. Each distribution is actually a superposition of two narrower distributions, one defined by the angle with respect to the light fragment and the other by the angle with respect to the heavy fragment. The sum of the most probable angles of emission with respect to the light and to the heavy fragments is slightly larger than 180°, reflecting the fact that the momentum of a 16-MeV α particle increases the angle between the fragments by 4.5° compared to binary fission, or slightly more than 2° per fragment. The angular distribution of α particles in the spontaneous fission of ^{252}Cf is given in Fig. XIV-5.

Kinetic energy spectra have been measured for most of the light charged particles emitted in fission. Experimental difficulties associated with particle identification and with α particles associated with natural radioactivity limit the energy region which can be studied. Some experimental values of the most probable energies of various light particles for several fissioning systems are

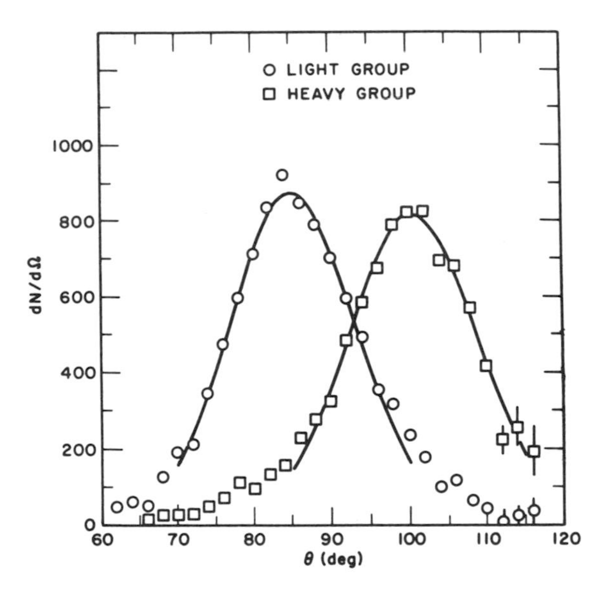

Fig. XIV-5. Angular distribution of all α particles with energy greater than 11 MeV. (□) Angular distribution of α particles measured in coincidence with heavy fission fragments; (○) angular distribution of α particles measured in coincidence with light fission fragments. [From Fluss *et al.* (1973).]

TABLE XIV-2

Most Probable Energies for Light Particles Accompanying Thermal Neutron-Induced Fission of ^{233}U and Spontaneous Fission of ^{252}Cf

	Most probable energy			
	^{233}U+n		^{252}Cf	
Particle	Vorobiev *et al.* (1969a)	Whetstone and Thomas (1967)	Cosper *et al.* (1967)	Gazit *et al.* (1970)
^{1}H			8	
^{2}H	8.4		8	
^{3}H	8.4	8	8	
^{3}He		17		
^{4}He	16.3	16	16	
^{6}He	11.5	13	12	
^{8}He	9.7		10	
^{7}Li	15.8			
^{8}Li	14.4		20	18.5
^{9}Li	12.0			
Be	17		~26	19
C				23

given in Table XIV-2. The most probable energy increases with the atomic number of the particle. For the helium isotopes, the most probable energy decreases with increasing mass, as illustrated in Fig. XIV-6.

A correlation between the average kinetic energy of the α particle and the angle of emission has been noted. The kinetic energy of the α particle increases as the angle changes away from the most probable angle. Conversely, this effect can be described as a broadening of the angular distribution for α particles with energies considerably above the average energy. A correlation of the most probable angle with mass ratio has been observed also. We defer further discussion of these results until comparison can be made with trajectory calculations.

There is evidence (Piasecki *et al.*, 1970) that a very small fraction (~1%) of the α particles are emitted at angles so close to 0° or 180° with respect to the light fragment that they must come from the poles of the fragments rather than from the neck region. The kinetic energies of these α particles are considerably greater than the energies of those emitted perpendicular to the fragment direction. This can be understood from the fact that both fragments are exerting a Coulomb force in the same direction, whereas alphas born in the

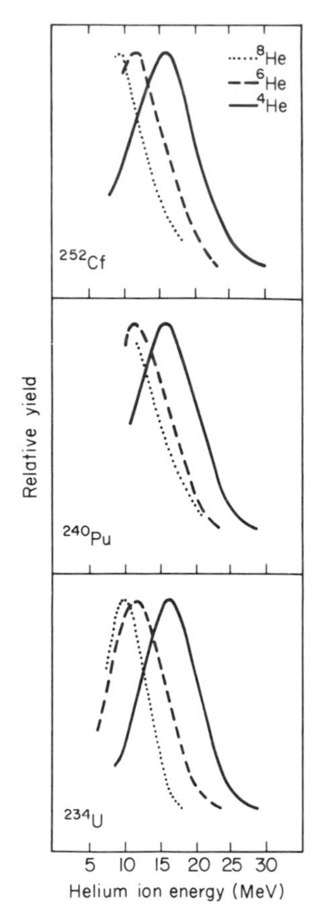

Fig. XIV-6. Helium-ion energy spectra emitted in the fission of ^{234}U (Vorobiev *et al.*, 1969a), ^{240}Pu (Krogulski *et al.*, 1969), and ^{252}Cf (Cosper *et al.*, 1967). The decrease in mean energies with increasing mass of the emitted ion is comparable for the three systems. [From Halpern (1971).]

neck region have forces exerted by the two fragments which have large components in opposite directions tending to cancel each other.

The angular distribution of the light charged particles suggests that the large majority are emitted with low kinetic energy in the region between the fragments. The Coulomb field of the fragments accelerates the particles in the direction perpendicular to the fragment axis, as illustrated in Fig. XIV-7. The fact that the energy required to release an α particle between the fragments is quite high suggests that it is the rapid snap of the neck which concentrates sufficient energy on the particle to emit it. Thus the particle is born at scission with a fairly modest initial kinetic energy. The final energy and angular

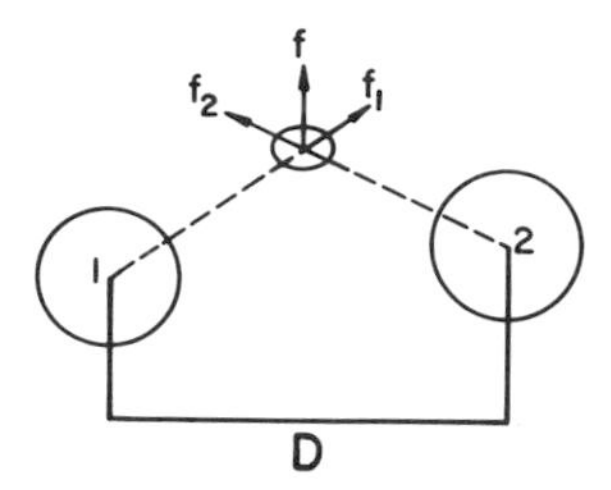

Fig. XIV-7. Schematic illustration of Coulombic forces of fragments on the light charged particle.

direction of the α particle depend not only on its initial energy and direction but also rather sensitively on the time evolution of the fragments' Coulomb field. To relate the final observed properties of the emitted particle to those of the initial system it is necessary to study trajectory calculations which link the two situations.

A number of trajectory calculations have been made. They generally assume three classical point charges interacting by Coulomb forces. The resulting differential equations are of the form $M_i\, d^2X_{ij}/dt^2 = F_{ij}$ where M_i is the mass of the ith particle, X_{ij} is the jth spatial coordinate of the ith particle, and F_{ij} is the jth component of the force on the ith particle. These are integrated numerically until a time when the Coulomb potential has become negligible. The final energies of the three particles as well as their mutual angles are obtained for a particular choice of initial conditions at the time of release of the light particle. For each of the three bodies there are three spatial variables describing the position of the center of mass and three conjugate momenta, making a total of 18 variables. This number of variables can be reduced immediately to 12 by subtraction of three variables for conservation of linear momentum and three variables for the conservation of angular momentum. The specification of the remaining parameters differs slightly in the various calculations which have been published. Recognizing that (at least for spontaneous fission) there is no preferred direction for the fission axis in space, we can reduce the number of variables to 10. In most calculations it has been further assumed that the momenta of the three particles lie in a plane, leaving a two-dimensional problem with seven remaining free parameters. In view of the large number of independent variables, considerable care in exploring the initial conditions is necessary before drawing definite conclusions from trajectory calculations. There are also undoubtedly restraints on the values certain parameters can have after other parameters have been chosen. For example, the distribution in starting positions must be chosen in a way consistent with the uncertainty principle for a given choice of the initial energy. An example of several trajectories is shown in Fig. XIV-8.

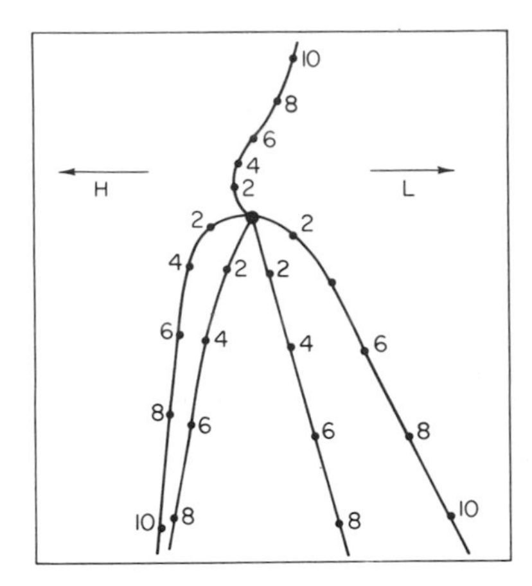

Fig. XIV-8. Early portions of computed trajectories of α particles moving in the α particle fragment plane. The α particles are assumed to start at a point displaced from the fragment axis with a moderate initial energy. The initial interfragment spacing is 20.5 F. The black dots show where the α particles are at different times. All times here are expressed in units of 1.7×10^{-22} sec. Six α particles were started in directions spaced 60° apart. Only five trajectories are seen because the α particle emitted into the upper right-hand corner struck the light fission fragment and was presumably absorbed. It is seen how very strongly focusing (at right angles to the fragments) the fragment Coulomb field is. [From Halpern (1971).]

At this point we will try to summarize the dependence of the final energy and angle on the remaining free parameters to which these properties are most sensitive. These parameters are identified in Fig. XIV-9.

Figure XIV-10 shows the final α-particle energy E_α as a function of its initial

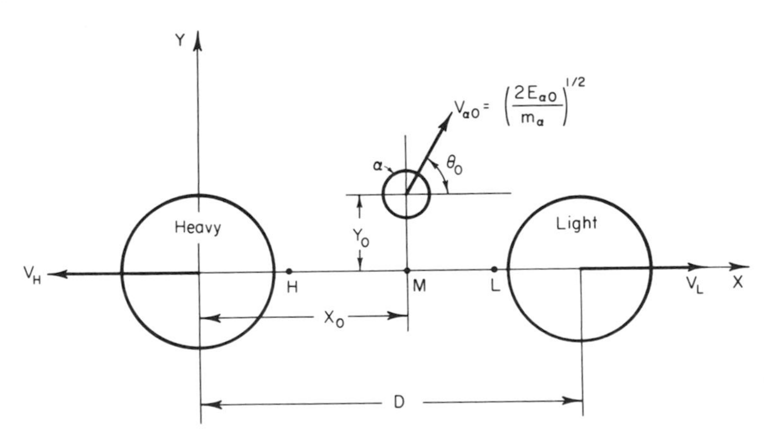

Fig. XIV-9. Diagram identifying the initial parameters of trajectory calculations. [From Boneh *et al.* (1967).]

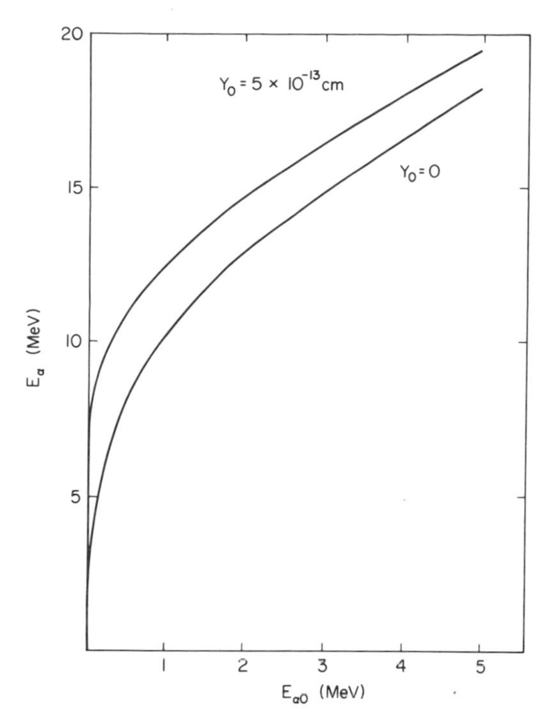

Fig. XIV-10. The final value of the α-particle energy E_α as a function of its initial energy $E_{\alpha 0}$ and its initial displacement from the fragment axis Y_0. The initial fragment energies were approximately $\frac{1}{4}$ of their final kinetic energies in this example. [From Boneh *et al.* (1967).]

energy $E_{\alpha 0}$ for a starting position on the fission axis ($Y_0 = 0$) and for a starting position displaced 5 F from the fission axis ($Y_0 = 5$ F). It is seen that a small variation in the initial α-particle energy gives rise to a much larger change in the final α-particle energy. The origin of this amplification effect is that the Coulomb field experienced during the acceleration of the α particle depends on the speed of the α particle relative to that of the fragments. If the α particle has a high initial velocity, it receives essentially the full acceleration possible from the fragments' field. On the other hand, if the initial α-particle velocity is low compared to that of the fragments, the fragments sneak away before the α particle can get far enough from the fission axis to feel the full force of the fragments' Coulomb potential. A considerable broadening of the final α-particle energy distribution may result from a rather narrow initial α-particle energy distribution. Figure XIV-10 also shows that, except for very small initial α-particle energies, the dependence of the final energy on Y_0 is rather weak, considering the extreme values chosen in this example. The final α-particle energy is somewhat more sensitive to a displacement along the fission axis, generally increasing when the α particle originates from a region of higher potential energy closer to one of the fragments. The final α-particle energy is also somewhat dependent on the initial angle of emission, decreasing due to reflection for initial angles significantly different from 90° with respect to a fragment. Lastly, the final α-particle energy E_α is nearly a linearly decreasing function of both the fragment separation at scission and the velocity of the fragments at scission.

The final angle θ_α is surprisingly insensitive to both the initial emission angle θ_0 and the initial energy $E_{\alpha 0}$, as illustrated in Fig. XIV-11. The final angle is, however, somewhat more sensitive to the position of origin along the fragment axis, especially for higher values of $E_{\alpha 0}$.

We will now give some examples of comparisons of trajectory calculations with experimental energy and angular distributions. For the particular calculations compared here the masses and charges of the fragments were assigned fixed values corresponding to the most probable mass split in ^{252}Cf. The initial emission angle was assumed isotropic, and a distribution in both the initial energies and initial positions X_0 was assumed. The average initial light particle energy and the fission-fragment total energy were chosen to reproduce the experimental α-particle energy spectrum. The resulting energies, $\bar{E}_{\alpha 0} = 2$ MeV and $E_{f0} = 8$ MeV, were used to calculate the energy and angular spectra for a

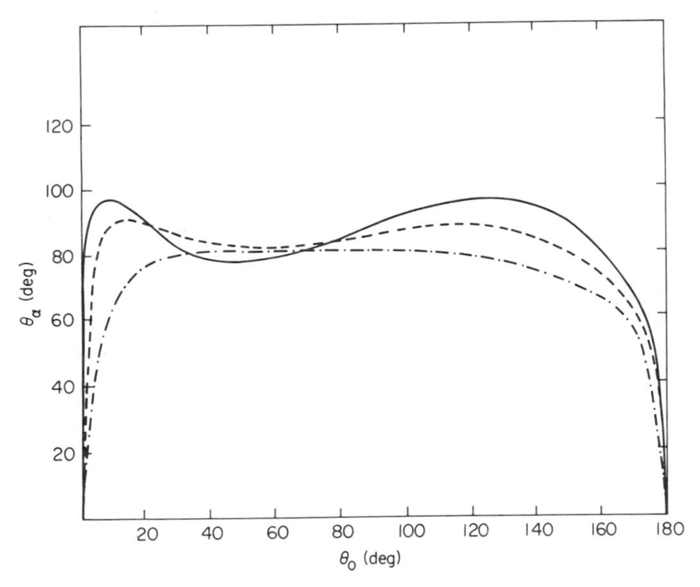

Fig. XIV-11. Dependence of the final angle θ_α as a function of the initial angle θ_0 for three values of the initial energy $E_{\alpha 0}$ ((–·–) 0.5 MeV, (---) 3.0 MeV, (—) 6.0 MeV). [From Boneh *et al.* (1967).]

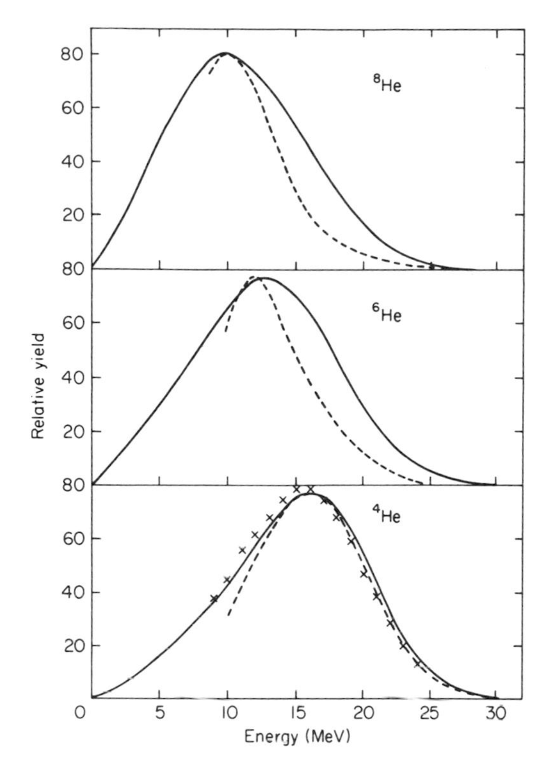

Fig. XIV-12. Comparison of calculated (—) and measured (---, ×) energy spectra for ^{4}He, ^{6}He, and ^{8}He. (---) Data of Cosper *et al.* (1967); (×) data of Raisbeck and Thomas (1968). [From Raisbeck and Thomas (1968).]

variety of particles. The energy spectra of the helium isotopes are shown in Fig. XIV-12. The observed decrease in most probable energy with increasing mass is reproduced quite well. The origin of this effect is that for a given initial energy the heavier helium isotopes are moving more slowly, allowing the fission fragments to move farther away before the heavier helium isotope is fully accelerated. The energy spectra for protons and tritons are compared with the calculations in Fig. XIV-13. The angular distributions for some hydrogen and helium isotopes are compared in Fig. XIV-14. Better fits to the α-particle angular distribution have been obtained by increasing the dispersion in starting positions (Musgrove, 1971; Rajagopalan and Thomas, 1972). The angular distributions were calculated for a fixed mass ratio, hence, are expected to be somewhat narrower than the experimental distributions, which contain all mass ratios. With the exception of the protons, the experimental and calculated energy and angular distributions are in satisfactory agreement.

The hydrogen ion distributions appear to be somewhat anomalous. The

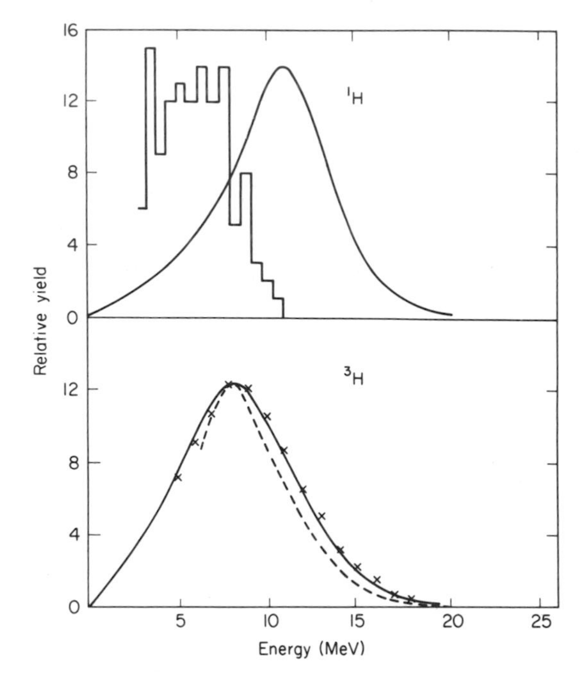

Fig. XIV-13. Comparison of calculated (—) and measured energy spectra for ^1H and ^3H. The experimental ^1H data (step curve) were obtained at 90°. (×) Data of Raisbeck and Thomas (1968); (---) data of Cosper *et al.* (1967). [From Raisbeck and Thomas (1968).]

energy distributions do not show the decrease in most probable energy with increasing light particle mass that was observed for the helium isotopes and is expected from the trajectory calculations if the starting energy is the same for all light particles. The comparison in Fig. XIV-13 suggests that the anomaly is primarily associated with the protons, although the deuterons also seem to be affected somewhat, as their energy spectrum is essentially the same as the tritons' (Fig. XIV-15) rather than shifted to the expected higher energies.

Halpern (1971) has suggested that the lower than expected proton energies may be related to the relatively greater mobility of such a light particle. The protons can move fairly large distances during the surface collapse time associated with their emission. This will result in a larger emission volume, which could result in both a broader angular distribution and lower average final energies due to the possibility of starting from regions where the Coulomb potential is not as large as for particles constrained by smaller emission volumes to the neck region. However, to obtain a significant lowering of the average final energies requires a very large emission volume. The result of moving the starting point from the Coulomb saddle to a point 5 F away from the Coulomb saddle for α particles is to increase the final kinetic energy by an

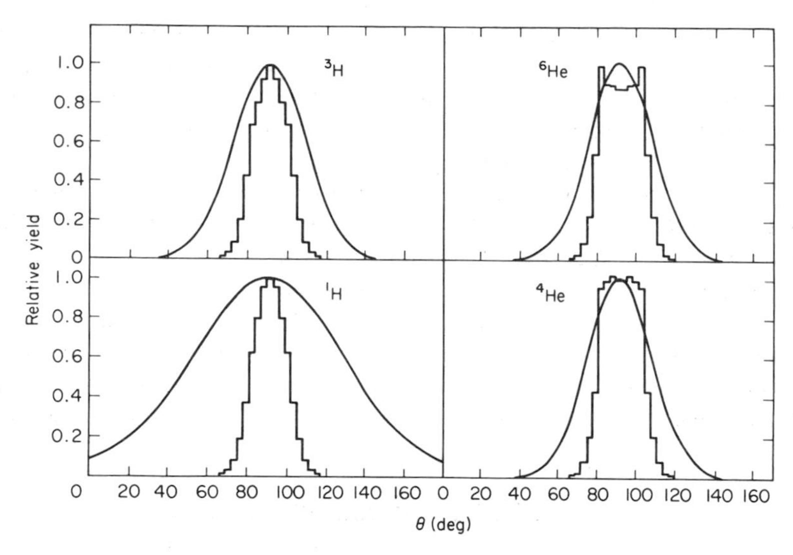

Fig. XIV-14. Comparison of calculated (step curves) and measured (—) angular distributions. The apparent separation of the calculated ^{6}He peak into two parts is a result of events being associated with either the light or the heavy mass peak. The calculated values are for the most probable mass ratio, whereas the measured values are integrated over all mass ratios. [From Raisbeck and Thomas (1968).]

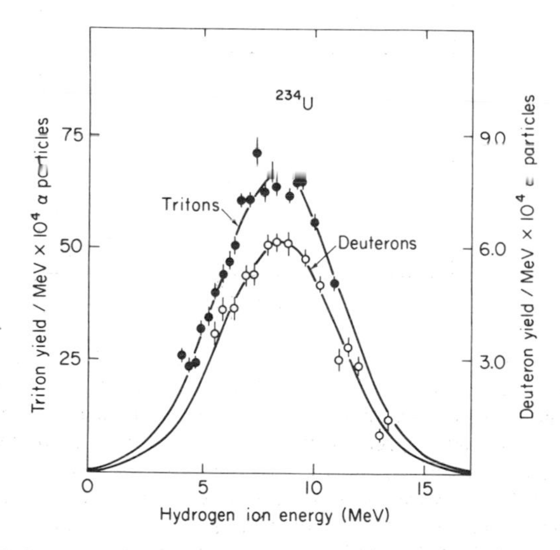

Fig. XIV-15. Hydrogen ion spectra from the fission of ^{234}U (Vorobiev *et al.*, 1969a). The mean energies for the hydrogen isotopes are the same, whereas for the helium isotopes (see Fig. XIV-6) they are different. [From Halpern (1971).]

amount illustrated in Fig. XIV-10. This result is a consequence of the weakening of the cancellation of the Coulomb forces on the light particle as the starting point moves off axis. If the α-particle trajectory calculations are indicative of the results expected for protons, a large contribution from points considerably farther than 5 F from the fragment axis must occur in order to get a sufficient lowering of the average final energy.

The dependence of the most probable angle on mass ratio has been measured (Fraenkel, 1967) and found to be relatively weak in the case of ^{252}Cf. [In an early report (Fraenkel and Thompson, 1964) of a dramatic variation with fragment mass, an important kinematic recoil correction was neglected.] The relatively weak dependence of the most probable angle on the mass ratio is to be contrasted with the rather strong dependence given by the trajectory calculations if the α particle is started midway between the fragment centers. This comparison is illustrated in Fig. XIV-16a. The implication is that as R increases

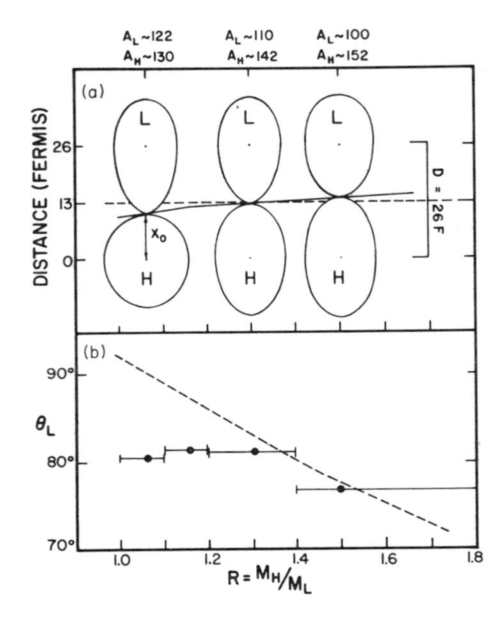

Fig. XIV-16. (a) Most probable angle of emission of α particles with respect to the light fragment plotted as a function of mass ratio. The data points [taken from Fraenkel (1967)] are plotted at the average mass ratio for which α-accompanied fission occurs within the mass ratio bin. (---) Angle expected if the α particle always started from a point midway between the fragment centers. (b) (—) Dependence of X_0, the distance of the starting point from the center of the heavy fragment on R, required to account for the dependence of the most probable angles on mass ratio. An attempt has also been made to indicate the average relative distortions of the light and heavy fragments for each mass ratio interval. Larger variations in starting positions and fragment distortions would be expected if narrower mass ratio intervals were considered.

from near symmetric fission ($R \sim 1$) to rather asymmetric fission ($R \sim 1.5$), the emission point moves from a region close to the heavy fragment to one closer to the light fragment, as illustrated in Fig. XIV-16b. This dependence of fragment distortion on fragment mass is similar to that deduced from the dependence of the neutron emission probability on fragment mass, and can be understood in terms of the nuclear shell structure of the fragments (Terrell, 1962; Vandenbosch, 1963). For mass ratios with heavy fragments close to the $A = 132$ double closed shell the starting point is closer to the heavy fragment, indicating a more spherical nascent heavy fragment than nascent light fragment at scission. For the more probable mass ratios corresponding to fragments away from closed shells the fragment distortions implied are more comparable. At the largest mass ratios where the light fragment is approaching the $N = 50$ closed shell a more spherical nascent light fragment than heavy fragment is implied. The observed approximate insensitivity of the emission angle to the mass ratio has been qualitatively reproduced in calculations by Fong (1970). The results of recent measurements of the dependence of the α-particle emission angle on mass ratio for thermal neutron fission of ^{235}U are somewhat contradictory, as illustrated in Fig. XIV-17. This aspect of fission deserves further experimental study, as the emission angle dependence on mass ratio may provide important information on the relative distortions of the two fragments at scission.

Let us turn now to the motivation for the trajectory calculations and ask what we learn about the scission configuration from the principal features of

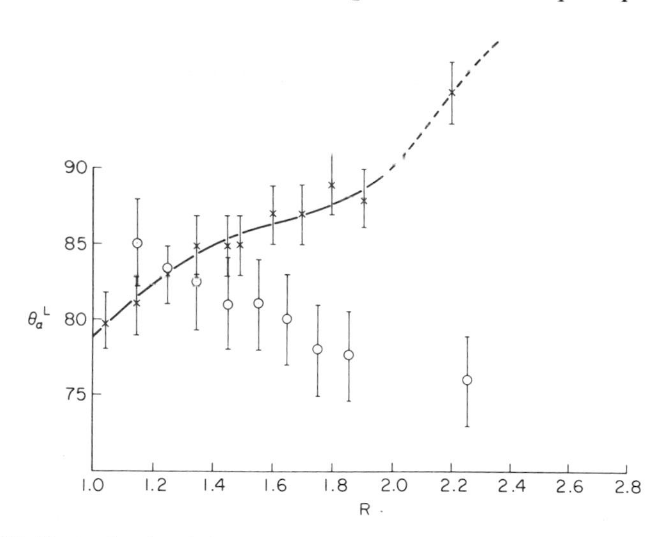

Fig. XIV-17. Angle of emission of α particle with respect to light fragment as a function of mass ratio for thermal neutron fission of ^{235}U as reported by Carles *et al.* (1969) and Gazit *et al.* (1969). (×) Results of Carles *et al.*; (o) results of Gazit *et al.*

the energy and angular distributions of the light particle. A number of independent calculations attempting to reproduce the characteristics of α particle-accompanied fission of ^{252}Cf have been reported. A summary of the implied values of some of the variables describing the initial conditions is given in Table XIV-3. It can be seen that there is considerable agreement on the interfragment distance D at the instance of scission. Of particular interest are the initial kinetic energies of the fragments at scission. The first trajectory calculations indicated already that these initial kinetic energies were very large; much larger, for example, than the values assumed in the statistical theory of fission. Unfortunately there is considerable disagreement in the various estimates of the initial fragment energies at scission, as can be seen from Table XIV-3. The initial energies in the calculation of Fong are based on a statistical model rather than a fit to the data. The primary concern in Fong's work was the correlation of the most probable α-particle angle with fragment mass, and no energy or angular distributions are presented. The final α-particle energy comes out too high, while the final total kinetic energy of the fragments comes out much too low in this calculation. There has been a recent attempt (Vitta, 1971) to rescue the failure of the statistical model initial conditions to reproduce the α-particle angular distribution by postulating delayed emission. Such an assumption restores the result of the previous analyses, i.e., that the fragments are moving with appreciable energies at the time the α particle is released. Whether the release time is the time of scission or some later time is not an issue which can be decided on the basis of trajectory calculation comparisons. It is difficult, however, to imagine a mechanism, particularly a mechanism consistent with a statistical view of fission, for introducing an emission delay.

The uniqueness of the determination of the initial kinematic variables $E_{\alpha 0}$, E_{f0}, and D has been explored by Boneh *et al.* (1967). They found, as shown in

TABLE XIV-3

Summary of Interfragment Distance D, Average α-Particle Initial Kinetic Energy $E_{\alpha 0}$, and Total Fragment Energy E_{f0} at Scission for Various Analyses of the Data for ^{252}Cf

D (fermi)	$E_{\alpha 0}$ (MeV)	E_{f0} (MeV)	Ref.
27	4.4	40	Halpern (1963)
25	3	35	Boneh *et al.* (1967)
26.7	4.5	63	Katase (1968)
21.5	2	8	Raisbeck and Thomas (1968), Rajagopalan and Thomas (1972)
24	3	28	Nardi *et al.* (1969)
~24	0.5	0.5	Fong (1970)
26	2	40	Krogulski and Blocki (1970)
23.7	2.75	25	Musgrove (1971)

TABLE XIV-4

Six Sets of Starting Conditions for $E_{\alpha 0} = 0.2$–5.0 MeV[a]

$E_{\alpha 0}$ (MeV)	D (10^{-13} cm)	E_{f0} (MeV)	E_α (MeV)	θ_L (deg)	E_f (MeV)
0.2	20.7	0	15.2	85	167
0.5	21.0	1.8	15.9	85	166
1.0	21.7	7.3	16.1	85	166
2.0	23.5	19.8	16.0	85	166
3.0	26.0	34.4	15.5	86	167
5.0	30.0	52.3	16.2	87	169

[a] The separation distance D and initial fragment energy E_{f0} were adjusted to satisfy the conditions $\bar{E}_f \simeq 167$ MeV and $\bar{E}_\alpha \simeq 15.9$ MeV at $\theta_L \simeq 85°$. The values of the other initial parameters are $R = 1.4$, $Y_0 = 0$, $\theta_0 = 90°$. [From Boneh *et al.* (1967).]

Table XIV-4, that the experimentally observed most probable α-particle energy and angle, and the total fragment kinetic energy, can be reproduced by initial conditions spanning a large region of $E_{\alpha 0}$. The three parameters $E_{\alpha 0}$, E_{f0}, and D are interrelated, so that determination of any one of the parameters fixes the values of the other two. To further specify the initial values it is necessary to appeal to the characteristics of the experimental observations beyond those corresponding to the most probable energies and angle. Two features of the data appear to be most relevant here. One is the width of the angular distribution. For very low values of the initial energy $E_{\alpha 0}$ the calculated angular distribution becomes too narrow. The second feature of the experimental results which bears on this question is the correlation between the final α-particle and total fragment kinetic energies. These quantities are negatively correlated, with the strength of the anticorrelation increasing with decreasing initial α-particle energy $E_{\alpha 0}$. Although a significant anticorrelation is observed experimentally, comparison with calculations again indicates that $E_{\alpha 0}$ must be larger than 1 MeV. The validity of both of these conclusions is, however, somewhat dependent on certain assumptions concerning the distribution of the other dynamical variables.

The situation may be summarized by saying that the comparison of trajectory calculations with experiment provides considerable evidence that in α particle-accompanied fission the fragments have considerable kinetic energy at the time of release of the α particle. On the basis of the mechanism for the process discussed in an earlier section, the release time must be very close to the time of scission. If we accept the position that the kinetic energy of the fragments at scission is rather considerable for α accompanied fission, we can still question whether this conclusion is applicable to binary fission. There is, however, considerable evidence that α particle accompanied fission resembles binary fission fairly closely in a number of respects, such as the mass distribution and the dependence of the total kinetic energy on mass ratio. On the other hand it is clear that α particles are emitted only for those scission configurations for which there is a larger than average distortion of the nascent fragments, hence, a larger than average fragment separation D. The initial kinetic energy at scission for unusually stretched configurations may be greater than for more typical distortions. A careful study of the dependence of both the light particle energy and angular distributions on the final total fragment kinetic energy might lead to a correlation of initial fragment velocity with initial fragment separation which could be extrapolated to the separations typical of binary fission.

5. Dependence of the Probability of Alpha-Particle-Accompanied Fission on Mass Division

We remarked earlier that light particle accompanied fission is typical with respect to the kinetic energy and mass ratio distributions. The qualitative similarity of the binary and ternary mass distributions for ^{235}U is illustrated in Fig. XIV-18. Mass yield curves for binary and for ternary fission have been measured for other fissioning systems, including spontaneous fission of ^{252}Cf, α particle-induced fission of ^{238}U, and proton-induced fission of ^{226}Ra and ^{230}Th. The binary and ternary mass distributions are qualitatively similar in all cases; the results for ^{226}Ra are illustrated in Fig. XIV-19.

There has been considerable interest in going beyond the qualitative similarities of binary and ternary fission to look for detailed differences between the mass distributions for binary and α-accompanied fission. In Fig. XIV-18 a displacement of the ternary relative to the binary yields is shown for the heavier light fragments and the heavier heavy fragments, while the yields of the lighter light fragments and lighter heavy fragments overlap in binary and ternary fission. In particular, the question is sometimes asked, Which fragment did the α particle come from? The phrasing of the question is actually prejudicial against an alternative possibility that the α particle is formed from nucleons drawn from both fragments. The dependence of the α-particle emission probability on fragment mass depends critically on the assumption made with regard to the number of nucleons given up by each fragment in the process of α emission. Halpern (1965, 1971) has emphasized that it is not possible to attribute the difference between the binary and ternary mass distributions uniquely to a mass-dependent emission probability nor to a mass dependence of the number of nucleons contributed to the α particle by a particular fragment. This point is illustrated in Fig. XIV-20. It is also not possible to distinguish experimentally whether, for example, in a particular mass split, three nucleons came from the heavy fragment and one from the

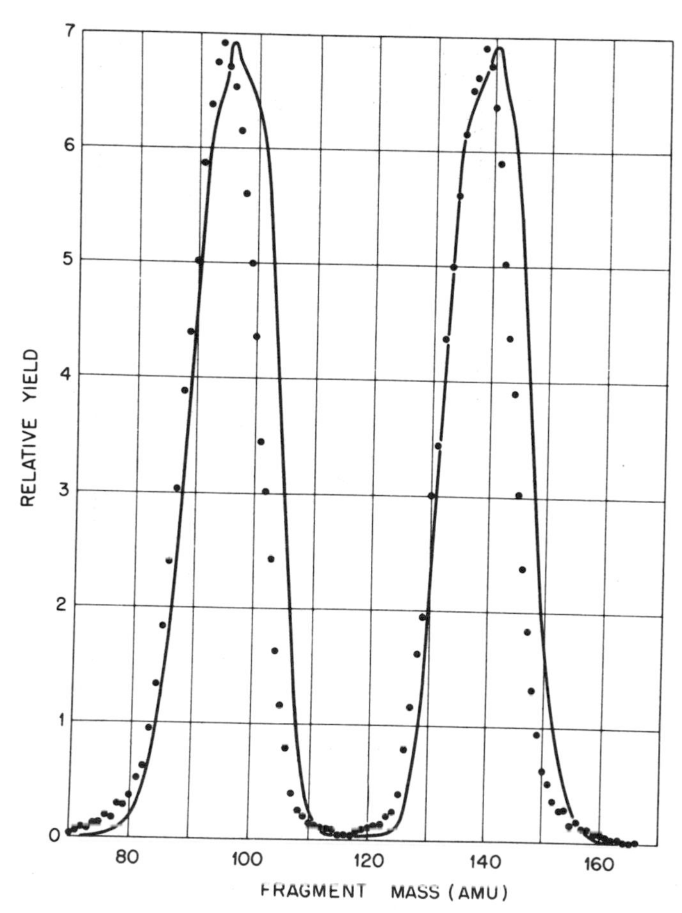

Fig. XIV-18. Relative mass yields for fragments from ^{235}U thermal neutron-induced binary (—) and ternary (●) fission. Yields are normalized at the peaks. [From Schmitt *et al.* (1962).]

light, or whether all of the nucleons comprising the α particle came from the heavy fragment in $\frac{3}{4}$ of the events and from the light fragment in $\frac{1}{4}$ of the events.

A better way of learning something about where the α particle came from is to ask which kinds of fragments are missing energy rather than which kinds of fragments are missing "mass." A fragment that has used up some of its deformation energy to release an α particle would be expected to emit fewer neutrons. By comparing the neutron yield as a function of fragment mass for binary and for ternary fission, the energy loss as a function of fragment mass can be deduced. Such an experiment has been performed for ^{252}Cf (Nardi and Fraenkel, 1970). The results show that those fragments which have a larger

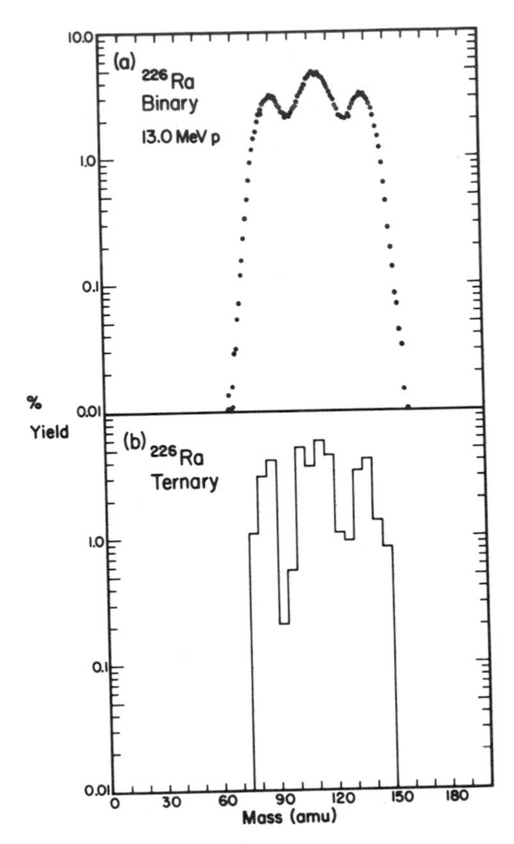

Fig. XIV-19. Comparison of mass yield curves for binary and charged particle-accompanied fission of ^{226}Ra. [From Perry (1970).]

than average deformation at scission (as indicated by their ν values in binary fission) also have larger than average energy loss in α-accompanied fission. Thus the dependence of the energy loss on fragment mass exhibits a sawtooth behavior similar to that of $\nu(A)$, with the heavier members of both the light and the heavy groups having the larger decrease in energy for α-accompanied fission as compared to binary fission. This conclusion is essentially independent of whether we assume that the nucleons given up by each fragment to form the α particle have come from one fragment or the other fragment or have come from both fragments.

In summary, the neutron yield data show that α-particle emission occurs at the expense of the excitation energy of those fragments having the largest deformation energy at scission, and the mass yield data are consistent with α-particle emission occurring at the expense of the yield of those fragments having the largest deformation energy at scission.

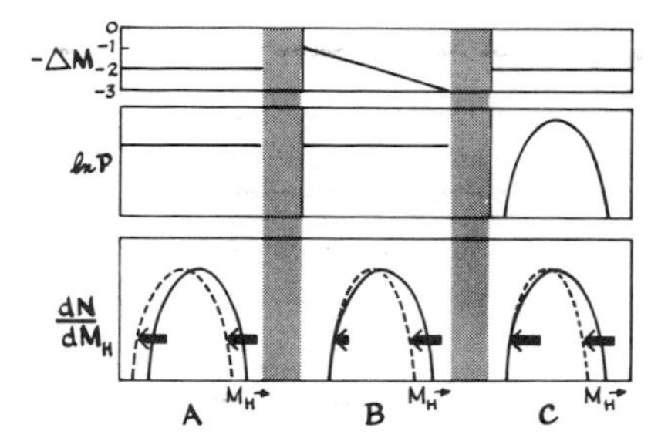

Fig. XIV-20. Idealized comparisons of binary and ternary fragment mass distributions. The ternary distribution (– – –) can be generated from the binary distribution if we can specify for each binary mass division (i) the chance P that an α particle will be emitted and (ii) the mean number of nucleons ΔM that this α particle draws from, say, the heavy fragment (it therefore draws $4-\Delta M$ nucleons from the light one). The three columns show the ternary mass distributions expected on the basis of the observed binary distribution (solid curves in bottom row) and three sets of assumptions: (A) The probability P for emitting an α particle is independent of the mass division and each fragment contributes 2 mass units ($\Delta M = 2$) to the α particle. Then the ternary mass distribution is simply shifted 2 mass units below the binary, contrary to observation. (B) P is as before, but ΔM varies linearly with fragment mass. (C) ΔM is as in (A) but P varies as shown with fragment mass. The expected mass distributions in ternary fission are indistinguishable for columns (B) and (C). These ternary distributions resemble those actually observed and we see that there is no unique way to interpret them in terms of ΔM and P. [From Halpern (1971).]

This might suggest that for symmetric mass divisions in thermal neutron fission of ^{235}U there would be an enhanced probability for emission of α particles. This is based on the neutron yields and kinetic energy release for symmetric divisions in binary fission, which indicate that both fragments are unusually distorted at scission. Any such enhancement is expected to disappear as the excitation energy is increased and the fragment mass dependence of the energy stored in deformation is washed out due to the dissolution of shell effects. Thus the absence of significant enhancement in the case of energetic particle-induced fission is not too surprising (Loveland *et al.*, 1967).

6. Energy Balance in Alpha-Accompanied Fission

The difference between the total fragment kinetic energy in binary fission and in α-accompanied fission

$$\bar{E}_{k_B} - \bar{E}_{k_T}$$

is reported to be 12.8 MeV for ^{252}Cf (Nardi and Fraenkel, 1970). This is somewhat larger than can be accounted for simply on the basis of the reduction of the Coulomb potential due to loss of 2 units of charge from one of the fragments. The indication is that the average velocity of the fragments at distances equal to or greater than the scission distance for α-accompanied fission is less than that for binary fragments at the *same separation*. Adding the average kinetic energy of the α particle, 16 MeV, to that of the fragments in α-accompanied fission, we obtain a total kinetic energy of the three particles that is 3 MeV greater than that in binary fission. The number of neutrons emitted in α-accompanied fission of ^{252}Cf is 0.65 neutron less than in binary fission (Nardi and Fraenkel, 1970; Piekartz *et al.*, 1970). Using a value of $(d\nu/dE)^{-1} = 8.7$ MeV, corresponding to an average of the values for binary and ternary fission, the indication is that the excitation energy of the fragments is 5.7 MeV less in the case of α-accompanied fission than in the case of binary fission. The total γ-ray energy in binary and ternary fission appears to be the same to within less than 1 MeV (Adamov *et al.*, 1967; Ajitanand, 1969). Thus the total energy release in α-accompanied fission is 2.5 MeV less than in the case of binary fission. This is somewhat less than might be expected on the basis of an average α-binding energy of about 5 MeV. This discrepancy can perhaps be attributed, at least in part, to a greater exploitation in ternary fission than in binary fission of the larger available energy release for fission into fragments with an even number of neutrons or protons. Feather (1970) has concluded, from a comparative analysis of the K x-ray probabilities for binary and ternary fission (Watson *et al.*, 1967; Watson, 1969), that the yield of even-Z x rays is relatively slightly greater in α-accompanied fission than in binary fission.

B. Fission in Which Three Fragments of Comparable Mass Are Produced

The possibility that a heavy excited nucleus might divide into three fragments of comparable mass has been considered since soon after the discovery of fission. Early experiments with emulsions showed that if the process occurred, it was fairly rare. Unambiguous identification of infrequent ternary fission events is difficult because of the possibility of a binary fragment scattering off an atom of the target or emulsion material, and the recoil being mistaken for a ternary fragment.

More recently a rather extensive instrumental investigation of events in which each of three particles is detected in time coincidence in separate detectors has been reported by Muga and collaborators (Muga, 1965, 1967; Muga and Rice, 1969; Muga *et al.*, 1967). For thermal neutron fission of uranium and plutonium isotopes a relative frequency of ternary to binary events of between one in 10^5 and one in 10^6 is reported. There appears to be a "grouping" of events; for one group the light fragment mass is near 30 and for the other group the light fragment mass is in the mass range 50–60. Attempts to identify these light fragments in radiochemical investigations have been unsuccessful (Stoenner and Hillman, 1966; Prestwood and Bayhurst, 1966). The yields of approximately 16 radionuclides between $A = 30$ and $A = 60$ have been shown to be several orders of magnitude smaller than implied by the detector measurements. Thus the ternary process must be very selective if the results of the two experiments are to be reconciled.

More recently it has been suggested that the detector results can be reinterpreted as the consequence of scattering of a binary fragment in the target or target backing (Steinberg *et al.*, 1970). The likelihood of such a contribution is very dependent on the angles between the detectors, the angular acceptance of the detectors, and the finite extension of the source. The possibility of scattering had been considered earlier by Muga and rejected on the basis of a number of tests, including the insensitivity of the ternary-to-binary ratio to target thickness. Further experimental work will be required to resolve these discrepancies. A smaller source extension and better angular definition of the detectors would be expected to reduce the scattering contribution greatly, but the practical difficulty in obtaining many events in a more restricted geometry for this rare process must be recognized.

Although radiochemical attempts to identify very light fragments which might arise from ternary fission have been negative for low energy fission, positive results for fragments with masses between $A = 20$ and $A = 40$ have been obtained at higher excitation energies (Iyer and Cobble, 1966, 1968). The yields of ^{28}Mg and ^{38}S relative to binary fission are strongly dependent on the excitation energy, as illustrated in Fig. XIV-21. It is clear that any extrapolation of these yields to the excitation energies of the experiments of Muga *et al.* would predict yields several orders of magnitude lower than observed.

A fairly clear separation between the light fragments attributed to ternary fission and the fragments from normal binary fission is indicated in the mass yield curve illustrated in Fig. XIV-22. As a check on the possibility that the $A = 20$ to $A = 40$ products might originate in very asymmetric binary fissions, it is of interest to measure the yield of the expected heavy complementary fragments. Unfortunately the measurements reported thus far do not adequately test this possibility. The upper limits for Au and Pb nuclides indicated in Fig. XIV-22 were apparently calculated on the basis of the assumption that their efficiency for recoiling from the thick targets (1-mil uranium) was equal to that for their complementary partners. If they are binary fragments, their kinetic energies are expected to be $\frac{1}{5}$ or less of that of their complementary light fragment. Thus it is impossible to exclude the possibility of binary fission on the basis of the presently available information on yields of possible complementary fragments.

There is some information on the kinetic energy of these light products in the case of ^{28}Mg. By comparing the amount of ^{28}Mg escaping to catcher foils in

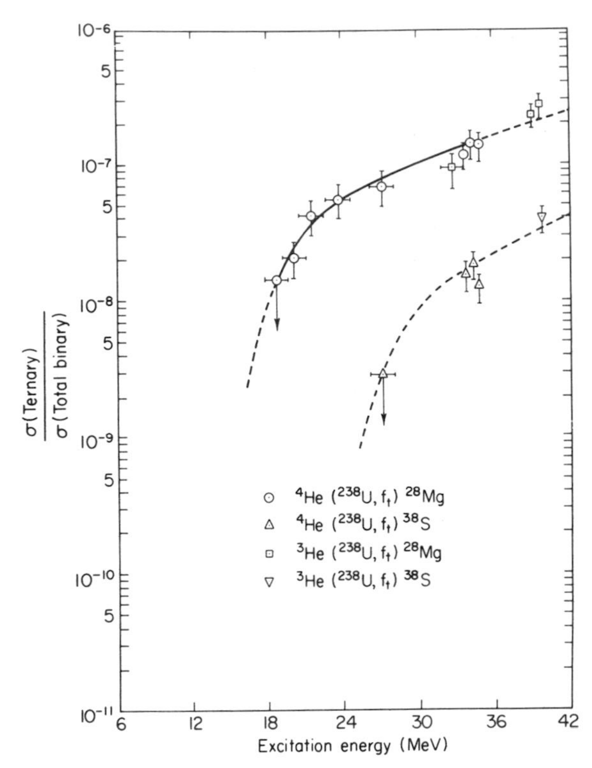

Fig. XIV-21. The ratio of ^{28}Mg and ^{38}S yield to total binary yield as a function of excitation energy. [From Iyer and Cobble (1966).]

front of and behind the target with that in the target itself, a range in uranium of 12 mg/cm^2 has been determined for α particle induced fission of ^{238}U (Iyer and Cobble, 1968). A rather surprising result is that although the range is independent of α-particle bombarding energy between 40 and 118 MeV, the range for 30-MeV ^{3}He-induced fission is approximately 25% lower than for ^{4}He-induced fission (MacMurdo and Cobble, 1969). Estimates of the kinetic energy implied by the range measurements for α-induced fission vary between 40 and 100 MeV, with 50 MeV a fairly likely value (Iyer and Cobble, 1966, 1968). The expected result from binary fission is approximately 70 MeV.

If the products with $A = 20$–40 arise from ternary fission, the decrease in yield with mass number is qualitatively consistent with the trend established by the lighter ($A = 4$–16) particles (see Fig. XIV-4). A more quantitative comparison is not possible in view of the relative insensitivity of the light particle (α) yields to excitation energy, as compared to the strong dependence observed for the heavier particles. The greater sensitivity of the heavier particle yields

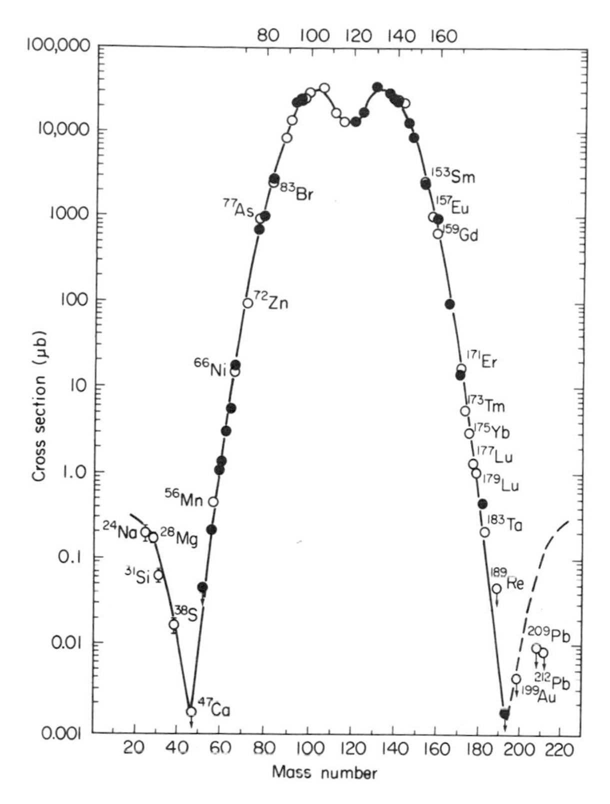

Fig. XIV-22. Fission mass yield distribution for ^{238}U excited by 30.6-MeV ^{3}He ions. (---) Heavy mass yields expected from the reflection of the light mass yield curve from hypothetical binary events. [From MacMurdo and Cobble (1969).]

to excitation energy may simply reflect the greater energy cost for producing heavier fragments.

There is quite clear evidence for fission into three comparable fragments for nuclei which have both higher excitation energies and higher Z^2/A values (Fleischer *et al.*, 1966; Karamyan *et al.*, 1967; Perelygin *et al.*, 1969). For 400-MeV argon ions on thorium the ratio of ternary to binary fission has reached 0.03. There are two possible mechanisms by which three comparable mass fragments can be produced in heavy ion-induced fission of heavy element targets. In a "true" ternary event two necks are formed essentially instantaneously, leading to three fragments. In the second mechanism, designated "cascade" fission (Karamyan *et al.*, 1967), the heavy fragment produced in normal binary fission may have sufficient excitation energy to subsequently fission also. These processes are indicated schematically in Fig. XIV-23. It is possible to make a fairly quantitative estimate of the probability of cascade

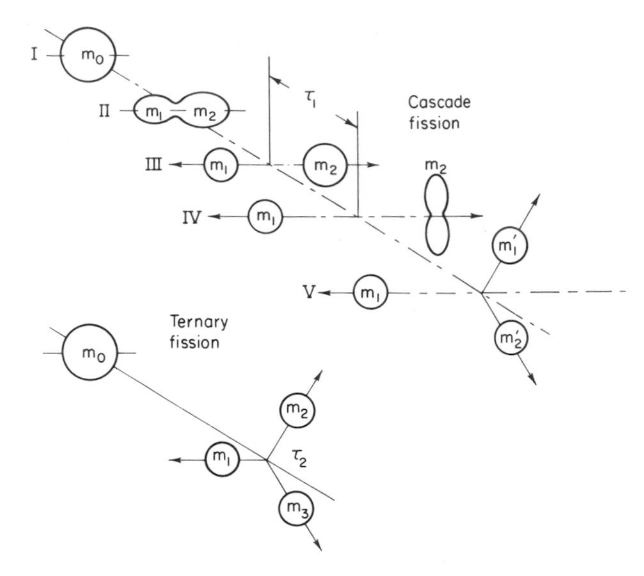

Fig. XIV-23. Schematic representation of two possible mechanisms of fission into three fragments. [From Muzychka *et al.* (1968).]

fission (Muzychka *et al.*, 1968). The probability of cascade fission can be written as

$$P_{3\mathrm{f}} = \sum_{A_{\mathrm{bin}}} P_E(A) W_{\mathrm{f}}^{E}(A)$$

where $P_E(A)$ is the relative yield of the (heavy) binary fragment with mass A at a given excitation energy of the compound nucleus, and $W_{\mathrm{f}}^{E}(A)$ is the probability of fission of a heavy fragment with mass A and excitation energy E. The energy E is obtained by assuming that the available excitation energy in the initial fission is distributed among the fragments in proportion to their masses. The fission yields $P_E(A)$ are given by a Gaussian parameterization of the observed dependence on initial excitation energy and Z^2/A. The probability of fission of the binary fragment can be expressed by

$$W_{\mathrm{f}}^{E} = \sum_i (\Gamma_{\mathrm{f}i}/(\Gamma_{\mathrm{f}i}+\Gamma_{\mathrm{n}i})) \prod_{k=0}^{i} \Gamma_{\mathrm{n}k}/(\Gamma_{\mathrm{n}k}+\Gamma_{\mathrm{f}k})$$

where the sum i is over the number of neutrons emitted prior to fission. In practice, only the first two or three emissions need be considered. The fission and neutron widths Γ_{f} and Γ_{n} are also functions of energy and can be evaluated using Eqs. (VII-7). The level density parameters in these equations are chosen so as to reproduce the known dependence of $\Gamma_{\mathrm{f}}/\Gamma_{\mathrm{n}}$ in the appropriate mass region. The results of such a calculation are compared with the experimental

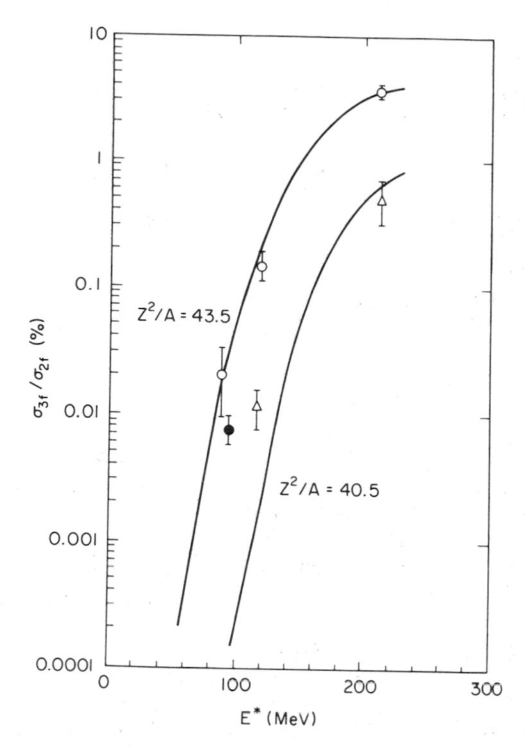

Fig. XIV-24. Comparison of experimental probabilities for ternary fission relative to binary fission with that calculated for the cascade mechanism (—). Experimental points: (○) $^{238}U+{}^{40}Ar$, (●) $^{238}U+{}^{22}Ne$, (△) $Pb+{}^{40}Ar$. [From Karamyan *et al.* (1967).]

data in Fig. XIV-24. The comparison indicates that cascade fission can account for all of the ternary fission observed except for systems with both low Z^2/A and low excitation energy.

References

Adamov, V. M., Drapchinskii, L. V., Kovalenko, S. S., Petrzhak, K. A., and Tyutyugin, I. I. (1967). *Sov. J. Nucl. Phys.* **5**, 30.

Ajitanand, N. N. (1969). *Nucl. Phys. A* **133**, 625.

Andreev, V. N., Nedopekin, V. G., and Rogov, V. I. (1969). *Sov. J. Nucl. Phys.* **8**, 22.

Blocki, J., Chwaszczewska, J., Dakowski, M., Krogulski, T., Piasecki, E., Sowinski, M., Stegner, A., and Tys, J. (1969). *Nucl. Phys. A* **127**, 495.

Boneh, Y., Fraenkel, Z., and Nebenzahl, I. (1967). *Phys. Rev.* **156**, 1305.

Bowman, H. R., Milton, J. C. D., Thompson, S. G., and Swiatecki, W. J. (1962). *Phys. Rev.* **126**, 2120.

Carles, C. Asghar, M., Doan, T. P., Chastel, R. and Signanbieux, C. (1969). *Proc. IAEA Symp. Phys. Chem. Fission, 2nd, Vienna, 1969*, p. 130. IAEA, Vienna.
Cosper, S. W., Cerny, J., and Gatti, R. C. (1967). *Phys. Rev.* **154**, 1193.
Dakowski, M., Chwaszczewska, J., Krogulski, T., Piasecki, E., and Sowinski, M. (1967). *Phys. Lett. B* **25**, 213.
Feather, N. (1969). *Proc. IAEA Symp. Phys. Chem. Fission, 2nd, Vienna, 1969*, p. 83. IAEA, Vienna.
Feather, N. (1970). *Phys. Rev. C* **1**, 747.
Fleischer, R. L., Price, P. B., Walker, R. M., and Hubbard, E. L. (1966). *Phys. Rev.* **143**, 943.
Fluss, M. J., Kaufman, S. B., Steinberg, E. P., and Wilkins, B. D. (1973). *Phys. Rev., C* **7**, 353.
Fong, P. (1970). *Phys. Rev. C* **2**, 735.
Fraenkel, Z. (1967). *Phys. Rev.* **156**, 1283.
Fraenkel, Z., and Thompson, S. G. (1964). *Phys. Rev. Lett.* **13**, 438.
Fuller, R. W. (1962). *Phys. Rev.* **126**, 684.
Gazit, Y., Katase, A., Ben-David, G., and Moreh, R. (1969). *Proc. IAEA Symp. Phys. Chem. Fission, 2nd, Vienna, 1969*, p. 130. IAEA, Vienna.
Gazit, Y., Nardi, E., and Katcoff, S. (1970). *Phys. Rev. C* **1**, 2101.
Halpern, I. (1963). CERN Rep. 6812/P, unpublished.
Halpern, I. (1965). *Proc. IAEA Symp. Phys. Chem. Fission, Salzburg, 1965*, **2**, p. 369. IAEA, Vienna.
Halpern, I. (1971). *Annu. Rev. Nucl. Sci.* **21**, 245.
Iyer, R. H., and Cobble, J. W. (1966). *Phys. Rev. Lett.* **17**, 541.
Iyer, R. H., and Cobble, J. W. (1968). *Phys. Rev.* **172**, 1186.
Kapoor, S. S., Ramanna, R., and Rama Rao, P. N. (1963). *Phys. Rev.* **131**, 283.
Karamyan, S. A., Kuznetsov, V. I., Oganesyan, Yu. Ts., and Penionzhkevich, Yu. E. (1967). *Sov. J. Nucl. Phys.* **5**, 684.
Katase, A. (1968). *J. Phys. Soc. Jap.* **25**, 933.
Krogulski, T., and Blocki, J. (1970). *Nucl. Phys. A* **144**, 617.
Krogulski, T., Chwaszczewska, J., Dakowski, M., Piasecki, E., Sowinski, M., and Tys, J. (1969). *Nucl. Phys. A* **128**, 219.
Loveland, W. D., Fairhall, A. W., and Halpern, I. (1967). *Phys. Rev.* **163**, 1315.
MacMurdo, K. W., and Cobble, J. W. (1969). *Phys. Rev.* **182**, 1303.
Milton, J. C. D., and Fraser, J. S. (1965). *Proc. IAEA Symp. Phys. Chem. Fission, Salzburg, 1965*, **2**, p. 39. IAEA, Vienna.
Muga, M. L. (1965). *Proc. IAEA Symp. Phys. Chem. Fission, Salzburg, 1965*, **2**, p. 409. IAEA, Vienna.
Muga, M. L. (1967). *Phys. Rev. Lett.* **18**, 404.
Muga, M. L., and Rice, C. R. (1969). *Proc. IAEA Symp. Phys. Chem. Fission, 2nd, Vienna, 1969*, p. 107. IAEA. Vienna.
Muga, M. L., Rice, C. R., and Sedlacek, W. A. (1967). *Phys. Rev.* **161**, 1266.
Musgrove, A. R. de L. (1971). *Austral. J. Phys.* **24**, 129.
Muzychka, Yu. A., Oganesyan, Yu. Ts., Pustylnik, B. I., and Flerov, G. N. (1968). *Sov. J. Nucl. Phys.* **6**, 222.
Nagy, L., Nagy, T., and Vinnay, I. (1969). *Sov. J. Nucl. Phys.* **8**, 257.
Nardi, E., and Fraenkel, Z. (1970). *Phys. Rev. C* **2**, 1156.
Nardi, E., Boneh, Y., and Fraenkel, Z. (1969). *Proc. IAEA Symp. Phys. Chem. Fission, 2nd, Vienna, 1969*, p. 143. IAEA, Vienna.
Natowitz, J. B., Khodai-Joopari, A., Alexander, J. M., and Thomas, T. D. (1968). *Phys. Rev.* **169**, 993.

Nobles, R. A. (1962). *Phys. Rev.* **126**, 1508.
Perelygin, V. P., Shadieva, N. H., Tretiakova, S. P., Boos, A. H., and Brandt, R. (1969). *Nucl. Phys. A* **127**, 577.
Perry, D. G. (1970). Thesis, Univ. of Washington, Seattle, unpublished.
Piasecki, E., Dakowski, M., Krogulski, T., Tys. J., and Chwaszczewska, J. (1970). *Phys. Lett. B* **33**, 568.
Piekartz, H., Blocki, J., Krogulski, T., and Piasecki, E. (1970). *Nucl. Phys. A* **146**, 273.
Prestwood, R. J., and Bayhurst, B. P. (1966). *Amer. Chem. Soc. Meeting, 151st, Pittsburgh, Pennsylvania, 1966.*
Raisbeck, G. M., and Thomas, T. D. (1968). *Phys. Rev.* **172**, 1272.
Rajagopalan, M., and Thomas, T. D. (1972). *Phys. Rev. C* **5**, 2064.
Roy, J. C. (1961). *Can. J. Phys.* **39**, 315.
Schmitt, H. W., Neiler, J. H., Walter, F. J., and Chetham-Strode, A. (1962). *Phys. Rev. Lett.* **9**, 427.
Skarsvåg, K., and Bergheim, K. (1963). *Nucl. Phys.* **45**, 72.
Steinberg, E. P., Wilkins, B. D., Kaufman, S. B., and Fluss, M. J. (1970). *Phys. Rev. C* **1**, 2046.
Stoenner, R. W., and Hillman, M. (1966). *Phys. Rev.* **142**, 716.
Terrell, J. (1962). *Phys. Rev.* **127**, 880.
Thomas, T. D., and Whetstone, S. L., Jr. (1966). *Phys. Rev.* **144**, 1060.
Vandenbosch, R. (1963). *Nucl. Phys.* **46**, 129.
Vitta, P. B. (1971). *Nucl. Phys. A* **170**, 417.
Vorobiev, A. A., Seleverstov, D. M., Grachov, V. T., Kondurov, I. A., Nikitin, A. M., Yegorov, A. I., and Zalite, Yu. K. (1969a). *Phys. Lett. B* **30**, 332.
Vorobiev, A. A., Grachev, V. T., Komar, A. P., Kondurov, I. A., Nikitin, A. M., and Seleverstov, D. M. (1969b). *Sov. J. At. Energy* **27**, 713.
Watson, R. L. (1969). *Phys. Rev.* **179**, 1109.
Watson, R. L., Bowman, H. R., and Thompson, S. G. (1967). *Phys. Rev.* **162**, 1169.
Whetstone, S. L., Jr., and Thomas, T. D. (1967). *Phys. Rev.* **154**, 1174.

Author Index

A

Adamov, V. M., 400, *405*
Ajitanand, N. N., 400, *405*
Alam, G. D., 76, *77*, 320, 321, 322, *333*
Aleksandrov, B. M., 46, 47, *77*
Alexander, J. M., 378, *406*
Aley, R., 262, *287*, 297, *303*, 348, 349, 354, *355*
Alstad, J., 318, *334*
Amiel, S., 337, *355*
Andersen, B. L., 166, 167, *176*
Andreev, V. N., 379, *405*
Apalin, V. F., 346, 347, *355*
Armbruster, P., 326, *334*, 358, 359, 360, 363, 364, *373*
Asghar, M., 85, 87, 88, 89, 91, *107*, 393, *406*
Asplund-Nilsson, I., 338, *355*
Attree, P., 338, *356*
Auchampaugh, G. F., 91, *107*, 219, 220, *252*
Auerbach, E. H., 144, 161, *176*
Axel, P., 119, *176*

B

Baba, H., 212, *215*
Babenko, Yu. A., 219, *252*
Bacharach, S. L., 61, *80*
Back, B. B., 39, *42*, 61, *77*, 166, 167, 169, 172, 173, *176*, *177*
Baerg, A. P., *116*, 130, 139, *177*, 221, *253*
Balagna, J. P., 295, *303*, 306, 307, *333*
Bang, J. M., 58, *80*, 166, 167, *176*
Bannister, F. J., 46, *80*
Barclay, F. R., 46, *77*
Bardeen, J., 197, *215*
Barnes, R. F., 47, *78*, *79*, 305, 306, 316, *334*
Baron, E., 343, *355*
Bartholomew, R. M., *116*, 130, 139, *177*
Barton, D. M., 47, *77*
Bassichis, W. A., 30, *42*
Bate, G. L., 189, 192, 193, 195, 205, 206, 207, *215*
Baybarz, R. D., 219, *253*
Bayhurst, B. P., 89, 94, 95, *107*, 401, *407*
Behkami, A. N., 111, 140, 145, 146, 148, 149, 150, 151, 152, 153, 154, 156, 157, 162, 163, *177*, *178*, 184, 187, *215*
Bemis, C. E., 338, *356*
Ben-David, G., 393, *406*
Bergen, D. W., 91, 92, *107*, 219, *252*
Bergheim, K., 381, *407*
Bes, D., 25, *42*
Bishop, C. J., 239, 241, 242, 243, *252*, 262, *287*, 297, *303*, 348, 349, 354, *355*
Bjørnholm, S., 59, 61, 62, 65, 73, *77*, *79*, 96, *107*, 157, *177*
Blann, H. M., 332, *334*
Blatt, J. M., 123, *177*, *252*
Blinov, M. V., *355*
Block, R. C., 89, 95, *108*

Blocki, J., 379, 394, 400, *405*, *406*, *407*
Blons, J., 89, *107*
Bloomquist, C. A., 305, 306, 316, *334*
Bohr, A., 7, *13*, 94, *107*, 109, 112, *177*
Bohr, N., 3, *13*, 16, *42*, *77*, *252*
Boldeman, J. W., 338, 341, 344, *355*
Bolsterli, M., 4, *13*, 36, 37, *43*, *258*, 276, 278, 280, *287*, *288*
Bondelid, R. O., 212, 213, 214, *215*
Bondorf, J. P., 39, *42*, *77*, 169, 172, 173, *176*, *177*
Boneh, Y., 386, 387, 388, 394, 395, *405*, *406*
Boos, A. H., 403, *407*
Borggreen, J., 61, 62, 65, *77*
Bowman, C. D., 47, *77*, 89, 91, *107*, *108*, 220, *252*
Bowman, H. R., 336, 338, 342, 353, 354, *355*, 359, 367, 369, 370, 371, *372*, *373*, 381, 400, *405*, *407*
Boyer, K., 163, 169, *178*
Brack, B., 58, 62, *77*
Brack, M., 3, *13*, 36, *43*, 129, *178*, 278, 280, *288*
Brandt, R., 47, 48, *77*, *79*, 403, *407*
Brenner, D. S., 61, *77*
Britt, H. C., 39, *43*, 61, 65, 73, 76, *77*, 90, *107*, 164, 166, 169, 170, 171, 172, 173, *176*, 220, *252*, 256, *258*, 298, 299, *303*
Brown, F., 46, *80*, *116*, 130, 139, *177*, 221, *253*
Brunhart, G., 85, 87, *108*
Burnett, D. S., 217, 236, *252*, 255, *258*
Burnett, S. C., 61, 65, 73, 76, *77*, 169, 173, *177*
Butler, J. P., 46, 47, 48, *77*, *79*

C

Caldwell, J. T., 47, *77*
Cao, M. G., 89, *107*
Carles, C., 393, *406*
Cauvin, B., 84, *107*
Cerny, J., 379, 383, 384, 389, *406*
Ceulemans, H., 87, *108*
Chamberlain, O., 46, *77*
Chan, G., 241, 242, *253*
Chase, D. M., 139, *178*
Chastel, R., 393, *406*
Chaudhry, R., 189, 193, 205, *215*, 217, 235, *252*
Cheifetz, E., 239, *252*, 338, *355*, 369, 370, 371, *373*
Chetham-Strode, A., 397, *407*
Chikov, N., 90, 91, 95, *108*
Chistyakov, L. V., 47, *80*, 219, *253*
Choppin, G. R., 47, 48, *77*, *78*, 338, *355*
Chwaszczewska, J., 379, 383, 384, *405*, *406*, *407*
Clarke, K. M., 120, *177*
Clarkson, J. E., 236, *253*
Cobble, J. W., 217, 236, *253*, 255, *258*, 313, 317, *333*, 401, 402, 403, *406*
Cohen, S., 4, 9, *13*, 18, 19, 20, 21, 22, 36, *43*, 207, 208, *215*, 263, 274, *288*
Colby, L. J., Jr., 313, 317, *333*
Colvin, D. W., 338, *355*
Conde, H., 338, *355*
Cooper, L. N., 197, *215*
Coryell, C. D., 369, *373*
Cosper, S. W., 379, 383, 384, 389, *406*
Cowan, G. A., 89, 94, 95, *107*
Cowper, G., 46, 47, 48, *79*
Cox, S., 359, *373*
Cramer, J. D., 58, *77*, 87, 91, *107*, 152, 153, 169, 176, *177*, 220, *252*, 256, *258*
Crespi, V. C., 242, *253*
Croall, I. F., 298, 301, 302, *303*, 369, *372*
Cuninghame, J. G., 298, 301, 302, *303*

D

Dabbs, J. W. T., 83, 84, 85, *107*
Dakowski, M., 338, *355*, 379, 383, 384, *405*, *406*, *407*
Dalhsuren, B., 71, *77*
Dalton, A. W., 338, 341, 344, *355*
Damgaard, J., 3, *13*, 58, 62, *77*, 280, *288*
Davey, W. G., 153, 219, 223, *252*
Davies, W. G., 173, 174, *178*
Decowski, P., 242, *252*
Demidov, A. M., 366, *372*
Derrien, H., 89, *107*
Diamond, H., 47, *78*, *79*
Diven, B. C., 338, *355*
Doan, T. P., 393, *406*
Drapchinskii, L. V., 400, *405*
Druin, V. A., 46, 47, 48, 59, *77*, *79*, *80*
Dubrovina, S. M., 47, *80*, 219, *253*
Duffy, G. J., 121, *177*
Dupzyk, R. J., 48, *78*
Dyachenko, P. P., 344, *356*

E

Earwaker, L. G., 156, *177*
Eastwood, T. A., 46, 47, 48, *77*, *79*
Eccleshall, D., 169, *177*
Edwards, R. R., 46, *79*
Eggerman, C., 84, 89, *107*
Eidens, J., 326, *334*
Eismont, V. F., 336, *355*
Ekberg, K., 338, *356*
Elwyn, A. J., 61, *77*
Engelbrecht. C. A., 121, *177*
Erkkila, B. H., 61, 65, 73, 76, *77*, 173, *177*
Eskola, K., 48, *78*, *79*
Eskola, P., 48, *78*, *79*
Evans, J. E., 48, *78*, 91, *107*

F

Fairhall, A. W., 314, 315, *334*, 378, 399, *406*
Farrar, H., 318, *319*, *333*
Farrell, J. A., 90, 91, 93, *107*, 122, *177*, 219, *252*
Farwell, G. W., 46, *77*
Feather, N., 400, *406*
Ferguson, A. T. G., 61, *77*
Ferguson, R. L., 76, *77*, 320, 321, 322, 325, 326, *333*, *334*, 348, *356*, 363, 364, 365, 366, *373*
Fermi, E., 1, *13*
Feshbach, H., 144, *177*
Fields, P. R., 46, 47, 48, *77*, 305, 306, 316, *334*, 359, *373*
Fillmore, F. L., 338, *355*
Fiset, E. O., 4, *13*, 36, 37, *43*, 276, 278, 280, *287*, *288*
Fleischer, R. L., 46, *78*, 403, *406*
Flerov, G. N., 46, 48, 61, 63, 71, *77*, *78*, *79*, 404, *406*
Fluss, M. J., 382, 401, *406* *407*
Flynn, K. F., 210, *215*, 305, 306, 307,311, 312, 313, 314, 315, 316, *333*, *334*
Fomenko, D. E., 358, *373*
Fomichev, V. A., 59, *80*
Fomoshkin, E. F., 219, *252*
Fong, P., 3, *13*, 261, 283, *288*, *334*, 393, 400, *406*
Ford, G. P., 295, *303*, 306, 307, *333*
Fox, J. P., 61, *79*
Fraenkel, Z., 239, *252*, 386, 387, 388, 392, 394, 395, 397, 399, 400, *405*, *406*
Frankel, S., 19, *43*
Fraser, J. S., 3, *13*, 89, *107*, 169, 173, 174, *178*, 336, 337, 346, *356*, 363, *373*, 381, *406*
Frehaut, J., 343, 344, 351, *355*, *356*
Fried, S., 47, *78*
Friedman, A. M., 46, 47, *77*, *78*, 359, *373*
Frisch, O. R., 2, *13*, *14*
Fuller, R. W., 376, *406*
Fullwood, R. R., 91, *107*
Fultz, S. C., 47, *77*, 91, *107*, 220, *252*
Funshtein, V. B., 219, *252*

G

Galbraith, W., 46, *77*
Galin, J., 239, *252*
Gangrsky, Yu. P., 61, 63, 70, 71, *77*, *78*
Gatti, R. C., 47, 48, *77*, *79*, 188, *215*, 217, 236, *252*, 255, *258*, 379, 383, 384, 389, *406*
Gauriau, J., 343, 344, *356*
Gavrilov, K. A., 48, 70, 71, *78*, *79*
Gazit, Y., 379, 383, 393, *406*
Gerling, E. K., 46, *78*
Geroy, J., 338, *356*
Ghiorso, A., 46, 47, 48, *77*, *78*, *79*, *80*
Gibbs, W. R., 39, *43*, 164, 166, 169, 171, 172, *177*
Gilmore, J. S., 89, 94, 95, *107*
Gindler, J. E., 3, *13*, 46, *77*, 120, *177*, 195, 205, *215*, 293, *303*
Glendenin, L. E., 210, *215*, 305, 306, 307, 308, 311, 312, 313, 314, 315, 316, 319, 321, 325, 326, 327, 329, *333*, *334*, 367, *372*
Glover, K. M., 46, *77*
Götz, U., 41, *43*
Gordon, G. E., 191, *215*. *303*, 369, *373*
Grachov, V. T., 379, 383, 384, 391, *407*
Gray, J., Jr., 47, *79*
Green, A. E. S., 17, 21, *43*
Greiner, W., 275, 276, 277, *288*
Griffin, H. C., 319, 321, 326, 327, 329, *334*, 367, *372*
Griffin, J. J., 38, 39, 41, *43*, 50, 58, *78*, 123, 164, 166, 169, 171, 172, *177*, 197, 201, 202, *215*, 286, 287, *288*
Gritsyuk, Yu. N., 346, 347, *355*
Grochulski, W., 242, *252*
Grogdev, B. A., 47, *80*
Groshev, L. V., 366, *372*
Grover, J. R., 242, *252*, 360, *373*
Gunther, H., 330, 331, 332, *334*
Gursky, J. C., 298, 299, *303*

Gustafson, C., 25, 31, 34, 35, 36, 39, 40, *43*, 58, *80*
Gutnikova, E. K., 219, *252*

H

Hagebø, E., 318, *334*
Hahn, O., 1, 2, *13*
Haines, E. L., 212, *215*
Hall, G. R., 46, *77*
Hall, W. S., 90, *107*, 169, *177*
Halperin, J., 338, *356*
Halpern, I., 3, *13*, 56, *78*, 181, 212, *215*, 217, 234, 239, 241, 242, 243, *252*, 262, *287*, 297, *303*, 328, *334*, 348, 349, *354*, 362, *372*, 375, 378, 379, 384, 386, 390, 391, 394, 396, 399, *406*
Hanna, G. C., 47, *78*, 338, *356*
Harkness, A. L., 46, *77*, 226, *253*
Harvey, B. G., 47, 48, *77*, *78*, 338, *355*
Harvey, J. W., *303*
Hasse, R. W., 282, *288*
Hauser, W., 144, *177*
Heffner, R. H., 62, *78*
Henderson, D. J., 47, *79*, 224, *253*
Henkel, R. L., 196, *215*
Henley, E. M., 39, 40, *43*, 50, *79*
Heske, M., 87, 88, 91, *108*
Heunemann, D., 61, 63, *80*
Hicks, D. A., 338, *355*, *356*
Higgens, G. H., 46, 47, *78*
Hill, D. L., 57, *78*, 120, *177*, 260, 263, 267, *288*
Hillman, M., 242, *252*, 401, *407*
Hirsch, A., 46, 48, *78*
Hiskes, J. R., 244, 245, *252*
Hockenbury, R. W., 87, 89, *108*
Hoff, R. W., 47, 48, *77*, *78*, 91, *107*
Hoffman, D. C., 295, *303*, 306, 307, *333*
Hoffman, M. M., 361, *372*
Horwitz, E. P., 47, *78*, 305, 306, 316, *334*
Howerton, R. J., 218, *252*
Hubbard, E. L., 403, *406*
Huizenga, J. R., 3, *13*, 42, *43*, 46, 47, *77*, *78*, 88, *108*, 120, 140, 145, 146, 148, 149, 150, 151, 152, 153, 154, 156, 157, 162, 163, 164, 169, 170, 172, 175, *177*, *178*, 184, 187, 188, 189, 190, 192, 193, 194, 195, 197, 201, 202, 204, 205, 206, 207, 210, 211, *215*, 217, 221, 222, 225, 226, 227, 231, 232, 235, 238, 241, 242, 247, 250, 251, *252*, *253*, 301, *303*, 318, *334*, 339, *356*
Hulet, E. K., 47, 48, *78*, 295, *303*, 307, *334*
Hunter, J. B., 338, *355*
Hyde, E. K., 3, *14*, *43*, 46, 48, *78*, *252*

I

Ignatyuk, A. V., 39, *43*, 154, *177*, 285, 286, *288*
Inglis, D. R., 33, *43*
Ippolitov, V. T., 219, *252*
Ise, J. I., Jr., 338, *355*, *356*
Ivanov, O. I., 363, *372*
Iyer, R. H., 401, 402, *406*

J

Jackson H. G., 46, 47, 48, *77*, *79*
Jackson, J. D., 63, *78*, 223, *253*
Jägare, S., 63, 76, *79*
Jaffey, A. H., 46, 48, *78*, 338, *356*
James, G. D., *108*, 156, *177*
Janeva, N., 90 91, 95, *108*
Jared, R. C., 369, 370, 371, *373*
Jéki, L., 363, *372*
Jensen, A. S., 3, *13*, 280, *288*
Jensen, R. C., 314, *334*
Johansson, S. A. E., *43*, 52, *79*, 129, *177*, 364, *372*
John, W., 295, *303*, 307, *334*
Jones, M., 46, 47, 48, *79*
Jorgenson, A. B., 61, *79*
Jungclaussen, H., 59, 61, 63, *78*
Junken, K., 41, *43*

K

Kahn, A. M., 124, *177*
Kalashnikova, V. I., 338, *356*
Kapitza, S. P., 118, 126, 127, 128, 130, 132, *178*
Kapoor, S. S., 203, 208, 212, *215*, 241, *253*, 367, *372*, 381, *406*
Karamyan, S. A., 403, 405, *406*
Karnaukhov, V. A., 59, 61, 62, 65, *77*, *80*
Kataria, S. K., 203, 208, *215*, 241, *253*
Katase, A., 393, 394, *406*
Katcoff, S., 379, 383, *406*
Katz, L., *116*, 130, 139, *177*, 221, *253*
Kaufman, S. B., 401, *407*
Kazarimov, N. M., *355*
Kelly, F. R., 47, *79*
Kennedy, R. C., 39, 40, *43*, 50, *79*

Kerman, A. K., 30, *42*
Keyworth, G. A., 91, *108*, 219, *253*
Khan, A. M., 123, 124, *177*
Khan, N. K., 70, 71, *78*
Khanh, N. C., 71, *77*
Kharisov, I. F., 59, *78*
Khlebnikov, G. I., 46, *77*
Khodai-Joopari, A., 217, 236, *253*, 255, *258*, 378, *406*
Kiker, W. E., 318, *334*
Kinderman, E. M., 46, *79*
King, T. J., 87, 89, *108*
Klema, E. D., 162, 163, *177*
Kluge, G., 363, *372*
Knight, J. D., 295, *303*, 306, 307, *333*
Knobeloch, G. W., 89, 94, 95, *107*
Knowles, J. W., 123, 124, *177*
Kolesov, I. V., 48, *79*
Kondurov, I. A., 379, 383, 384, 391, *407*
Konecny, E., 61, 63, *80*, 314, 330, 331, 332, *334*, 348, *356*
Koontz, P. G., 47, *77*
Kovacs, I., 61, *78*
Kovalenko, S. S., *355*, 400, *405*
Kowalski, S. B., *116*, 130, 139, *177*
Krisyuk, I. T., *355*
Krivokhatskii, L. S., 46, 47, *77*
Krogulski, T., 379, 383, 384, 394, 400, *405*, *406*, *407*
Kroshkin, N. I., 338, *356*
Kuiken, R., 84, *108*
Kuroda, P. K., 46, *79*
Kushnir, Yu. A., 363, *372*
Kutikov, I. E., 346, 347, *355*
Kutsaeva, L. S., *79*
Kuzminov, B. D., 169, *178*, 344, *356*
Kuznetsov, V. I., 48, *79*, 403, 405, *406*
Kvitek, J., 96, *108*
Kyz'minov, B. D., *79*

L

Laidler, J. B., 46, *80*
Lajtai, A., 363, *372*
Lamm, I. L., 25, 31, 34, 35, 39, 40, *43*
Lamphere, R. W., 111, 140, *153*, 175, *177*
Lang, D. W., 354, *356*
Lark, N. L., 59, 61, *79*
Larsh, A. E., 46, 47, 48, *78*, *80*, 191, *215*
Larsson, S. E., 41, *43*
Lasarev, Yu. A., 48, 71, *77*, *78*, *79*, 338, *355*
Latimer, R., 48, *80*
Lawrence, J. N. P., 18, *43*
Lazarev, Yu. A., *see* Lasarev, Yu. A.
Leachman, R. B., 76, *77*, 345, *356*
Lebedev, V. I., 338, 346, 347, *355*, *356*
Ledergerber, T., 36, *43*, 71, 72, 73, *79*, 129, *178*, 278, 280, *288*
Lefort, M., 239, *252*
Lerner, J. L., 47, *78*, 338, *356*
Lessler, R. M., 226, *253*
Limkilde, P., 61, *80*
Lin, W., 28, *43*
Lindner, M., *253*
Lobanov, Yu. V., 48, *78*, *79*
Lottin, A., 96, *108*
Lougheed, R. W., 295, *303*, 307, *334*
Lounsbury, M., 46, *77*
Love, T. A., 360, *372*
Loveland, W. D., 140, 146, 149, 151, 152, 153, 154, 165, 166, 169, *177*, *178*, 300, *303*, 369, *372*, 378, 399, *406*
Lukyanov, A. A., 106, *108*
Lutsenko, V. N., 366, *372*
Lyashchenko, N. Ya, 9, *14*, 18, 20, *44*, 207, 208, *215*
Lynn, J. E., 61, 65, 73, 76, *77*, 92, 104, 105, 106, *108*, 154, 156, 173, *177*

M

McHugh, J. A., 333, *334*
MacMurdo, K. W., 402, 403, *406*
McNally, J. H., 219, *253*
Magnusson, L. B., 47, *79*
Maienschein, F. C., 359, 360, *372*, *373*
Maier-Leibnitz, H., 358, 359, 360, 363, 364, *373*
Malkin, L. Z., 46, 47, *77*, *79*
Mang, H. J., 371, 372, *373*
Manuel, O. K., 305, *333*
Marcinkowski, A., 242, *252*
Markov, B. N., 59, 61, 63, 70, 71, *77*, *78*
Martin, H. C., 338, *355*
Martologu, N., 61, *78*
Maslov, A. N., 219, *252*
Meadows, J. W., 151, 162, 163, *177*, *178*
Mech, J., 46, *79*
Mehta, G. K., 89, 95, 96, *108*
Mehta, M. K., 61, 67, *80*
Meitner, L., 2, *14*
Melkonian, E., 89, 95, 96, *108*

Merritt, J. S., 46, *77*
Metag, V., 61, *79*, *80*
Metropolis, N., 19, *43*
Metta, D. N., 47, *78*, *79*
Michaudon, A., 84, 85, 87, 88, 89, 91, 96, *107*, *108*
Michel, M. C., 333, *334*
Migdal, A. B., 201, *215*
Migneco, E., 87, 89, *107*, *108*
Mikaelyan, L. A., 346, 347, *355*
Mikheev, V. L., 46, 59, *79*, *80*
Miller, L. G., 87, 89, *108*
Milsted, J., 47, *78*, *79*
Milton, J. C. D., 3, *13*, 169, 173, 174, *178*, 336, 342, 346, 353, 354, *355*, *356*, 363, *373*, 381, *405*, *406*
Minor, M. M., 212, 213, 214, *215*
Möller, P., 25, 34, 35, 36, 37, 39, 40, *43*, 58, *80*, 129, *177*, 273, 274, 275, 278, 280, 282, *288*, 317, *334*
Moldauer, P. A., 121, 143, *177*
Moore, M. S., 87, 89, 91, *107*, *108*, 219, *253*
Moreh, R., 393, *406*
Moretto, L. G., 184, 187, 188, *215*, 241, 242, *253*, 255, *258*
Morrissey, J. A., 61, *80*
Mosel, U., 40, *43*, 241, 243, *253*, 275, 276, 277, *288*
Mottelson, B. R., 25, *43*, 302, *303*
Muga, M. L., 400, *406*
Musgrove, A. R., de L., 389, 394, *406*
Muzychka, Yu. A., 34, 36, *43*, 404, *406*
Myers, W. D., 4, 6, *14*, 29, 30, 37, 38, *43*, 49, 50, *79*, 240, *253*

N

Nagy, L., 378, *406*
Nagy, T., 61, *78*, 378, *406*
Nakahara, H., *303*
Nardi, E., 379, 383, 394, 397, 399, 400, *406*
Natowitz, J. B., 378, *406*
Nebenzahl, I., 386, 387, 388, 394, 395, *405*
Nedopekin, V. G., 379, *405*
Neiler, J. H., 294, 295, 296, *303*, 321, *334*, 397, *407*
Nemilov, Yu. A., 219, *252*, 255, *258*
Nesteron, V. G., *79*
Nethaway, D. R., 325, 326, *334*
Newton, J. O., 50, 54, *79*
Nicholson, W. J., 212, *215*, 217, 234, *252*
Niday, J. B., *303*
Nifenecker, H., 351, *356*
Nikitin, A. M., 379, *407*
Nilsson, B., 25, 31, 34, 35, 39, 40, *43*, 58, *80*
Nilsson, S. G., 25, 26, 31, 33, 34, 35, 36, 37, 39, 40, 41, *43*, *44*, 53, 57, 58, 67, 69, *79*, *80*, 129, *177*, 273, 274, 275, 278, 280, 282, *288*, 317, *334*
Nix, J. R., 4, *13*, 22, 23, 24, 36, 37, 41, *43*, 57, 58, 67, 75, *77*, *79*, *80*, 152, 153, *177*, *258*, 263, 264, 265, 266, 267, 268, 270, 271, 272, 274, 276, 278, 280, *287*, *288*, 352, *356*, 371, 372, *373*
Nobles, R. A., 377, 378, *407*
Nörenberg, W., 371, 372, *373*
Norris, A. E., 328, *334*
Northrop, J. A., 163, 169, *178*
Norton, J. L., 4, *13*, 36, 37, *43*, *258*, 276, 278, 280, *288*
Novoselov, G. V., 219, *252*
Nurmia, M., 47, 48, *78*, *79*, *80*

O

Oganesian, Yu. Ts., *see* Oganesyan, Yu. Ts.
Oganesyan, Yu. Ts., 48, *78*, *79*, 338, *355*, 403, 404, 405, *406*
Okolovich, V. N., *303*, 342, *356*
Orth, C. J., 338, *356*
Otozai, K., 151, *178*
Otroschenko, G. A., *see* Otroshenko, G. A.
Otroshenko, G. A., 39, 42, 47, *77*, *80*, 169, 172, 173, *176*, *177*, 219, *253*
Ouvry, F., 343, *355*

P

Panin, V. L., 219, *252*
Pappas, A. C., 318, *334*
Parker, G. W., 83, 84, 85, *107*
Pashkevich, V. V., 34, 36, 37, 41, *43*, 129, *178*, 276, 278, 279, 280, 281, *288*
Passell, L., 85, 87, *108*
Pattenden, N. J., 84, *108*, 338, *356*
Pauli, H. C., 3, *13*, 36, 41, *43*, 58, 62, 71, 72, 73, *77*, *79*, 103, *108*, 129, *178*, 278, 280, *288*
Paya, D., 85, 87, 88, 89, 91, 96, 98, *107*, *108*
Pedersen, J., 39, *42*, 59, 61, 62, *77*, *78*, *79*, 169, 172, 173, *176*, *177*, *178*
Peelle, R. W., 359, 360, *372*, *373*
Pelekhov, V. I., 366, *372*
Penionzhkevich, Yu. E., 403, 405, *406*

Perelygin, V. P., 46, 48, *77*, *79*, 403, *407*
Perey, F. G. J., 144, 161, *176*
Perfilov, N. A., 46, *79*
Perry, D. G., 315, *334*, 398, *407*
Peter, J., 239, *252*
Petrov, G. A., 363, 364, *373*
Petrzhak, K. A., 46, 47, *77*, 400, *405*
Phillips, L., 47, 48, *77*, *79*
Piasecki, E., 379, 383, 384, 400, *405*, *406*, *407*
Piekartz, H., 400, *407*
Pik-Pichak, G. A., 244, *253*
Plasil, F., 76, *77*, 169, 170, 171, *177*, 217, 236, 246, 248, *252*, *253*, 255, *258*, 320, 321, 322, *333*, 348, *356*
Pleasonton, F., 363, 364, 365, 366, *373*
Pleva, Y. S., 363, 364, *373*
Pleve, A. A., 59, 61, 63, *78*, *80*
Plotko, V. M., 48, *79*
Podgurskaya, A. V., 46, *79*
Poenaru, D., 61, *78*
Polikanov, S. M., 59, 60, 61, 63, 68, 70, 71, *78*, *79*, *80*
Poluboyarinov, Yu. F., 48, *79*
Poortmans, F., 87, *108*
Popov, N. A., 9, *14*, 18, 20, *44*, 207, 208, *215*
Popov, Yu. P., 96, *108*
Postma, H., 84, *108*
Preston, M. A., 111, *178*
Prestwood, R. J., 89, 94, 95, *107*, 401, *407*
Price, P. B., 46, *78*, 217, 236, *252*, 255, *258*, 403, *406*
Primack, J. R., 25, *43*
Prokhorova, L. I., *79*
Protopopov, A. N., 359, *373*
Pustylnik, B. I., 404, *406*
Pyle, R. V., 338, *355*, *356*

Q

Quaheim, B. J., 48, *78*
Quill, L. L., 1, *14*

R

Rabotnov, N. S., 118, 126, 127, 128, 130, 132, 154, *177*, *178*
Rae, E. R., *108*
Ragnarsson, I., 41, *43*
Raisbeck, G. M., 217, 236, *253*, 255, *258*, 378, 379, 389, 390, 391, 394, *407*
Rajagopalan, M., 389, 394, *407*
Ramamurthy, V. S., 203, 208, *215*, 241, *253*
Ramanna, R., 381, *406*
Rama Rao, P. N., 381, *406*
Ramler, W. J., 224, 226, *253*
Rasmussen, B., 39, *42*, *77*, 169, 172, 173, *176*, *177*
Rasmussen, J. O., 188, *215*, 367, 369, 370, 371, *372*, *373*
Ratcliffe, C. T., 369, *373*
Rau, F. E. W., 360, *373*
Reed, R., 95, *108*
Reisdorf, W., 319, 321, 326, 327, 329, *334*
Reising, R. F., 205, 206, 207, *215*
Repnow, R., 61, *79*, *80*
Reynolds, C. A., 85, 87, *108*
Rice, C. R., 400, *406*
Rickey, F. A., Jr., 90, *107*, 169, *177*
Ripon, P., 89, *107*
Rizzo, G. T., 156, 157, *178*
Roberts, J. H., 140, 145, 146, 148, 149, 150, 151, 152, 153, 154, *177*
Robinson, H. P., 47, *78*
Roeckl, E., 326, *334*
Rogov, V. I., 379, *405*
Rose, M. E., 111, *178*
Rouse, R. A., 328, *334*
Routi, J., 255, *258*
Roy, J. C., 379, *407*
Rud, V. I., 47, 48, *77*
Rudy, C. R., 61, *80*, 369, *373*
Russo, P. A., 61, 67, *80*
Ryabov, Y. V., 90, 91, 95, 96, *108*

S

Sailor, L. V., 85, 87, *108*
Salwin, A. E., 212, 213, 214, *215*
Sanatini, S., 338, *356*
Sanche, M., 84, *107*
Sarantites, D. G., 369, *373*
Sauter, G. D., 89, 91, *108*
Scharnweber, D., 275, 276, 277, *288*
Schermer, R. I., 85, 87, *108*
Schmitt, H. W., 76, *77*, 294, 295, 296, *303*, 314, 318, 320, 321, 322, *333*, *334*, 338, 348, *356*, 363, 364, 365, 366, *373*, 397, *407*
Schrieffer, J. R., 197, *215*
Schuman, R. P., 46, 47, 48, *77*, *79*
Schwartz, R. B., 89, *107*
Seaborg, G. T., 46, 47, *78*, 191, *215*
Sedlacek, W. A., 400, *406*
Segrè, E., 46, *77*, *80*

Seleverstov, D. M., 379, 383, 384, 391, *407*
Selitskii, Yu. A., 219, *252*, 255, *258*
Sergachev, A. I., 344, *356*
Sezon, M., 61, *78*
Shadieva, N. H., 403, *407*
Shaker, M. O., 104, 106, *108*
Shapiro, F. L., *108*
Shaw, R. W., Jr., 239, 241, 242, 243, *252*, 262, *287*, 297, *303*, 348, 349, 354, *355*
Shepherd, B. J., 362, *372*
Shev, R., 338, *356*
Shigin, V. A., 47, *80*, 219, *253*
Shiryaev, B. M., 359, *373*
Shoaf, M. L., 313, 317, *333*
Shore, F. J., 85, 87, *108*
Shubko, V. M., 47, *80*, 219, *253*
Shum, Y. S., 369, *372*
Siegert, G., 330, 331, 332, *334*
Sighanbienv, C., 393, *406*
Sik, D. J., 90, 91, *108*
Sikkeland, T., 47, 48, *78*, *79*, *80*, 191, 212, *215*, 236, 247, 250, 251, *253*
Silbert, M. G., 91, 92, *107*
Silva, R., 47, 48, *79*
Simmons, J. E., 196, *215*
Simpson, F. B., 87, 89, 91, *107*, *108*
Sistemich, K., 326, *334*
Siwek, K., 242, *252*
Sjoblom, R. K., 47, *78*, 305, 306, 316, *334*, 359, *373*
Skarsvåg, K., 362, *373*, 381, *407*
Skliarevskii, V. V., 358, *373*
Skobelev, N. K., 46, 48, 48, 59, *77*, *79*, *80*
Sletten, G., 59, 60, 61, 68, *77*, *79*, *80*
Smirenkin, G. N., 39, *43*, *79*, 118, 126, 127, 128, 130, 132, 135, 154, *177*, *178*, *303*, 342, *356*, 363, *372*
Smirenko, L. D., 344, *356*
Smith, A., 359, *373*
Sobiczewski, A., 25, 34, 35, 36, 39, 40, *43*, 51, 58, *80*
So, Don Sik, 95, *108*
Soldatov, A. S., 117, 118, 126, 127, 128, 130, 132, 135, *178*
Soleilhac, M., 343, 344, 351, *355*, *356*
Sowerby, M. G., 338, *355*
Sowinski, M., 379, 384, *405*, *406*
Specht, H. J., 61, 63, *80*, 169, 173, 174, *178*, 358, 359, 360, *373*
Srinivasan, B., 305, *333*
Starfelt, N., 338, *355*
Stavinsky, V., 104, 106, *108*
Stegner, A., 379, *405*
Steiger-Shafrir, N. H., 236, *253*
Stein, W. E., 61, 65, 73, 76, *77*, 173, *177*
Steinberg, E. P., 382, 401, *406*, *407*
Stella, R., 241, 242, *253*
Stenholm-Jensen, A., 58, 62, *77*
Stepanov, E. P., 358, *373*
Stephan, C., 169, *178*
Stepien, W., 40, *43*
Stevens, C. M., 47, *79*
Stoenner, R. W., 401, *407*
Stokes, R. H., 39, *43*, 163, 164, 166, 169, 171, 172, *177*, *178*
Stoughten, R. W., 338, *356*
Strassmann, F., 1, 2, *13*
Strutinsky, V. M., 3, 5, 9, *13*, *14*, 18, 20, 21, 22, 27, 28, 29, 30, 34, 36, *43*, *44*, 58, 62, 67, *77* *80*, 96, 103, *107*, *108*, 157, *177*, 181, 207, 208, *215*, 263, 265, 280, *288*, 362, 371, *373*
Stubbins, W. F., 91, *107*
Studier, M. H., 46, *77*
Subbotin, V. G., 59, *80*
Swanson, H. H., 62, *78*
Swiatecki, W. J., 4, 6, 9, *13*, *14*, 18, 19, 20, 21, 22, 29, 30, 36, 37, 38, *43*, *44*, 48, 49, 50, *79*, *80*, 207, 208, *215*, 217, 236, 240, 246, 248, *252*, *253*, 255, *258*, 263, 264, 274, *288*, 332, 334, 336, 342, 352, 353, 354,*355*, *356*, 371, 372, *373*, 381, *405*
Szymanski, Z., 25, 34, 35, 36, 39, 40, *43*, 58, *80*

T

Tarrago, X., 239, *252*
Taschek, R. F., 338, *355*
Temperley, J. K., 61, *80*
Ter-Akopyan, G. M., 59, *80*
Terrell, J., 289, *303*, 338, 339, 340, 346, 347, *355*, *356*, 393, *407*
Tesmer, J. R., 61, 67, *80*
Theobald, J. P., 87, 89, *107*, *108*
Theus, R. B., 212, 213, 214, *215*
Thomas, T. D., 318, 320, 331, *334*, 360, *373*, 377, 378, 379, 380, 383, 389, 390, 391, 394, *406*, *407*
Thompson, M. C., 338, *356*
Thompson, S. G., 46, 47, 48, *77*, *78*, *79*, 181, 201, *215*, 217, 236, *252*, 255, *258*, 336, 338,

342, 353, 354, *355*, 359, 367, 369, 370, 371, *372*, *373*, 381, 392, 400, *405*, *406*, *407*
Tomlinson, R. H., 318, *319*, *333*
Tretiakova, S. P., 48, *78*, *79*, 403, *407*
Tretjakova, S. P., 61, *78*
Trochon, E. J., 96, *108*
Troutner, D. E., 325, 326, *334*
Tsang, C. F., 25, 30, 33, 34, 35, 39, 40, *42*, *43*, 58, *80*, 280, 282, *288*, 317, *334*
Tsipenyuk, Yu. M., 61, *78*, 118, 126, 127, 128, 130, 132, 135, *178*
Tuerpe, D. R., 30, *42*
Turkevich, A., *253*
Turner, L. A., 3, *14*
Tys, J., 379, 383, 384, *405*, *406*, *407*
Tyutyugin, I. I., 400, *405*

U

Unik, J. P., 61, *80*, 164, 169, 170, 172, 175, *177*, *178*, 211, *215*, 293, 298, 300, 301, 302, *303*, 318, 319, 321, 326, 327, 329, *334*, 367, *372*
Urin, M. G., 51, 58, *80*
Usachev, L. N., 118, 126, 127, 128, 130, 132, *178*

V

Valatin, J. G., 302, *303*
Val'skii, G. V., 363, 364, *373*
Vandenbosch, R., 59, 61, 63, 67, 74, *80*, 88, *108*, 120, 133, 140, 164, 165, 166, 169, 175, *177*, *178*, 190, 202 211, *215*, 217, 221, 222, 224, 225, 226, 227, 231, 232, 235, 238, 239, 241, 242, 243, 247, *252*, *253*, 262, *287*, 297, *303*, 318, 320, 331, *334*, 339, 348, 349, 354, *355*, *356*, 369, *373*, 393, *407*
Vilcov, I., 61, *78*
Vilcov, N. 61, *78*
Vinnay, I., 61, *78*, 378, *406*
Viola, V. E., Jr., *80*, 212, 213, 214, *215*, 236, *253*, 292, 293, *303*
Vitta, P. B., 394, *407*
von Brentano, P. 61, *79*, *80*
von Gunten, H. R., 305, *334*
Vorobeva, V. G., 344, *356*
Vorobiev, A. A., 379, 383, 384, 391, *407*
Vorotnikov, P. E., 47, *80*, 219, *253*

W

Wahl, A. C., 311, 316, 324, 325, 326, *334*
Walker, G. E., 67, 75. *79*
Walker, R. M., 403, *406*
Walsh, R. L., 344, *355*
Walter, F. J., 83, 84, 85, *107*, 294, 295, 296, *303*, 321, *334*, 397, *407*
Warhanek, H., 190, 202, *215*, 369, *373*
Wartena, J. A., *107*
Watson, R. L., 367, *372*, 400, *407*
Watt, D. E., 46, *80*
Weber, J., 61, 63, *80*
Wegner, H. E., 298, 299, *303*
Weigmann, H., 87, 88, 91, *108*
Weinstein, S., 95, *108*
Weisskopf, V. F., 123, *177*, *252*
Wesolowski, J. J., 295, *303*, 307, *334*
Westcott, C. H., 338, *356*
Westgaard, L., 61, 62, 65, *77*
Wheeler, J. A., 3, *13*, *14*, 16, *42*, 50, 54, 57, *77*, *78*, *80*, 94, *108*, 110, 120, *177*, *178*, 184, *215*, *252*, 260, 263, *288*
Whetstone, S. L., Jr., 377, 379, 380, 383, *407*
Whitehouse, W. J., 46, *77*
Wilets, L., 3, *14*, 30, 39, 40, *42* *43*, 50, 58, *79*, *80*, 93, *108*, 139, *178*, 261, 283, *288*
Wilhelmi, Z., 242, *252*
Wilhelmy, J. B., 280, 282, *288*, 317, *334*, 369, 370, 371, *373*
Wilkins, B. D., *80*, 164, 169, 170, 172, *177*, *178*, 382, 401, *406*, *407*
Williams, C. W., 318, *334*
Williams, F. C., Jr., 241, 242, *253*
Williams, J. C., 328, *334*
Williamson, C. F., 362, *372*
Willis, H. H., 369, *372*
Wing, J., 224, 226, *253*, *334*
Winter, J., 87, 88, 91, *107*, *108*
Winter, L., 330, 331, 332, *334*
Wolf, K. L., 55, 59, 61, 67, 74, *80*, 165, 166, 169, *178*
Wolfsberg, K., 325, 326, *334*
Wong, C. Y., 3, *13*, 58, 62, *77*, *80*, 280, *288*
Wycech, S., 25, 34, 35, 36, 39, 40, *43*, 58, *80*

Y

Yates, M. J. L., 169, *177*
Young, T. E., 91, *107*
Yuen, G., 156, 157, *178*

Z

Zamgatnia, Y. S., 338, *356*
Zaretsky, D. F., 51, 58, *80*
Zhagrov, E. A., 255, *258*
Zobel, W., 360, *372*

Subject Index

A

Adiabatic models, 259–283
α-Particle emission in fission, 374–400
  angular distribution of, 382
  dependence on fragment mass, 396
  dependence upon fragment kinetic energy, 399
  dependence upon initial excitation energy, 376
  energy distribution of, 382
  mechanism of, 375
  polar emission of, 383
  trajectory calculations for, 385–396
Angular correlation between fission fragments, 212–214
Angular distribution of fission fragments, 109–215
  from aligned nuclei, 83
  dependence on fragment mass, 209–212
  effect of target spin, 186
  in fission induced by charged particles of moderate energy, 179–215
  in fission induced by MeV neutrons, 139–163
  from nuclei excited in direct reactions, 163–176
  pairing effects on, 197–203
  in photofission near threshold, 112–139
  statistical theory of, 180–189
Angular momentum effects, on fission probability, 244–250
Angular momentum of fragments, 368–372
Asymmetry in fission, *see* Mass distribution in fission

B

Barriers, 15–44, 254–258
  barrier penetration, 51–59, 75
  curvature, 22, 57
  double-humped, 73, 96, 257, 258
  from fitting isomer data, 73
  heights, summary, 255, 258
  odd-even effects on, 38, 50, 54
  of superheavy elements, 34
  systematics of, 37–40
  types of excitations, 40–42
β Particles emitted per fission 323, 326
BCS theory, 31, 197
Breit–Wigner formula for resonances in neutron capture and fission, 82

C

Cascade fission, 403
Central neutrons, 336, 381, *see also* Neutrons, scission
Chain yields 309–311, *see also* Mass distribution in fission
Charge distribution in fission, 322–333
  distribution around most probable charge, 323
  equal charge displacement hypotheses, 325
  excitation energy dependence, 331–333
  independent yield, 323

low-energy fission, 322–331
maximum energy release hypothesis, 327, 328
minimum potential energy hypothesis, 328
most probable charge, 323
physical measurements, 325–327
radiochemical results, 323–325
relation of mass equations to, 327–329
unchanged charge distribution (UCD)-hypothesis, 324
Competition between neutron emission and fission, 216–253
angular momentum effects on, 244–250
experimental data concerning, 216–226
theoretical expressions for, 227–250
Conversion electrons from fission, 367
Coriolis force, 55
Coulomb energy, 16, 17, 289, 378
Coulomb forces, 371, 374
Cross sections, 82, 216–225
Cumulative yield, 323, *see also* Charge distribution in fission

D

Deformation energy
of fissioning nucleus, 17–37
of fragments, 347–351
Delayed neutrons, 337
Direct fission, 250–252, *see also* Angular distribution of fission fragments from nuclei excited in direct reactions
Discovery of fission, 1–3

E

Emission of central neutrons, *see* Scission neutrons
Energy, Coulomb, *see* Coulomb energy
Energy
specialization, 54
threshold, *see* Barriers
surface, 15–17
Energy release, 294, 329
Expansion about spherical shape, 17

F

Fermi gas model, 198, 233
Fissility parameter, 16
Fissionability parameter, *see* Fissility parameter
Fluctuations in fission widths, 92, 106
Fong model, 283
Fragment Anisotropy, *see* Angular distribution of fission fragments

G

γ-Ray emission in fission, 357–373
angular distribution of, 361–363, 370
dependence on fragment mass, 363–366
energy spectra, 359–360
enhancement of emission due to angular momentum, 359
multipolarity of, 359, 362
number of photons per fission, 359
population of ground state rotational bands, 369
time-scale for emission, 357, 358, 367

H

Hartree–Fock, 30

I

Inertial mass parameter, 58, 283, 286
Intermediate structure, 96
Irrotational flow, 58
Isomeric state fission, 59–76
Isomeric yield ratios of fission products, 368

K

K quantum number, 41, 54, 110
Kinetic energy of fragments, 267–273, 289–303
dependence on angular momentum, 301
dependence on charge division, 303
dependence on excitation energy, 300
dependence on fission fragment mass, 294–300
dependence on resonance spin, 95
widths of total kinetic energy distributions, 269, 271, 293, 298–300

L

Level density, 229, 233, 240, 284, 354
spin dependence of, 180–181
Light particles emitted in fission, 374–400
energies of, 383, 389
relative yields of different light particles, 378–381
Liquid drop model, 15–24
distortion energy, 15–24
nuclear mass, 15
potential energy, 15–24
saddle shapes, 204–209
surface energy, 16, 17

M

Mapping of potential energy, 17–34, 263–265, 273–282
Mass asymmetry, 22, 263–287, 307–317, *see also* Mass distribution in fission
  dependence on resonance spin, 94
Mass distribution in fission, 259–287, 304–321
  theories of, 259–287
  fine structure, 317–320
  particle-induced fission, 307
  spontaneous fission. 305, 320
  widths of, 271
Mass equation, 15
Mass parameters, *see* Inertial mass parameters
Mass yield, *see* Mass distribution in fission
Models
  adiabatic, 259–283
  collective, 40–42
  independent particle, 24–27
  kinetic dominance, 286–287
  liquid drop, 15–24, 263–273
    dynamics, 263–273
    statics, 15–24
  Nilsson, 28–35
  shell, 273, *see also* Shell effects
  statistical, 283–286
  Strutinsky hybrid, 27–34
Moment of inertia, 62, 180, 205
Most probable charge, 323
Multiple-chance fission, 221–222, 196–197

N

Neutron-induced fission, 139–163, 218–223
Neutrons from fission, 335–356
  angular distributions of neutrons, 336
  correlation between yields from complementary fragments, 351
  delayed, 337
  dependence on fragment mass, 345–351
  dependence of $\bar{\nu}$ upon excitation energy, 340
  dependence on resonance spin, 95
  emission probability distribution, 337–340
  emission time, 335
  energy spectrum, 344, 352
  number of neutrons emitted per fission, 337, 338
  scission, 336, 381
Neutron resonance fission, 81–107
  average widths, 88–92, 102
  γ-ray multiplicities, 85–87
  interference between levels, 82
  intermediate structure, 96
  kinetic energy release, 95
  mass asymmetry, 94
  $(n,\gamma f)$ process, 104–107
  neutron multiplicity, 95
  neutron scattering, 82
  nuclear orientation, 83
  polarized neutrons through polarized targets, 85
  ratio of ternary to binary, 95
  spin determination, 81–88, 89
  width distributions, 92–94, 106
Nilsson states, 28–35, 67
Nonadiabaticity, 261, 283

P

Pairing, 31, 197
  deformation dependence of, 38–40
Penetrability formulas, 51–59, 154
Photofission, 112–139
  angular distribution of fragments from, 112–139
  cross sections for, 119–130
Potential energy surface, 17–37
Primary fission fragment, 304
Probability of fission, 51, 145, 216–253
Prompt neutrons, *see* Neutrons from fission

R

Resonances, 81, 98, 103, 154

S

Saddle point, 17–22, 34, 36, 110, 203, 275
  effective moment of inertia, 181, 197, 203–209, 249
  rotational energy, 249
  thermodynamic temperature at, 180, 198
Scission
  deformation, 380, 394
  energy release, 289–303
  time of, 262, 267
Secondary fission fragments, 304
Semiempirical mass formula, 15
Shape isomers, *see* Spontaneously fissionable isomers
Shell effects
  dependence of potential energy surface on, 27–34, 273–280

dependence of width on, 240–243
of fragments, 276, 285, 346–351
on level densities, 235, 354
Single particle model, 24–27
Specialization energy, 54, 67
Spontaneous fission, 45–80
ground-state half-lives, 46
isomeric-state half-lives, 60, 61
odd–even effects, 49
Spontaneously fissionable isomers, 59–76
excitation energy of, 63
excitation functions of, 63, 71
γ-ray decay of, 62, 67
half-lives, 60, 61
mass and kinetic energy distributions of, 76, 320–322
moment of inertia of, 61–63
population of, 63, 75
Statistical models, 283–286
Strong coupling, 101, 102
Strutinsky method, 27–37
extension to reflection-asymmetric saddle-point shapes, 36–37
Sudden approximation, 261
Superconductivity, *see* Pairing, BCS theory
Superheavy elements, 34
Surface energy, 16, 17
curvature dependence of, 20–22

T

Ternary fission, 374–407, *see also* α-Particle emission in fission. Light particles emitted in fission
Threshold energy, *see* Barriers
Total kinetic energy, *see* Kinetic energy of fragments
Transition state nucleus
energy needed to reach, *see* Barriers
even–even, 40–42, 112, 159, 169
moments of inertia, 203–209
shape of, 203
study of low-lying levels in, 40, 109–176
Two-center potentials, 275
Two-mode hypothesis, 298

V

Vibrations, 41
Viscosity, 260, 273, 371

W

Weak coupling, 101
Widths, 81, 216
average, 88, 102, 227, 232
distributions, 92, 106
neutron, 227

X

X rays in coincidence with fission fragments, 367